FUNCTIONAL INTEGRATION AND PARTIAL DIFFERENTIAL EQUATIONS

BY

MARK FREIDLIN

ANNALS OF MATHEMATICS STUDIES
PRINCETON UNIVERSITY PRESS

Annals of Mathematics Studies

Number 109

FUNCTIONAL INTEGRATION AND PARTIAL DIFFERENTIAL EQUATIONS

BY

MARK FREIDLIN

PRINCETON UNIVERSITY PRESS

PRINCETON, NEW JERSEY

1985

Clothbound editions of Princeton University Press books are printed on acid-free paper, and binding materials are chosen for strength and durability. Paperbacks, while satisfactory for personal collections, are not usually suitable for library rebinding

ISBN 0-691-08354-1 (cloth)
ISBN 0-691-08362-2 (paper)

Library of Congress Cataloging in Publication data will be found on the last printed page of this book

Printed in the United States of America
by Princeton University Press, 41 William Street
Princeton, New Jersey

CONTENTS

PREFACE

With every second-order elliptic differential operator L, one can associate a family of probability measures in the space of continuous functions on the half-line. This family of measures forms the Markov process corresponding to the operator L. If one knows some properties of the operator L, it is possible to draw conclusions about properties of the Markov process. And conversely, studying the Markov process one can obtain new information concerning the differential operator.

This book considers problems arising in the theory of differential equations. Markov processes (or the corresponding families of measures in the space of continuous functions) are here only a tool for examining differential equations. As a rule, the necessary results from the theory of Markov processes are given without proof in this book. We restrict ourselves to commentaries clarifying the meaning of these results. There are already excellent books where these results are set forth in detail, and we give references to these works.

The probabilistic approach makes many problems in the theory of differential equations very transparent; it enables one to carry out exact proofs and discover new effects. It is the latter—the possibility of seeing new effects—which seems to us the most significant merit of the probabilistic approach.

This book is intended not only for mathematicians specializing in the theory of differential equations or in probability theory but also for specialists in asymptotic methods and functional analysis. The book may also be of interest to physicists using functional integration in their research.

The two years I have spent writing this book were very hard, I would even say desperate, for me and my family. And I am glad to be able to thank my colleagues for their support. I have been happy to see convincing evidence of the high moral standards of many colleagues. I especially wish to express my gratitude to E. B. Dynkin for his constant attention and concern about all our problems.

Finally, I must say that this book would never be brought into the world without the enormous labor of my wife, Valeria Freidlin, in her editing, translating and retyping the manuscript. I feel even awkward about thanking her for this labor; in essence, she was my co-author.

MARK FREIDLIN

Functional Integration

and Partial

Differential Equations

INTRODUCTION

It was known long ago that there is a close relation between the theory of second-order differential equations and Markov processes with continuous trajectories. As far back as 1931, the parabolic equations for transition probabilities were written down in the article of Kolmogorov [1]. Still earlier, these equations on the theory of Brownian motion appeared in physics literature (Einstein [1]). It was also established that the mean values of some functionals of the trajectories of diffusion processes (as functions of an initial point) are the solutions of boundary value problems for the corresponding elliptic differential equations.

For a long time the connection between Markov processes and differential equations was used mainly in one direction: from the properties of the solutions of differential equations, some or other conclusions on Markov processes were made. Meanwhile, probabilistic arguments in problems of the theory of differential equations played at best the role of leading reasoning. This may be explained by lack of direct probabilistic methods for studying diffusion processes. Even the construction of such a process with given characteristics was carried out with the help of the existence theorems for the corresponding parabolic equations.

For the last quarter of a century the situation has changed in an essential way. The rapid development of direct probabilistic methods for examining Markov processes allowed one to construct and study them without turning to partial differential equations. Conversely, the construction and analysis of the trajectories of the corresponding diffusion process via direct probabilistic methods, enabled the solutions of differential equations to be constructed and the properties of these solutions to be examined.

It is not for the first time that such a situation arises in the theory of differential equations. For example, recall the mutual relations between differential equations and the calculus of variations. Originally, the differential equations served as the means of seeking solutions of extremal problems. With the development of the direct methods in the calculus of variations, the possibility appeared of constructing and studying the solutions of differential equations as the extremals of the corresponding functionals. Similar mutual relations have now been established between the theory of differential equations and that of diffusion processes.

Speaking somewhat inaccurately, one can say that, in the theory of second-order parabolic and elliptic differential equations, the trajectories of diffusion processes play the same part as characteristics do for first-order equations. Just as the theory of characteristics makes first-order equations geometrically descriptive, the probabilistic considerations make transparent many problems arising in the theory of second-order elliptic and parabolic equations.

Sometimes the probabilistic methods play the role of a tool for deriving delicate analytical results. Sometimes they are a basis for the extension of some analytical theory. However, in my view, the greatest value of such an approach consists in its visualization which turns this approach into an especially helpful instrument for discovering new effects, for a deeper qualitative understanding of the classical objects of mathematical analysis.

Among the tools of the direct probabilistic research of diffusion processes, one should, first of all, mention stochastic differential equations. The theory of such equations, originating in the works of Berstein, was basically founded by Ito and (independently) by Gihman, and then has been developed by a number of mathematicians. The stochastic integral introduced by Ito, Ito's formula, and the generalizations of these notions play the central part in the whole theory. The present state of the theory of stochastic differential equations is described in the monograph of Ikeda and Watanabe [2]; references to the original works can be found there too.

As another important factor permitting the direct study of diffusion processes, one should mention the convenient general concept of Markov process and Markov family introduced by Dynkin [1], [3] as well as the detailed analysis of the strong Markov property. The wide use of the theory of one-parameter semi-groups due to Feller is also worthwhile noting.

The theory of martingales serves as a highly suitable instrument for examining Markov processes (see Doob [1], Delacherie and Meyer [1]).

The transformations of Markov processes, in particular, those involving an absolutely continuous change of measure in the space of trajectories, are also very useful tools which enable one, in a transparent and explicit fashion, to understand the effects of potential terms and first order terms. This leads to an understanding of the affects of these terms on the behavior of the solution of the differential equation.

The last ten to fifteen years have seen a development of limit theorems for random processes—central limit theorem type results as well as theorems on the asymptotics of probabilities of large deviations. In particular, the counterpart of the asymptotic Laplace method for functional integrals pertains to the results of that kind. These results proved to be highly useful in a great number of problems in differential equations which have waited long to be solved.

The application of the probabilistic methods for examining differential equations is usually based on the representation of the solution of these equations as the mean value of some functional of the trajectories of a proper diffusion process. The mean value of a functional of the trajectories of a random process may be written down as the integral of the corresponding functional on the space of functions with respect to the measure in this space induced by the random process. This is why such representations of solutions are sometimes called the representations in the form of functional integrals.

The construction of the diffusion process corresponding to the differential operator

$$L = \frac{1}{2} \sum_{i,j=1}^{r} a^{ij}(x) \frac{\partial^2}{\partial x^i \partial x^j} + \sum_{i=1}^{r} f^i(x) \frac{\partial}{\partial x^i} \tag{1}$$

with the non-negative definite matrix $(a^{ij}(x))$, is carried out with the help of stochastic differential equations. The Wiener process W_t, the simplest of the non-trivial Markov processes serves as a starting point.

By a Wiener process (one-dimensional), we mean a random process $W_t = W_t(\omega)$, $t \geq 0$, having independent increments and continuous trajectories (with probability 1), and for which $EW_t = 0$, $EW_t^2 = t$ (E being the mathematical expectation sign).

It is established that such a process does exist and its finite-dimensional distributions are Gaussian. In particular, for every $t > 0$, the random variable $W_t(\omega)$ has the density function $(2\pi t)^{-1/2} \exp\left\{-\frac{x^2}{2t}\right\}$, $-\infty < x < \infty$. This process is connected, in the closest way, with the operator $\frac{1}{2}\frac{d^2}{dx^2}$ and with the simplest heat conduction equation. For instance, the solution of the Cauchy problem

$$\frac{\partial u(t,x)}{\partial t} = \frac{1}{2}\frac{\partial^2 u(t,x)}{\partial x^2}, \quad u(0,x) = g(x), \tag{2}$$

may be represented in the form

$$u(t,x) = \frac{1}{\sqrt{2\pi t}} \int_{-\infty}^{\infty} g(x+y) e^{-\frac{y^2}{2t}} dy = Eg(x + W_t).$$

This assertion is checked by direct substitution into equation (2). Just as any random process, the Wiener process induces a measure in the space of functions. In the present case, it is a measure in the space of continuous functions on the half-line $t \geq 0$ with the values in R^1. This measure is referred to as the Wiener measure. It plays the principal role in all the questions to be considered in this book. The first construction of this measure was published by Wiener in 1923 [1]. Later on the Wiener process and the Wiener measure have been studied in detail.

An ordered collection of r independent Wiener processes $(W_t^1, \cdots, W_t^r) = W_t$ is termed an r-dimensional Wiener process. Such a process is connected with the Laplace operator in R^r. What process corresponds to the operator L in (1)? Let us assume for a moment that the coefficients of the operator are constant: $a^{ij}(x) = a^{ij}$, $b^i(x) = b^i$. Denote by $\sigma = (\sigma^i_j)$ a matrix such that $\sigma\sigma^* = (a^{ij})$ and consider the family of random processes

$$X_t^x = \sigma W_t + bt + x, \; x \in R^r, b = (b^1, \cdots, b^r), t \geq 0 . \tag{3}$$

It is not difficult to find the distribution function of the Gaussian process X_t^x and then to check that $u(t,x) = E\, g(X_t^x)$ is the solution of the Cauchy problem

$$\frac{\partial u}{\partial t} = L\, u(t,x) , \quad u(0,x) = g(x) , \tag{4}$$

for any continuous bounded function $g(x)$. Therefore, the random process (3) is associated with the operator L with constant coefficients.

It is natural to expect that, in the vicinity of every point $x \in R^r$, the process corresponding to the operator L with variable (sufficiently smooth) coefficients, must behave just as the process corresponding to the operator with the constant coefficients frozen at this point x. On the basis of this reasoning, for the family of the processes X_t^x corresponding to the operator L with variable coefficients, we obtain the differential equation

$$d\, X_t^x = \sigma(X_t^x)\, d\, W_t + b(X_t^x) dt, \; X_0^x = x , \tag{5}$$

where the matrix $\sigma(x)$ is such that $\sigma(x)\sigma^*(x) = (a^{ij}(x))$, $b(x) = (b^1(x), \cdots, b^r(x))$.

If the trajectories of the Wiener process were differentiable functions or at least had bounded variation, then equation (5) could be treated within the framework of the usual theory of ordinary differential equations. But, with probability 1, the trajectories of the Wiener process have infinite variation on every non-zero time interval. Therefore, equation (5) should be given a meaning. Ito's construction is most convenient for this. This construction is given in the beginning of Chapter I.

One can demonstrate that, under mild assumptions on the coefficients, equation (5) has a unique solution X_t^x. The random functions X_t^x, $x \in R^r$, together with the corresponding probability measure, form a Markov family connected with the operator L. A solution of Cauchy's problem (4) may be written in the form $u(t,x) = E\, g(X_t^x)$.

A solution of Dirichlet's problem for the operator L may also be represented in the form of the mathematical expectation of some functional of the process X_t^x. For example, if D is a bounded domain in R^r with a smooth boundary ∂D and the operator L does not degenerate for $x \in D \cup \partial D$, then the solution of the Dirichlet problem

$$L\,u(x) = 0, x \in D\, ;\; u(x)|_{\partial D} = \psi(x)\;, \tag{6}$$

where $\psi(x)$ is a continuous function on ∂D, may be written as follows

$$u(x) = E\,\psi(X^x_{\tau^x})\,. \tag{7}$$

Here $\tau^x = \inf\{t : X_t^x \notin D\}$ is the first exit time of the process X_t^x from the domain D.

If the term with a potential v is added to the operator L, then the solutions of various problems for the operator $L + v$ may also be represented in terms of the trajectories of the process X. For example, the solution of the Cauchy problem

$$\frac{\partial u(t,x)}{\partial t} = L\,u(t,x) + v(x)\,u(t,x),\; u(0,x) = g(x) \tag{8}$$

is given by the Feynman-Kac formula

$$u(t,x) = E\,g(X_t^x)\exp\left\{\int_0^t v(X_s^x)\,ds\right\}\,. \tag{9}$$

Notice that equation (5) may be looked upon as the mapping of the space $C_{0,\infty}(R^r)$ of continuous functions on the half-line with values in

R^r, into itself: $I: W_. \to X_.^x$. This mapping is defined a.e. with respect to the Wiener measure in $C_{0,\infty}(R^r)$. The value of X_t^x at time t is defined as a functional of the Wiener trajectory in the interval $[0,t]$ which depends on x as a parameter: $X_t^x = I^x(W_s, 0 \le s \le t)(t)$. This mapping allows formulae (7) and (9) to be rewritten in the form of integrals with respect to the Wiener measure.

Chapter I describes the construction and properties of the Wiener process. The necessary information on stochastic integrals, stochastic differential equations, and Markov processes and their transformations is given here. Some limit theorems for random processes are included as well. In particular, we provide the definition and properties of the action functional related to the Laplace type asymptotics for functional integrals. In short, Chapter I introduces those notions and methods which are necessary for the direct probabilistic analysis of processes (measures in the space of functions) connected with differential operators.

Today there are a number of monographs presenting these results in detail. Also, in this book, random processes are a tool rather than an object of research themselves. For this reason the results of Chapter I are, as a rule, cited without proof. We restrict ourselves to short comments and references.

In Chapter II, the formulas representing the solutions of differential equations in the form of functional integrals (in the form of the mean values of the functionals of the trajectories of the corresponding processes) are studied. Besides formulas (7) and (9), this chapter gives representations for the solutions of the second boundary value problem as well as some other problems. The behavior of random processes as $t \to \infty$ is a traditional subject of probability theory. This is closely related to problems concerning the stabilization, as $t \to \infty$, of the solutions of Cauchy's problem as well as of mixed problems. It is also related to the statement of boundary valued problems in unbounded domains. These questions are also considered in Chapter II. Speaking somewhat inaccurately, one can say that a solution of the first boundary value problem is unique if and only if the trajectories of the corresponding diffusion process leave the domain D with probability 1. Hence the question of the correct statement of the problem in an unbounded domain is closely related to the behavior of the trajectories as $t \to \infty$. If, with positive probability, the trajectories go to infinity without hitting the boundary of the domain, then supplementary conditions at infinity are required to single out the unique solution. For example, the Wiener process in R^2 does not run to infinity, and so the solution of the exterior Dirichlet problem for the Laplace operator in R^2 is unique in the class of bounded functions. Meanwhile, the Weiner process in R^r, for $r \geq 3$, goes to infinity with positive probability, and hence, when considering the exterior Dirichlet problem for the Laplace equation in these spaces, one must in addition define the value of the limit of the solution at infinity. In the case of equations of a more general form, "the boundary at infinity" may have a more complicated structure. Everything depends on the final (i.e. as $t \to \infty$) behavior of the trajectories of the corresponding diffusion process.

Probabilistic methods have proved to be greatly effective in examining degenerate elliptic and parabolic equations. Chapter III is devoted to these questions. If the coefficients are Lipschitz continuous, then existence and uniqueness theorems are valid for equation (5) regardless of any degeneration of the diffusion matrix $(a^{ij}(x))$. This enables one to examine the peculiarities of the statement of boundary value problems for degenerate equations. In particular, the behavior of the corresponding process near the boundary points is in exact agreement with where and how the boundary conditions will be taken. After the process corresponding to the operator has been constructed, it is not difficult to prove the existence theorem and to clarify uniqueness conditions. The generalized solution is described in the form of functional integral (7). This allows one to examine its local properties. Under broad assumptions, the generalized solution turns out to be Hölder continuous. In order to ensure Lipschitz continuity or smoothness, one should make some special assumptions. Chapter III clarifies the conditions under which the generalized solution is smooth and gives an example illustrating the importance of these conditions. Roughly speaking, the smoothness of the generalized solution is due to the relation between the rate of scattering of the trajectories of system (5) starting from close points and the first eigenvalue (generalized) of the boundary value problem. The rate of scattering of the trajectories is defined by a number which is expressed in terms of the Lipschitz constant of the coefficients of equation (5).

The results of Chapter III, besides being interesting on their own, serve as a basis for Chapter IV where elliptic equations with small parameter in higher derivatives are dealt with. The analysis of how the solutions of boundary value problems depend on these parameters reduces to the following two questions: first, to analyzing the dependence of the trajectories of "ordinary" equation (5) on these parameters, and then to examining the dependence of the functional integral on the parameters contained in the integrand. Here the dependence on the parameters may be understood in a rather broad sense. This may be the dependence on the initial point—in this way Chapter III studies the modulus of continuity and the smoothness of the generalized solutions. This may also be the dependence on various parameters involved in the operator $L + v(x)$. Here, for example, belongs the problem on the behavior of the solutions of the equations with fast oscillating coefficients and various versions of the averaging principle.

The fact that equations (5) are not sensitive to degenerations makes the probabilistic approach especially suitable in problems with small parameter in higher derivatives. Consider the Dirichlet problem in a bounded domain D:

$$
\begin{aligned}
L^{\varepsilon}u^{\varepsilon}(x) = (L_0 + \varepsilon L_1)\, u^{\varepsilon}(x) = \frac{1}{2} \sum_{i,j=1}^{r} A^{ij}(x) \frac{\partial^2 u^{\varepsilon}}{\partial x^i \partial x^j} + \sum_{i=1}^{r} B^i(x) \frac{\partial u^{\varepsilon}}{\partial x^i} + \\
+ \frac{\varepsilon}{2} \sum_{i,j=1}^{r} a^{ij}(x) \frac{\partial^2 u^{\varepsilon}}{\partial x^i \partial x^j} = 0, \; x \in D; \; u^{\varepsilon}(x)|_{\partial D} = \psi(x),
\end{aligned}
\tag{10}
$$

where $\psi(x)$ is a continuous function on ∂D. We admit that the small parameter may precede not all the second-order derivatives and thus the operator L_0 also may, generally speaking, involve terms with second-order derivatives.

The random process corresponding to the operator L^ε may be constructed with the help of the stochastic differential equations

$$\begin{aligned} dX_t^{\varepsilon,x} &= \sigma(X_t^{\varepsilon,x})\,dW_t + B(X_t^{\varepsilon,x})\,dt + \\ &+ \sqrt{\varepsilon}\,\tilde{\sigma}(X_t^{\varepsilon,x})\,d\tilde{W}_t,\ X_0^{\varepsilon,x} = x \in R^r,\ \varepsilon \geq 0\,, \end{aligned} \tag{11}$$

where $\sigma(x)\sigma^*(x) = (A^{ij}(x))$, $\tilde{\sigma}(x)\tilde{\sigma}^*(x) = (a^{ij}(x))$, and W_t and $\tilde{W}_t$ are independent Wiener processes. For $\varepsilon = 0$, equation (11) defines the random functions $X_t^{0,x}$, $x \in R^r$, $t \geq 0$, corresponding to the operator L_0.

From equation (11), one can easily deduce that

$$\lim_{\varepsilon \downarrow 0} P\{\sup_{0 \leq t \leq T} |X_t^{\varepsilon,x} - X_t^{0,x}| > \delta\} = 0 \tag{12}$$

for any $T > 0$, $\delta > 0$.

Denote by $\tau^{\varepsilon,x} = \inf\{t : X_t^{\varepsilon,x} \notin D\}$, $\varepsilon > 0$, the first exit time of the process $X_t^{\varepsilon,x}$ from the domain. The behavior of $\tau^{\varepsilon,x}$ as $\varepsilon \downarrow 0$ is an important characteristic of problem (10). If one supposes that, with probability 1, the trajectories of the degenerate process $X_t^{0,x}$, $x \in D$, leave the domain D in a finite time and, moreover, behave in a rather regular way near the boundary, then it is not difficult to conclude from (12) that $\tau^{\varepsilon,x}$ has a finite limit as $\varepsilon \downarrow 0$, $\lim_{\varepsilon \downarrow 0} u^\varepsilon(x) = u_0(x)$ exists, does not depend on the perturbating operator L_1 and is a unique solution of the equation $L_0 u_0(x) = 0$, $x \in D$, with the corresponding boundary conditions. This is the simplest case. If there are no second-order derivatives in the operator L_0, then we have the known result due to Levinson [1].

If $\tau^{\varepsilon,x}$ grows like ε^{-1} or faster (as $\varepsilon \downarrow 0$), then the limit behavior of $u^\varepsilon(x)$ already depends, generally speaking, on perturbations. For example, §4.3 considers the case when the operator L_0, in a sense, does not help, but does not hinder the trajectories $X_t^{\varepsilon,x}$ from hitting ∂D either. Here, under some extra conditions, hitting the boundary, and thereby $\lim_{\varepsilon \downarrow 0} u^{\varepsilon}(x)$ are controlled by a certain operator which is obtained from L_1 by means of averaging with respect to a measure specified by the operator L_0. If this averaged operator vanishes, then $\lim_{\varepsilon \downarrow 0} u^{\varepsilon}(x)$ is defined by the subsequent approximation which is of the central limit theorem nature.

Next, Chapter IV discusses the case when the operator L_0, in a sense, hinders the process $X_t^{\varepsilon,x}$ from leaving D. In these problems, $\tau^{\varepsilon,x}$ grows very fast as $\varepsilon \downarrow 0$, approximately like $\exp\{\text{const.}\ \varepsilon^{-1}\}$. The case is typical when there are no second-order derivatives in L_0 and the field $B(x) = (B^1(x), \cdots, B^r(x))$ is such that its integral curves everywhere cross the boundary ∂D of the domain D from the outside toward the interior. Here the exit from the domain is defined by the large deviations of the process $X_t^{\varepsilon,x}$ from $X_t^{0,x}$, and the result is formulated and established via the action functional.

In the last section of Chapter IV, a problem is treated where the small parameter precedes the terms of first order, but due to the presence of degenerations, these terms become the main ones. This section sets forth results of the averaging principle type and of large deviations type.

The last three chapters are devoted to the analysis of quasi-linear equations. Chapter V goes into the question of the existence "in the large" (that is for all $t \geq 0$) of a continuous solution of Cauchy's problem and of some mixed problems. The results of this chapter are based on transformations of Markov processes leading to an absolutely continuous change of measure. The last section of Chapter V is devoted to the analysis, as $t \to \infty$, of the solutions of one class of quasi-linear systems admitting a simple probabilistic interpretation.

Chapters VI and VII consider various generalizations of the Kolmogorov-Petrovskii-Piskunov equation [1]

$$\frac{\partial u(t,x)}{\partial t} = \frac{D}{2}\frac{\partial^2 u}{\partial x^2} + f(u),\ u(0,x) = \begin{cases} 1, x \leq 0 \\ 0, x > 0\ . \end{cases} \tag{13}$$

As the function $f(u)$, one can, for example, take $u(1-u)$ or $u(u-\mu)(1-u)$. It is known that, for large t, the solution of problem (13) behaves as a wave with some profile $V(\xi)$, $-\infty < \xi < \infty$, travelling with some velocity c^* from left to right: $u(t,x) \approx V(x-c^*t)$, $t \to \infty$. A small parameter may be introduced in the problem so that the wave (considered in a rough preliminary approximation) may become a step travelling with the velocity c^*. Such a consideration enables one to generalize the problem widely and examine a number of new effects, such as the appearance of "new sources" in space non-homogeneous media, the wave propagation at the expense of boundary conditions, and some others. This chapter also discusses the question of going over from the description of the wave propagation via equations of type (13) to the axiomatic theory of excitable media. The last section considers the problem of wave propagation in some systems of differential equations. As the basic apparatus in this and in the next chapters, we use asymptotic bounds of the Laplace type for functional integrals.

Chapter VII, the last one, examines the behavior, as $t \to \infty$, of the solution of equation (13) type in which the non-linear term has the form $f = f(x,u)$. As $f(x,u)$, we take either a function periodic in x, or a random field homogeneous in x. In either case the notion of wave propagation velocity is introduced (strictly speaking, the wave itself does not exist, though). This velocity is expressed in terms of some spectral characteristics of the operators.

Today a number of reviews and monographs are available where there are some applications of function integration and probabilistic methods in analysis, differential equations, and physics (e.g. Kac [1], [2], Feynman and Hibbs [1], Freidlin [7], Dynkin and Yushkevich [1], McKean [1], Friedman [2], Wentzell and Freidlin [2], Simon [1]). The overlap of this book with the above monographs is not large. However, to make this book self-contained we had to include some results which are contained in the foregoing works.

We note that in this book there are no results on potential theory. Special monographs are already available on this subject (Blumenthal and Getoor [1], Meyer [1]). The problems connected with analyzing the spectra for second-order operators are also barely mentioned in this book. The materials on these questions may be found in the works of Kac [2, 3], Simon [1], Wentzell and Freidlin [2].

In conclusion we will explain how formulas and theorems are numbered. For example, Theorem 3.2.1 is the first theorem in the second section in Chapter III. Inside Chapter III, it is written as Theorem 2.1 only. Formulas are numbered in a similar fashion. Figures are numbered consecutively within each chapter.

Chapter I

STOCHASTIC DIFFERENTIAL EQUATIONS AND RELATED TOPICS

§1.1 *Preliminaries*

As is customary in mathematical probability theory, we start with a *probability space* $(\Omega, \mathcal{F}, P)$. Here Ω is an arbitrary set which is interpreted as the space of elementary events. The second component $\mathcal{F}$ is a *σ-field* of subsets of the space Ω, i.e. the system of subsets of the space Ω containing Ω itself and being closed with respect to unions and intersections in finite or countable numbers, as well as with respect to the operation of taking the complement. The elements of the *σ-field* $\mathcal{F}$ are called *events*. The third component of the probability space, P, is a *probability measure* on the σ-field $\mathcal{F}$, i.e. it is a non-negative, countably additive function defined on $\mathcal{F}$ and such that $P(\Omega) = 1$.

A function $\xi(\omega)$ on Ω with real values, for which $\{\omega : \xi(\omega) < x\} \in \mathcal{F}$ for any $x \in (-\infty, \infty)$, is called a *random variable*.

Given a set X with some σ-field $\mathcal{B}$ of its subsets (a *measurable space* $(X, \mathcal{B})$), one can define a random variable $\xi(\omega)$ with values in X. Indeed, it is required that the function $\xi(\omega)$ be $(\mathcal{F}, \mathcal{B})$-measurable: $\{\omega : \xi(\omega) \in B\} \in \mathcal{F}$ for any $B \in \mathcal{B}$.

A probability measure $\mu(D) = P\{\xi(\omega) \in D\}$, $D \in \mathcal{B}$ is termed the *distribution* of the random variable $\xi(\omega)$.

If a space X is equipped with a topology, then the minimal σ-field containing all open sets, is called a *Borel σ-field* of the topological space X. The Borel σ-field in the Euclidean space R^r is denoted by $\mathcal{B}^r$; $(\mathcal{F}, \mathcal{B}^r)$-measurable (or briefly, $\mathcal{F}$-measurable) functions on Ω with values in R^r, are termed r-dimensional random variables. The mathematical expectation of a random variable $\xi(\omega)$ will be denoted by $E\xi$:

$$E\xi = \int_{\Omega} \xi(\omega)\, P(d\omega) .$$

To every r-dimensional random variable ξ, there is a corresponding *characteristic function* $f_\xi(\lambda)$: $f_\xi(\lambda) = E \exp\{i(\lambda, \xi)\}$, $\lambda \in R^r$.

If a characteristic function $f_\xi(\lambda)$ has the form

$$f_\xi(\lambda) = \exp\left\{-\frac{1}{2}\,(Q(\lambda-m), \lambda-m)\right\}, \ \lambda \in R^r ,$$

for some $m \in R^r$ and $Q = (q^{k\ell})$, $k, \ell = 1,2,\cdots,r$, then the random variable ξ is called *Gaussian*. Here $m = (m^1, \cdots, m^r)$ is the mathematical expectation of $\xi = (\xi^1, \cdots, \xi^r)$, and $(q^{k\ell}) = Q$ is the matrix of covariances: $q^{k\ell} = E(\xi^k - m^k)(\xi^\ell - m^\ell)$, $m^k = E\xi^k$. Here and henceforth we denote by $(\cdot\,,\cdot)$ the Euclidean scalar product. We remind that $q^{k\ell} = 0$ if and only if the components ξ^k and ξ^ℓ of a Gaussian random variable are independent. The class of Gaussian random variables is closed with respect to linear transformations and with respect to limit passage.

Suppose that in the space Ω, there is a σ-subfield $\mathcal{Y}$ of the underlying σ-field $\mathcal{F}: \mathcal{Y} \subseteq \mathcal{F}$. By a *conditional expectation* $E(\xi\,|\mathcal{Y})$ of a one-dimensional random variable ξ, we mean a $\mathcal{Y}$-measurable function on Ω for which the equality

$$\int_{\Lambda} E(\xi\,|\mathcal{Y})\, P(d\omega) = \int_{\Lambda} \xi(\omega)\, P(d\omega)$$

holds for any $\Lambda \in \mathcal{Y}$. These conditions define the conditional expectation in a unique way up to a set of zero measure. The equalities between conditional expectations are all fulfilled almost surely (P-a.s. or P-a.e.).

We shall list some basic properties of conditional expectations:

1. $E(\xi \mid \mathcal{Y}) \geq 0$, if $\xi \geq 0$.

2. $E(\xi + \eta \mid \mathcal{Y}) = E(\xi \mid \mathcal{Y}) + E(\eta \mid \mathcal{Y})$

if the summands on the right-hand side exist.

3. $E(\xi\eta \mid \mathcal{Y}) = \xi E(\eta \mid \mathcal{Y})$ if $E\xi\eta$ and $E\eta$ are defined and ξ is $\mathcal{Y}$-measurable.

4. Let $\mathcal{Y}_1$ and $\mathcal{Y}_2$ be two σ-fields such that $\mathcal{Y}_1 \subseteq \mathcal{Y}_2 \subseteq \mathcal{F}$. Then $E(\xi \mid \mathcal{Y}_1) = E(E(\xi \mid \mathcal{Y}_2) \mid \mathcal{Y}_1)$.

5. Suppose that a random variable ξ does not depend on a σ-field $\mathcal{Y}$, that is $P\{(\xi \in D) \cap \mathcal{A}\} = P\{\xi \in D\} P(\mathcal{A})$ for any Borel set $D \in R^1$ and $\mathcal{A} \in \mathcal{Y}$. Then $E(\xi \mid \mathcal{Y}) = E\xi$ provided $E\xi$ exists.

Let $\chi_{\mathcal{A}}(\omega)$ be the indicator of a set $\mathcal{A} \in \mathcal{F}$, i.e. $\chi_{\mathcal{A}}(\omega) = 1$ for $\omega \in \mathcal{A}$ and $\chi_{\mathcal{A}}(\omega) = 0$ for $\omega \in \Omega \setminus \mathcal{A}$. Then the random variable $E(\chi_{\mathcal{A}}(\omega) \mid \mathcal{Y})$ is said to be the conditional probability of the event $\mathcal{A}$ with respect to the σ-field $\mathcal{Y}$ and is denoted by $P(\mathcal{A} \mid \mathcal{Y})$.

A family of r-dimensional random variables $\xi_t(\omega)$ depending on a real parameter $t \in T \subset (-\infty, \infty)$ is called a *random process*. Thus, for every fixed $t = t_0$, we obtain a random variable $\xi_{t_0}(\omega)$. For a fixed $\omega = \omega_0$ we obtain a function of t which is called a trajectory or a sample function of the process $\xi_t(\omega)$.

The totality of distributions $\mu_{t_1, \dots, t_n}$ of random variables $(\xi_{t_1}, \dots, \xi_{t_n})$ for various $n = 1,2,3, \cdots$ and $t_1, \cdots, t_n \in T$, is termed the *family* of *finite-dimensional distributions* of the process $\xi_t(\omega)$. In this book, we shall, as a rule, consider random processes having parameter t which varies over the half-line $[0, \infty)$ or over the interval $[0, T_1]$ and having trajectories which are continuous with probability 1. Such processes are, in fact, defined in a unique way by their finite-dimensional distributions.

If the finite-dimensional distributions of a process $\xi_t(\omega)$, $t \in [0, \infty)$, are all Gaussian ones, then the process is called a *Gaussian random process*. The finite-dimensional distributions of such a process are all defined in a unique way by two functions—by the mathematical expectation $m(t) = E\xi_t$ and by the correlation function $R(s,t) = E(\xi_s - m(s))(\xi_t - m(t))$.

If $\xi_t = (\xi_t^1, \cdots, \xi_t^r)$ is an r-dimensional random process, then $m(t) = (E\xi_t^1, \cdots, E\xi_t^r)$; $R(s,t) = (R^{ij}(s,t))$, $R^{ij}(s,t) = E(\xi_s^i - m^i(s))(\xi_t^j - m^j(t))$.

With every random process $\xi_t(\omega)$, $t \in T$, one can associate an increasing system of σ-fields $\mathcal{F}_{\leq t} = \mathcal{F}_{\leq t}^{\xi} = \sigma(\xi_s, s \leq t)$ where $\sigma(\xi_s, s \leq t)$ is the minimal σ-field with respect to which the random variables ξ_s are all measurable for $s \leq t$, $s \in T$. Sometimes we shall also consider the σ-fields $\mathcal{F}_{\geq t} = \mathcal{F}_{\geq t}^{\xi} = \sigma(\xi_s, s \geq t)$. We shall use the notation $E(\eta|\xi_s, s \in S)$ for the conditional expectation of the random variable η with respect to the σ-field generated by the random variables ξ_s for $s \in S$ (i.e. the minimal σ-field with respect to which the random variables ξ_s are all measurable for $s \in S$).

All the objects introduced here and their properties, as well as other elementary information from probability theory, which is assumed to be known to the reader, are considered in detail in many courses in the theory of stochastic processes (see, e.g. Doob [1], Gihman and Skorohod [1], Wentzell [1]).

§1.2 *The Wiener measure*

In this section, the Wiener measure in the space of continuous functions will be constructed. This measure is connected with the Laplace operator and with the simplest heat operator. Solutions of some problems for such operators admit a representation in the form of integrals of appropriate functionals with respect to this measure. In many respects, the measures connected with general second-order elliptic operators are similar to the Wiener measure, and it is convenient to construct such measures proceeding from the Wiener measure.

Because the Wiener measure is so important, we will describe a few constructions for it and list its basic properties. As a rule, we shall drop proofs restricting ourselves to brief comments on what needs to be proved. The detailed proofs of these properties are available in many manuals and monographs, so the reader is referred to the corresponding literature.

Let $C_{0,T} = C_{0,T}(R^1)$ be the space of continuous functions on $[0,T]$, $T < \infty$, with values on the real line R^1; $C_{0,\infty} = C_{0,\infty}(R^1)$ being the space of continuous functions on $[0,\infty)$. We put $C^0_{0,T} = \{\phi \in C_{0,T} : \phi_0 = 0\}$, $0 < T \le \infty$. The spaces C_{0T} and C^0_{0T} will be thought of as equipped with the uniform convergence topology (uniform convergence on every finite interval whenever $T = \infty$).

As the σ-field in the spaces $C^0_{0,T}$, $0 < T \le \infty$, one can take the Borel σ-field $\mathcal{B}_{0,T}$, that is the minimal σ-field containing all open sets of the space $C^0_{0,T}$.

Suppose we are given a measurable space $(X, \mathcal{B})$, i.e. there is a set X, with a σ-field of subsets $\mathcal{B}$ being defined in it. How can one put a measure on this space? There are several ways of assigning a measure. Presumably, it is the easiest way to define a measure with the aid of a density function with respect to some standard measure defined on the same σ-field $\mathcal{B}$. For example, let $X = R^1$ be the real line equipped with the σ-field of Borel sets. Then, as the standard measure, it is sometimes natural to take the Lebesgue measure; that is, a measure on the line, unique up to scale factor and invariant with respect to translations. For instance, Gaussian measure on the line is specified by the density function

$$p(x) = \frac{1}{\sigma\sqrt{2\pi}} \exp\left\{-\frac{(x-a)^2}{2\sigma^2}\right\}$$

with respect to the Lebesgue measure, where a and σ are real parameters.

So far we have had no measure in the space $C^0_{0,T}$. It should be noted that it is not easy to define some non-trivial measure in this space, and there is no measure which is invariant with respect to translations on $C^0_{0,T}$. Therefore, this way does not fit as a starting point, but we shall bear it in mind for later constructions.

Another way of defining a measure consists of prescribing it with the help of some mapping. Let a measurable space $(Y, \mathcal{B}')$ and a probability measure $P'(A)$, $A \in \mathcal{B}'$ be already available. Suppose we are given a measurable mapping $f: (Y, \mathcal{B}') \to (X, \mathcal{B})$; that is, a mapping for which $f^{-1}(B) \in \mathcal{B}'$ for $B \in \mathcal{B}$. Then this mapping induces the probability measure $P(B) = P'(f^{-1}(B))$ on the σ-field $\mathcal{B}$ in X. And this way is already suitable for constructing the Wiener measure in $C^0_{0,T}$. We shall provide this construction later on in this section.

The third way of defining a measure is with the help of passage to the limit. Namely, it is possible to construct a sequence of measures μ_n in $C^0_{0,T}$ being described in a comparatively simple manner, and then to consider the limit of μ_n as $n \to \infty$. For example, limit in the sense of weak convergence of measures is convenient. Recall that a sequence of measures μ_n in $C^0_{0,T}$ converges weakly to a measure μ if

$$\int_{C^0_{0,T}} f(x)\,\mu_n(dx) \to \int_{C^0_{0,T}} f(x)\,\mu(dx)\,, \quad n \to \infty\,,$$

for any continuous bounded functional $f(x)$ on the space $C^0_{0,T}$.

In this way, the Wiener measure may be constructed in $C^0_{0,T}$ as well.

Finally, we recall another approach widely used in probability theory—Kolmogorov's extension of measure. According to this method, a measure for some collection of relatively simple sets must be defined, and then it must be continued, by the countable additivity property, onto the smallest σ-field containing the original collection of sets. Of course, in doing so one must demonstrate that such an extension will not face obstacles and will give a measure, countably additive on this minimal σ-field. If one wants to get a measure on the σ-field which has been set beforehand, then, in addition, one should make certain that this σ-field is contained in the minimal σ-field generated by the simple sets.

We begin by outlining the construction of the Wiener measure via the last procedure.

So, first of all, a collection of "simple" sets in the space of continuous functions must be indicated. Let $0 < t_1 < t_2 < \cdots < t_n$, where n is any positive integer. Moreover, let $\Gamma_1, \Gamma_2, \cdots, \Gamma_n$ be intervals of the real line (not necessarily different). We will denote by $\Pi^{t_1,\ldots,t_n}_{\Gamma_1,\ldots,\Gamma_n}$ the following set in the space $C^0_{0,\infty}$ (Fig. 1):

$$\Pi^{t_1,\ldots,t_n}_{\Gamma_1,\ldots,\Gamma_n} = \{\phi \in C^0_{0,\infty} : \phi_{t_1} \in \Gamma_1, \cdots, \phi_{t_n} \in \Gamma_n\}.$$

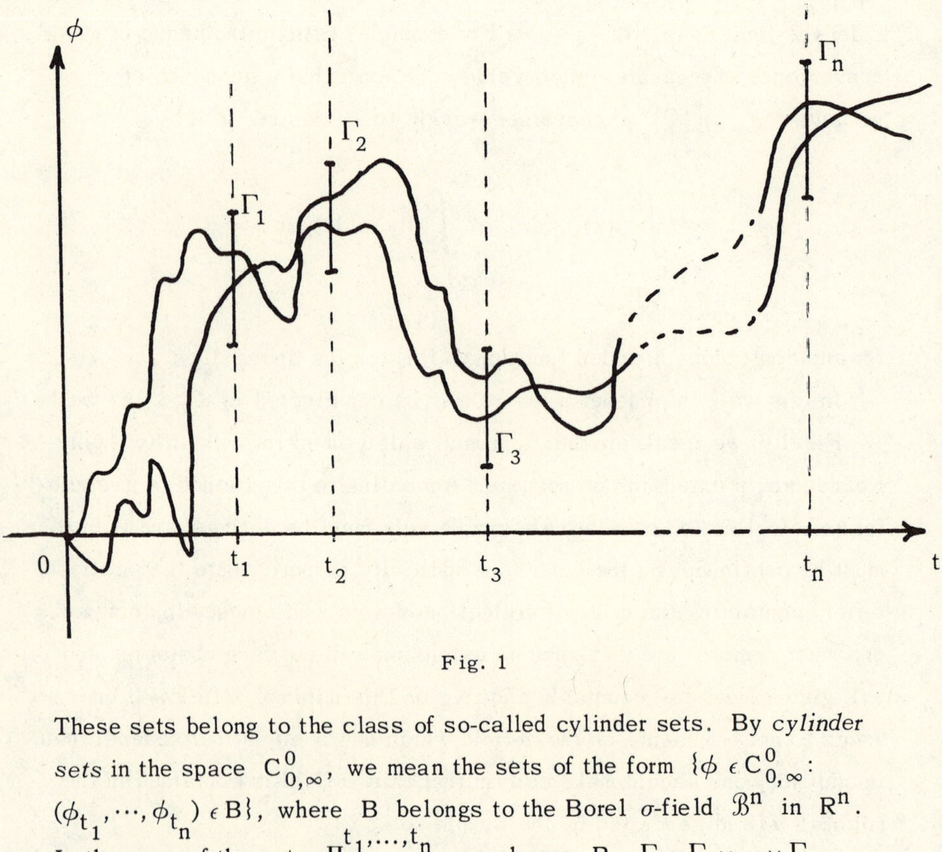

Fig. 1

These sets belong to the class of so-called cylinder sets. By *cylinder sets* in the space $C^0_{0,\infty}$, we mean the sets of the form $\{\phi \in C^0_{0,\infty} : (\phi_{t_1}, \cdots, \phi_{t_n}) \in B\}$, where B belongs to the Borel σ-field $\mathcal{B}^n$ in R^n. In the case of the sets $\Pi^{t_1,\ldots,t_n}_{\Gamma_1,\ldots,\Gamma_n}$ we choose $B = \Gamma_1 \times \Gamma_2 \times \cdots \times \Gamma_n$.

Now let us introduce a measure on the sets $\Pi^{t_1,\ldots,t_n}_{\Gamma_1,\ldots,\Gamma_n}$. We designate

$$p(t,x,y) = \frac{1}{\sqrt{2\pi t}} e^{-\frac{(x-y)^2}{2t}}, x,y \in R^1, t > 0, \tag{2.1}$$

and put

$$\mu(\Pi^{t_1,\ldots,t_n}_{\Gamma_1,\ldots,\Gamma_n}) = \int_{\Gamma_1} dy_1 \cdots \int_{\Gamma_n} dy_n p(t_1,0,y_1)\, p(t_2-t_1,y_1,y_2) \cdots p(t_n-t_{n-1},y_{n-1},y_n)\,. \tag{2.2}$$

Notice that if, for some i, Γ_i coincides with the whole space R^1, then

$$\Pi^{t_1,\ldots,t_{i-1},t_i,t_{i+1},\ldots,t_n}_{\Gamma_1,\ldots,\Gamma_{i-1},R^1,\Gamma_{i+1},\ldots,\Gamma_n} = \Pi^{t_1,\ldots,t_{i-1},t_{i+1},\ldots,t_n}_{\Gamma_1,\ldots,\Gamma_{i-1},\Gamma_{i+1},\ldots,\Gamma_n}\,.$$

In other words, the superscript t_i and the corresponding interval $\Gamma_i = R^1$ may be omitted. Hence, one and the same set in the space of functions may be written in the form $\Pi^{t_1,\ldots,t_n}_{\Gamma_1,\ldots,\Gamma_n}$ with different number of indices.

Formula (2.2) for the measure of this set can also be written in various ways. Thus, for our definition of the measure of "simple" sets to be correct, it is necessary that, if $\Pi^{t_1,\ldots,t_n}_{\Gamma_1,\ldots,\Gamma_n} = \Pi^{t'_1,\ldots,t'_m}_{\Gamma'_1,\ldots,\Gamma'_m}$ then the corresponding values of the measure defined by formula (2.2) coincide. This property is referred to as the compatibility of a family of distributions. It is possible to check that formula (2.2) does define a compatible family of distributions. This compatibility comes from the Kolmogorov-Chapman equation

$$p(s+t,x,y) = \int_{-\infty}^{\infty} p(s,x,z)\, p(t,z,y)\, dz \tag{2.3}$$

which is fulfilled for function (2.1).

It is easily seen that the function $p(t,x,y)$ defined by (2.1) also has the following properties:

$$p(t,x,y) = p(t,0,y-x), \int_{-\infty}^{\infty} x^{\alpha} p(h,0,x) dx = const \times h^{\alpha/2}, \qquad (2.4)$$

where α is an arbitrary positive number.

By using the compatibility of the distributions given by formula (2.2) and properties (2.4), it is now possible to prove that the function defined for simple sets by formula (2.2) may be extended to a measure on the minimal σ-field of subsets of the space $C_{0,\infty}^0$ which contains all possible $\Pi_{\Gamma_1,\ldots,\Gamma_n}^{t_1,\ldots,t_n}$ (see e.g. Ito and McKean [1], Wentzell [1]). This minimal σ-field will be denoted by $\mathfrak{N}_0^{\infty}$, $\mathfrak{N}_0^t$ will designate the σ-subfield of the σ-field $\mathfrak{N}_0^{\infty}$ generated by the sets $\Pi_{\Gamma_1,\ldots,\Gamma_n}^{t_1,\ldots,t_n}$ for arbitrary natural numbers n and $0 < t_1 < t_2 < \cdots < t_n \leq t$.

What kind of sets belong to the σ-fields $\mathfrak{N}_0^t$ and $\mathfrak{N}_0^{\infty}$? Generally speaking, the σ-fields $\mathfrak{N}_0^t$ and $\mathfrak{N}_0^{\infty}$ are rather extensive. For example, let us consider the set:

$$\mathfrak{A} = \{\phi \in C_{0,\infty}^0 : f_s < \phi_s < g_s \text{ for } s \in [0,t]\},$$

where f_s, g_s, $s \in [0,t]$ are arbitrary continuous functions on the segment $[0,t]$, $f_s < g_s$, $f_0 < 0 < g_0$. Since we deal with continuous functions,

$$\mathfrak{A} = \bigcup_{n=1}^{\infty} \bigcap_{s_i \in \Lambda_{0,t}} \left\{ f_{s_i} + \frac{1}{n} < \phi_{s_i} < g_{s_i} - \frac{1}{n} \right\},$$

where $\Lambda_{0,t}$ denotes the set of rational numbers in the segment $[0,t]$. Therefore, the set $\mathfrak{A}$ may be represented in the form of a union of the intersections of a countable number of simple sets, and thus $\mathfrak{A} \in \mathfrak{N}_0^t$.

Hence it appears clear that the σ-field $\mathfrak{N}_0^{\infty}$ contains the Borel σ-field $\mathfrak{B}_{0,\infty}$. It is possible to make sure that the opposite inclusion is also valid, and therefore, $\mathfrak{B}_{0,\infty} = \mathfrak{N}_0^{\infty}$. For $0 < t_1 < \cdots < t_n \leq T$, the sets $\Pi_{\Gamma_1,\ldots,\Gamma_n}^{t_1,\ldots,t_n}$ may be projected into the space $C_{0,T}^0$. The σ-field generated by these sets in $C_{0,T}^0$ will also be called $\mathfrak{N}_0^T$. It is readily checked that this σ-field in $C_{0,T}^0$ coincides with the Borel σ-field $\mathfrak{B}_{0,T}$.

We proceed to present some other examples of sets from $\mathfrak{N}_0^T$ and $\mathfrak{N}_0^{\infty}$.

Let $x > 0$, $\tau_x(\phi) = \min\{t : \phi_t = x\}$, i.e. $\tau_x(\phi)$ is the first moment when $\phi \in C^0_{0,\infty}$ hits the level x; if the set $\{t : \phi_t = x\}$ is empty (the function ϕ never attains the level x), then we set $\tau_x(\phi) = +\infty$. Consider the set $\mathfrak{D} = \{\phi \in C^0_{0,\infty} : \tau_x(\phi) \le t\}$. It is clear that $\mathfrak{D} = \bigcap_{n=1}^{\infty} \bigcup_{s_i \in \Lambda_{0,t}} \left\{\phi \in C^0_{0,t} : \phi_{s_i} > x - \frac{1}{n}\right\}$ which implies that $\mathfrak{D} \in \mathfrak{N}^t_0$. It is not difficult to verify that the set $\mathfrak{E} = \{\phi \in C^0_{0,\infty} : \tau_x(\phi) < \tau_x(-\phi)\}$ of the trajectories reaching the point x before the point $-x$, belongs to the σ-field $\mathfrak{N}^\infty_0$. Throughout this book, we will often consider the first hitting times of closed sets.

So, in the space $C^0_{0,T}$ there is a σ-field $\mathfrak{N}^T_0$ which coincides with the Borel σ-field $\mathfrak{B}_{0,T}$ of this space (if $T = \infty$, then $\mathfrak{N}^\infty_0 = \mathfrak{B}_{0,\infty}$).

The measure on the σ-field $\mathfrak{N}^T_0$, $0 < T \le \infty$, defined by equality (2.2) for the "simple" sets $\Pi^{t_1,\dots,t_n}_{\Gamma_1,\dots,\Gamma_n}$, is called the *Wiener measure.*

The first mathematically correct construction of such a measure and the analysis of its basic properties was brought about by N. Wiener [1] in 1923.

REMARK. The statement that the function given by (2.2) for simple sets may be extended to a measure on the σ-field $\mathfrak{N}^T_0$ in the space $C^0_{0,T}$, is based on two different results. The first of them is Kolmogorov's general theorem on the extension of a measure in the space $H_{0,T}$ of all functions on $[0,T]$ with values in R^1. The second result is concerned with the conditions under which this measure is concentrated on the space of continuous functions. In order to clarify the meaning of these results, let us consider the "simple" sets in $H_{0,T}$:

$$\tilde{\Pi}^{t_1,\dots,t_n}_{\Gamma_1,\dots,\Gamma_n} = \{\phi \in H_{0,T} : \phi_{t_1} \in \Gamma_1, \cdots, \phi_{t_n} \in \Gamma_n\}.$$

Here n is any natural number, $\Gamma_1, \cdots, \Gamma_n$ are intervals on the line, $0 < t_1 < \cdots < t_n \le T$. For the simple sets $\tilde{\Pi}^{t_1,\dots,t_n}_{\Gamma_1,\dots,\Gamma_n} \subset H_{0,T}$ we will define the function

$$\tilde{\mu}(\tilde{\Pi}_{\Gamma_1,\ldots,\Gamma_n}^{t_1,\ldots,t_n}) =$$

$$= \int_{\Gamma_1} dy_1 \cdots \int_{\Gamma_n} dy_n p(t_1,0,y_1)\, p(t_2-t_1,y_1,y_2) \cdots p(t_n-t_{n-1},y_{n-1},y_n), \tag{2.5}$$

where $p(t,x,y)$ is specified by equality (2.1).

Just as in the case of the simple sets in $C^0_{0,T}$, one and the same simple set in $H_{0,T}$ may be described in different ways, for example, with a different number of indices. However, the function $\tilde{\mu}$ depends only on the set $\tilde{\Pi}_{\Gamma_1,\ldots,\Gamma_n}^{t_1,\ldots,t_n} \subset H_{0,T}$ and not on how it is written down. This follows from the Kolmogorov-Chapman equation (2.3). By the theorem on the extension of measure, the function $\tilde{\mu}$, originally defined only on simple sets in $H_{0,T}$, may be extended to a measure on the minimal σ-field $\tilde{\mathfrak{N}}_0^T$ in $H_{0,T}$ which contains all simple sets $\tilde{\Pi}_{\Gamma_1,\ldots,\Gamma_n}^{t_1,\ldots,t_n}$, $0 < t_1 < t_2 < \cdots < t_n < T$.

The explicit form of the function $p(t,x,y)$ is not essential for the extension of $\tilde{\mu}$ to a measure on $\tilde{\mathfrak{N}}_0^T$ to exist. For example, the function $p_1(t,x,y) = \dfrac{t}{\pi[t^2+(y-x)^2]}$ also satisfies the Kolmogorov-Chapman equation. If in equalities (2.5), $p(t,x,y)$ is replaced by $p_1(t,x,y)$, then we have some new function $\tilde{\mu}_1$ on the simple sets in $H_{0,T}$. This function may be extended to a measure on the σ-field $\tilde{\mathfrak{N}}_0^T$ in $H_{0,T}$ as well.

Suppose now that in equalities (2.2), p_1 is substituted for p. If these modified equalities are used to define the function $\mu_1(\Pi_{\Gamma_1,\ldots,\Gamma_n}^{t_1,\ldots,t_n})$ for simple sets in $C^0_{0,T}$, then this function may no longer be extended to a measure on $\mathfrak{N}_0^T$. In order that one could perform the extension to a measure on the σ-field $\mathfrak{N}_0^T$ in the space $C^0_{0,T}$, it is sufficient to check that the outer measure of the set $C^0_{0,T}$ in $H_{0,T}$ is equal to 1 ($C^0_{0,T}$ does not belong to $\tilde{\mathfrak{N}}_0^T$, thus the measure $\tilde{\mu}(C^0_{0,T})$ is not defined):

$$\inf\{\tilde{\mu}(\mathfrak{A}) : C^0_{0,T} \subset \mathfrak{A} \in \tilde{\mathfrak{N}}^T_0\} = 1 .$$

It turns out that, for the last equality to be fulfilled, simple sufficient conditions may be given. Namely, for the outer measure of the set $C^0_{0,T}$ to be 1, and therefore, in order that one could extend μ from simple sets in $C^0_{0,T}$ to a measure on $\mathfrak{N}^T_0$, it suffices that, for some $\alpha, \beta, c > 0$

$$\int_{-\infty}^{\infty}\int_{-\infty}^{\infty} |x_1-x_2|^\alpha \mu_{t_1,t_2}(dx_1 \times dx_2) \le c|t_1-t_2|^{1+\beta} , \tag{2.6}$$

where $\mu_{t_1,t_2}(\Gamma_1 \times \Gamma_2) = \mu(\Pi^{t_1,t_2}_{\Gamma_1,\Gamma_2})$. The above assertion is the Kolmogorov theorem on the existence of a continuous modification. In the case of the function $p(t,x,y)$ such constants α, β, c exist. This follows, for example, from (2.4) with $\alpha = 4$. Therefore, one can make the extension of the measure defined by (2.2) for simple sets in $C^0_{0,T}$ to a measure on the σ-field $\mathfrak{N}^T_0$. For μ_1, introduced previously, such an extension in $C^0_{0,T}$ is impossible.

If the compatibility condition and bound (2.6) holds, then one can arrange the proof of the existence of an extended measure immediately in the space $C^0_{0,T}$ (see Ito and McKean [1]), without dividing it into these two stages.

The notion of the Wiener measure is closely related to that of Wiener random process. A random process $W_t(\omega)$, $t \ge 0$, on a probability space $(\Omega, \mathcal{F}, P)$ is called a *Wiener process* if its trajectories are continuous with probability 1, $P\{W_0 = 0\} = 1$, and the finite-dimensional distributions are given by

$$P\{W_{t_1} \in \Gamma_1, \cdots, W_{t_n} \in \Gamma_n\} =$$
$$= \int_{\Gamma_1} dy_1 \cdots \int_{\Gamma_n} dy_n\, p(t_1,0,y_1)\, p(t_2-t_1,y_1,y_2) \cdots p(t_n-t_{n-1},y_{n-1},y_n) \tag{2.7}$$

where $p(t,x,y)$ is defined by equality (2.1).

Formula (2.7) implies that W_t is a Gaussian process. By using (2.7) it is not difficult to calculate that $EW_t = 0$, and the correlation function $R(s,t)$ of the process W_t has the form $R(s,t) = EW_sW_t = s \wedge t$. Note that the correlation function and the expectation of a Gaussian process determine, in a unique way, its finite-dimensional distribution functions. Thus, the Wiener process could be defined as a mean zero Gaussian process with continuous trajectories having the correlation function $R(s,t) = s \wedge t$.

Every one-dimensional random process $X_t(\omega)$, $t \geq 0$, $\omega \in \Omega$ whose trajectories are continuous with probability 1, induces a mapping $\Omega \to C_{0,\infty} : \omega \to X_{\cdot}(\omega)$. This mapping induces a probability measure on the σ-field[1] $\mathcal{N}_0^\infty$ in $C_{0,\infty}$. Comparing (2.7) with (2.2) we draw the conclusion that the process $W_t(\omega)$ induces the Wiener measure in the space $C^0_{0,\infty}$.

On the other hand, given the space $C^0_{0,\infty}$ with the Wiener measure μ_W on the σ-field $\mathcal{N}_0^\infty$, one can take $(C_{0,\infty}, \mathcal{N}_0^\infty, \mu_W)$ as a probability space, and define the random process $W_t(\omega) = \phi_t$, $t \in [0,\infty)$, for $\omega = \phi$. By virtue of (2.2) this process is a Wiener process. Therefore, the trajectories of the Wiener process $W_t(\omega)$ are simply elements of the space $C^0_{0,\infty}$. The elements of the σ-field $\mathcal{N}_0^t$ are events defined by the motion of the Wiener process W_s for $s \in [0,t]$.

The construction of Wiener measure and the examination of its properties is in essence equivalent to the construction and examination of the properties of Wiener process.

Let us consider other constructions of the Wiener measure. We will construct the Wiener measure in $C^0_{0,1}$. Afterwards, it is not hard to define the Wiener measure in $C^0_{0,T}$ for any $T > 0$.

---

[1]We preserve the notation $\mathcal{N}_0^t$, $0 \leq t \leq \infty$, for the σ-field of all cylinder sets of $C_{0,t}$ (not only of $C^0_{0,t}$), i.e. $\mathcal{N}_0^t$ is the σ-field generated by the sets $\{\phi \in C_{0,t} : (\phi_{t_1}, \ldots, \phi_{t_n}) \in B\}$ for any natural n, any $0 \leq t_1 < t_2 < \cdots < t_n \leq t$, and $B \in \mathcal{B}^n$.

Suppose we are given a probability space $(\Omega, \mathcal{F}, P)$ and a sequence of independent mean zero Gaussian random variables on it having variance $1: \xi_0, \xi_1, \xi_2, \cdots$. Such a probability space may be constructed, for example, by taking an infinite product of lines equipped with Borel σ-fields and the standard Gaussian measure. We will define the mapping $(\Omega, \mathcal{F}) \to (C^0_{0,1}, \mathcal{N}^1_0)$ in such a fashion that in $C^0_{0,1}$ the Wiener measure will be induced.

Let us consider the series

$$\phi_s(\omega) = \xi_0 s + \frac{\sqrt{2}}{\pi} \sum_{n=1}^{\infty} \sum_{k=2^{n-1}}^{2^n-1} \xi_k \frac{\sin k\pi s}{k}, \quad 0 \leq s \leq 1 . \tag{2.8}$$

This series converges uniformly on $[0,1]$ with probability 1 (see, e.g. Ito and McKean [1]).

Thus, formula (2.8) gives the mapping $f: \Omega \to C^0_{0,1}$. This mapping is readily checked to be measurable, i.e. $f^{-1}(B) \in \mathcal{F}$ for $B \in \mathcal{N}^1_0$. Since the sum of independent Gaussian random variables is also a Gaussian random variable, and since the limit of Gaussian random variables is also a Gaussian random variable, we conclude that the process $\phi_s(\omega)$ is Gaussian. Now let us verify that it is a Wiener process. As was said above, for this it suffices to check that $E\phi_s = 0$, $E\phi_s\phi_t = s \wedge t$. The first of these equalities comes from the fact that series (2.8) may be integrated with respect to the measure $P(d\omega)$ on Ω termwise. Noting that $E\xi_i\xi_j = \delta_{ij}$ we conclude from (2.8) that

$$E\phi_s\phi_t = st + \sum_{k=1}^{\infty} \frac{2}{\pi^2} \frac{\sin k\pi s \sin k\pi t}{k^2}, \quad s,t \in [0,1] .$$

We leave to the reader to make certain that the expression on the right-hand side is equal to $s \wedge t$. Hence, $\phi_s(\omega)$ is a Wiener process. The mapping $f: \Omega \to C^0_{0,1}$ induces the Wiener measure in $C^0_{0,1}$.

Finally we mention another way of constructing the Wiener measure ensuring a transparent image of the nature of the set of functions this measure is concentrated on.

We shall consider a symmetric random walk over a lattice on the line. Let a particle (being at time t at a point kh) jump to one of the neighboring points $(k-1)h$ or $(k+1)h$ with equal probabilities at time $t+\Delta$. Here $h > 0$ is the lattice spacing, $\Delta > 0$ is a time interval between sequential jumps, $k = 0,1,2,\cdots$. Let us denote by $X_{k\Delta}^{\Delta,h} = X_{k\Delta}^{\Delta,h}(\omega)$, $k = 0,1,2,\cdots$, the trajectory of this particle starting from zero at time $t = 0$. Clearly, $X_{k\Delta}^{\Delta,h}$ is a random sequence. Let $(\Omega, \mathcal{F}, P)$ be the probability space this sequence is defined on. We will introduce the random broken lines $X_s^{\Delta,h}$ $s \in [0,1]$, consisting of the segments which connect sequential points $(k\Delta, X_{k\Delta}^{\Delta,h}(\omega))$ and $((k+1)\Delta, X_{(k+1)\Delta}^{\Delta,h})$, $k = 0,1,\cdots,\left[\frac{1}{\Delta}\right], \left[\frac{1}{\Delta}\right]+1$. The random broken lines $X_s^{\Delta,h}$ induce in $C_{0,1}^0$ certain measure $\mu^{\Delta,h}$ concentrated on the broken lines with vertices at the points $(k\Delta, \ell h)$, $k = 0,1,2,\cdots$; $\ell = 0,\pm 1,\pm 2,\cdots$. For any integers $k_1 < k_2 < \cdots < k_n$ and any intervals $\Gamma_1, \Gamma_2, \cdots, \Gamma_n \subset R^1$, the value of the measure $\mu^{\Delta,h}$ of the simple set

$$\{X_{\cdot}^{\Delta,h} \in C_{0,1}^0 : X_{k_1\Delta}^{\Delta,h} \in \Gamma_1, X_{k_2\Delta}^{\Delta,h} \in \Gamma_2, \cdots, X_{k_n\Delta}^{\Delta,h} \in \Gamma_n\}$$

may be written down explicitly via binomial probabilities.

Now let $\Delta, h \downarrow 0$ in such a way that $h^2\Delta^{-1} = 1$. Then the family of measures $\mu^{\Delta,h}$ proves to converge weakly to the Wiener measure in the space $C_{0,1}^0$. The proof of this statement may be decomposed into two stages. First, one must prove that the measures $\mu^{\Delta,h}$ of simple sets in $C_{0,1}^0$ converge to the Wiener measure of these sets. This convergence is an implication of a version of the central limit theorem. Secondly, one must verify that the family of measures $\mu^{\Delta,h}$, $h^2 = \Delta$, $\Delta \downarrow 0$ is weakly compact. This implies that the measures $\mu^{\Delta,h}$ converge weakly to the Wiener measure. For a detailed proof, see Donsker [1], Ito and McKean [1].

A more strong statement is also true (see Knight [1], Ito and McKean [1]): one can find a probability space $(\Omega, \mathcal{F}, P)$ and a family of random walks $X_{k\Delta}^{\Delta,h}(\omega)$, $\omega \in \Omega$, such that the broken lines $X_s^{\Delta,h}(\omega)$ drawn by these walks, with probability 1, converge uniformly on $[0,1]$ to the continuous functions $X_s = X_s(\omega)$. The random process $X_s(\omega)$ is a Wiener process.

So, the functions the Wiener measure is concentrated on, may be imagined as the limits of random broken lines $X_s^{\Delta,h}$ for $\Delta = h^2$ as $\Delta \downarrow 0$. Whence, via the properties of symmetric random walk over a lattice on the line, we can obtain the properties of the Wiener process. For example, the fact that, with probability 1, random walk returns to zero an infinite number of times, implies that, for almost all trajectories of the Wiener process $W_s(\omega)$, the set $\{s \in [0,1], W_s(\omega) = 0\}$ is a perfect one (i.e. a closed set every point of which is a limit point for this set).

So, suppose we are given a Wiener process $W_t(\omega)$, $t \geq 0$, on the probability space $(\Omega, \mathcal{F}, P)$. We will list some important properties of the Wiener process (and, thereby, of the Wiener measure as well) in addition to those mentioned previously.

We will show that the Wiener process has independent increments, i.e. for any $0 < t_1 < t_2 < \cdots < t_n$, the random variables $W_{t_1}, W_{t_2} - W_{t_1}, \cdots, W_{t_n} - W_{t_{n-1}}$ are independent. As W_t is a Gaussian process, all the n differences are jointly Gaussian distributed. Hence, to prove the independence of these random variables, it is sufficient to check that they are uncorrelated, i.e. that $E(W_{t_{k+1}} - W_{t_k})(W_{t_{i+1}} - W_{t_i}) = 0$ for $i \neq k$. The last equality may be easily checked by remembering that $EW_s W_t = s \wedge t$. It is possible to demonstrate that the Wiener process is the only random process having continuous, with probability 1, trajectories and independent increments for which $EW_t = 0$, $EW_t^2 = t$.

From the independence of increments, we can readily deduce the so-called *Markov property* of the Wiener process which may be stated as follows: for any $T > 0$, the random process $\widetilde{W}_t = W_{T+t}(\omega) - W_T(\omega)$ does not depend on the σ-field $\mathcal{F}^W_{\leq T} = \sigma(W_s, s \in [0,T])$[2] and is a Wiener process.

The proof of this statement is immediate from the independence of increments and from the following observation: for any $0 \leq t_1 < t_2 < \cdots < t_n < T$, the σ-fields $\sigma(W_{t_1}, \cdots, W_{t_n})$ and $\sigma(W_{t_1}, W_{t_2} - W_{t_1}, \cdots, W_{t_n} - W_{t_{n-1}})$ coincide.

May we substitute a random variable for T? Will the process $\widetilde{W}_t = W_{t+\tau} - W_\tau$ be a Wiener process if the time $\tau = \tau(\omega)$ itself depends on the trajectory? One can show that one cannot take an arbitrary random variable. For example, let $\theta = \sup\{t \in [0,1] : W_t = 0\}$. By the definition of θ, the process $\widetilde{W}_t = W_{t+\theta} - W_\theta$ does not change the sign in the interval $(0, 1-\theta)$ with probability 1. For a Wiener process, such a behavior (no zeroes in some interval $(0, 1-\theta)$) has probability 0. For instance, this follows from the foregoing Knight's result asserting that the Wiener trajectories may be represented in the form of the limit of symmetric walks over a lattice, if one takes into account that symmetric random walk returns to zero infinitely many times. Therefore, $\widetilde{W}_t = W_{t+\theta} - W_\theta$ is not a Wiener process. However, an important class of random variables $\tau(\omega)$ does exist for which the process $W_{t+\tau} - W_\tau$ is a Wiener process. These are the so-called Markov times.

A random variable $\tau(\omega)$ which is allowed to take non-negative values and the value $+\infty$ is said to be a *Markov time*,[3] whenever for every $t \geq 0$ the event $\{\tau(\omega) \leq t\}$ belongs to $\mathcal{F}^W_{\leq t} = \sigma(W_s, 0 \leq s \leq t)$.

In other words, a Markov time is one whose occurrence may be known without any information about what will happen after this time. This property is emphasized by another term for the Markov time—a random variable independent of future. In particular, the random variable θ introduced above (the last time, before 1, when the Wiener trajectory

---

[2] We will remind that if $\mathfrak{A}$ is a collection of random variables on a space $(\Omega, \mathcal{F}, P)$ then $\sigma(\mathfrak{A})$ is the minimal σ-field with respect to which the variables of the class $\mathfrak{A}$ are all measurable.

---

[3] To be more exact, a Markov time with respect to the expanding system of the σ-fields $\mathcal{F}^W_{\leq t}$. Later on we shall introduce a more general notion of Markov time.

visits zero) is not a Markov time. On the other hand, $\tau(\omega) = \inf\{t : W_t = a\}$ –(the first hitting time to the point a)–is a Markov time.

It is easy to ascertain the elementary properties of the Markov times. For instance, if τ_1 and τ_2 are Markov times, then $\tau_1 \wedge \tau_2$ and $\tau_1 \vee \tau_2$ are also Markov times. The variables $\tau_1 + 1$ and $2\tau_1$, for example, are Markov times as well; but the variables $\frac{\tau_1}{2}$ and $\tau_2 - 1$ are not, in general, Markov times.

A σ-field $\mathcal{F}^W_{\leq\tau} \subset \mathcal{F}$ is associated with every Markov time $\tau(\omega)$. Namely, an event $\mathcal{A} \in \mathcal{F}$ belongs to $\mathcal{F}^W_{\leq\tau}$ if and only if, for every $t \geq 0$, the inclusion $\mathcal{A} \cap \{\tau \leq t\} \in \mathcal{F}^W_{\leq t}$ holds. It is not difficult to check that the collection of such events actually forms a σ-field. Intuitively, this σ-field is the collection of events which are defined by the behavior of the Wiener process W_s up to the time $\tau(\omega)$.

It is possible to prove that if $\tau(\omega)$ is a Markov time and $P\{\tau(\omega) < \infty\} = 1$, then the random process $W_t = W_{t+\tau} - W_\tau$ does not depend on the σ-field $\mathcal{F}^W_{\leq\tau}$ and is a Wiener process (see Dynkin [1], Hunt [1], Ito and McKean [1]). This property is called the *strong Markov property* of a Wiener process.

Up to now we have been dealing with the Wiener trajectories starting solely from zero. In what follows it will be suitable for us to consider a family of Wiener processes starting from various points $a \in R^1 : W^a_t = a + W_t$. Then the Markov property may be interpreted like this: the process $\widetilde{W}^{W_T}_t = W_{T+t}$ is a new Wiener process starting from the point W_T, at time $t = 0$ and conditionally independent of the behavior of the process W_t for $t \leq T$, given W_T.

The strong Markov property may be applied for deducing the helpful relation (see, e.g. Ito and McKean [1]):

$$P\{\max_{0\leq s\leq t} W_s > a\} = 2P\{W_t > a\}.$$

In fact, by denoting $\tau_a = \min\{t : W_t = a\}$ we arrive at

$$P\{\max_{0\leq s\leq t} W_s > a\} = P\{\tau_a \leq t\} = P\{\tau_a \leq t, W_t < a\} + \\ + P\{\tau_a \leq t, W_t \geq a\}.$$

The terms on the right-hand side of this equality are equal. This is immediate from the strong Markov property and from the symmetry of the distribution of W_s with respect to zero. Noting this we derive

$$P\{\max_{0\le s\le t} W_s > a\} = P\{\tau_a \le t\} = 2P\{\tau_a \le t, W_t \ge a\} =$$

$$= 2P\{W_t \ge a\} = \frac{2}{\sqrt{2\pi t}} \int_a^\infty e^{-\frac{x^2}{2t}} dx = \sqrt{\frac{2}{\pi}} \int_{\frac{a}{\sqrt{t}}}^\infty e^{-\frac{y^2}{2}} dy . \tag{2.9}$$

From (2.9) we conclude that $P\{\tau_a < \infty\} = \lim_{t\to\infty} P\{\tau_a < t\} = 1$, but $E\tau_a = \infty$.

The joint distribution of W_t and $\max_{0\le s\le t} W_s = \zeta_t$ may also be computed (see Ito and McKean [1]). The density function $p_{W,\zeta}(x,y)$ of this two-dimensional random variable is the following

$$p_{W,\zeta}(x,y) = \left(\frac{2}{\pi t^3}\right)^{\frac{1}{2}} (2y-x)\, e^{-\frac{(2y-x)^2}{2t}} , t \ge 0, y \ge 0, x \le y .$$

Now we dwell briefly on the local properties of Wiener trajectories. By remembering that the increments of Wiener process are independent, it is not difficult to prove that, for every fixed t, almost all Wiener trajectories W_t are not differentiable. Paley, Wiener, and Zygmund [1] established a stronger result: Wiener trajectories are nowhere differentiable with probability 1.

With probability one, the trajectories of Wiener process are Hölder continuous with exponent $\frac{1}{2} - \varepsilon$ for every $\varepsilon > 0$, and do not satisfy the Hölder condition with the exponent $\frac{1}{2}$. Wiener trajectories have infinite variation in every non-empty interval. For the proof of all these assertions, the reader is referred to Ito and McKean [1], Chapter 1.

The properties of symmetry and self similarity of the Wiener process play a highly significant role.

It is obvious that if $W_t(\omega)$ is a Wiener process, $W_0(\omega) = 0$, then the process $\tilde{W}_t = -W_t(\omega)$ is also a Wiener process. No matter what the positive number α, the process $\tilde{\tilde{W}}_t = \alpha W_{t/\alpha^2}$ is a Wiener process as well. Indeed, the process $\tilde{\tilde{W}}_t$ has continuous trajectories and is a Gaussian process; $E\tilde{\tilde{W}}_t = \alpha E W_{t/\alpha^2} = 0$. The correlation function of the process $\tilde{\tilde{W}}_t$ has the form: $E\tilde{\tilde{W}}_s\tilde{\tilde{W}}_t = \alpha^2 E W_{s/\alpha^2} W_{t/\alpha^2} = \alpha^2\left(\frac{s}{\alpha^2} \wedge \frac{t}{\alpha^2}\right) = s \wedge t$. As was remarked above, these properties characterize the Wiener process.

There is another interesting transformation which preserves Wiener processes. Let W_t be a Wiener process. The process

$$\hat{W}_t = \begin{cases} 0, & t = 0, \\ t W_{1/t}, & t > 0 \end{cases} \tag{2.10}$$

is a Wiener process too. In order to prove this, we again make use of the fact that a mean zero Gaussian process with continuous trajectories and the correlation function $R(s,t) = s \wedge t$, is a Wiener process. The continuity of the function $\hat{W}_t$ may be broken at zero only, but one can see that for any $\varepsilon > 0$

$$\lim_{t \downarrow 0} \frac{|\hat{W}_t|}{t^{(1/2)-\varepsilon}} = 0 \quad \text{P-a.s.}$$

This can be checked, for example, with the law of iterated logarithm for $t \to \infty$ (see below). Thus, the functions $\hat{W}_t$ are continuous at zero. Taking into consideration that $E\hat{W}_t = 0$, $E\hat{W}_s\hat{W}_t = st\left(\frac{1}{s} \wedge \frac{1}{t}\right) = s \wedge t$, we deduce that $\hat{W}_t$ is a Wiener process. This transformation enables us to determine the properties of a Wiener process as $t \to 0$, by studying its properties as $t \to \infty$, and vice versa.

There is an interesting property, the so-called law of iterated logarithm which holds for almost all the Wiener trajectories: namely, with probability 1.

$$\overline{\lim_{t\to\infty}} \frac{W_t}{\sqrt{2t \ln \ln t}} = 1\,, \quad \underline{\lim}_{t\to\infty} \frac{W_t}{\sqrt{2t \ln \ln t}} = -1\,.$$

The proof of this statement may be found in Ito and McKean [1]. Using transformation (2.10), the law of the iterated logarithm becomes the local law of the iterated logarithm:

$$\overline{\lim_{t\downarrow 0}} \frac{W_t}{\sqrt{2t \ln \ln t^{-1}}} = 1\,, \quad \underline{\lim}_{t\downarrow 0} \frac{W_t}{\sqrt{2t \ln \ln t^{-1}}} = -1$$

almost surely. The local law of the iterated logarithm, in particular, implies that a Wiener process starting from zero, returns to the point 0 with probability 1 in an arbitrarily small interval $(0,h)$, $h > 0$. We have used this property when stating that the variable $\theta = \sup\{t \in [0,1], W_t = 0\}$ is not a Markov time. Furthermore the law of the iterated logarithm implies that the set of zeroes of a Wiener trajectory is unbounded with probability 1 as $t \to \infty$.

Notice that the Lebesgue measure of the set $\{t : W_t = 0\}$ is zero with probability 1. Really, the time $\Lambda_\Gamma(t)$, a trajectory spends in the set $\Gamma \subset R^1$ up to time t, may be written in the form $\Lambda_\Gamma(t) = \int_0^t \chi_\Gamma(W_s)\,ds$ where $\chi_\Gamma(x)$ is the indicator of the set Γ. From this we conclude that $E\,\Lambda_\Gamma(t) = \int_0^t E\,\chi_\Gamma(W_s)\,ds = \int_0^t \int_\Gamma p(s,0,y)\,dy\,ds$. In particular, if Γ has zero measure, then $E\,\Lambda_\Gamma(t) = 0$ and therefore $\Lambda_\Gamma(t) = 0$ with probability 1. Many interesting properties of the Wiener process are available in the books of Levy [1] and Ito and McKean [1].

A collection of r independent Wiener processes $(W_t^1, \cdots, W_t^r) = W_t$, $t \geq 0$, defined on a probability space $(\Omega, \mathcal{F}, P)$, is said to be an *r-dimensional Wiener process*. The measure in the space $C_{0,\infty}(R^r)$ of continuous functions on $[0,\infty)$ with values in R^r, induced by the process $W_t(\omega)$ in $C_{0,\infty}(R^r)$, is called a *Wiener measure in* $C_{0,\infty}(R^r)$.

A family of σ-fields $\mathcal{F}_{\leq t}^W = \sigma(W_s^1, \cdots, W_s^r\,;\ s \in [0,t])$ is associated with the process W_t. Corresponding to this family of σ-fields (as to any increasing family of σ-fields) there is a collection of Markov times: namely, non-negative random variables $\tau(\omega)$ for which $\{\tau(\omega) \leq t\} \in \mathcal{F}^W_{\leq t}$, $t \geq 0$. This concept of a Markov time, if considered in connection with the one-dimensional process W_t^1, is broader than that introduced before, because the σ-fields $\mathcal{F}^W_{\leq t}$ are broader than $\mathcal{F}^{W^1}_{\leq t}$. For these new Markov times the strong Markov property is fulfilled as well: $W_{\tau+t} - W_\tau = \widetilde{W}_t$ is an r-dimensional Wiener process independent of events belonging to the σ-field $\mathcal{F}^W_{\leq \tau} = \{\mathcal{A} \in \mathcal{F} : \mathcal{A} \cap \{\tau \leq t\} \in \mathcal{F}^W_{\leq t}$ for any $t \geq 0\}$.

We will note some differences in the behavior of r-dimensional Wiener processes as we vary the dimension r (these distinctions follow from the properties of the one-dimensional Wiener process and from the independence of the components $(W_t^1, \cdots, W_t^r)$). As we observed, after any time t, a Wiener process in R^1 will hit zero and, hence, will hit any other point of the line as well. For $r \geq 2$, however, with probability 1, the Wiener trajectory will never hit any fixed point of R^r. But for $r = 2$, with probability 1, the Wiener trajectory does hit any open set after any fixed time. For $r > 2$, this is not the case. What is more, one can prove that, for $r \geq 3$, $\lim_{t\to\infty} |W_t| = \infty$ a.s..

Finally, we remark that the Wiener process in R^r is invariant with respect to rotations: if W_t is an r-dimensional Wiener process, then $\widetilde{W}_t = Q W_t$ is also a Wiener process for any orthogonal matrix Q. This assertion follows from the definition of W_t and from the properties of the Gaussian distribution. The family of Wiener processes $W_t^x = x + W_t$ in R^r is invariant with respect to the group of all rigid motions of the space.

Concluding this section it is worthwhile to draw the reader's attention to the close connection between the Wiener measure (or process) and the Laplace operator[4] Δ.

---

[4]More precisely, this is the operator $\frac{1}{2}\Delta$ rather than Δ. This is seen in considering Cauchy's problem. In the case of Dirichlet's problem for the homogeneous equation $\Delta u = 0$, this difference is of course imperceptible.

The simplest example of this connection is the following: the mean value $E\,g(W_t^x) = u(t,x)$ (here $g(x)$ is a continuous bounded function in R) is a solution of Cauchy's problem for the simplest heat conduction equation:

$$\frac{\partial u}{\partial t} = \frac{1}{2}\Delta u\,, \quad \lim_{t \downarrow 0} u(t,x) = g(x)\,.$$

This is straightforward since according to the definition of W_t^x

$$u(t,x) = E\,g(W_t^x) = \int_{R^r} g(y)\frac{1}{(\sqrt{2\pi t})^r} e^{-\frac{|x-y|^2}{2t}}\,dy\,.$$

Now we shall show how the Wiener process is linked with Dirichlet's problem for the Laplace operator.

Suppose we are given a bounded domain $D \subset R^r$ with a boundary ∂D, the domain D, for simplicity, being assumed convex. Consider the Dirichlet problem in the domain D :

$$\Delta u(x) = 0\,,\ x \in D\,, \quad \lim_{x \to x_0 \in \partial D} u(x) = \psi(x_0)\,, \tag{2.11}$$

where $\psi(x)$ is a continuous function defined on ∂D. We will show how the function $u(x)$ which is a solution of problem (2.11) can be represented in the form of the average value of a proper functional of the Wiener trajectory, or (what is the same) in the form of an integral with respect to the Wiener measure.

Let W_t be a Wiener process in R^r. We will introduce the Markov time $\tau_D^x = \tau_D^x(\omega) = \inf\{t : x + W_t \in \partial D\}$ —the first hitting time to the boundary of the domain D starting from $x \in D \cup \partial D$. Let us show that $P\{\tau_D^x < \infty\} = 1$ for arbitrary $x \in D$. In fact, let $W_t^{1,x}$ be the first component of $W_t^x = W_t + x$ and let a number a be such that $D \subset \{x \in R^r; x^1 < a\}$. We shall denote by $\tau_a^{1,x}$ the first hitting time of the one-dimensional process $W_t^{1,x}$ to the point $a : \tau_a^{1,x} = \inf\{t : W_t^{1,x} = a\}$. It is clear that $\tau_D^x < \tau_a^{1,x}$ so the finiteness of τ_D^x comes from the fact that $P\{\tau_a^{1,x} < \infty\} = 1$.

The last equality results from (2.9). At the time τ_D^x, the point $W^x_{\tau_D^x}$ belongs to ∂D, thus one can consider $\psi(W^x_{\tau_D^x})$ and the function

$$V(x) = E\,\psi(W^x_{\tau_D^x}) . \tag{2.12}$$

Let us show that the function $V(x)$ is a solution of Dirichlet's problem (2.11).

First we will make sure that the function $V(x)$ defined by (2.12) satisfies the boundary conditions, i.e. $\lim_{x \to x_0 \in \partial D} V(x) = \psi(x_0)$. To that end, let us draw the line of support Γ (for the sake of visualization, we shall speak of the two-dimensional case) through the point $x_0 \in \partial D$ (see Fig. 2). Since the Wiener process is invariant with respect to the group of movements of the plane, the point x_0 may be considered as coinciding with the origin and the line Γ —with the x^2-axis. Let $\tau_0^{1,x}$ be the first hitting time of the first component $W_t^{1,x}$ of the Wiener process W_t^x to the point 0. Then noting equality (2.9) we obtain

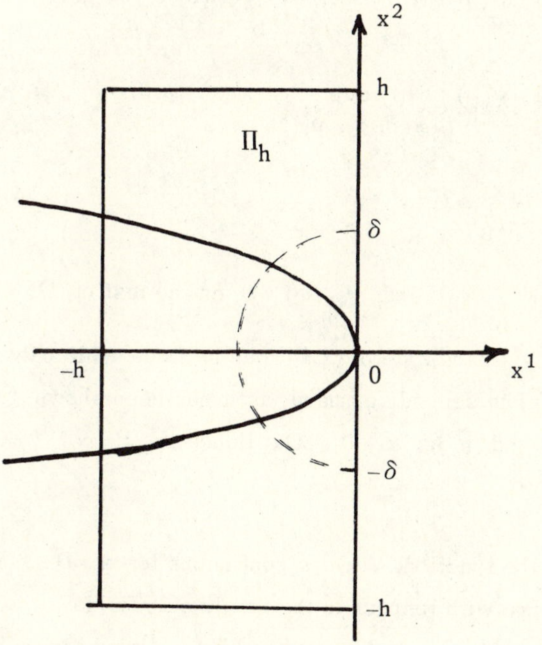

Fig. 2

$$P\{\tau_D^x > t\} < P\{\tau_0^{1,x} > t\} = \frac{2}{\sqrt{2\pi}} \int_0^{|x^1|t^{-1/2}} e^{-\frac{y^2}{2}} dy \to 0 \qquad (2.13)$$

as $x^1 \to 0$. Formula (2.9) also readily implies that, for any fixed $h > 0$, the probability of leaving the rectangle Π_h shown in Fig. 2 before time t across the horizontal sides or across the side lying on the line $x^1 = -h$, tends to zero as $t \downarrow 0$ uniformly in the set of the initial points $x \in D \cap U_\delta(0)$, where $U_\delta(0) = \{x : |x| < \delta\}$. This remark along with (2.13) implies that $P\{W^x_{\tau^x_D} \in \gamma_h\} \to 1$ as $x \to 0$, where $\gamma_h = \partial D \cap \Pi_h$. Now it is easy to verify that $V(x)$ takes on the prescribed boundary values:

$$|V(x) - \psi(0)| = |E(\psi(W^x_{\tau^x_D}) - \psi(0))| \le E|\psi(W^x_{\tau^x_D}) - \psi(0)|\chi_{\gamma_h} +$$

$$+ E|\psi(W^x_{\tau^x_D}) - \psi(0)|(1 - \chi_{\gamma_h}) \le \sup_{x \in \partial D, |x-0| \le 2h} |\psi(x) - \psi(0)| +$$

$$+ 2 \max_{y \in \partial D} |\psi(y)| \times P\{W^x_{\tau^x_D} \notin \gamma_h\}. \qquad (2.14)$$

Here $\chi_{\gamma_h}(\omega) = 1$ for $\omega \in \{W^x_{\tau^x_D} \in \gamma_h\}$ and $\chi_{\gamma_h}(\omega) = 0$ on the rest of Ω. The first summand on the right-hand side of (2.14) can be made arbitrarily small by making h small. The second summand, as it has been shown above, tends to zero for a fixed h as $x \to 0 \in \partial D$. Hence,

$$\lim_{x \to x_0 \in \partial D} V(x) = \psi(x_0).$$

Now we will show that the function $V(x)$ is continuous for $x \in D$. First let $\bar{\tau}^x$ be a Markov time with respect to the family of σ-fields $\mathcal{F}^W_{\le t}$, which satisfies $P\{\bar{\tau}^x \le \tau^x_D\} = 1$. Let $\eta = \eta(\omega) = W^x_{\bar{\tau}^x}$. Using the strong Markov property one can write down the following chain of equalities:

$$V(x) = E\,\psi(W^x_{\tau^x_D}) = E(E(\psi(\widetilde{W}^{\eta}_{\tau^{\eta}_D})|\eta)) = \tag{2.15}$$

$$= EV(\eta) = EV(W^x_{\hat{\tau}^x}) \; ,$$

where $\widetilde{W}_t$ is a Wiener process independent of W_t. Let us introduce the Markov time $\hat{\tau} = \hat{\tau}^{x,y} = \tau^x_D \wedge \tau^y_D$, $x, y \in D$. Using (2.15) leads to

$$V(x) - V(y) = E(V(W^x_{\hat{\tau}}) - V(W^y_{\hat{\tau}})) \; . \tag{2.16}$$

One of the points $W^x_{\hat{\tau}}$ or $W^y_{\hat{\tau}}$ belongs to ∂D, and the distance of the other point from D is bounded by $|x-y|$, because the trajectory $W^y_{\hat{\tau}}$ is a translation of $W^x_{\hat{\tau}}$ by the vector $(y-x)$. Hence, on account of the foregoing bounds near the boundary we have:

$$|V(W^x_{\hat{\tau}}) - V(W^y_{\hat{\tau}})| \to 0 \quad \text{as} \quad |x-y| \to 0 \quad \text{a.s.}$$

Noting that the function $V(x)$ is bounded, we deduce from (2.16) that $|V(x) - V(y)| \to 0$ as $|x-y| \to 0$.

Let us show that the function $V(x)$ given by (2.12) has the mean-value property: for any circle K having its center at a point $x \in D$ and lying entirely in $D \cup \partial D$, the function $V(x)$ satisfies

$$V(x) = \int_K V(y)\, m(dy) \tag{2.17}$$

where $m(dy)$ is a uniform distribution on the circle with the condition $m(K) = 1$. Indeed, let us put $\tau^x_K = \min\{t : W^x_t \in K\}$; that is, the first hitting time of the circle K starting from $x \in D$. The random variable τ^x_K is a Markov time satisfying $P\{\tau^x_K < \tau^x_D\} = 1$, and by (2.15)

$$V(x) = EV(W^x_{\tau^x_K}) \; .$$

Further, remembering that the Wiener process is invariant with respect to rotations one can conclude that when starting from the center of the circle K, the distribution at the first exit time from K is uniform: that is, $P\{W^x_{\tau^x_K} \in \gamma\} = m(\gamma)$ for every arc $\gamma \subset K$. Therefore,

$$V(x) = EV(W^x_{\tau^x_K}) = \int_K V(y)\, m(dy) .$$

So, we have shown that $V(x)$ is a continuous function which takes on the prescribed boundary values and has the mean-value property. As is known in the theory of differential equations, a solution $u(x)$ of Dirichlet's problem (2.11) possesses these properties too. Consequently, to prove that $u(x) = V(x)$, it remains to note that there is but one continuous function taking the value $\psi(x)$ on ∂D and satisfying the mean-value property. Indeed, suppose on the contrary, that there are two such functions and that $a(x)$ is their difference. For $x \in \partial D$, the function $a(x)$ vanishes identically and satisfies the mean-value property as well. Assume that $a(x)$ attains its maximum for $x = x_0 \in D$. Let us draw a circle with center at x and with radius equal to the distance from x to ∂D (Fig. 3).

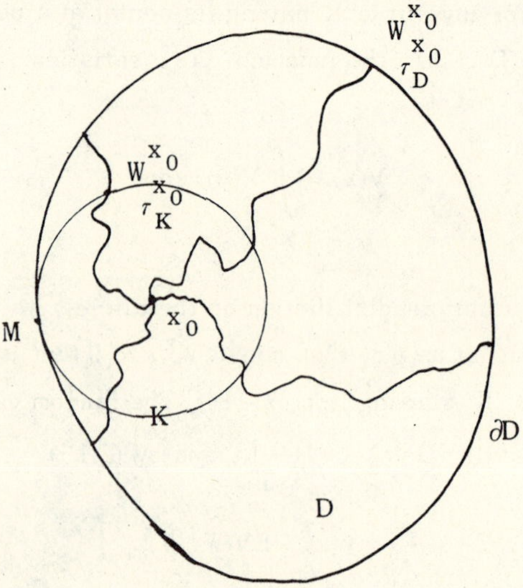

Fig. 3

Using the continuity of $a(x)$ and the mean-value theorem yields that $a(x_0) = a(M) = 0$ (M is shown in Fig. 3). Thus, $a(x) \equiv 0$ for $x \in D$ and hence $u(x) \equiv V(x)$.

So we have established that the solution of Dirichlet's problem can be represented in the form

$$u(x) = E\,\psi(W^x_{\tau^x_D}) . \tag{2.18}$$

Intuitively this means that, for the value of $u(x)$ at any point $x \in D$ to be determined, we are to let out Wiener trajectories from the point x, then to watch where they first hit ∂D and what the boundary values are at the points of the first hitting ∂D, and then to average these values over all the trajectories $W^x_t(\omega)$.

§1.3 *Stochastic differential equations*

Our next goal now is to construct a family of measures in $C_{0,\infty}(R^r)$ (or, equivalently, a family of random processes with continuous trajectories) which corresponds to an elliptic, possibly degenerate, second-order differential operator of general form. As we have seen in the previous section, a family of Wiener processes W^x_t is associated with the operator $\frac{1}{2}\Delta$. Now suppose that we are given a second-order operator

$$\tilde{L} = \frac{1}{2}\sum_{i,j=1}^{r} a^{ij}\frac{\partial^2}{\partial x^i \partial x^j} + \sum_{i=1}^{r} b^i \frac{\partial}{\partial x^i}$$

with constant coefficients and non-negative characteristic form: $\sum a^{ij}\lambda_i\lambda_j \geq 0$, $a^{ij} = a^{ji}$. Consider the family of random processes

$$\tilde{X}^x_t = x + \sigma W_t + bt \tag{3.1}$$

where $x \in R^r$ is an initial point, and σ is a real matrix[5] such that

[5]The existence of such a matrix follows from the fact that the matrix (a^{ij}) is non-negative.

$\sigma \cdot \sigma^* = (a^{ij})$, $b = (b^1, \cdots, b^r)$. Knowing the density function of the random variable W_t it is easy to evaluate the distribution of $\tilde{X}_t^x$ and to make sure that the function $u(t,x) = E\, g(\tilde{X}_t^x)$ is a solution of Cauchy's problem

$$\frac{\partial u}{\partial t} = \tilde{L}u\,, \quad \lim_{t \downarrow 0} u(t,x) = g(x)\,,$$

where $g(x)$, $x \in R^r$, is a bounded continuous function. Therefore, the Gaussian process $\tilde{X}_t^x$ and the corresponding measure in $C_{0,\infty}(R^r)$ are related in a natural way to the operator $\tilde{L}$.

Now consider the operator of general form

$$L = \frac{1}{2} \sum_{i,j=1}^{r} a^{ij}(x) \frac{\partial^2}{\partial x^i \partial x^j} + \sum_{i=1}^{r} b^i(x) \frac{\partial}{\partial x^i}, \quad \sum_{i,j=1}^{r} a^{ij}(x) \lambda i \lambda j \geq 0\,,$$

with variable coefficients. The question suggests itself: what stochastic process X_t^x is related to this operator L? If the coefficients of the operator possess some continuity properties, then it is natural to expect that, near the point x, such a process X_t^x is close to the process $\tilde{X}_t^x$ defined in (3.1) with σ and b held fixed to their values at point x. In other words, the desired process must satisfy the differential equation

$$dX_t^x = \sigma(X_t^x)\, dW_t + b(X_t^x) dt\,, \quad X_0^x = x\,, \tag{3.2}$$

where $b(x) = (b^1(x), \cdots, b^r(x))$, $\sigma(x)\sigma^*(x) = (a^{ij}(x))$.

Later on we will see that actually such is the case, but first we must make sense of equation (3.2). To make the meaning of equation (3.2) more precise, we will follow the ideas of K. Ito. Namely, we integrate it from 0 to t, taking into account the initial condition:

$$X_t^x - x = \int_0^t \sigma(X_s^x) dW_s + \int_0^t b(X_s^x) ds\,. \tag{3.3}$$

If the matrix σ does not depend on x, then the existence and uniqueness of a solution of such an equation for a fixed trajectory W_s, follow from the corresponding results for ordinary differential equations (non-stochastic), whenever one assumes, for example, that the function b(x) satisfies a Lipschitz condition. In the general case, the situation is more complicated: here the first integral in (3.3) must be given a meaning. But this is not easy to do, because the Wiener trajectories W_s have with probability one infinite variation on every interval.

We proceed now to construct Ito's stochastic integral and describe its properties (see, e.g. Gihman and Skorohod [1], [2], McKean [1], Wentzell [1]). The general outline is as follows (now W_s is supposed to be a one-dimensional Wiener process). To begin with, we define $\int_a^b f(s,\omega)\,dW_s$ for the "simple" real-valued functions $f(s,\omega)$, $s \in [a,b]$. Here it turns out that $\int_a^b f(s,\omega)\,dW_s = \eta(\omega)$ is a random variable such that $E|\eta(\omega)|^2 = \int_a^b E|f(s,\omega)|^2 ds$. Therefore, to every simple function $f(s,\omega)$, the stochastic integral is a random variable $\eta(\omega)$. We introduce the Hilbert space $H^2([a,b]\times\Omega)$ of functions with the norm $\|f\|_{H^2} = \int_a^b E|f(s,\omega)|^2 ds$ and the Hilbert space $L^2(\Omega)$ of the random variables $\eta(\omega)$, $\|\eta\|_{L^2} = E|\eta(\omega)|^2$. Then integration becomes a linear isometric mapping of the set $\widetilde{H}^2_{a,b}$ of simple functions from $H^2([a,b]\times\Omega)$ into $L^2(\Omega)$. This mapping can be extended in a continuous way to the closure of the set $\widetilde{H}^2_{a,b} \subset H^2([a,b]\times\Omega)$ with the isometry being preserved. Such a continuation defines an integral on the closure of the set of simple functions which turns out to be a sufficiently extensive set. Now, we proceed with the details.

Suppose we are given a one-dimensional Wiener process W_t, $t \geq 0$, and an increasing family of σ-fields $\mathcal{N}_t$, $t \geq 0$. We assume that $\mathcal{F}^W_{\leq t} \subset \mathcal{N}_t$ and that the increments $W_s - W_t$ do not depend on the σ-field $\mathcal{N}_t$ for $s > t$. Such a family of σ-fields $\mathcal{N}_t$ will be called a *family adapted* to the Wiener process W_t.

A function $f(t,\omega)$ is called a *simple step-function* (or a step function independent of future) if there are points $a = t_0 < t_1 < t_2 < \cdots < t_n = b$ such that $f(t,\omega) = f(a,\omega)$ for $t \in [a, t_1)$; $f(t,\omega) = f(t_1,\omega)$ for $t \in [t_1, t_2); \cdots; f(t,\omega) = f(t_{n-1},\omega)$ for $t \in [t_{n-1}, b)$, where the random variables $f(t_i,\omega)$ are assumed to be $\mathcal{N}_{t_i}$-measurable and $E|f(t_i,\omega)|^2 < \infty$. The collection of all such functions will be denoted by $\widetilde{H}^2_{a,b}$.

For the step-functions independent of future we will define the stochastic integral as:

$$I(f) = \int_a^b f(s,\omega)\,dW_s = \sum_{i=0}^{n-1} f(t_i,\omega)(W_{t_{i+1}} - W_{t_i}) .$$

It is clear that $I(f)$ is a linear mapping of $\widetilde{H}^2_{a,b} \subset H^2([a,b]\times\Omega)$ into $L^2(\Omega)$. Since the variable $f(t,\omega)$ is $\mathcal{N}_{t_i}$-measurable, and $W_{t_{i+1}} - W_{t_i}$ is independent of the σ-field $\mathcal{N}_{t_i}$, we have: $E\,I(f) = \sum_{i=0}^{n-1} E\,f(t_i,\omega)E(W_{t_{i+1}} - W_{t_i}) = 0$. Let us check that this mapping is isometric:

$$\|I(f)\|^2_{L^2} = E|I(f)|^2 = E\left|\sum_{i=0}^{n-1} f(t_i,\omega)(W_{t_{i+1}} - W_{t_i})\right|^2 =$$

$$= E\left[\sum_{i=0}^{n-1} |f(t_i,\omega)|^2 (W_{t_{i+1}} - W_{t_i})^2\right] + E\left[\sum_{i\neq j} f(t_i,\omega)f(t_j,\omega)(W_{t_{i+1}} - W_{t_i})(W_{t_{j+1}} - W_{t_j})\right] . \tag{3.4}$$

The second summand on the right-hand side of (3.4) vanishes. In fact, if for example, $j > i$, then $W_{t_{j+1}} - W_{t_j}$ and $f(t_i,\omega)f(t_j,\omega)(W_{t_{i+1}} - W_{t_i})$ are independent. Thus, $E\,f(t_i,\omega)f(t_j,\omega)(W_{t_{i+1}} - W_{t_i})(W_{t_{j+1}} - W_{t_j}) = E\{f(t_i,\omega) \times f(t_j,\omega)(W_{t_{i+1}} - W_{t_i})\}E(W_{t_{j+1}} - W_{t_j}) = 0$. Noting that $W_{t_{i+1}} - W_{t_i}$ does not depend on $f(t_i,\omega)$ and $E(W_{t_{i+1}} - W_{t_i})^2 = t_{i+1} - t_i$, the first summand in (3.4) can be reduced to the form:

$$\sum_{i=0}^{n-1} E|f(t_i,\omega)|^2(W_{t_{i+1}}-W_{t_i})^2 = \sum_{i=0}^{n-1} E|f(t_i,\omega)|^2(t_{i+1}-t_i) =$$

$$= \int_a^b E|f(t,\omega)|^2 dt = \|f\|_{H^2} .$$

Hence, $I(f)$ is an isometric mapping $\widetilde{H}^2_{a,b} \to L^2(\Omega)$. Now if f belongs to the closure $\overline{H}^2_{a,b}$ of $\widetilde{H}^2_{a,b}$ in the space $H^2([a,b]\times\Omega)$, then we define

$$\int_a^b f(s,\omega)\,dW_s = \lim_{n\to\infty} I(f_n) ,$$

where $f_n \in \widetilde{H}^2_{a,b}$ and $\|f-f_n\|_{H^2} \to 0$ as $n\to\infty$. Since the operator $I(f)$ is isometric, such a definition is correct.

Denote by $\mathcal{B}_{a,t}$ the Borel σ-field in $[a,t]$, and let $\mathcal{B}_{a,t}\times N_t$ be the minimal σ-field in the product $[a,t]\times\Omega$ which contains all the sets $A\times B$ for $A\in\mathcal{B}_{a,t}$, $B\in N_t$. A random function $f(s,\omega)$ is called *progressively measurable (or independent of future)*, if for every $t\in[a,b]$ the function $f(s,\omega)$ is measurable in (s,ω) on the set $[a,t]\times\Omega$ with respect to the σ-field $\mathcal{B}_{a,t}\times N_t$.

It turns out that the progressively measurable functions, for which $\int_a^b E|f(s,\omega)|^2 ds < \infty$, all belong to the space $\overline{H}^2_{a,b}$.

We will list the basic properties of the stochastic integral ($f(s,\omega)$, $g(s,\omega)\in\overline{H}^2_{a,b}$):

1) $$\int_a^b [\alpha f(s,\omega)+\beta g(s,\omega)]\,dW_s = \alpha\int_a^b f(s,\omega)\,dW_s + \beta\int_a^b g(s,\omega)\,dW_s ;$$

2) $$E\left[\int_a^b f(s,\omega)\,dW_s \,\middle|\, N_a\right] = 0$$

$$3)\quad E\left[\int_a^b f(s,\omega)\,dW_s \cdot \int_a^b g(s,\omega)\,dW_s \,|\, N_a\right] = E\left[\int_a^b f(s,\omega)\,g(s,\omega)\,ds \,|\, N_a\right],$$

in particular,

$$E\left[\left(\int_a^b f(s,\omega)\,dW_s\right)^2 |\, N_a\right] = E\left[\int_a^b f^2(s,\omega)\,ds \,|\, N_a\right].$$

These properties are readily verified for simple functions, and then can be extended to all functions in $\overline{H}^2_{a,b}$. We will remark that the stochastic integral is defined up to a set of zero measure, and these equalities are all fulfilled P-a.s. in the space Ω. Similarly, one can define the integral $\int_a^\infty f(s,\omega)\,dW_s$ for the functions from $H^2([0,\infty)\times\Omega)$ independent of future. The stochastic integral can be defined for a slightly broader class of functions than the set $\overline{H}^2_{a,b}$ (see, e.g. Gihman and Skorohod [1]), but here the space $\overline{H}^2_{a,b}$ will be quite sufficient for our purposes.

Now let us consider stochastic integrals with a variable upper limit of integration. We will denote by $\chi_t(s)$ the function equal to 1 for $s \le t$, and equal to zero elsewhere. If $f(s,\omega) \in \overline{H}^2_{a,b}$, then $\chi_t(s) f(s,\omega) \in \overline{H}^2_{a,b}$ for every $t \in [a,b]$. For $t \in [a,b]$, we will set

$$\int_a^t f(s,\omega)\,dW_s = \int_a^b \chi_t(s)\,f(s,\omega)\,dW_s\,.$$

Since for every $t \in [a,b]$ the right-hand side is defined up to a set of zero measure, there is some ambiguity in the definition of the left-hand side. It can be shown that, for every t, the right-hand sides may be defined in such a way that the stochastic integral on the left-hand side is a continuous function of the upper limit of integration for a.a. ω. It is this continuous version that will always be meant without special mentioning.

We will list some more properties of the stochastic integral as a function of the upper limit of integration $(f \in \overline{H}^2_{0,T})$

4) (Kolmogorov's generalized inequality)

$$P\left\{\max_{0\leq t\leq T} \left| \int_0^t f(s,\omega)\,dW_s \right| > c\right\} \leq \frac{1}{c^2}\int_0^T E f^2(s,\omega)\,ds \;;$$

5) $$E \max_{0\leq t\leq T} \left| \int_0^t f(s,\omega)\,dW_s \right|^2 \leq 4 \int_0^T E f^2(s,\omega)\,ds$$

6) $$P\left\{\max_{0\leq t\leq T}\left[\int_0^t f(s,\omega)\,dW_s - \frac{\alpha^2}{2}\int_0^t f^2(s,\omega)\,ds\right] > \beta\right\} \leq e^{-\alpha\beta}.$$

These very useful properties of stochastic integrals follow from the fact that the random process $\eta_t = \int_0^t f(s,\omega)\,dW_s$ together with the family of σ-fields N_t is a martingale.

We remind the reader that a random process η_t, $t \in T \subset R^1$, is called a *martingale* with respect to an increasing family of σ-fields N_t, $t \in T$, if

a) $\eta_t(\omega)$ is N_t-measurable for every $t \in T$;

b) $E|\eta_t| < \infty$ for $t \in T$;

c) $\eta_s = E(\eta_t | N_s)$ a.s. for every $s, t \in T$, $s \leq t$.

If, in place of c), the condition

c′) $\eta_s \leq E(\eta_t | N_s)$ a.s. for every $s, t \in T$, $s \leq t$

is fulfilled, then the process η_t is called a *submartingale*.

If, in place of c), the condition

c″) $\eta_s \geq E(\eta_t | N_s)$ a.s. for every $s, t \in T$, $s \leq t$

is fulfilled, then η_t is called a *supermartingale*.

A Wiener process W_t together with the increasing family of the σ-fields N_t adapted to W_t is an example of a martingale. Another example is the stochastic integral $\eta_t = \int_0^t f(s,\omega)\,dW_s$ together with the same family of σ-fields N_t.

One can give simple conditions in "martingale terms" which uniquely determine the Wiener process. Namely, the following assertion (Lévy's theorem) holds:

Suppose that $(X_t, \mathcal{N}_t)$, $t \geq 0$, is a martingale (with respect to an increasing family of σ-fields $\mathcal{N}_t$), which is continuous with probability 1 and satisfies the conditions

$$E((X_t - X_s)^2 | \mathcal{N}_s) = t - s \quad \text{for} \quad t \geq s, \quad P\{X_0 = 0\} = 1 .$$

Then X_t is a Wiener process (adapted to the family of σ-fields $\mathcal{N}_t$).

The proof is, for instance, available in Doob [1]. The multidimensional version of this theorem is also valid:

If $X_t^1, X_t^2, \cdots, X_t^r$, $t \geq 0$, are continuous martingales with respect to the increasing family of σ-fields $\mathcal{N}_t$, $P\{X_0^i = 0\} = 1$ for $i = 1, \cdots, r$, and $E((X_t^i - X_s^i)(X_t^j - X_s^j)|\mathcal{N}_s) = \delta^{ij} \cdot (t-s)$, for $t \geq s$, where δ^{ij} is the Kronecker delta, then $(X_t^1, \cdots, X_t^r)$ is a Wiener process in R^r.

Inequality (6) follows from the fact that the random process $\xi_t = \exp\{\int_0^t f(s,\omega)\, dW_s - \frac{1}{2}\int_0^t f^2(s,\omega)\, ds\}$ is a supermartingale with respect to the family of σ-fields $\mathcal{N}_t$, for $f \in \overline{H}^2_{0,T}$ (McKean [1]).

We shall utilize the convergence theorem for supermartingales as well. Let ξ_t be a non-negative supermartingale with respect to the increasing family of σ-fields $\mathcal{N}_t$, $t \in [0, \infty)$. Suppose that its trajectories are continuous with probability 1. Then, with probability 1, $\lim_{t \to \infty} \xi_t$ exists. The detailed theory of martingales is set forth in the monograph of Doob [1] and in a more modern fashion, in Delacherie et Meyer [1].

Sometimes, one has to consider stochastic integrals with a random time as the upper limit of integration. Let τ be a Markov time with respect to the family of σ-fields $\mathcal{N}_t$. We denote by $\chi_\tau(s)$ the function equal to 1 for $s \leq \tau$ and equal to zero for $s > \tau$. Suppose that $\chi_\tau(s) f(s,\omega) \in \overline{H}^2_{0,\infty}$. Then the equality $\int_0^\tau f(s,\omega)\, dW_s = \int_0^\infty \chi_\tau(s) f(s,\omega)\, dW_s$ holds. In this case, $E\int_0^\tau f(s,\omega)\, dW_s = 0$. In particular, $\chi_\tau(s) f(s,\omega) \in \overline{H}^2_{0,\infty}$, if $|f(s,\omega)| < c < \infty$ a.s., $f(s,\omega)$ is progressively measurable and $E\,\tau < \infty$.

Another property of stochastic integrals should be mentioned (see e.g. McKean [1]). Suppose that $f(s,\omega) \in \overline{H}^2_{0,t}$ for any $t \geq 0$ and consider the stochastic integral $\int_0^t f(s,\omega)\,dW_s$. Then one can find a Wiener process $\widetilde{W}_t$ such that

$$\int_0^t f(s,\omega)\,dW_s = \widetilde{W}_{\int_0^t f^2(s,\omega)\,ds}, \quad t \geq 0 .$$

This equality means that the process $Z_t = \int_0^t f(s,\omega)\,dW_s$ may be obtained from some Wiener process $\widetilde{W}_t$ via a random time change.

Suppose that W_t is an r-dimensional Wiener process adapted to the increasing family of σ-fields N_t. Moreover, let $\Phi(s,\omega) = (\phi^i_d(s,\omega))$ be a matrix having r columns whose elements belong to the space $\overline{H}^2_{0,t}$. Then $\int_0^t \Phi(s,\omega)\,dW_s$ can be defined in a natural way.

Levy's theorem implies the useful

COROLLARY. *If the matrix* $a(t,\omega) = \Phi(t,\omega)\Phi^*(t,\omega)$ *is orthogonal, then* $\zeta_t = \int_0^t \Phi(s,\omega)\,dW_s$ *is a Wiener process adapted to the family of σ-fields* N_t. *The dimension of the process* ζ_t *is equal to the number of rows in the matrix* $\Phi(t,\omega)$.

The following formula by Ito [1] plays an essential role in the theory of stochastic differential equations (see also McKean [1]). Let a function $u(t,x)$, $t \geq 0$, $x \in R^r$, have two continuous derivatives in x and one in t, the derivatives $\frac{\partial u}{\partial x^i}$ being bounded. If

$$Y_t = y + \int_0^t \Phi(s,\omega)\,dW_s + \int_0^t \Psi(s,\omega)\,ds\,, \quad \Phi(s,\omega) = (\Phi^i_j(s,\omega)),$$
$$\Psi(s,\omega) = (\Psi^i(s,\omega)) \tag{3.5}$$

then the following equality is valid:

$$u(t,Y_t)-u(0,y)=\int_0^t \sum_{i,j}\frac{\partial u}{\partial x^i}(s,Y_s)\,\Phi^i_j(s,\omega)\,dW^j_s +$$

$$+\int_0^t \left[\frac{1}{2}\sum_{i,j,k}\frac{\partial^2 u}{\partial x^i \partial x^j}(s,Y_s)\,\Phi^i_k(s,\omega)\,\Phi^j_k(s,\omega)+\sum_i \frac{\partial u}{\partial x^i}(s,Y_s)\,\Psi^i(s,\omega)+\frac{\partial u}{\partial t}(s,Y_s)\right] ds\,.$$

This equality looks awkward, but it has a highly transparent meaning. We will explain it in the one-dimensional case. The increment of the process Y_t in a small time dt has the form:

$$dY_t=\Phi(t,\omega)\,dW_t+\Psi(t,\omega)\,dt\,.$$

By Taylor's formula:

$$du(t,Y_t)=\frac{\partial u}{\partial t}(t,Y_t)\,dt+\frac{\partial u}{\partial y}(t,Y_t)\,dY_t+\frac{1}{2}\frac{\partial^2 u}{\partial y^2}(t,Y_t)\,(dY_t)^2+\cdots=$$

$$=\frac{\partial u}{\partial t}(t,Y_t)\,dt+\frac{\partial u}{\partial y}(t,Y_t)\,\Phi(t,\omega)\,dW_t+\frac{\partial u}{\partial y}(t,Y_t)\,\psi(t,\omega)\,dt+ \tag{3.6}$$

$$+\frac{1}{2}\frac{\partial^2 u}{\partial y^2}\Phi^2(t,\omega)\,(dW_t)^2+\cdots\,.$$

Here we denoted by the dots the terms containing factors of the form $(dW_t)^3$, or $(dW_t)^2\,dt$, or $(dt)^2\,dW_t$, or $(dt)^2$. The expectation of each of these factors is the quantity of a higher order of smallness than dt. After integration, these summands give zero contribution. This statement is, roughly speaking, a version of the law of large numbers. Here, it is helpful to bear in mind that the increments of the Wiener process are independent. Noting that $E(dW_t)^2=dt$, we conclude that after integration the last term on the right-hand side in (3.6), becomes $\frac{1}{2}\int_0^t \frac{\partial^2 u}{\partial y^2}(s,Y_s)\,\Phi^2(s,\omega)\,ds$. Lastly, integrating (3.6) from zero to t we obtain:

$$u(t,Y_t) - u(0,y) = \int_0^t \frac{\partial u}{\partial y}(s,Y_s)\,\Phi(s,\omega)\,dW_s +$$

$$+ \int_0^t \left[\frac{\partial u}{\partial t}(s,Y_s) + \frac{\partial u}{\partial y}(s,Y_s)\,\Psi(s,\omega) + \frac{1}{2}\,\frac{\partial^2 u}{\partial y^2}(s,Y_s)\,\Phi^2(s,\omega)\right] ds\,.$$

This formula corresponds to (3.5) in the one-dimensional case.

Certain generalizations of Ito's formula are possible. For example, it is sufficient to assume that

$$E \int_0^t \sum_{j=1}^r \left[\sum_{i=1}^r \frac{\partial u}{\partial x^i}(s,Y_s)\,\Phi_j^i(s,\omega)\right]^2 ds < \infty$$

rather than suppose first derivatives to be bounded. This condition ensures the existence of the stochastic integral in Ito's formula. The requirement for the functions $u(t,x)$ to be smooth may be weakened by allowing discontinuities of first derivatives in y on a set in which the process Y_t spends zero time.

It should be noted that stochastic integrals can be defined not only with respect to the Wiener process but also with respect to every martingale which is regular enough. A generalized Ito formula is available for stochastic integrals with respect to such martingales (see Gihman and Skorohod [3], v. 3). This generalized formula is useful in studying diffusion processes as well.

Now we return to equation (3.2). By a solution of this equation, we mean a function $X_t = X_t^x(\omega)$, $t \geq 0$, $\omega \in \Omega$, $x \in R^r$, which is progressively measurable (with respect to the family of σ-fields $\mathcal{N}_t$), continuous in t with probability 1, and which satisfies equation (3.3).

It is established (see e.g. Gihman and Skorohod [1], McKean [1], Wentzell [1]) that if the functions $\sigma_j^i(x)$ and $b^i(x)$ satisfy the Lipschitz condition

$$\sum_{i,j=1}^{r} |\sigma_j^i(x) - \sigma_j^i(y)| + \sum_{i=1}^{r} |b^i(x) - b^i(y)| \le K|x-y|, x,y \in R^r, K < \infty ,$$

then, for every $x \in R^r$, a unique solution $X_t^x(\omega)$ of equation (3.2) exists. This solution may be derived as the limit of successive approximations:

$$X_t^{(0)} \equiv x, X_t^{(n)} = x + \int_0^t \sigma(X_s^{(n-1)})\, dW_s + \int_0^t b(X_s^{(n-1)})\, ds .$$

If the coefficients of equation (3.2) or the initial point depend on some parameter α, then the solution $X_t^{x,\alpha}$ also depends on this parameter. Just as in the theory of ordinary (non-stochastic) differential equations, it is possible to study the functions $X_t^{x,\alpha}$ continuity or differentiability in this parameter. In doing so, the continuity and differentiability may be regarded in different ways, for instance, in the mean or with probability 1. General results of this kind and references are available in the works of Gihman and Skorohod [2], Wentzell and Freidlin [2]. Below we will discuss in more detail some special cases of this problem.

By analogy with ordinary differential equations, the existence of a solution of equation (3.2) can also be ensured under weaker requirements on the coefficients. It suffices, e.g. to require the coefficients to be bounded and continuous (cf. Peano's theorem for ordinary differential equations). For details the reader is referred to Stroock and Varadhan [1]. However, known examples from the theory of ordinary differential equations show that unless we assume non-degeneracy (see below), uniqueness does not hold if the coefficients have a modulus of continuity which is essentially worse than the Lipschitz one. For example, if the coefficients satisfy at best a Hölder condition with an exponent smaller than 1, the solution is not unique. However, if $\det(\sigma(x) \cdot \sigma^*(x)) \neq 0$, then the solution is unique under weak requirements concerning regularity of the coefficients (Stroock and Varadhan [1]).

In the case of bad coefficients, the notion of the so-called weak solution of equation (3.2) is introduced. We shall not strive for minimal hypotheses on the coefficients. As a rule, the coefficients will usually be assumed to satisfy a Lipschitz condition (for results on the existence and uniqueness of the strong and weak solutions of stochastic differential equations, see Stroock and Varadhan [1]).

We shall again return to studying the properties of the solution X_t^x of equation (3.2) in the next section. Now we will remark that one can consider equations of (3.2) or (3.3) type with the coefficients depending on t:

$$X_t - x = \int_s^t \sigma(u, X_u)\, dW_u + \int_s^t b(u, X_u)\, du . \tag{3.7}$$

This equation has a unique solution if, for example, the functions $\sigma_j^i(u,x)$, $b^i(u,x)$, $u \geq 0$, $x \in R^r$, $i,j = 1, 2, \cdots, r$, are bounded and continuous in the pair of variables (u,x) and are Lipschitz continuous in x. A solution $X_t = X_t^{s,x}(\omega)$ of equation (3.7) depends on $s > 0$ and $x \in R^r$. We shall consider such equations, non-homogeneous in time, when dealing with differential operators which have coefficients depending on both time and space. Clearly, such nonhomogeneous equations may be reduced to the homogeneous case, provided time is thought of as a new space variable.

Finally, we shall consider stochastic equations of the form:

$$X_t - x = \int_0^t \sigma[X_u, 0 \leq u \leq s]\, dW_s + \int_0^t b[X_u, 0 \leq u \leq s]\, ds , \tag{3.8}$$

with the coefficients σ_j^i, b^i, $(i,j = 1, \cdots, r)$, being non-anticipating functionals. This means that the values of the coefficients at the time s are random variables measurable with respect to the σ-field $\mathcal{N}_s$. If the coefficients of equation (3.8) are bounded and Lipschitz continuous:

$$|\sigma_j^i[\phi_s, 0 \le s \le t] - \sigma_j^i(\psi_s, 0 \le s \le t)| \le K \max_{0 \le s \le t} |\phi_s - \psi_s| ,$$

$$|f^i[\phi_s, 0 \le s \le t] - f^i[\psi_s, 0 \le s \le t]| \le K \max_{0 \le s \le t} |\phi_s - \psi_s| ,$$

then equation (3.8) has a unique solution $X_s^x(\omega)$ continuous with probability 1. This solution can be constructed via successive approximations similar to those used in the case of equation (3.3). For every s, the random variables X_s^x are $\mathcal{N}_s$-measurable.

§1.4 *Markov processes and semi-groups of operators*

Now consider equation (3.2) with bounded coefficients satisfying a Lipschitz condition. For the solution X_t^x of this equation, the following relation

$$X_t^x - X_s^x = \int_s^t \sigma(X_u^x)\,dW_u + \int_s^t b(X_u^x)\,du \tag{4.1}$$

is fulfilled for $s \le t$. From (4.1) it follows that, for a fixed X_s^x, the function X_t^x, $t \ge s$, is defined in a unique way by the increments of the Wiener process after the time s. These increments do not depend on the σ-field $\mathcal{N}_s$ and therefore

$$P\{X_t^x \in \Gamma | \mathcal{N}_s\} = P\{X_t^x \in \Gamma | X_s^x\} \tag{4.2}$$

for $0 \le s \le t$, $\Gamma \subset \mathcal{B}^r$.

Relation (4.2) is called the *Markov property*. Its intuitive meaning is the following. If the position of a trajectory X_s^x at a time s is known, then no information on the behavior of the process before time s adds any supplementary data about the behavior of the process after the time s. For each x, the random functions $X_t^x(\omega)$, $t \ge 0$, for which (4.2) is fulfilled are called Markov random functions (Dynkin [3]).

The stochastic differential equation (3.2) defines a family of Markov random functions $X_t^x(\omega)$ for all possible initial points $x \in R^r$. The following relations are valid (see, e.g. Dynkin [3]):

(i) $P\{X_t^x \in \Gamma\} = p(t,x,\Gamma)$ is a $\mathcal{B}^r$-measurable function of $x \in R^r$.

(ii) $p(0,x,R^r \setminus x) = 0$

(iii) If $s \in [0,t]$, $\Gamma \in \mathcal{B}^r$ then $P\{X_t^x \in \Gamma | N_s\} = p(t-s, X_s^x, \Gamma)$ P-a.s.

Consider the pair (X_t^x, P), where X_t^x, $x \in R^r$, is the collection of Markov random functions for which properties (i)-(iii) are fulfilled and P is the corresponding measure on Ω. Such a pair was called by Dynkin [1] a *Markov family*. The measurable space $(R^r, \mathcal{B}^r)$ is called the *state space* of this Markov family, and $p(t,x,\Gamma)$ is called the *transition function*.

Of course, it is possible to consider Markov random functions and Markov families whose state space differs from $(R^r, \mathcal{B}^r)$. Any measurable space $(\mathcal{E}, \mathcal{B})$ can be taken as the state space. In this case, one should assume that the functions $X_t^x(\omega)$ are defined for every $x \in \mathcal{E}$ and take values in $\mathcal{E}$. Equality (4.2) and properties (i)-(iii) are to be fulfilled for every $\Gamma \in \mathcal{B}$. Here by N_s one can mean any increasing family of σ-fields in Ω such that $\{X_u^x \in \Gamma\} \in N_s$ for $u \leq s$, $\Gamma \in \mathcal{B}$, $x \in \mathcal{E}$. Such a family of random functions is referred to as Markov family in the state space $(\mathcal{E}, \mathcal{B})$ adapted to the family of σ-fields N_t.

We observe that the stochastic differential equations which are used to define a Markov family (X_t^x, P) in $(R^r, \mathcal{B}^r)$ determine certain relations between the trajectories starting from different points $x \in R^r$. Namely, we construct trajectories starting from neighboring points $x,y \in R^r$ using one and the same underlying Wiener process. As we shall see later on, this leads to the following effect: trajectories $X_t^x(\omega)$ and $X_t^y(\omega)$ will be close in a certain sense if the initial points are close. This is easily seen when observing the Wiener family W_t^x, where trajectories starting from different points can be derived out of each other by translation (see page 42). This additional structure which is not involved in the general concept of Markov is very suitable, and we shall utilize it repeatedly.

Sometimes it is more convenient to use the notion of Markov process which is close to that of Markov family.

To define the Markov process corresponding to stochastic differential equation (4.1) (and to the corresponding Markov family (X_t^x, P)), we shall consider the measurable space $(C_{0,\infty}(R^r), \mathfrak{N})$, where $\mathfrak{N}$ is the σ-field of all the cylinder sets of the space $C_{0,\infty}(R^r)$. We shall define on the σ-field $\mathfrak{N}$ a family of probability measures P_x, $x \in R^r$, by putting

$$P_x\{\mathfrak{A}\} = P\{X_\cdot^x \in \mathfrak{A}\},$$

where $\mathfrak{A} \in \mathfrak{N}$, and $X_\cdot^x$ is a solution of equation (3.2) with the initial condition $X_0^x = x$. Let us define the random function $X_t(\phi)$ on the measurable space $(C_{0,\infty}(R^r), \mathfrak{N})$:

$$X_t(\phi) = \phi_t \quad \text{for} \quad \phi \in C_{0,\infty}(R^r).$$

The collection $(X(\phi), P_x)$ of all random functions $X_t(\phi)$ and all probability measures P_x on the σ-field $\mathfrak{N}$ is called the *Markov process* corresponding to equation (3.2).

Certainly, one can define a Markov process corresponding to any Markov family (X_t^x, P) and not only to ones constructed from stochastic differential equations. If the state space of a family (X_t^x, P) is a topological space $\mathcal{E}$ equipped with the σ-field of Borel sets $\mathcal{B}$, and if the trajectories X_t^x are continuous a.s., then the corresponding process $(X_t(\phi), P_x)$ is defined on the space $C_{0,\infty}(\mathcal{E})$ of all continuous functions on $[0, \infty)$ with values in $\mathcal{E}$. As the σ-field in $C_{0,\infty}(\mathcal{E})$, we choose the σ-field $\mathfrak{N}$ of cylinder sets. One should put $X_t(\phi) = \phi_t$ for $\phi_\cdot \in C_{0,\infty}(\mathcal{E})$, and $P_x(\mathfrak{A}) = P\{X_\cdot^x \in \mathfrak{A}\}$ for $\mathfrak{A} \in \mathfrak{N}$, $x \in \mathcal{E}$.

If the trajectories of the family (X_t^x, P) are not a.s. continuous, similar formulas define the corresponding process on the space $H_{0,\infty}(\mathcal{E})$ of all functions on $[0, \infty)$ with values in $\mathcal{E}$. As the σ-field in $H_{0,\infty}(\mathcal{E})$, one should take the σ-field of cylinder sets in $H_{0,\infty}(\mathcal{E})$.

The state space $(\mathcal{E}, \mathcal{B})$ of a family (X_t^x, P) is called the state space of the process (and, unless it leads to misunderstanding, the space $\mathcal{E}$ itself will be called the state space without indicating the σ-field $\mathcal{B}$).

The transition function of a family (X_t^x, P) is also called the transition function of the process.

The above cited notions of Markov family and Markov process are special cases of the corresponding notions in the monograph by Dynkin [3].

We have defined a Markov process as some object corresponding to a Markov family. From the properties of Markov family, one can deduce that, for $0 \le t \le u$, $x \in \mathcal{E}$, $\Gamma \in \mathcal{B}$, the equality

$$P_x\{X_u \in \Gamma | \mathcal{N}_t\} = p(u-t, X_t, \Gamma)$$

holds P_x-a.s.. Here $\mathcal{N}_t$ is the minimal σ-field in the space $C_{0,\infty}(\mathcal{E})$ (or in $H_{0,\infty}(\mathcal{E})$, when trajectories are discontinuous), generated by the sets $\{\phi_s \in B\}$ for $s \le t$, $B \in \mathcal{B}$.

This equality is called the *Markov property of the process* (X_t, P_x).

There are a number of equivalent formulations of the Markov property. To cite them, we introduce (following Dynkin [1]) a family of shift operators θ_t, $t \ge 0$, acting on $\Omega = H_{0,\infty}(\mathcal{E})$. The action of θ_t on a trajectory $\phi \in H_{0,\infty}(\mathcal{E})$ is defined by the equalities:

$$\theta_t X_s(\phi) = X_{t+s}(\phi), \quad s \ge 0 .$$

The shift operator θ_t acts on random variables according to the formula

$$\theta_t \xi(\phi) = \xi(\theta_t \phi) .$$

If $\tau = \tau(\phi)$ is a non-negative random variable, one can in a similar way define the shift operator θ_τ.

The formulation of the Markov property to be used in the sequel, is as follows (Dynkin [3]). Suppose that a random variable $\eta(\phi)$ is $\mathcal{N}$-measurable and bounded and let $\xi(\phi)$ be $\mathcal{N}_t$-measurable and $E_x|\xi| < c < \infty$ for all $x \in \mathcal{E}$. Then

$$E_x \xi \theta_t \eta = E_x \xi E_{X_t} \eta . \tag{4.3}$$

Therefore, according to our definition, a Markov process (X_t, P_x) is essentially a family of probability measures P_x, $x \in \mathcal{E}$, in the space of trajectories which possesses certain properties. In particular, for the processes constructed via the stochastic differential equation (3.2), this is a family of measures in the space $C_{0,\infty}(R^r)$ of continuous functions.

In what follows we shall consider both Markov families and the Markov processes corresponding to these families. In essence, the distinction here is only in the notations: if the trajectory X_t^x has the index x, then, the Markov family is considered; if the probability measure P_x has the index x, then the corresponding Markov process is considered. The expectation with respect to the measure P_x is denoted by E_x. As the state space, we shall usually take the Euclidean space R^r or a closed region in R^r together with the σ-field of Borel sets.

Let us denote by B the Banach space of bounded measurable functions on the measurable space $(\mathcal{E}, \mathcal{B})$. The norm in this space is defined by the equality: $\|f\| = \sup_{x \in \mathcal{E}} |f(x)|$. With every Markov family (X_t^x, P) in the state space $(\mathcal{E}, \mathcal{B})$ (or with the Markov process (X_t, P_x) in this space) one can associate a family of linear continuous operators T_t, $t \geq 0$, acting in the space B according to the formula:

$$(T_t f)(x) = E f(X_t^x) = E_x f(X_t) = \int_{\mathcal{E}} f(y)\, p(t,x,dy) .$$

Since $p(t,x,\Gamma)$ is a probability measure in $\Gamma \in \mathcal{B}$, we conclude that the operators T_t do not increase the norm and transform non-negative functions into non-negative ones: $\|T_t f(x)\| \leq \|f\|$, $T_t f(x) \geq 0$, provided $f(x) \geq 0$. The Markov property (4.3) yields:

$$T_{t+s} f(x) = E_x f(X_{t+s}) = E_x \theta_t f(X_s) = E_x(E_{X_t} f(X_s)) = T_t(T_s f)(x) ,$$

i.e. the operators T_t form a *semi-group* $T_{t+s} = T_t T_s$. In particular, let us suppose that as $f(x)$ we have chosen the indicator function $\chi_\Gamma(x)$.

Then $T_t \chi_\Gamma(x) = E_x \chi_\Gamma(X_t) = p(t,x,\Gamma)$, and the semi-group property can be written down in the form:

$$p(s+t,x,\Gamma) = \int_{\mathcal{E}} p(s,x,dy)\, p(t,y,\Gamma) .$$

This relation is called the *Kolmogorov-Chapman equation.* We will illustrate how the operators T_t act in the simplest examples.

Suppose that in equation (3.2), $\sigma(x)$ vanishes identically. Hence, the Markov family reduces to the family of deterministic trajectories of the dynamical system:

$$d\, X_t^x = b(X_t^x)\, dt$$

with the various initial conditions $X_0^x = x \in R^r$. In this case, the action of the operator T_t on a function is a translation of its argument along the trajectory of the dynamical system. Namely, $T_t f(x) = f(X_t^x)$, where X_t^x is the trajectory with the initial condition $X_0^x = x$. The trajectory X_t^x is not random, so no expectation needs to be written. In particular, if $b(x) = b$ is a constant vector, then the corresponding dynamical system describes a movement with a constant speed b, and $T_t f(x) = f(x+bt)$.

An alternate example. Let $b(x)$ vanish identically, and $\sigma(x)$ be the unit matrix. Then $X_t^x = W_t^x$ is an r-dimensional Wiener process. In this case

$$T_t f(x) = E f(x+W_t) = \int_{R^r} f(x+y) \frac{1}{(2\pi t)^{r/2}} e^{-\frac{|y|^2}{2t}} dy . \tag{4.4}$$

The fact that the operators (4.4) form a semi-group is a well-known property of solutions of Cauchy's problem for the simplest heat equation $\frac{\partial u}{\partial t} = \frac{1}{2} \Delta u(t,x)$.

Suppose that as the state space one has chosen a topological space $\mathcal{E}$ equipped with Borel σ-field. Henceforth we shall deal solely with this case. We consider the space $C = C_{\mathcal{E}}(R^1)$ of bounded continuous functions on $\mathcal{E}$.

A Markov *family* (X_t^x, P) (Markov process (X_t, P_x), semi-group T_t, or transition function $p(t,x,\Gamma)$) is said to be a *Feller* one, whenever the operators T_t transform the space C into itself: $T_t f \in C$ for $f \in C$, $t \geq 0$.

Let us show that a stochastic differential equation with Lipschitz coefficients generates a Feller Markov family. For this, we need a few simple lemmas which will be useful for other purposes as well.

LEMMA 4.1. *Suppose that* $m(t)$, $0 \leq t \leq T$, *is a non-negative function for which the inequality*

$$m(t) \leq c + a \int_0^t m(s)\,ds, \quad 0 \leq t \leq T,$$

holds for some $c, a > 0$. *If also* $\int_0^t m(s)\,ds < \infty$, *then* $m(t) \leq c \exp\{(a,t)\}$ *for every* $t \in [0,T]$.

This lemma is a special case of Gronwell's lemma the proof of which can be found in many books, so we drop it.

LEMMA 4.2. *Suppose that the coefficients in equation (3.2) are bounded*: $|\sigma_j^i(x)| \leq M$, $|b^i(x)| \leq M$; $i,j = 1, \cdots, r$; $x \in R^r$. *Then one can find a constant* $c = c(M,r)$ *such that*

$$E \sup_{0 \leq t \leq T} |X_t^x - x|^2 \leq c(T+T^2).$$

Proof. Evidently, it is sufficient to obtain such a bound separately for each component $X_t^{i,x}$ of the process X_t^x. The equation for $X_t^{i,x}$ yields:

$$\sup_{0\le t\le T} |X_t^{i,x} - x^i|^2 \le 2T^2 \sup_x |b^i(x)|^2 + 2\left[\sum_{j=1}^{r} \max_{0\le t\le T} \left| \int_0^t \sigma_j^i(X_s^x)\, dW_s^j \right|\right]^2 .$$

To estimate the stochastic integrals, we shall apply Property 5 (for the properties of stochastic integrals, see the preceding section). Since the coefficients are bounded, we have

$$E \sup_{0\le t\le T} |X_t^{i,x} - x|^2 \le 2T^2M^2 + 8T\, r^2M^2 .$$

This bound clearly implies the claim of Lemma 4.2. □

LEMMA 4.3. *Suppose that the coefficients of equation (3.2) are bounded and Lipschitz continuous:*

$$\sum_{i,j=1}^{r} |\sigma_j^i(x) - \sigma_j^i(y)| + \sum_{i=1}^{r} |b^i(x) - b^i(y)| \le K\, |x-y| .$$

Let X_t *and* Y_t *be the solutions of (3.2) with the initial conditions* $X_0 = x$ *and* $Y_0 = y$ *respectively.*

Then there exists a constant c *determined by the Lipschitz constant* K *such that*

$$E|X_t - Y_t|^2 \le |x-y|^2 e^{ct} , \tag{4.5}$$

and for every $\delta > 0$, $T > 0$

$$\lim_{|x-y|\to 0} P\{ \sup_{0\le s\le T} |X_s - Y_s| > \delta \} = 0 .$$

For the proof of Lemma 4.3, we shall apply Ito's formula to the function $\rho^2(X_t, Y_t) = \sum_{i=1}^{r} (X_t^i - Y_t^i)^2$:

$$\rho^2(X_t,Y_t) = \rho^2(x,y) + 2\int_0^t \sum_{i=1}^r \left[(X_s^i - Y_s^i)\sum_{j=1}^r (\sigma_j^i(X_s) - \alpha_j^i(Y_s))\,dW_s^j\right] +$$

$$+\, 2\int_0^t \sum_{i=1}^r (X_s^i - Y_s^i)(b^i(X_s) - b^i(Y_s))\,ds + \int_0^t \sum_{i,j=1}^r (\sigma_j^i(X_s) - \sigma_j^i(Y_s))^2\,ds\,. \tag{4.6}$$

Notice that the first derivatives of the function $\rho^2(x,y) = |x-y|^2$ are not bounded. Nevertheless, Ito's formula may be used since Lemma 4.2 implies the finiteness of the expectation needed to ensure that the stochastic integrals in (4.6) exist (see page 53).

Remembering that the coefficients $\sigma_j^i(x)$, $b^i(x)$ are Lipschitz continuous, from (4.6) we conclude that

$$E\,\rho^2(X_t,Y_t) \le \rho^2(x,y) + c\int_0^t E\,\rho^2(X_s,Y_s)\,ds$$

for some $c < \infty$. By Lemma 4.1, this inequality implies the first assertion of Lemma 4.3.

Next, observe that

$$X_t^i - Y_t^i = x^i - y^i + \int_0^t [b^i(X_s) - b^i(Y_s)]\,ds + \int_0^t \sum_{j=1}^r (\sigma_j^i(X_s) - \sigma_j^i(Y_s))\,dW_s^j\,.$$

Thus, to prove the second assertion of Lemma 4.3, it suffices to make sure that

$$P\{\sup_{0\le t\le T} \left|\int_0^t [\sigma_j^i(X_s) - \sigma_j^i(Y_s)]\,dW_s^j\right| > \delta^1\} \to 0\,,$$

$$P\{\int_0^t |b^i(X_s) - b^i(Y_s)|\,ds > \delta^1\} \to 0$$

for every $\delta^1 > 0$ and $i,j = 1,2,\cdots,r$ as $|x-y| \to 0$. The first of these relations comes from the Kolmogorov inequality for stochastic integrals (Property 4) and from bound (4.5):

$$P\left\{ \sup_{0 \le t \le T} \left| \int_0^t [\sigma_j^i(X_s) - \sigma_j^i(Y_s)] dW_s^j \right| > \delta^1 \right\} \le$$

$$\le (\delta^1)^{-2} k^2 \int_0^T E|X_s - Y_s|^2 ds \to 0, \quad |x-y| \to 0 .$$

The second relation is an implication of the Chebyshev inequality and bound (4.5). □

LEMMA 4.4. *If the coefficients of equation (3.2) are bounded and Lipschitz continuous, then the corresponding Markov family* (X_t^x, P) *is a Feller one.*

Proof. In fact, let $f \in C$, $\varepsilon > 0$. For the prescribed ε we shall pick a δ such that $|f(a) - f(b)| < \varepsilon$ provided $|a-b| < \delta$, $|a| + |b| < \frac{1}{\varepsilon}$. Then putting $X_t = X_t^x$, $Y_t = X_t^y$ and using Chebyshev's inequality, we arrive at

$$|T_t f(x) - T_t f(y)| \le E|f(X_t) - f(Y_t)| \le 2\|f\| \Big(P\{|X_t - Y_t| > \delta\} +$$

$$+ P\{|X_t| + |Y_t| > \tfrac{1}{\varepsilon}\} \Big) + \varepsilon \le 2\|f\| \Big(\frac{1}{\delta^2} E|X_t - Y_t|^2 +$$

$$+ 2\varepsilon^2 (E|X_t|^2 + E|Y_t|^2) \Big) + \varepsilon .$$

The right-hand side of the last inequality may be made arbitrarily small by means of selecting ε and $|x-y|$ small enough. This results from Lemmas 4.2 and 4.3. Therefore, the function $(T_t f)(x)$ is continuous in x, and (X_t^x, P) is a Feller family. □

In §1.2 the strong Markov property of the Wiener family was dealt with. Now we are going to formulate the strong Markov property for a general

Markov family. To that end, let us consider a Markov family (X_t^x, P) adapted to the increasing family of σ-fields N_t. To formulate the strong Markov property of this Markov family, one should consider times τ, Markovian with respect to the family of σ-fields N_t, and the σ-field N_τ consisting of events A such that $A \cap \{\tau(\omega) < t\} \in N_t$ for all $t \geq 0$. We will remind that a random time τ, is Markov with respect to the family of σ-fields N_t, if it is a random variable taking non-negative values and $+\infty$ for which the event $\{\tau(\omega) < t\}$ belongs to N_t for every $t \geq 0$.

A family (X_t^x, P) is said to be *strong Markov*, if for any Markov time $\tau(\phi)$ $P\{X_{\tau+t}^x \in \Gamma | N_\tau\} = p(t, X_\tau^x, \Gamma)$ a.s. on the set $\Omega^1 = \{\omega \in \Omega, \tau(\omega) < \infty\}$ for every $\Gamma \in \mathcal{B}$.

The intuitive meaning of the strong Markov property is like this. Once the position of the trajectory X_t^x at the Markov time τ is known, its future behavior does not depend on events defined by the motion of the process before the time τ (it is these events which form the σ-field N_τ).

Let us formulate the strong Markov property for a Markov process (X_t, P_x) in the state space $(\mathcal{E}, \mathcal{B})$. One associates with this process the non-decreasing family of σ-fields $\mathcal{N}_t$, $t \geq 0$, in the space $C_{0,\infty}(\mathcal{E})$ generated by the sets $\{\phi \in C_{0,\infty}(\mathcal{E}) : \phi_s \in B\}$ for $s \leq t$, $B \in \mathcal{B}$ (or in the space $H_{0,\infty}(\mathcal{E})$ if the corresponding trajectories are discontinuous).

By a Markov time with respect to a family $\mathcal{N}_t$, we mean a random variable $\tau(\phi)$ taking non-negative values and $+\infty$ such that $\{\tau(\phi) \leq t\} \in \mathcal{N}_t$ for any $t \geq 0$.

Denote by $\mathcal{N}_\tau$ the totality of sets $A \in \mathcal{N}$ such that $A \cap \{\tau \leq t\} \in \mathcal{N}_t$. It is easily checked that $\mathcal{N}_\tau$ is a σ-field.

A process (X_t, P_x) is said to be a *strong Markov process* if for any Markov time $\tau(\phi)$

$$P_x\{X_{\tau+t} \in \Gamma | \mathcal{N}_\tau\} = p(t, X_\tau, \Gamma)$$

P_x a.s. for any $x \in \mathcal{E}$, $t \geq 0$, $\Gamma \in \mathcal{B}$, on the set $\{\phi \in C_{0,\infty}(\mathcal{E}) : \tau(\phi) < \infty\}$.

There are a number of useful implications of the strong Markov property of processes and families. In particular, the following property will be used repeatedly: Let (X_t, P_x) be a strong Markov process in the state space $(\mathfrak{S}, \mathfrak{B})$, $f(x)$, $x \in \mathfrak{S}$, be a bounded $\mathfrak{B}$-measurable function, and suppose that τ_1, τ_2 are finite Markov times such that $P_x\{\tau_1 \leq \tau_2\} = 1$, $x \in \mathfrak{S}$, τ_2 being the first hitting time to some set in $\mathfrak{S}$. Then

$$E_x f(X_{\tau_1 + \theta_{\tau_1}\tau_2}) = E_x E_{X_{\tau_1}} f(X_{\tau_2}) . \tag{4.7}$$

Here θ_{τ_1} is the shift operator.

The first hitting time to a closed set gives an example of a Markov time for a Markov family (or process) derived by equation (3.2). A detailed examination of the strong Markov property has been carried out by Dynkin in [1] (see also, e.g. Wentzell [1]). In particular, a Feller Markov family whose trajectories are continuous with probability 1 and the corresponding Markov process are shown to be strong Markovian there. Therefore, by Lemma 4.4, the strong Markov property holds for the Markov family derived via the stochastic differential equation (3.2) with bounded Lipschitz continuous coefficients and for the corresponding Markov process.

The *infinitesimal operator* A of the semi-group T_t is defined by the equality

$$Af = \lim_{t \downarrow 0} \frac{T_t f - f}{t} .$$

Here the limit is understood in the sense of convergence in the norm: $\|Af - \frac{1}{t}(T_t f - f)\| \to 0$ as $t \downarrow 0$. The operator A is also called the infinitesimal operator of the corresponding Markov family or Markov process. The *domain* of the definition of this operator is a linear space denoted by D_A. Let us set $B_0 = \{f \in B : \lim_{t \downarrow 0} \|T_t f(x) - f\| = 0\}$. It is easy to make certain that B_0 is a closed linear subspace of the space B. One can prove (see, e.g. Dynkin [3]) the manifold D_A to be dense everywhere in B_0.

A transition function $p(t, x, \Gamma)$ in a topological state space $\mathfrak{S}$ equipped with the Borel σ-field $\mathfrak{B}$ is said to be *stochastically continuous* whenever $\lim_{t \downarrow 0} p(t, x, \mathfrak{S} \setminus U) = 0$ for every $x \in \mathfrak{S}$ and every neighborhood U of the point x.

If the transition function is stochastically continuous, then the infinitesimal operator defines all finite-dimensional distributions of the Markov process in a unique way. A stochastically continuous transition function is readily shown to correspond to a Markov family generated by equation (3.2) with bounded coefficients.

Suppose that $f \in D_A$. Then $\|T_t f - f\| \to 0$ as $t \downarrow 0$. We also have that

$$\frac{T_{t+h}f - T_t f}{h} = \frac{T_h T_t f - T_t f}{h} = T_t \frac{T_h f - f}{h}. \tag{4.8}$$

Since $f \in D_A$, it follows that the last expression in the chain of equalities (4.8) converges in the norm to $T_t Af$ as $h \downarrow 0$. Therefore, the limits of the other ratios exist too; in particular, $T_t f \in D_A$ and $AT_t f = T_t Af$. We also note that the left derivative in t of $T_t f$ exists and is equal to $T_t Af$ as well:

$$\left\| \frac{-1}{h}(T_{t-h}f - T_t f) - T_t Af \right\| \le$$

$$\le \|T_{t-h}(-h)^{-1}(f - T_h f) - T_{t-h}Af\| + \|T_{t-h}Af - T_t Af\| \le$$

$$\le \left\| \frac{1}{h}(T_h f - f) - Af \right\| + \|T_{t-h}Af - T_t Af\|,$$

as $h \downarrow 0$. Here we have used the fact that T_t is a contraction semi-group, $f \in D_A$, and also that $Af \in B_0$ (it is easy to check that if $f \in D_A$, then $Af \in B_0$). Hence, if $f \in D_A$, the function $u_t(x) = T_t f(x)$ is a solution of the next abstract Cauchy's problem:

$$\frac{du_t}{dt} = Au_t, \quad \lim_{t \downarrow 0} u_t(x) = f(x). \tag{4.9}$$

In the theory of semi-groups (see, e.g. Hille and Phillips [1], Dynkin [3]), it is established that the solution of problem (4.9) is unique in the class of functions $v_t(x)$ such that $\|v_t\| \le c_1 e^{\alpha t}$, where c_1 and α are arbitrary constants.

Recall that the stochastic differential equation

$$dX_t^x = \sigma(X_t^x)dW_t + b(X_t^x)dt\,, \quad X_0^x = x \in R^r \tag{4.10}$$

is connected with the differential operator

$$L = \frac{1}{2}\sum_{i,j=1}^{r} a^{ij}(x)\frac{\partial^2}{\partial x^i \partial x^j} + \sum_{i=1}^{r} b^i(x)\frac{\partial}{\partial x^i}\,,$$

where $(a^{ij}(x)) = \sigma(x)\sigma^*(x)$. The following result clarifies this connection.

THEOREM 4.1. *Suppose that the coefficients* $\sigma^i_j(x)$ *and* $b^i(x)$, $i,j = 1,\cdots,r$, *are bounded and Lipschitz continuous. Also suppose that the function* $f(x)$ *has bounded, uniformly continuous partial derivatives up to the second order inclusively. Then* $f(x)$ *belongs to the domain* D_A *of the infinitesimal operator* A *of the Markov family* (X_t^x, P) *generated by equation (4.10) and moreover* $Af = Lf$.

Proof. Let us apply Ito's formula to the function $f(X_t^x)$:

$$f(X_t^x) - f(x) = \int_0^t (\nabla f(X_s^x), \sigma(X_s^x)dW_s) + \int_0^t Lf(X_s^x)ds\,. \tag{4.11}$$

Here $\nabla f(x)$ is the gradient of the function $f(x): \nabla f = \left(\frac{\partial f}{\partial x^1}, \cdots, \frac{\partial f}{\partial x^r}\right)$; $(\nabla f, \sigma dW)$ being the scalar product of two vectors. Let us take the expectation of equality (4.11). Noting that the expectation of the stochastic integral is zero, we derive

$$Af(x) = \lim_{t\downarrow 0}\frac{1}{t}(Ef(X_t^x) - f(x)) = \lim_{t\downarrow 0}\frac{1}{t}\int_0^t Lf(X_s^x)ds = Lf(x)\,.$$

The convergence in the norm is ensured by the fact that the function $Lf(x)$ is bounded and uniformly continuous. □

Therefore, the domain D_A of the infinitesimal operator of the Markov family (x_t^x, P) at least contains sufficiently smooth functions.

From equation (4.10), remembering the properties of stochastic integrals we deduce immediately the following relations for the Markov random functions $X_t^x = (X_t^{1,x}, \cdots, X_t^{r,x})$.

$$\lim_{t\downarrow 0} \frac{1}{t} E(X_t^{i,x} - x^i) = \lim_{t\downarrow 0} \frac{1}{t} \int_0^t E\, b^i(X_s^x)\, ds = b^i(x)\,,$$

$$\lim_{t\downarrow 0} \frac{1}{t} E(X_t^{i,x} - x^i)(X_t^{j,x} - x^j) = \lim_{t\downarrow 0} \frac{1}{t} \int_0^t E \sum_{k=1}^r \sigma_k^i(X_s^x)\, \sigma_k^j(X_s^x)\, ds =$$

$$= \lim_{t\downarrow 0} \frac{1}{t} \int_0^t E\, a^{ij}(X_s^x)\, ds = a^{ij}(x)\,.$$

The coefficients $a^{ij}(x) = \sum_k \sigma_k^i(x) \sigma_k^j(x)$ characterize the intensity of jiggle of the trajectory X_t^x near the point x, and they are called *diffusion coefficients*.

The collection of functions $b^1(x), \cdots, b^r(x)$ is called the *drift*. Note that $(b^1, \cdots, b^r)$ which are the coefficients of the first order derivatives in the operator L, do not form a vector. When we change variables, the derivatives of the diffusion coefficients are a part of expression for the coefficients of the first order derivatives in new coordinates.

The Markov process $X = (X_t, P_x)$ constructed via stochastic differential equation (4.10), is said to be a *diffusion process governed by the operator* L *in the space* R^r.

Sometimes we shall consider the Markov process $X^D = (X_t^D, P_x^D)$ obtained from the process $X = (X_t, P_x)$ by means of stopping at the first exit time from some domain $D \subset R^r$, that is $\tau_D = \inf\{t : x_t \notin D\}$ and the corresponding Markov family $(X_t^{D,x}, P)$. The trajectories $X_t^{D,x}$ can be constructed via the stochastic equation

$$X_t^{D,x} - x = \int_0^{t\wedge\tau_D^x} \sigma(X_s^{D,x})\, dW_s + \int_0^{t\wedge\tau_D^x} b(X_s^{D,x})\, ds ,$$

$$\tau_D^x = \inf\{t : X_t^{D,x} \notin D\}, \quad x \in D \cup \partial D .$$

The operation of stopping a Markov process has been studied by Blumenthal [1] and Dynkin [3].

For a Feller Markov process, Dynkin has introduced the operator $\mathfrak{A}$ (Dynkin [1]) which proved to be suitable. This local operator acts according to the formula

$$\mathfrak{A} f(x) = \lim_{U \downarrow x} \frac{E_x f(X_{\tau_U}) - f(x)}{E_x \tau_U} ,$$

where U is a neighborhood of the point x, $\tau_U = \inf\{t : X_t \notin U\}$ is the first exit time from U, and the limit is taken over all the possible neighborhoods contracting to the point x.

Given a Feller Markov process with continuous trajectories, the operator $\mathfrak{A}$ is defined at least for continuous functions $f(x)$ belonging to the domain D_A of the infinitesimal operator A, and $\mathfrak{A}f = Af$. Let us suppose, in addition, that the state space of the process is compact. We will consider continuous functions belonging to the domain of the definition of the operator $\mathfrak{A}$ and such that $\mathfrak{A}f(x)$ is continuous. Then every such function belongs to D_A, and $\mathfrak{A}f(x) = Af(x)$.

Another semi-group of operators denoted by U_t, $t \geq 0$, can be associated with the transition function $p(t, x, \Gamma)$ (and, therefore, with the corresponding Markov family and process in the state space $(\mathcal{E}, \mathcal{B})$ as well). Let us denote by $\mu(\Gamma)$, $\Gamma \in \mathcal{B}$, countably additive functions of sets having finite variation. The above-mentioned semi-group of operators U_t acts in the Banach space V of these functions $\mu(\cdot)$. The norm $\|\mu\|^*$ in this space is defined as the total variation of the function of set

$\mu(\cdot)$. The semi-group U_t acts by the formula

$$(U_t\mu)(\Gamma) = \int_{\mathcal{E}} p(t, x, \Gamma)\,\mu(dx)\,.$$

It is easy to check that the operators U_t and T_t are mutually adjoint:

$$\int_{\mathcal{E}} T_t f(x)\,\mu(dx) = \int_{\mathcal{E}} f(x)\,(U_t\mu)\,(dx)\,,\ f \in B,\ \mu \in V\,.$$

Now let us consider a Markov random function X_t^ξ which starts not a fixed point $x \in \mathcal{E}$, but at a random point ξ having the distribution μ: $P\{\xi \in \Gamma\} = \mu(\Gamma)$, $\Gamma \in \mathcal{B}$. Then the distribution of X_t^ξ at the time t is as follows:

$$P\{X_t^\xi \in \Gamma\} = \int_{\mathcal{E}} \mu(dx)\ p(t,x,\Gamma) = (U_t\mu)(\Gamma)$$

Thereby, the semi-group U_t describes the evolution of the distribution of the random function X_t^ξ.

A measure μ on the state space $(\mathcal{E}, \mathcal{B})$ is called an invariant measure of a Markov process (Markov family) with the transition function $p(t,x,\Gamma)$ if $U_t\mu = \mu$ for all $t \geq 0$. If $\mu(\mathcal{E}) = 1$, then the invariant measure $\mu(\cdot)$ is also called a *stationary distribution.*

It is clear that if the initial point ξ has a stationary distribution μ, then the distribution of X_t^ξ does not change with time. A linear combination of two invariant measures with non-negative coefficients is an invariant measure too. Therefore, the collection of all invariant measures of a Markov process forms a cone. The infinitesimal operator of the semi-group U_t is denoted by A^*. It is easy to see that every invariant measure μ belongs to the domain of the definition of the operator A^*, and $A^*\mu = 0$.

The converse is also true, namely: every measure μ obeying the equation $A^*\mu = 0$ is an invariant measure.

Suppose that $p(t, x, \Gamma)$ is the transition function of a Markov family in R^r constructed by equation (4.10). Then the operator A^* is an extension of the operator

$$L^*u = \frac{1}{2} \sum_{i,j=1}^{r} \frac{\partial^2}{\partial x^i \partial x^j} (a^{ij}(x)\, u(x)) - \sum_{i=1}^{r} \frac{\partial}{\partial x^i} (b^i(x)\, u(x))$$

(which is formally adjoint to the operator L) in the following sense: if $\mu(\Gamma) = \int_\Gamma m(x)\, dx$, $\Gamma \in \mathcal{B}^r$ and the function $m(x)$ is smooth enough, then $(A^*\mu)(\Gamma) = \int_\Gamma L^*m(x)\, dx$. In particular, if $L^*m = 0$, $m \geq 0$, then $m(x)$ is the density function of the invariant measure for the corresponding Markov process.

§1.5 *Measures in the space of continuous functions corresponding to diffusion processes*

Consider the differential operator

$$L = \frac{1}{2} \sum_{i,j=1}^{r} a^{ij}(x) \frac{\partial^2}{\partial x^i \partial x^j} + \sum_{i=1}^{r} b^i(x) \frac{\partial}{\partial x^i} .$$

We assume that the coefficients of this operator are bounded, $b^i(x)$ is Lipschitz continuous, and a matrix $\sigma(x) = (\sigma^i_j(x))$, exists, $i = 1, \cdots, r$, $j = 1, \cdots, \ell$, with Lipschitz continuous coefficients such that $\sigma(x)\sigma^*(x) = (a^{ij}(x))$.

Let $W_t(\omega)$ be an ℓ-dimensional Wiener process; $\mathcal{N}_t$, $t \geq 0$, being a non-decreasing family of σ-fields in the space Ω adapted to the process W_t.

Consider the stochastic differential equation

$$dX^x_t = \sigma(X^x_t)\, dW_t + b(X^x_t)\, dt\, , \; X^x_0 = x \in R^r . \tag{5.1}$$

The fact that the matrix $\sigma(x)$ is now not necessarily square does not affect the existence and uniqueness of the solution of problem (5.1): if the coefficients are bounded and Lipschitz continuous, then, for any $x \in R^r$, a unique, a.s. continuous solution $X_t^x(\omega)$ of equation (5.1) exists, and $X_t^x(\omega)$ are $\mathcal{N}_t$-measurable random variables. The pair (X_t^x, P) makes up a Markov family adapted to the σ-fields $\mathcal{N}_t$, $t \geq 0$.

The representation $(a^{ij}(x)) = \sigma(x) \cdot \sigma^*(x)$ is not unique. Hence, one can associate with the operator L different Markov families, for example, by selecting the matrices $\sigma(x)$ with differing number of columns. Let us show that the distributions in the space of functions $C_{0,\infty}(R^r)$ coincide for all such random functions X_t^x.

LEMMA 5.1. *Suppose that* X_t^x *and* $\tilde{X}_t^x$ *are unique solutions of the stochastic differential equations*

$$dX_t^x = \sigma(X_t^x)\,dW_t + b(X_t^x)\,dt\,,\ X_0^x = x\ ;$$
$$d\tilde{X}_t^x = \tilde{\sigma}(\tilde{X}_t^x)\,d\tilde{W}_t + b(\tilde{X}_t^x)\,dt\,,\ \tilde{X}_0^x = x\ ,$$

respectively, and let $\sigma(x)\sigma^*(x) = \tilde{\sigma}(x) \cdot \tilde{\sigma}^*(x)$. *Then the distributions in the space* $C_{0,\infty}(R^r)$ *induced by the random functions* X_t^x *and* $\tilde{X}_t^x$ *coincide.*

Proof. Without loss of generality, the matrices $\sigma(x)$ and $\tilde{\sigma}(x)$ may be thought of as having not only the same number of rows, but of columns as well. If this is not the case, then one of the matrices may be supplemented with zero columns. The Wiener processes W_t and $\tilde{W}_t$ may also be assumed to be of equal dimension.

Since $\sigma(x)\sigma^*(x) = \tilde{\sigma}(x)\tilde{\sigma}^*(x)$, the scalar products of the row vectors of the matrix $\sigma(x)$ and the scalar products of the corresponding row vectors of the matrix $\tilde{\sigma}(x)$ are equal. Hence, there exists an orthogonal matrix $\theta(x)$ such that $\tilde{\sigma}(x) = \sigma(x)\,\theta(x)$. This reasoning leads to the equality

$$d\tilde{X}_t^x = \tilde{\sigma}(\tilde{X}_t^x)\,d\tilde{W}_t + b(\tilde{X}_t^x)\,dt = \sigma(\tilde{X}_t^x)\,\theta(\tilde{X}_t^x)\,d\tilde{W}_t + b(\tilde{X}_t^x)\,dt\,. \tag{5.2}$$

By the corollary of Levy's Theorem cited in §1.3, $\tilde{\tilde{W}}_t = \int_0^t \theta(\tilde{X}_s^x)\, d\tilde{W}_s$ is a Wiener process. Thus, (5.2) yields

$$d\tilde{X}_t^x = \sigma(\tilde{X}_t^x)\, d\tilde{\tilde{W}}_t + b(\tilde{X}_t^x)\, dt\ .$$

Therefore, the equation for $\tilde{X}_t^x$ has the same $\sigma(x)$ and $b(x)$ as that for X_t^x has; it only involves the other Wiener process $\tilde{\tilde{W}}$. These equations define the same mapping of the space $C_{0,\infty}(R^\ell)$ into $C_{0,\infty}(R^r) : W_\cdot \to X_\cdot^x$, and induce the same measure in $C_{0,\infty}(R^r)$. □

Therefore, different Markov families (X_t^x, P) may correspond to the same operator L (in particular, those defined on different probability fields), however, the distributions in the space $C_{0,\infty}(R^r)$ induced by the random functions $X_t^x(\omega)$ coincide. Whence, it follows that a unique canonical Markov process (X_t, P_x) corresponds to every operator L. Namely, P_x, $x \in R^r$, is the measure in $C_{0,\infty}(R^r)$ induced by the random functions $X_t^x(\omega)$. These measures also may be considered on the space $C_{0,t}(R^r)$ for $t < \infty$.

In particular, let P_x^W denote the family of measures corresponding to the Wiener Markov process which are obtained out of the standard Wiener measure μ_W on $C_{0,\infty}^0(R^r)$ by means of translation: the measure P_x^W is concentrated on the functions of $C_{0,\infty}(R^r)$ which are equal to x for $t = 0$, $P_x^W(\mathfrak{A}) = \mu_W(\mathfrak{A}_{-x})$ where $\mathfrak{A}_{-x} = \{\phi \in C_{0,\infty}(R^r) : \phi_\cdot + x \in \mathfrak{A}\}$.

The measures P_x corresponding to a non-degenerating operator L are similar to the Wiener measure P_x^W in many respects, if considered in the space $C_{0,t}(R^r)$, $t < \infty$.

Just as the Wiener measure, this measure has all the space $C_{0,t}^x(R^r) = \{\phi \in C_{0,t}(R^r) : \phi(0) = x\}$, as its support, i.e., for any $\delta > 0$, the δ-neighborhood of any function $\phi \in C_{0,t}^x(R^r)$ has a positive P_x-measure. The measure P_x, just as the Wiener one, is concentrated on the functions satisfying a Hölder condition with the exponent $\frac{1}{2}$ at no point and satisfying a Hölder condition with the exponent $\frac{1}{2} - \varepsilon$ for any $\varepsilon > 0$.

If the operator L degenerates, then of course the corresponding measures P_x can differ vastly from the Wiener measures. For example, if $a^{ij}(x) \equiv 0$ for $i,j = 1, \cdots, r$, then the corresponding measure P_x is concentrated on a single function of $C_{0,t}(R^r)$ —namely, the solution of the equation $\dot{X}_t^x = b(X_t^x)$, $X_0^x = x$.

Let (X_t, P_x) be the Markov process corresponding to the operator L. We remind that this process is defined on the probability space $(C_{0,\infty}(R^r), \mathcal{N}, P_x)$, $\mathcal{N}$ being the σ-field of cylinder sets in $C_{0,\infty}(R^r)$. The σ-field generated by the sets $\{\phi \in C_{0,\infty}(R^r) : \phi_s \in B\}$, $s \le t$, $B \in \mathcal{B}^r$, is denoted by $\mathcal{N}_t$.

Formula (4.11) implies that, for every $x \in R^r$ and for every function $f : R^r \to R^1$, having bounded, continuous first- and second-order derivatives, the random process

$$Z_t = f(X_t) - \int_0^t Lf(X_s)\,ds \tag{5.3}$$

is a martingale with respect to the increasing family of σ-fields $\mathcal{N}_t$ and the measure P_x.

This property may be taken as the definition of the Markov process corresponding to the operator L. Namely, the random process $X_t(\phi)$ defined by the equality $X_t(\phi) = \phi_t$ on the probability space $(C_{0,\infty}(R^r), \mathcal{N}, P_x)$ is called the *solution of the martingale problem for the operator* L, if, for any function $f : R^r \to R^1$, having bounded, continuous first- and second-order derivatives, the process Z_t defined by (5.3) is a martingale with respect to the increasing family of σ-fields $\mathcal{N}_t$ (and the measure P_x). Sometimes, by the solution of martingale problem, one means the measure P_x itself with respect to which the process Z_t is a martingale. Therefore, formula (4.11) implies that the Markov process (X_t, P_x) gives the solution of martingale problem for every $x \in R^r$.

The martingale problem has been studied in detail recently (see Stroock and Varadhan [1], there are also other references there). Under very weak assumptions on the coefficients of the operator L, the existence of the solution of this problem is established, and the conditions ensuring uniqueness are established. In particular, it is demonstrated that if the coefficients of the operator L obey the conditions formulated at the beginning of this section, then the martingale problem has a unique solution. Thus, the Markov process corresponding to the operator L may be defined as the solution of the martingale problem for the operator L.

Such an approach makes the theory more symmetric. Here no process (or any measure in $C_{0,\infty}(R^r)$) is particularly singled out; processes all have "equal rights" (as we saw, in the classical theory due to Ito, the Wiener process plays the special role). The martingale approach enables one directly to find the connection between differential operators and processes (families of measures in $C_{0,\infty}(R^r)$). Such an approach allows one, in a more natural way, to generalize the theory to the case of integro-differential operators which correspond to processes with jumps; it also permits one to study processes corresponding to operators with bad coefficients. The fact that, under minor extra assumptions, the martingale problem has a unique solution, is highly useful when proving limit theorems for random processes. We shall return to this point later on.

Now we shall go into the question of the absolute continuity of the measures in the space $C_{0,t}(R^r)$ corresponding to various random functions. The first important result in this direction is the Cameron-Martin theorem [1] concerns absolutely continuous shift transformations of the Wiener measure. Together with the r-dimensional random function W_t, $t \in [0,T]$, we will consider the random function $X_t = W_t + \phi_t$, $\phi \in C_{0,T}(R^r)$. Let μ_W and μ_X be the corresponding measures in the space $C_{0,T}(R^r)$. Clearly, for the measures μ_X and μ_W to be absolutely continuous, it is necessary that $\phi_0 = 0$. It is also clear that the function ϕ_t should have sufficiently good modulus of continuity: unless the function ϕ_t satisfies a Holder condition with an exponent $a \leq \frac{1}{2}$ the measures μ_X and μ_W are no longer absolutely continuous. It turns out that the measures μ_X

and μ_W are absolutely continuous, if and only if $\int_0^T |\dot{\phi}_s|^2 ds < \infty$, $\phi_0 = 0$. In this case, the density $\frac{d\mu_X}{d\mu_W}$ has the form

$$\frac{d\mu_X}{d\mu_W} = \exp\left\{\int_0^T (\dot{\phi}_s, dW_s) - \frac{1}{2}\int_0^T |\dot{\phi}_s|^2 ds\right\}. \tag{5.4}$$

We shall clarify this formula in the one-dimensional case. Let

$$\zeta_t[W] = \zeta_t = \int_0^t \dot{\phi}_s dW_s - \frac{1}{2}\int_0^t \dot{\phi}_s^2 ds,\ a_t = \int_0^t \dot{\phi}_s^2 ds < \infty .$$

It is easy to see that the random variables $\int_0^t \dot{\phi}_s dW_s$ and W_{a_t} have the same distribution. Hence,

$$\int_0^T E \exp\{2\zeta_t\} dt = \int_0^T E \exp\{2W_{a_t} - a_t\} dt \le TEe^{2W_{a_T}} < \infty .$$

Therefore, Ito's formula can be applied to the function e^{ζ_t}:

$$e^{\zeta_t} = 1 + \int_0^t e^{\zeta_s} \dot{\phi}_s dW_s .$$

The integrand in the non-stochastic integral vanished. The last equality implies that e^{ζ_t} is a martingale with respect to the family of the σ-fields $\mathcal{N}_t$, and $E\exp\{\zeta_t\} = \int_{C_{0,T}(R^1)} \exp\{\zeta_T(W.)\} d\mu_W = 1$. Hence the positive function $\exp\{\zeta_T(W.)\}$ can serve as a probability density function with respect to the Wiener measure on the space $C_{0T}(R^1)$. We will define the new measure on the σ-field $\mathcal{N}_T$:

$$\tilde{\mu}(A) = \int_A \exp\{\zeta_T(W.)\} d\mu_W,\ A \in \mathcal{N}_T .$$

To prove the Cameron-Martin assertion, it suffices to check that $\mu(A) = \mu_X(A)$ or in other words, to verify that the random process $\tilde{X}_t$, $t \in [0,T]$ defined by the equality $\tilde{X}_t(\psi_.) = \psi_t$ on the probability space $(C_{0T}(R^1), \mathcal{N}_T, \tilde{\mu})$, $\psi_. \in C_{0T}(R^1)$, has the same distribution as the process $X_t = W_t + \phi_t$. For this, it is sufficient to check that $\tilde{W}_t = \tilde{X}_t - \phi_t$ is a Wiener process. Probably, the simplest way of doing this is that via Lévy's theorem (see §1.3). We leave this to the reader (see Girsanov [1], though).

This work of Girsanov (see also Gihman and Skorohod [3]) contains the generalization of the Cameron-Martin result too. Consider the random functions X_t and Y_t satisfying the stochastic differential equations

$$X_t = x + \int_0^t \sigma(u,X_u)\,dW_u + \int_0^t b(u,X_u)\,du\,,$$

$$Y_t = x + \int_0^t \sigma(u,Y_u)\,dW_u + \int_0^t [b(u,Y_u)+c(u,Y_u)]\,du$$

respectively, $x \in R^r$, $t \in [0,T]$. The coefficients of these equations are assumed bounded.

Denote by μ_X and μ_Y the measures in $C_{0,T}(R^r)$ induced by the random functions X_t and Y_t. We shall suppose that the system of linear algebraic equations $\sigma(u,x)\phi(u,x) = c(u,x)$ has a bounded solution $\phi(u,x)$, $u \in [0,T]$, $x \in R^r$. Then the measures μ_X and μ_Y are absolutely continuous with respect to each other, and

$$\frac{d\mu_Y}{d\mu_X}(X_.) = \exp\left\{\int_0^T (\phi(u,X_u), dW_u) - \frac{1}{2}\int_0^T |\phi(u,X_u)|^2\,du\right\}. \quad (5.5)$$

In particular, if the diffusion matrix $(a^{ij}(t,x)) = \sigma(t,x)\sigma^*(t,x)$ is uniformly non-degenerate, (i.e. $\sum_{i,j=1}^r a^{ij}(t,x)\lambda_i\lambda_j > a\sum_1^r \lambda_i^2$ for some $a > 0$, any real

$\lambda_1, \cdots, \lambda_r$ and $x \in R^r$, $t > 0$), then the measures μ_X and μ_Y are absolutely continuous with respect to each other for any bounded $c(t,x)$. If the matrix $(a^{ij}(t,x))$ degenerates, then the absolute continuity holds only for functions $c(t,x)$ having a special form.

The work of Girsanov [1] considers also a more wide class of stochastic equations including, in particular, equations of (3.8) type.

Measures corresponding to processes with different diffusion matrices are, however, not absolutely continuous. For example, the measures $\mu_{\sigma W}$ corresponding to the one-dimensional processes $X_t^\sigma = \sigma W_t$, are singular with respect to each other for every different $\sigma \geq 0$, in the sense that for any σ_1 one can indicate a set $\mathfrak{A} \subset C_{0T}^0(R^1)$ such that $\mu_{\sigma_1 W}(\mathfrak{A}) = 1$ and $\mu_{\sigma_2 W}(\mathfrak{A}) = 0$ for any $\sigma_2 \neq \sigma_1$. This is straightforward, for example, from the local law of interated logarithm: for a.a. (with respect to the measure $\mu_{\sigma W}$) functions $\phi \in C_{0T}^0(R^1)$ the following relation holds

$$\lim_{t \downarrow 0} \frac{|\phi_t|}{\sqrt{2t \ln \ln t^{-1}}} = \sigma .$$

Up to now we have only discussed the continuity of measures in the space of functions $C_{0,T}(R^r)$ on a finite time interval. The measures in $C_{0,\infty}(R^r)$ corresponding to the processes with the same diffusion matrices and different drifts are, generally speaking, no longer absolutely continuous. For example, as is easily seen from the law of iterated logarithm as $t \to \infty$, the trajectories of the one-dimensional process $X_t = W_t + t$ run to infinity with probability 1 as $t \to \infty$, whereas the Wiener trajectories, after any time t, return to zero with probability 1.

Now we will remind some results on the weak convergence of measures in $C_{0,T}(R^r)$. A sequence of measures $\mu^{(n)}$ on the space $C_{0,T}(R^r)$ is said to *converge weakly* to a measure μ as $n \to \infty$, if for any continuous bounded functional $F(\phi_.)$ on the space $C_{0,T}(R^r)$

$$\lim_{n \to \infty} \int_{C_{0,T}(R^r)} F(\phi_.)\, d\mu^{(n)} = \int_{C_{0,T}(R^r)} F(\phi_.)\, d\mu .$$

If a set A of measures on $C_{0,T}(R^r)$ is such that every sequence of measures in A has a weakly convergent subsequence, then the set A is called *relatively compact in the weak convergence topology* or *relatively weakly compact*. The proof of the weak convergence of the sequence of measures is usually divided into two stages: first, one proves the relative compactness of this family of measures; secondly, one checks that the family $\{\mu^{(n)}\}$ has only one limit point.

To check the relative weak compactness of a family of measures, one can use the following criterion (Prohorov [1], see also Gihman and Skorohod [1]). Suppose that $c, \alpha, \beta > 0$ exist such that, for any measure μ from the family of measures A on the space $C_{0,T}(R^r)$,

$$\mu(\phi \in C_{0,T}(R^r) : |\phi_0| > c) = 0, \qquad \int_{C_{0,T}(R^r)} |\phi_{t+h} - \phi_t|^\alpha d\mu < ch^{1+\beta} \tag{5.6}$$

for every t, $t+h \in [0,T]$. Then the family of measures A is relatively compact in the weak convergence topology.

A sequence of random processes $\xi_t^{(n)}(\omega)$, $t \in [0,T]$, defined on, generally speaking, different probability spaces $(\Omega^{(n)}, \mathcal{F}^{(n)}, P^{(n)})$ and such that their trajectories belong to $C_{0,T}(R^r)$, is *weakly convergent*, if the sequence of measures corresponding to these processes in $C_{0,T}(R^r)$ converges weakly. If the corresponding family of measures is relatively weakly compact, then the family of processes $\{\xi_t^{(n)}\}$ is termed *weakly compact*. The compactness criterion (5.6), in terms of random processes, is as follows: a family $\{\xi_t^{(n)}\}$ is weakly compact if there are $c, \alpha, \beta > 0$ such that

$$P^{(n)}\{|\xi_0^{(n)}| > c\} = 0, \; E^{(n)}|\xi_{t+h}^{(n)} - \xi_t^{(n)}|^\alpha < ch^{1+\beta},$$

$$t, t+h \in [0,T].$$

LEMMA 5.2. *Given fixed* $T > 0$ *and* $x \in R^r$, *denote by* $G_M^{x,T}$ *the set of random functions* X_t^x, $0 \le t \le T$, $X_0^x = x$, *which are the solutions of equation (5.1) for all possible coefficients* $\sigma_j^i(x)$ *and* $b^i(x)$, *bounded in absolute value by the constant* $M < \infty$. *Then the set* $G_M^{x,T}$ *is weakly compact in* $C_{0T}(R^r)$.

For the proof, we shall make use of the above criterion. Let us show that for some $c > 0$

$$E|X_t^x - X_s^x|^4 \leq c|t-s|^2 . \tag{5.7}$$

Remembering that $X_0^x = x$, we can get from (5.7) the claim of the lemma. To prove (5.7), it is sufficient to check that for a one-dimensional Wiener process W_t and, for every bounded, progressively measurable, real-valued function $f(u,\omega)$, $|f(u,\omega)| < M < \infty$,

$$E\left(\int_s^t f(u,\omega)\,dW_u\right)^4 \leq \text{const.}\,(t-s)^2 . \tag{5.8}$$

Let us compute $(\int_s^t f(u,\omega)\,dW_u)^4$ via Ito's formula:

$$\left(\int_s^t f(u,\omega)\,dW_u\right)^4 = 4\int_s^t\left(\int_s^{t_1} f(u,\omega)\,dW_u\right)^3 f(t_1,\omega)\,dW_{t_1} +$$

$$+\,6\int_s^t\left(\int_s^{t_1} f(u,\omega)\,dW_u\right)^2 f^2(t_1,\omega)\,dt_1 .$$

We shall take the expectation of both sides of this equality and use the properties of stochastic integrals:

$$E\left(\int_s^t f(u,\omega)\,dW_u\right)^4 \leq 6M^2\int_s^t\int_s^{t_1} Ef^2(u,\omega)\,du\,dt_1 \leq 3M^4(t-s)^2 .$$

Thus, we obtained inequality (5.8), and thereby the proof of Lemma 5.2. □

To verify the uniqueness of the limit point of the family of measures $\{\mu^{(n)}\}$ in $C_{0,T}(R^r)$ (or of the corresponding family of processes $\xi_t^{(n)}$), one can prove that, for some class Π of "simple" sets from $C_{0,T}(R^r)$, the limits $\lim_{n\to\infty} \mu^{(n)}(A)$, $A \in \Pi$, exist. Certainly, this class Π of "simple" sets must be wide enough. For instance, the collection of cylinder sets in $C_{0,T}(R^r)$ may be taken as Π. The existence of $\lim_{n\to\infty} \mu^{(n)}(A)$ for the cylindrical sets A means the convergence of the finite-dimensional distributions of the processes $\xi_t^{(n)}$.

There is another way of verifying the uniqueness of the limit point. One can show that every limit measure is a solution of some martingale problem. If this problem has a unique solution, then the limit point is unique.

The weak compactness of the family of measures (processes) and the uniqueness of the limit point of this family imply weak convergence.

§1.6 *Diffusion processes with reflection*

The Markov processes constructed in §1.3 enable us to examine Cauchy's problem and boundary value problems with boundary conditions of the Dirichlet type. To analyze Neumann's problem and problems close to it, one must construct a diffusion process with reflection on the boundary of domain. Before describing the construction of a process with reflection in the general case, we will first construct a Wiener process on the non-negative semi-axis with reflection at zero. To be more exact, we shall construct the Wiener Markov family of random functions with reflection at zero. The corresponding process is constructed starting from this family in the usual way (see §1.4). We shall recall two constructions of such a family of random functions.

In the first place, it is possible to define a Wiener family of random functions on the half-line $R_+^1 = \{x \geq 0\}$ with reflection at zero by the formula: $X_t^x = |x+W_t|$, $x \geq 0$, where W_t is the standard Wiener process on the line. Noting that the transition density function $p(t,x,y)$ of the Wiener process has the property $p(t,x,y) = p(t,-x,-y)$, the random functions X_t^x are readily seen to form a Markov family of random functions in the state space R_+^1 equipped with the Borel σ-field.

Let a bounded function $f(x)$, $x \in R^1$, have uniformly continuous and bounded second-order derivatives. It is easy to verify that this function belongs to the domain D_A of the infinitesimal operator of the family (X_t^x, P) if and only if $\left.\frac{d^+f}{dx}\right|_{x=0} = \lim\limits_{x\downarrow 0} \frac{f(x)-f(0)}{x} = 0$. To see this, extend the function $f(x)$ to an even function $\tilde{f}$ on the entire line R and apply Ito's formula. If $\frac{d^+f}{dx}(0) \neq 0$, then, at $x = 0$, the function $\tilde{f}$ has a corner and

$$\lim_{t\downarrow 0} \left|\frac{Ef(X_t^0)-f(0)}{t}\right| = \left|\frac{d^+f}{dx}(0)\right| \lim_{t\downarrow 0} \frac{E|W_t|}{t} = +\infty .$$

On account of the known distribution of the random variable W_t one can evaluate

$$v(t,x) = Ef(X_t^x) = Ef(|x+W_t|) = \frac{1}{\sqrt{2\pi t}} \int_{-\infty}^{\infty} f(|x+y|)\, e^{-\frac{y^2}{2t}} dy$$

and, by direct substitution, one can check that $v(t,x)$ is a solution of the mixed problem

$$\frac{\partial v}{\partial t} = \frac{1}{2}\frac{\partial^2 v}{\partial x^2}, \quad t>0,\ x>0 ;$$

$$\lim_{t\downarrow 0} v(t,x) = f(x), \quad \frac{\partial v}{\partial x}(t,0) = 0 ;$$

the function $f(x)$ being assumed bounded and continuous.

It is possible to derive the stochastic differential equation for the random function $X_t^x = |x+W_t|$. To this end, let us formally apply Ito's formula to the function $|x+W_t|$:

$$X_t^x = |x+W_t| = x + \int_0^t \operatorname{sign}(x+W_s)\,dW_s + \frac{1}{2}\int_0^t \delta(x+W_s)\,ds ,$$

where $\delta(x)$ is the δ-function concentrated at zero.

Noting that $\int_0^t \operatorname{sgn}(x+W_s)\,dW_s = \widetilde{W}_t$ is a Wiener process (see the corollary of Lévy's theorem in §1.3), the last equation may be rewritten in the form

$$X_t^x - x = \widetilde{W}_t + \frac{1}{2}\int_0^t \delta(X_s^x)\,ds . \tag{6.1}$$

The second summand in (6.1), $\frac{1}{2}\int_0^t \delta(X_s^x)\,ds = \zeta_t^x$ is a continuous, non-decreasing random process, adapted (just as X_t^x) to the σ-fields N_t. The process ζ_t^x increases only on the set $\Lambda = \{t \ge 0 : X_t^x = 0\} = \{t \ge 0 : x + W_t = 0\}$. The Lebesgue measure of this set Λ is zero. The random process ζ_t^x is called *local time* at zero for the random function X_t^x. It is possible to prove (Ito and McKean [1]) that

$$\zeta_t^x = \lim_{\varepsilon\downarrow 0} \frac{1}{2\varepsilon}\int_0^t \chi_{[0,\varepsilon]}(X_s^x)\,ds$$

where $\chi_{[0,\varepsilon]}(x)$ is the indicator of the set $[0,\varepsilon] \in R_+^1$. Equation (6.1) is a special case of (6.4) and so we shall make the meaning of equation (6.1) more precise later on.

Another construction of the Wiener process with reflection at zero was put forward by Lévy [1]. The random functions $\widetilde{X}_t^x = x + W_t - \left[\min_{0\le s\le t}(x+W_s)\wedge 0\right]$ were shown by Lévy to coincide (in the sense of distributions in the space of trajectories) with the random functions X_t^x (see also Ito and McKean [1]). The process $\widetilde{\zeta}_t^x = -\left[\min_{0\le s\le t}(W_s+x)\wedge 0\right]$ is the local time at zero for the random function $\widetilde{X}_t^x$.

Both of these constructions admit generalization to the multi-dimensional case. Now we shall describe the construction of a diffusion process in the half-space $R^r_+ = \{(x^1, \cdots, x^r) \in R^r : x^1 \geq 0\}$ with reflection along the normal to the boundary, i.e. in the direction of the x^1-axis, and then we shall make remarks concerning the general case.

So, suppose that we are given the differential operator

$$L = \frac{1}{2} \sum_{i,j=1}^{r} a^{ij}(x) \frac{\partial^2}{\partial x^i \partial x^j} + \sum_{i=1}^{r} b^i(x) \frac{\partial}{\partial x^i}$$

defined for $x \in R^r_+$. The functions $a^{ij}(x)$ are assumed to be bounded and twice continuously differentiable up to the boundary, their first- and second-order derivatives being bounded. The functions $b^i(x)$ are supposed Lipschitz continuous. In addition, we assume that the operator L is uniformly non-degenerate in a neighborhood of the boundary of this half-space. Let $\sigma(x) = (\sigma^i_j(x))$ be a matrix defined for $x \in R^r_+$ whose elements are Lipschitz continuous and such that $\sigma(x)\sigma^*(x) = (a^{ij}(x))$ (on the existence of such a matrix, see §3.2).

Let us extend the functions $\sigma^i_j(x)$ and $b^i(x)$ onto the entire space R^r by putting for $x^1 < 0$

$$b^i(x^1, \cdots, x^r) = b^i(-x^1, x^2, \cdots, x^r) \quad \text{if} \quad i \neq 1 ,$$

$$b^1(x^1, \cdots, x^r) = -b^1(-x^1, x^2, \cdots, x^r) ;$$

$\sigma^i_j(x^1, \cdots, x^r) = \sigma^i_j(-x^1, \cdots, x^r)$ if $i, j \neq 1$, $\sigma^i_j(x^1, \cdots, x^r) = -\sigma^i_j(-x^1, \cdots, x^r)$ if at least one index i or j is equal to 1.

Now consider the stochastic differential equation in R^r

$$d\tilde{X}^x_t = \sigma(\tilde{X}^x_t)\, dW_t + b(\tilde{X}^x_t)\, dt, \; \tilde{X}^x_0 = x . \tag{6.2}$$

The coefficients of this equation, generally speaking, have discontinuities in the plane $x^1 = 0$. Nevertheless, for such an equation, due to the fact that the diffusion matrix does not degenerate near the plane $x^1 = 0$, one can prove existence and uniqueness theorems (Girsanov [2]) and construct the Markov family $(\tilde{X}_t^x, P)$ in the state space $(R^r, \mathcal{B}^r)$. Let us denote by ϕ the symmetry mapping with respect to the plane $x^1 = 0$. Our rule of extending the coefficients onto the entire space R^r implies that the transition function $p(t, x, \Gamma)$ of the family $(\tilde{X}_t^x, P)$ has the following property:

$$p(t, x, \Gamma) = p(t, \phi(x), \phi(\Gamma)) . \tag{6.3}$$

It results from (6.3) that, for $x \in R^r_+$, the random functions $X_t^x = (X_t^{1,x}, \cdots, X_t^{r,x}) = (|\tilde{X}_t^{1,x}|, \tilde{X}_t^{2,x}, \cdots, \tilde{X}_t^{r,x})$, $x \in R^r_+$, together with the measure P, form a Markov family in the state space $(R^r_+, \mathcal{B}^r_+)$ (here $\mathcal{B}^r_+$ is the σ-field of Borel subsets of R^r_+). The Markov process (X_t, P_x) corresponding to this family is the process in R^r_+ with reflection on the boundary in the direction of the x^1-axis. By analogy to the one-dimensional case, one can verify that the infinitesimal operator A of this process is defined, at least, for the smooth functions $f(x)$ for which $\left.\frac{\partial f}{\partial x^1}(x)\right|_{x^1=0} = 0$ and $Af(x) = Lf(x)$. This construction of a process with reflection is available in the article of Freidlin [2].

As in the one-dimensional case, stochastic equations may be written down directly for X_t^x. For the case of an arbitrary closed domain D and vector field $\gamma(x)$ on ∂D, this equation may be arranged in the form

$$\begin{gathered} dX_t^x = \sigma(X_t^x)\, dW_t + b(X_t^x)\, dt + \chi_{\partial D}(X_t^x)\gamma(X_t^x)\, d\xi_t^x , \\ X_0^x = x, \ \xi_0^x = 0 , \end{gathered} \tag{6.4}$$

where $\chi_{\partial D}(x)$ is the indicator of the boundary of the domain, W_t is a Wiener process in R^r adapted to the family of σ-fields N_t, $\gamma(x)$ is a vector-valued function defined for $x \in \partial D$; in the above case, $D = R^r_+$, $\partial D = \{x \in R^r : x^1 = 0\}$, $\gamma(x) = e = (1, 0, \cdots, 0)$.

By a solution of equation (6.4), we mean a pair of a.s. continuous processes X_t^x, ξ_t^x satisfying (6.4), adapted to the underlying family of σ-fields N_t and satisfying, with probability 1, the following conditions:

$X_t^x \in D$; ξ_t^x is a non-decreasing process which increases only for $t \in \Lambda = \{t : X_t^x \in \partial D\}$, Λ having Lebesgue measure zero a.s. We do not give now the detailed derivation of equations (6.4) from equations (6.2), because a more suitable way of constructing the solution of system (6.4) will be shown below, and moreover, the uniqueness of such a solution will be established.

Skorohod [1] had been the first to consider equations like (6.4) involving two unknown processes (X_t^x, ξ_t^x) for constructing Markov families with boundary conditions, and Watanabe [1] developed this approach in the general situation. The approach which we proceed to present, belongs to Anderson and Orey [1]. Let $D = R_+^r$ for a moment.

We will define the mapping $\Gamma : C_{0,\infty}(R^r) \to C_{0,\infty}(R_+^r)$ via the following equalities: for $\zeta = (\zeta_t^1, \cdots, \zeta_t^r) \in C_{0,\infty}(R^r)$, we put $\Gamma(\zeta) = \eta = (\eta_t^1, \cdots, \eta_t^r)$, where $\eta_t^i = \zeta_t^i$ for $i = 2, \cdots, r$, and $\eta_t^1 = \zeta_t^1 - \min_{0 \le s \le t} \zeta_s^1 \wedge 0$. The mapping $\xi : C_{0,\infty}(R^r) \to C_{0,\infty}(R^1)$ will be defined by the condition $\Gamma(\zeta) - \zeta = (\xi(\zeta), 0, 0, \cdots, 0)$. It is clear from this definition that $\Gamma_t(\zeta) = (\eta_t^1, \cdots, \eta_t^r)$ takes values in R_+^r and is a non-anticipating functional, i.e. the value of $\Gamma_t(\zeta)$ at time t is determined by the behavior of ζ_s for $s \le t$. Besides, $\xi_t(\zeta) = (\Gamma_t(\zeta))^1 - \zeta_t^1$ and it is also a non-anticipating functional. In addition, observe, that $\sup_{0 \le s \le t} |\Gamma_s(\zeta) - \Gamma_s(\tilde{\zeta})| \le 2 \sup_{0 \le s \le t} |\zeta_s - \tilde{\zeta}_s|$ and $|\Gamma_s(\zeta) - \Gamma_t(\zeta)| \le |\zeta_s - \zeta_t|$.

Consider now the stochastic equation

$$Y_t^x = x + \int_0^t \sigma(\Gamma_s(Y^x))\, dW_s + \int_0^t b(\Gamma_s(Y^x))\, ds . \tag{6.5}$$

Note that although the coefficients $\sigma(x)$ and $b(x)$ were originally defined in R_+^r, $\sigma(\Gamma_s(y))$ and $b(\Gamma_s(y))$ are now defined for the functions Y_s, $0 \le s < \infty$ taking values in the whole space R^r. From the properties of the mapping Γ and the functions $\sigma_j^i(x)$, $b^i(x)$ it follows that

$$\sup_{0\le s\le t} |\sigma^i_j(\Gamma_s(\zeta)) - \sigma^i_j(\Gamma_s(\tilde{\zeta}))| \le K \sup_{0\le s\le t} |\zeta_s - \tilde{\zeta}_s| ,$$
$$\sup_{0\le s\le t} |b^i(\Gamma_s(\zeta)) - b^i(\Gamma_s(\tilde{\zeta}))| \le K \sup_{0\le s\le t} |\zeta_s - \tilde{\zeta}_s| \tag{6.6}$$

for some $K < \infty$, $i,j = 1,\cdots,r$, $t \ge 0$. From (6.6) we conclude that equation (6.5) has a unique, continuous with probability 1 solution Y_t^x which is measurable in the variables (t,x) and, for every t, is measurable with respect to the σ-field N_t. This solution may be constructed with the aid of successive approximations. The proof of these statements does not differ, in essence, from the similar proof for equation (3.2) which defines diffusion process in R^r.

THEOREM 6.1. *Suppose that* Y_t^x, $t \ge 0$, *is a solution of equation (6.5) with bounded, Lipschitz continuous coefficients, and suppose that the matrix* $a^{ij}(x)) = \sigma(x)\sigma^*(x)$ *is non-degenerate for* $x \in R^r_+$. *We put* $X_t^x = \Gamma_t(Y^x)$, $\xi_t^x = \xi_t(Y^x)$. *Then the pair* (X_t^x, ξ_t^x) *is a solution of equation (6.4) for* $D = R^r_+$, $\gamma(x) = (1, 0, \cdots, 0)$. *This solution is unique.*

Proof. First we note that the function Y_t^x is continuous with probability 1, measurable in (t,x) and N_t-measurable for every t. Whence, taking into account that $\Gamma_t(Y^x)$ and $\xi_t(Y^x)$ are non-anticipating continuous functionals, one can deduce that the functions (X_t^x, ξ_t^x) possess these properties as well. That Y_t^x is a solution of equation (6.5) implies immediately that the pair (X_t^x, ξ_t^x) obeys equation (6.4). Further, we see from its definition that ξ_t^x never decreases and increases only on the set $\Lambda = \Lambda(\omega) = \{t \ge 0 : X_t^x \in \partial D\}$, where $\partial D = \{x \in R^r : x^1 = 0\}$. Now we shall show that the Lebesgue measure of the set $\Lambda = \Lambda(\omega)$ is zero a.s. For this, clearly, it is sufficient to prove that the set $\Lambda_T = \Lambda \cap [0,T]$ has measure zero. From the definition of X_t^x we see that $\Lambda_T = \{t \in [0,T] : Y_t^{1,x} = \min_{0\le s\le t} Y_s^{1,x} \wedge 0\}$, where $Y_s^{1,x}$ is the first component of the process Y_s^x. Recall that, by assumption, the matrix $\sigma(x)\sigma^*(x)$ is non-degenerate.

Therefore, according to the results of Girsanov [1], the measure induced in the space $C_{0T}(R^r)$ by the process $Y_t^{1,x}$ is absolutely continuous with respect to the measure induced in $C_{0T}(R^r)$ by the process with zero drift. Because of this, it suffices to consider the case $b^1(x) \equiv 0$ only. Now notice that, for $b^1 \equiv 0$, the first component $Y_t^{1,x} = \int_0^t \sum_{i=1}^r \sigma_i^1(\Gamma_s(Y^x))dW_s$ may be obtained from the one-dimensional Wiener process via the non-singular change of time: $Y_t^{1,x} = \tilde{W}_{\int_0^t a^{11}(\Gamma_s(Y^x))ds}$, where $a^{11}(x) = \Sigma[\sigma_i^1(x)]^2$. Hence the question reduces to the proof of the following fact: for the Wiener process $\tilde{W}_t$, the set $\tilde{\Lambda}_T = \{t \in [0,T] : \tilde{W}_{\int_0^t a^{11}(\Gamma_s(Y^x))ds} = \min_{0 \leq s \leq t} \tilde{W}_{\int_0^s a^{11}(\Gamma_{s_1}(Y^x))ds_1}\}$ has the Lebesgue measure zero with probability 1. If $\chi_{a,b}(x)$ denotes the indicator of the set $[a,b)$, $t_a^b = \int_0^T \chi_{a,b}(\tilde{W}_s)ds$, then using the explicit form of the density function of the process $\tilde{W}_t$, one can show that, for some $c < \infty$, $\Delta^{-1} \sum_{k=-\infty}^{\infty} E(t_{k\Delta}^{(k+1)\Delta})^2 < c$ for any $\Delta > 0$. Whence, by Chebyshev's inequality, $\lim_{\Delta \downarrow 0} P\{\sup_k t_{k\Delta}^{(k+1)\Delta} > \Delta^{1/3}\} = 0$. Thus, for any $\gamma = \gamma(\omega)$, the set $\{s : \tilde{W}_s(\omega) = \gamma(\omega)\}$ has Lebesgue measure zero a.s. In particular, putting $\gamma(\omega) = \inf_{0 \leq s \leq \bar{a}T} W_s$, with $\bar{a} = \sup_x a^{11}(x)$, we get that $\tilde{\Lambda}_T$ has measure zero a.s.

Therefore, the pair (X_t^x, ξ_t^x) meets all the requirements to be a solution of equation (6.4).

We turn to proving the uniqueness of the solution of equation (6.4). Suppose, on the contrary, that together with the solution $(X_t^x, \xi_t^x) = (X_t^{1,x}, \cdots, X_t^{r,x}; \xi_t^x)$, another solution $(\tilde{X}_t^x, \tilde{\xi}_t^x) = (\tilde{X}_t^{1,x}, \cdots, \tilde{X}_t^{r,x}; \tilde{\xi}_t^x)$ exists. We set $Y_t = (Y_t^1, \cdots, Y_t^r)$, $\tilde{Y}_t = (\tilde{Y}_t^1, \cdots, \tilde{Y}_t^r)$ where $Y_t^1 = X_t^{1,x} - \xi_t^x$, $Y_t^2 = X_t^{2,x}, \cdots, Y_t^r = X_t^{r,x}$; $\tilde{Y}_t^1 = \tilde{X}_t^{1,x} - \tilde{\xi}_t^x$, $\tilde{Y}_t^2 = \tilde{X}_t^{2,x}, \cdots, \tilde{Y}_t^r = \tilde{X}_t^{r,x}$.

It is easily seen that Y_t as well as $\tilde{Y}_t$ satisfies equation (6.5). However, since this equation has a unique solution, $Y_t = \tilde{Y}_t$ a.s. This implies that $X_t^{i,x} = \tilde{X}_t^{i,x}$ for $i = 2, 3, \cdots, r$. For $i = 1$, we have

$$X_t^{1,x} - \widetilde{X}_t^{1,x} = \xi_t^x - \widetilde{\xi}_t^x .$$

This equality implies that, with probability 1, both its sides vanish for $t \geq 0$. In fact, if at some time t the left-hand side is positive, then $X_t^{1,x} > 0$, and the right-hand side may only decrease or remain constant on some interval $[t, t + \alpha(\omega))$. On account of the continuity of all functions under consideration and noting that $X_0^{1,x} - \widetilde{X}_0^{1,x} = 0$, we conclude from the above observation that $X_t^x \leq \widetilde{X}_t^x$ for all $t \geq 0$. The converse inequality may be derived similarly. Therefore, $X_t^x = \widetilde{X}_t^x$ a.s. and thereby, the solution of equation (6.4) is unique. □

Relying on the uniqueness theorem, we conclude from (6.4) that X_t^x, $t \geq 0$, is a Markov random function. The random function ξ_t^x is called a *local time on the boundary* of domain. The collection of functions X_t^x, $x \in R_+^r$, together with the measure P forms a Markov family (X_t^x, P) (with respect to the system of σ-fields $\mathcal{N}_t$, $t \geq 0$, adapted to the underlying Wiener process W_t). The state space of this family is R_+^r. The corresponding Markov process (X_t, P_x) is called a *Markov process in* R_+^r *with (instantaneous) reflection on the boundary.*

Just as in Lemma 4.3, one can prove that

$$E\,|Y_t^a - Y_t^b|^2 \leq c(t)|a-b|^2;\ a,b \in R_+^r ,$$

where the function $c(t)$ depends on the Lipschitz constant of the coefficients of equation (6.5). By the properties of the transformation Γ,

$$E|X_t^a - X_t^b|^2 \leq 2\,c(t)|a-b|^2 . \tag{6.7}$$

We will denote by T_t the semi-group of operators corresponding to the process (X_t, P_x) in the state space $R_+^r : T_t f(x) = E_x f(X_t)$. The bound (6.7) implies that the operators T_t transfer the space $C_{R_+^r}(R^1)$ of bounded continuous functions on R_+^r into itself, i.e. the process (X_t, P_x) is a Feller process. Being a Feller Markov process with continuous trajectories, this process is strong Markov. Let us evaluate the infinitesimal operator A of the process (X_t, P_x) on smooth functions. Suppose that a function $f(x)$ belonging to $C_{R^r_+}(R^1)$ has uniformly continuous, bounded first- and second-order derivatives, and also that $\left.\frac{\partial f(x)}{\partial x^1}\right|_{x^1=0} = 0$. Then, remembering that $X_t^x = \Gamma_t(Y^x)$, it is not difficult to compute via Ito's formula that

$$\lim_{t\downarrow 0} \frac{T_t f(x) - f(x)}{t} = Lf(x) = \frac{1}{2} \sum_{i,j=1}^{r} a^{ij}(x) \frac{\partial^2 f(x)}{\partial x^i \partial x^j} + \sum_{i=1}^{r} b^i(x) \frac{\partial f(x)}{\partial x^i},$$

where $(a^{ij}(x)) = \sigma(x)\sigma^*(x)$. The assumptions on the function $f(x)$ imply that this convergence is uniform in $x \in R^r_+$, so that $f \in D_A$ and $Af(x) = Lf(x)$.

If $\frac{\partial f}{\partial x^1}(\bar{x}) \neq 0$ for some $\bar{x} = (0, x^2, \cdots, x^r)$, then $f(x) \notin D_A$. Indeed,

$$\lim \frac{T_t f(\bar{x}) - f(\bar{x})}{t} = \lim_{t\downarrow 0} \frac{1}{t}\left(\frac{\partial f}{\partial x^1}(\bar{x}) E_{\bar{x}} X_t^1 + O(t)\right), \tag{6.8}$$

where $O(t)$ denotes terms such that $|t^{-1}O(t)| < c < \infty$ for small t. Next, note that

$$E_{\bar{x}} X_t^1 = E(Y_t^{1,\bar{x}} - \min_{0\le s\le t} Y_s^{1,\bar{x}}) =$$

$$= -E\left(\min_{0\le s\le t} \int_0^s \sum_{i=1}^{r} \sigma_i^1(\Gamma_u(Y^{\bar{x}}))\, dW_u^i\right) + O(t) > \tag{6.9}$$

$$> E(\max_{0\le s\le at} \widetilde{W}_s) \sim c_1 \sqrt{t}$$

as $t\downarrow 0$, where c_1 is a positive constant; $0 < a < a^{11}(x) = \sum_{i=1}^{r} (\sigma_i^1(\bar{x}))^2$. Here we used that $\int_0^s \sum_1^r \sigma_i^1(\Gamma_u(Y^{\bar{x}}))\, dW_u^i$, $s \in [0,t]$, has the same distribution in $C_{0,t}(R^1)$ as $\widetilde{W}_{\int_0^s a^{11}(\Gamma_u(Y^{\bar{x}}))du}$ has, where $\widetilde{W}_t$ is some Wiener process. We also resorted to formula (2.9) for the distribution of $\max_{0 \le s \le t} W_s$. From (6.8) and (6.9) we conclude that

$$\lim_{t \downarrow 0} |(T_t f(\bar{x}) - f(\bar{x}))| = +\infty \quad \text{and hence} \quad f(x) \notin D_A .$$

REMARK 1. In this construction of a process with reflection, we have always been assuming only that $a^{11}(x) \ge a > 0$. Thus the assumption on the non-degeneracy of the operator L may be replaced by a less restrictive assumption on the non-degeneracy in the direction $x^1 : a^{11}(x) \ge a > 0$. It is sufficient to assume that this inequality holds only on the plane $x^1 = 0$.

So, we have constructed a process in the half-space governed by the operator L in the interior points and reflected in the direction of the x^1-axis on the boundary of the half space. Now we proceed to construct a process with reflection in an arbitrary domain.

Given a bounded domain $D \subset R^r$, suppose that the direction cosines of the vector of the inward normal $n(x) = (n_1(x), \cdots, n_r(x))$ to the boundary ∂D of the domain D are twice continuously differentiable functions. We will designate by $\gamma(x)$ the vector field on $\partial D : \gamma(x) = (\gamma_1(x), \cdots, \gamma_r(x))$. Let the functions $\gamma_i(x)$ have continuous derivatives up to the third order inclusively and $(\gamma(x), n(x)) > 0$ for $x \in \partial D$. The last condition means that the vectors $\gamma(x)$ are directed into the domain D, and the angle between the direction $\gamma(x)$ and the plane tangent to ∂D at the point x, is not equal to zero.

In the vicinity of every point $x \in \partial D$, a neighborhood $U(x)$ exists such that it may be covered by a coordinate system in which the vectors $\gamma(y)$, $y \in \partial D \cap U(x)$, will have the form $(1, 0, \cdots, 0)$, and the boundary ∂D in the new coordinates is represented by the equation $x^1 = 0$, and also $D \cap U(x) \subset \{x^1 > 0\}$. One can choose a finite covering of the boundary out of these neighborhoods $U(x)$; suppose they are $U_1, \cdots, U_n$. Without any loss of generality, one can assume here that, for any point $x \in \partial D$, there is a neighborhood $U_{k(x)}$ for which the Euclidean distance $\rho(x, R^r \setminus U_{k(x)})$

is larger than ρ_0, where ρ_0 is some positive constant, one and the same for all the points x. We will denote by K_i the coordinate system of the above described form in the domain U_i. The original coordinate system will be denoted by K_0. This coordinate system covers all the domain D. It will be convenient for us to denote $D = U_0$.

Now we sketch the construction of the solutions of equation (6.4) in the domain D (for details, see Anderson and Orey [1]). The local coordinate systems enable us to reduce this problem to constructing a process in a half-space with reflection in the direction of the x^1-axis. In fact, the going over to the new coordinates and the construction described above, allow us to construct solutions of equation (6.4) with the initial condition $x \in U_{k(x)}$ up to the first hitting time τ^x to the boundary of $U_{k(x)}$. If $k(x) = 0$, then it is an ordinary process in unbounded space. If $k(x) = 1, \cdots, n$, then it is a process with reflection. Then this construction is repeated for the starting point $X^x_{\tau^x}$ and the processes X^x_t and ξ^x_t are constructed up to the first exit time from $U_{k(X^x_{\tau^x})}$ and so on. Thanks to the fact that the coefficients of equation (6.4) are bounded and the fields $n(x)$, $\gamma(x)$ are smooth, one will need, with probability 1, only a finite number of steps for constructing the random function (X^x_t, ξ^x_t) on any finite time interval $[0,T]$.

Thereby, we obtain a solution (X^x_t, ξ^x_t) of problem (6.4) for all $t > 0$. This construction also implies that, first, the functions (X^x_t, ξ^x_t) are progressively measurable and continuous with probability 1; and secondly, the function ξ^x_t is non-decreasing and increases only on the set $\Lambda = \{t : X^x_t \in \partial D\}$. With probability 1, this set has measure zero, whenever the diffusion coefficient in the direction $\gamma(x)$, $x \in \partial D$, is strictly positive: $\Sigma\, a^{ij}(x)\gamma_i(x)\gamma_j(x) > 0$. From the properties of the process with reflection in the half-space, it also follows that (X^x_t, P) is a Feller Markov family. This family has the strong Markov property.

The random function ξ^x_t is referred to as a local time on boundary. Note that, when the vector field $\gamma(x)$ is multiplied by a positive function

$\alpha(x)$, the random functions X_t^x remain unchanged, but the local time changes. To avoid such an ambiguity, one can impose some normalizing condition on the field $\gamma(x)$. For example, one can assume that $(\gamma(x), n(x)) = 1$.

The considerations carried out in the case of the half-space, immediately lead to the conclusion that the infinitesimal operator A of the Markov family (X_t^x, P) is defined at least on the functions f(x) having continuous (in the domain D up to the boundary) first- and second-order derivatives for which $\left.\frac{\partial f(x)}{\partial \gamma(x)}\right|_{x \in \partial D} = 0$. For these functions, $Af(x) = Lf(x)$.

REMARK 2. The assumption that the domain is bounded, was made solely to enable us to draw the following two conclusions from the fact that the coefficients, the boundary, and the field $\gamma(x)$ are continuous and smooth: first, the conclusion on the uniform boundedness of the coefficients and their derivatives in the new coordinates; secondly, the conclusion on the existence of a $\rho_0 > 0$ and a covering with coordinate neighborhoods $U_1, U_2, \cdots, U_n$ such that every point $x \in D \cup \partial D$ may be covered with a neighborhood $U_{k(x)}$; $\rho(x, R^r \backslash U_{k(x)}) > \rho_0 > 0$. The cited construction also remains valid without changes in the case of unbounded domains, provided one postulates the existence of a covering of this domain via the coordinate neighborhoods with the listed properties.

In conclusion we will give Ito's generalized formula for functions of the process (X_t^x, ξ_t^x) defined by equation (6.4). We will restrict ourselves to the special case to be used in §2.5. Suppose that $f(t,x,y,z)$, $t \in [0,T]$, $x \in D \cup \partial D$; $y,z \in (-\infty, \infty)$, has continuous bounded derivatives $\frac{\partial f}{\partial t}, \frac{\partial f}{\partial x^i}, \frac{\partial f}{\partial y}, \frac{\partial f}{\partial z}, \frac{\partial^2 f}{\partial x^i \partial x^j}$, $i,j = 1, \cdots, r$. We put $Y_t = \int_0^t c(X_s^x)\, ds$ and $Z_t = \int_0^t \lambda(X_s^x)\, d\xi_s^x$ where $\lambda(x)$, $c(x)$ are continuous bounded functions. Then the following formula holds

$$f(t, X_t^x, Y_t, Z_t) - f(0, x, 0, 0) =$$

$$= \int_0^t \left(\frac{\partial f}{\partial t} + L_x f\right)(s, X_s^x, Y_s, Z_s)\,ds + \int_0^t \frac{\partial f}{\partial y}(s, X_s^x, Y_s, Z_s)\,c(X_s^x)\,ds +$$

$$+ \int_0^t \frac{\partial f}{\partial z}(s, X_s^x, Y_s, Z_s)\,\lambda(X_s^x)\,d\xi_s^x + \int_0^t (\nabla_x f(s, X_s^x, Y_s, Z_s), \gamma(X_s^x))\,d\xi_s^x .$$

The intuitive meaning of this formula is the same as in the case when we considered Ito's formula in unbounded space. The proof may be found, for example, in Gihman and Skorohod [3], v. 3, §3.

§1.7 *Limit theorems. Action functional*

Let $\xi_1, \cdots, \xi_n, \cdots$ be a sequence of independent random variables. In the classical theory of probability, an important role is played by questions concerning the asymptotic behavior of the sums $\sum_{k=1}^{n} \xi_k$ as $n \to \infty$. The simplest assertion of this kind is the law of large numbers: $\frac{1}{n} \sum_{k=1}^{n} \xi_k - \frac{1}{n} \sum_{k=1}^{n} E\xi_k \to 0$ in some sense or other (in probability, in the mean-square, or with probability 1).

A more precise bound is given by the central limit theorem: under small extra assumptions on the ξ_k,

$$P\left\{\frac{\sum_1^n (\xi_k - E\xi_k)}{\sqrt{\sum_1^n D\xi_k}} < x\right\} \to \frac{1}{\sqrt{2\pi}} \int_{-\infty}^{x} e^{-\frac{z^2}{2}}\,dz\ , \quad n \to \infty .$$

Here $D\xi_k$ denotes the variance of $\xi_k : D\xi_k = E\xi_k^2 - (E\xi_k)^2$. If x grows with n on the left-hand side of the above relation, then the central limit theorem implies only that the probability on the left-hand side tends to 1. In this case, a more exact bound may be obtained via the limit theorems for probabilities of large deviations. These theorems give (with some order of approximation) the asymptotics as $n \to \infty$ for the probabilities of the form

$$P\left\{\sum_{k=1}^{n} (\xi_k - E\xi_k) > an^{\alpha}\right\}, \quad \alpha > \frac{1}{2}.$$

Each of these problems has many modifications. For instance, many interesting and useful generalizations arise when going from sums of independent random variables to random processes.

We shall consider some of these generalizations. In the statement of some of them, the Doeblin condition plays an important role.

Suppose we are given a Markov process (X_t, P_x), a topological space $\mathcal{E}$ (with the Borel σ-field $\mathcal{B}$) serving as its state space.

Doeblin's condition is said to be fulfilled for this process if there exists a finite measure $\phi(\cdot)$, $\phi(\mathcal{E}) > 0$, on $\mathcal{B}$, and there are numbers $\varepsilon, t > 0$ such that $p(t,x,A) < 1 - \varepsilon$ for all $x \in \mathcal{E}$ and any Borel set A for which $\phi(A) < \varepsilon$.

For example, the Doeblin condition is fulfilled for a diffusion process in a bounded domain with reflection on the boundary, if the corresponding operator is elliptic, and its coefficients as well as the vector field (along which the reflection takes place) are regular enough.

Another example is given by a non-degenerate diffusion process on a torus or, generally, on any compact smooth manifold. Constructing such processes may be carried out by glueing together the trajectories of the corresponding processes on the different charts which taken together cover all the manifold (see, e.g. Ikeda and Watanabe [2]).

The following theorem holds. The proof is available in Chapter IV of the monograph by Doob [1].

THEOREM 7.1. *Suppose we are given a Markov process* (X_t, P_x) *in a state space* $(\mathcal{E}, \mathcal{B})$ *satisfying the Doeblin condition. Then a unique stationary distribution* μ *for the process* (X_t, P_x) *exists. For any bounded, Borel measurable function* $f(x)$ *on* $\mathcal{E}$,

$$\lim_{t\to\infty} \frac{1}{t} \int_0^t f(X_s)ds = \int_{\mathcal{E}} f(x)\,\mu(dx),\ P_x\text{-a.s. } \textit{for any } x \in \mathcal{E}\,;$$

$$|E_x f(X_t) - \int_{\mathcal{E}} f(x)\,\mu(dx)| \le c \cdot \sup_{x\in\mathcal{E}} |f(x)| \exp\{-at\}\,,$$

where c, a *are some positive constants (independent of* $f(x)$ *).*

As an example of an application of this theorem, we will investigate the behavior of a family of diffusion processes with periodic, quickly oscillating coefficients. Let the Markov family $(X_t^{\varepsilon,x}, P)$ in R^r, $\varepsilon > 0$, be defined by the stochastic differential equations

$$dX_t^{\varepsilon,x} = \sigma(\varepsilon^{-1}X_t^{\varepsilon,x})\,dW_t\,,\quad X_0^{\varepsilon,x} = x\,, \tag{7.1}$$

where W_t is a Wiener process, and $\sigma(x) = (\sigma_j^i(x))$, $x = (x^1, \cdots, x^r)$, is a matrix with Lipschitz continuous elements, periodic in each argument x^i with period 1.

Consider the mapping $\phi : R^r \to R^r$ defined by the formula

$$\phi(x^1, \cdots, x^r) = (\tilde{x}^1, \cdots, \tilde{x}^r)\,,\quad \tilde{x}^i = x^i (\text{mod } 1)\,,\quad 0 \le x^i < 1\,.$$

For such a mapping, the entire space R^r goes into an r-dimensional torus Π. It is not difficult to check that, since $(X_t^{1,x}, P)$ is a Markov family in R^r defined by equation (7.1) for $\varepsilon = 1$, the pair $(\tilde{X}_t^{1,x}, P)$, $x \in \Pi$, $\tilde{X}_t^{1,x} = \phi(X_t^{1,x})$, forms a Markov family in the state space Π. Together with the families $(X_t^{\varepsilon,x}, P)$ in R^r and $(\tilde{X}_t^{1,x}, P)$ in Π, we shall also consider the corresponding Markov processes $(X_t^{\varepsilon}, P_x^{\varepsilon})$ and $(\tilde{X}_t^1, \tilde{P}_x^1)$.

The matrix $(a^{ij}(x)) = \sigma(x)\sigma^*(x)$ is assumed positive definite. Then, for the process $(\tilde{X}_t^1, \tilde{P}_x^1)$ on the torus Π, the Doeblin condition is fulfilled, and by Theorem 7.1, a unique normalized invariant measure of the process $(\tilde{X}_t^1, \tilde{P}_x^1)$ exists. This measure has a density function $m(x)$, $x \in \Pi$, which is the unique positive solution of the equation $\sum_{i,j=1}^{r} \frac{\partial^2}{\partial x^i \partial x^j} (a^{ij}(x)m(x)) = 0$, $x \in \Pi$, for which $\int_\Pi m(x)\,dx = 1$. For any continuous function $f(x)$ on the torus

$$\tilde{P}_x^1 \left\{ \lim_{t\to\infty} t^{-1} \int_0^t f(\tilde{X}_s^1)\,ds = \int_\Pi f(x)\,m(x)\,dx \right\} = 1 .$$

Let us show that the family of processes $(X_t^\varepsilon, P_x^\varepsilon)$ in R^r converges (in the sense of weak convergence of the corresponding distributions in the space $C_{0,T}(R^r)$) to a diffusion process governed by the operator $\bar{L} = \frac{1}{2} \sum_{i,j=1}^{r} \bar{a}^{ij} \frac{\partial^2}{\partial x^i \partial x^j}$, $\bar{a}^{ij} = \int_\Pi a^{ij}(x)\,m(x)\,dx$, as $\varepsilon \downarrow 0$.

For this, first of all we note that the corresponding family of measures in $C_{0,T}(R^r)$ is weakly compact by Lemma 5.2. To prove the weak convergence, it suffices to check that the corresponding finite-dimensional distributions converge. We put $0 < t_1 < t_2 < \cdots < t_n \le T$. Note that the random variable $Z^\varepsilon = (X_{t_1}^{\varepsilon,x}, \cdots, X_{t_n}^{\varepsilon,x})$ has the same distribution as $Z_1^\varepsilon = (\varepsilon X^{1,x\varepsilon^{-1}}_{t_1/\varepsilon^2}, \cdots, \varepsilon X^{1,x\varepsilon^{-1}}_{t_n/\varepsilon^2})$. We will show that the distribution of Z_1^ε converges to the distribution of $(\bar{X}_{t_1}^x, \cdots, \bar{X}_{t_n}^x)$ as $\varepsilon \downarrow 0$, where $\bar{X}_t^x$ is defined by the equation

$$d\bar{X}_t^x = \bar{\sigma}\,dW_t,\ \bar{X}_0^x = x,\ \bar{\sigma} = (\bar{a}^{ij})^{1/2} . \tag{7.2}$$

Indeed, the characteristic function $f_\varepsilon(\lambda_1, \cdots, \lambda_n)$, $\lambda_k \in R^r$, of the vector Z_1^ε has the form

$$f_\varepsilon(\lambda_1, \cdots, \lambda_n) = E \exp\left\{ i\varepsilon \sum_{k=1}^{n} \left(\lambda_k, X^{1,x}_{\frac{t_k}{\varepsilon^2}} \right) \right\}.$$

We will make use of the Markov property of the process (X_t^1, P_x^1):

$$f_\varepsilon(\lambda_1,\cdots,\lambda_n) = E_x \exp\left\{ i\varepsilon \left(\lambda_1, X^1_{\frac{t_1}{\varepsilon^2}}\right)\right\} E_{X^1_{\frac{t_1}{\varepsilon^2}}} \exp\left\{ i\varepsilon\left(\lambda_2, X^1_{\frac{t_2-t_1}{\varepsilon^2}}\right)\right\} \times$$

$$\times \cdots \times E_{X^1_{\frac{t_{n-1}}{\varepsilon^2}}} \exp\left\{ i\varepsilon\left(\lambda_n, X^1_{\frac{t_n - t_{n-1}}{\varepsilon^2}}\right)\right\}. \tag{7.3}$$

Next, noting that $X_t^{1,\hat{x}} - \hat{x} = \int_0^t \sigma(X_s^{1,\hat{x}})\, dW_s$, we find that, there is a one-dimensional Wiener process $\widetilde{W}_t$ for which the following equality holds:

$$\varepsilon\left(\lambda_n, X^{1,\hat{x}}_{\frac{t_n - t_{n-1}}{\varepsilon^2}} - \hat{x}\right) = \varepsilon\left(\int_0^{(t_n - t_{n-1})\varepsilon^{-2}} \sigma(X_s^{1,\hat{x}})\, dW_s, \lambda_n\right)$$

$$= \widetilde{W}_{A_{t_n - t_{n-1}}} \tag{7.4}$$

where

$$A_t = \varepsilon^2 \int_0^{t\varepsilon^{-2}} (a(X_s^{1,\hat{x}})\lambda_n, \lambda_n)\, ds\ .$$

We have used here the following extension to several dimensions of the result discussed on page 50: let W_s be a Wiener process in R^r and put

$$X_t = \int_0^t \sigma(X_s)\,dW_s\,,$$

$$A_t = \int_0^t |\sigma(X_s)\lambda|^2 ds\,,$$

$$\tau_t = A_t^{-1}\,,$$

then $\widetilde{W}_t = (\lambda, x_{\tau_t})$ is a one-dimensional Brownian motion adapted to the σ-fields $\mathcal{F}_t = N_{\tau_t}$.

Now we resort to the law of large numbers. Since the elements of the matrix $(a^{ij}(x)) = a(x)$ are periodic, this matrix may be considered on the torus Π, and therefore, with probability 1,

$$\lim_{\varepsilon \downarrow 0} \varepsilon^2 \int_0^{(t_n - t_{n-1})\varepsilon^2} (a(X_s^{1,\hat{x}})\lambda_n, \lambda_n)\,ds = \lim_{\varepsilon \downarrow 0} \varepsilon^2 \int_0^{(t_n - t_{n-1})\varepsilon^{-2}} (a(\widetilde{X}_s^{1,\hat{x}})\lambda_n, \lambda_n)\,ds =$$

$$= (t_n - t_{n-1})(\bar{a}\lambda_n, \lambda_n), \quad \hat{x} \in \Pi\,.$$

Together with (7.4), this implies that, for any $\hat{x} \in R^r$,

$$\lim_{\varepsilon \downarrow 0} E_{\hat{x}} \exp\left\{ i\varepsilon\left(\lambda_n, X^1_{\frac{t_n - t_{n-1}}{\varepsilon^2}}\right)\right\} = E \exp\{i\widetilde{W}_{(t_n - t_{n-1})(\bar{a}\lambda_n, \lambda_n)}\} =$$

$$= \exp\left\{ -\frac{t_n - t_{n-1}}{2}(\lambda_n, \bar{a}\lambda_n)\right\}.$$

By using similar reasoning consecutively to each expectation in (7.3), we derive

$$\lim_{\varepsilon \downarrow 0} f_\varepsilon(\lambda_1, \cdots, \lambda_n) = \exp\left\{ i \sum_{k=1}^{n} (x, \lambda_k) - \right.$$

$$\left. - \frac{1}{2} \sum_{k=1}^{n} \left[(t_{n-k+1} - t_{n-k}) \left(\bar{a} \sum_{\ell=0}^{k-1} \lambda_{n-\ell}, \sum_{\ell=0}^{k-1} \lambda_{n-\ell} \right) \right] \right\}.$$

The function on the right-hand side is easily checked to be the characteristic function of the random vector $(\bar{X}_{t_1}^x, \cdots, \bar{X}_{t_n}^x)$, where $\bar{X}^x$ is determined by equation (7.2). The convergence of the characteristic functions implies the convergence of the distributions. Thus the processes $(X_t^\varepsilon, P_x^\varepsilon)$ converge weakly to the process defined by equation (7.2). Notice that instead of proving the convergence of the finite-dimensional distributions, one could prove that all the limit points of the family of the measures $\mu^{\varepsilon,x}$ in $C_{0,T}(R^r)$ corresponding to the random functions $X_t^{\varepsilon,x}$, are solutions of the martingale problem for the operator $\bar{L}$. Taking into account that this problem has a unique solution, we can conclude from the compactness of the family $\{\mu^{\varepsilon,x}\}$ that the measures $\mu^{\varepsilon,x}$ converge weakly to the solution of the martingale problem for $\bar{L}$ as $\varepsilon \downarrow 0$. In our case, the limit process is Gaussian. For such processes, the characteristic functions of the finite-dimensional distributions may be written down explicitly. Thus, using the martingale description of the limit process gives no essential advantages. If a limit process has a more complicated structure, then using the martingale approach may make the reasoning essentially simpler.

For example, let the matrix σ in equation (7.1) have the form $\sigma = \sigma\left(x, \frac{x}{\varepsilon}\right)$, $x \in R^r$, where $\sigma(x,y)$ is a bounded smooth matrix function, periodic in the variables y. Then the corresponding process $(X_t^\varepsilon, P_x^\varepsilon)$ in R^r converge weakly to a Markov process with zero drift and diffusion coefficients $\bar{\bar{a}}^{ij}(x)$ as $\varepsilon \downarrow 0$:

$$\bar{\bar{a}}^{ij}(x) = \int_{\Pi} a^{ij}(x,y)\, m_x(y)\, dy, \quad (a^{ij}(x,y)) = \sigma(x,y)\sigma^*(x,y) .$$

Here, for a fixed $x \in R^r$, $m_x(y)$ is a density function of the invariant distribution of the process $(\tilde{X}_t^1, \tilde{P}_t^1)$ on the torus corresponding to the operator

$$\tilde{L} = \frac{1}{2} \sum_{i,j=1}^{r} a^{ij}(x,y) \frac{\partial^2}{\partial y^i \partial y^j}, \quad y \in \Pi .$$

In this case, the limit process has, generally speaking, variable diffusion coefficients, and in order to prove the convergence, it is convenient to use the martingale description of the limit process. Namely, one should verify that every process $(\bar{\bar{X}}_t, \bar{\bar{P}}_x)$ which is a limit of some subsequence of the processes $(X_t^\varepsilon, P_x^\varepsilon)$, is a solution of the martingale problem for the operator $\bar{\bar{L}} = \frac{1}{2} \Sigma \bar{a}^{ij}(x) \frac{\partial^2}{\partial x^i \partial x^j}$. Then, on account of Lemma 5.2, the uniqueness of the solution of such a problem implies the convergence.

The weak convergence of the processes $(X_t^\varepsilon, P_x^\varepsilon)$, together with the results of the next chapter, leads to the conclusion that the solutions of Cauchy's problem as well as of a number of mixed problems for the equation

$$\frac{\partial u^\varepsilon(t,x)}{\partial t} = L^\varepsilon u^\varepsilon = \frac{1}{2} \sum_{i,j=1}^{r} a^{ij}\left(x, \frac{x}{\varepsilon}\right) \frac{\partial^2 u^\varepsilon}{\partial x^i \partial x^j}$$

converge to the solutions of the corresponding problems for the equation

$$\frac{\partial \bar{\bar{u}}}{\partial t} = \bar{\bar{L}} \bar{\bar{u}} ,$$

as $\varepsilon \downarrow 0$. It is also readily verified that the solution of Dirichlet's problem for the operator L^ε converges to the solution of the same problem for $\bar{\bar{L}}$ as $\varepsilon \downarrow 0$.

Here we made use of the law of large numbers for the process $(\tilde{X}_t^1, \tilde{P}_x^1)$ on the torus.

Now we consider the behavior as $\varepsilon \downarrow 0$ of the family of processes corresponding to the self-adjoint operators

$$\mathcal{L}^\varepsilon = \frac{1}{2} \sum \frac{\partial}{\partial x^i}\left(a^{ij}\left(\frac{x}{\varepsilon}\right)\frac{\partial}{\partial x^j}\right)$$

with sufficiently smooth periodic coefficients. For this, we shall need a central limit theorem type result. However, we shall get the required version of the central limit theorem via the law of large numbers. For brevity, we confine ourselves to the case when $r = 1$. It offers no difficulty to carry over the reasoning to the multi-dimensional case, although the result will no longer be so explicit. The trajectories of the Markov family $(X_t^{\varepsilon,x}, P)$ corresponding to the operator $\mathcal{L}^\varepsilon$, $r = 1$, satisfy

$$X_t^{\varepsilon,x} = x + \int_0^t \sigma(\varepsilon^{-1}X_s^{\varepsilon,x})\, dW_s + \frac{1}{2\varepsilon}\int_0^t a'(\varepsilon^{-1}X_s^{\varepsilon,x})\, ds\,, \tag{7.5}$$

where $\sigma(x) = \sqrt{a(x)}$, $a'(x) = \frac{da(x)}{dx}$, $a(x) = a^{11}(x)$. Since $a(x) = a(x+1)$, one can consider the Markov process $(\tilde{X}_t^1, \tilde{P}_x^1)$ on the circle S governed by the operator $\frac{1}{2}\frac{d}{dx}\left(a(x)\frac{d}{dx}\right)$. The corresponding invariant measure is the Lebesgue measure on S. As is easily seen from the law of large numbers for $(\tilde{X}_t^1, \tilde{P}_x^1)$ and from the properties of stochastic integrals, the first summand on the right-hand side of (7.5) converges weakly to $\sqrt{\bar{a}}\, W_t$ as $\varepsilon \downarrow 0$, where $\bar{a} = \int_0^1 a(x)dx$. If the second summand did not have the factor ε^{-1}, it would tend to zero by the law of large numbers, since $\int_0^1 a'(x)dx = 0$.

As we shall see later on, it is the factor ε^{-1} which ensures the asymptotic normality of the second summand as $\varepsilon \downarrow 0$. To describe the limit process and to prove the convergence, let us denote by $u(x)$ the periodic solution of the equation

$$\mathcal{L}^1 u(x) = \frac{1}{2}\frac{d}{dx}\left(a(x)\frac{du}{dx}\right) = -\frac{1}{2}\, a'(x)\,. \tag{7.6}$$

Since $\int_S a'(x)dx = 0$, such a solution exists and is unique up to an additive constant. In the one-dimensional case, the solution of problem

(7.6) may be written down explicitly, and we shall do this a little later. Let us apply Ito's formula to the function $\varepsilon u(\varepsilon^{-1}X_t^{\varepsilon,x})$:

$$\varepsilon u(\varepsilon^{-1}X_t^{\varepsilon,x}) - \varepsilon u(\varepsilon^{-1}x) =$$

$$= \int_0^t u'(\varepsilon^{-1}X_s^{\varepsilon,x})\,\sigma(\varepsilon^{-1}X_s^{\varepsilon,x})\,dW_s - \frac{1}{2\varepsilon}\int_0^t a'(\varepsilon^{-1}X_s^{\varepsilon,x})\,ds . \qquad (7.7)$$

From (7.5) and (7.7), it follows that

$$X_t^{\varepsilon,x} = x + \int_0^t [\sigma(\varepsilon^{-1}X_s^{\varepsilon,x}) + \sigma(\varepsilon^{-1}X_s^{\varepsilon,x})\,u'(\varepsilon^{-1}X_s^{\varepsilon,x})]\,dW_s - \qquad (7.8)$$

$$- \varepsilon\,[u(\varepsilon^{-1}X_s^{\varepsilon,x}) - u(\varepsilon^{-1}x)] .$$

One can find a Wiener process $\widetilde{W}(t)$ such that the first summand on the right-hand side of (7.8) induces in the space of trajectories the same measure as the process $\widetilde{W}(\int_0^t [\sigma+\sigma u']^2(\varepsilon^{-1}X_s^{\varepsilon,x})\,ds)$ does. By the law of large numbers, with probability 1, uniformly in all initial points,

$$\int_0^t [\sigma+\sigma u']^2(\varepsilon^{-1}X_s^{\varepsilon,x})\,ds \to \int_0^1 [\sigma(x)+\sigma(x)\,u'(x)]^2\,dx = \hat{a} ,$$

as $\varepsilon \downarrow 0$. Together with (7.8), this implies that the processes $(X_t^\varepsilon, P_x^\varepsilon)$ governed by the operators $\mathcal{L}^\varepsilon$ converge weakly to the process corresponding to the operator $\frac{1}{2}\hat{a}\frac{d^2}{dx^2}$ in R^1 as $\varepsilon \downarrow 0$. To compute $\hat{a}$, note that the solution of problem (7.6) has the form

$$u(x) = \int_0^x \frac{(\overline{a^{-1}})^{-1} - a(y)}{a(y)}\,dy ,$$

where $\overline{a^{-1}} = \int_0^1 a^{-1}(x)dx$. Then for $\hat{a}$ we get

$$\hat{a} = \int_0^1 a(x)[1+u'(x)]^2\,dx = \left(\int_0^1 a^{-1}(x)\,dx\right)^{-1} = (\overline{a^{-1}})^{-1} .$$

Whence, of course, one can easily deduce the convergence of the solutions of various problems for the equation $\frac{\partial u^\varepsilon}{\partial t} = \mathcal{L}^\varepsilon u(t,x)$ to the solutions of the equation $\frac{\partial \hat{u}}{\partial t} = \frac{1}{2}\,\hat{a}\,\frac{\partial^2 \hat{u}}{\partial x^2}$.

In the multi-dimensional case, the limit coefficients are expressed in terms of the solution of equations like (7.6). Equations with periodic coefficients were examined using probability methods by Freidlin [3], [13], Bensoussan, Lious and Papanicolaou [1], Nikunen [1]. Further results and references to analytical and probabilistic works may be found in Bensoussan et al [1], Zikov, Kozlov and Oleinik [1].

In the sequel, we shall study several theorems similar to the law of large numbers or the central limit theorem. Now we will turn to the limit theorems for the probabilities of large deviations. We shall dwell only on the rough (logarithmic) asymptotics of the probabilities of large deviations. Of course, far more precise theorems are available in the case of the sums of independent random variables. However, it is easier to extend the rough results to wide classes of random processes, and moreover, their corollaries are, in a sense, most interesting. First, we will formulate a result on large deviations for a family of measures in a finite-dimensional space.

Let μ^h be a family of probability measures in R^r, the parameter h varying in some set on the positive part of the real line which has 0 as its limit point. We put

$$H^h(\alpha) = \ln \int_{R^r} \exp\{(\alpha,x)\}\mu^h(dx) , \quad \alpha \in R^r .$$

From the definition of the function $H^h(\alpha)$, it is clearly convex. Applying Fatou's lemma, it is not difficult to show that $H^h(\alpha)$ is lower semi-continuous. Notice that the function $H^h(\alpha)$ may become infinite for some $\alpha \in R^r$, but $H^h(0) = 0$.

Let $\lambda(h)$, $h > 0$, be a real-valued function tending to $+\infty$ as $h \downarrow 0$. Suppose that for $\alpha \in R^r$, the limit

$$\lim_{h \downarrow 0} \lambda^{-1}(h) H^h(\lambda(h)\alpha) = H(\alpha)$$

exists. Clearly, the function $H(\alpha)$ just as $H^h(\alpha)$ is convex and $H(0) = 0$. We will denote by $L(\beta)$, $\beta \in R^r$, the Legendre transform of $H(\alpha)$:

$$L(\beta) = \sup_{\alpha} [(\alpha,\beta) - H(\alpha)] .$$

THEOREM 7.2. *Suppose that the function* $H(\alpha)$ *is finite in some neighborhood of the point* $\alpha = 0$, *is nowhere equal to* $-\infty$, *and is lower semi-continuous. Then, no matter what the* δ, γ, $s > 0$, *one can find* $h_0 > 0$ *such that, for* $h < h_0$,

$$\mu^h\{y : \rho(y, \Phi(s)) \geq \delta\} \leq \exp\{-\lambda(h)(s-\gamma)\} ,$$

where $\Phi(s) = \{\beta : L(\beta) \leq s\}$, *and* $\rho(\cdot,\cdot)$ *is the Euclidean metric in* R^r.

Moreover, suppose that the function $L(\beta)$ *is strictly convex*[6] *on a set of points forming an everywhere dense set in* $\{\beta : L(\beta) < \infty\}$. *Then, no matter what the* δ, $\gamma > 0$, $x \in R^r$, *one can find* $h_0 > 0$ *such that, for* $h < h_0$,

$$\mu^h\{y : \rho(y,x) < \delta\} \geq \exp\{-\lambda(h)(L(x)+\gamma)\} .$$

[6] A convex function $L(\beta)$ is strictly convex at a point β_0, whenever for some $\alpha \in R^r$

$$L(\beta) > L(\beta_0) - (\alpha, \beta-\beta_0)$$

for all $\beta \in R^r$, $\beta \neq \beta_0$. The strict convexity condition for the function $L(\beta)$ may be given in terms of the smoothness of the function $H(\alpha)$ conjugate to $L(\beta)$ (see Rockafellar [1]). For example, the conditions of Theorem 7.2 are fulfilled if the function $H(\alpha)$ is finite and differentiable for all $\alpha \in R^r$.

For instance, let $\xi_1, \xi_2, \cdots$ be a sequence of independent, identically distributed r-dimensional random variables. Suppose that the function $H_0(\alpha) = \ln E \exp\{(\alpha, \xi_i)\}$ is finite for all $\alpha \in R^r$. We will designate by $\mu^{\left(\frac{1}{n}\right)}$ a measure on the Borel σ-field $\mathcal{B}^r$ in R^r, defined by the equalities

$$\mu^{\left(\frac{1}{n}\right)}(B) = P\left\{\frac{1}{n}\sum_{i=1}^{n} \xi_i \in B\right\}, \quad B \in \mathcal{B}^r .$$

Then

$$H^{\left(\frac{1}{n}\right)}(\alpha) = \ln \int_{R^r} \exp\{(\alpha, x)\}\, \mu^{\left(\frac{1}{n}\right)}(dx) =$$

$$= \ln E \exp\left\{\left(\alpha, \frac{1}{n}\sum_{1}^{n} \xi_i\right)\right\} = n H_0\left(\frac{\alpha}{n}\right) .$$

Setting $\lambda\left(\frac{1}{n}\right) = n$ we get that $\lambda^{-1}\left(\frac{1}{n}\right) H^{\left(\frac{1}{n}\right)}\left(\lambda\left(\frac{1}{n}\right)\alpha\right) = H_0(\alpha)$. The finiteness of the function $H_0(\alpha)$ for all $\alpha \in R^r$ implies that it is differentiable. Thus, in our example, the conditions of Theorem 7.2 hold. The function $L(\beta) = \sup_{\alpha} [(\alpha, \beta) - H_0(\alpha)]$ is lower semi-continuous. It is not hard to deduce from Theorem 7.2 that if $\inf_{y > x} L(y) = \inf_{y \geq x} L(y)$, then

$$\lim_{n \to \infty} \frac{1}{n} \ln P\left\{\frac{1}{n}\sum_{i=1}^{n} \xi_i > x\right\} = -\inf_{y > x} L(y) .$$

This statement now more resembles an ordinary limit theorem for the probabilities of large deviations.

The proof of Theorem 7.2 can be found in Chapter V of the monograph by Wentzell and Freidlin [2]. In what follows, this theorem will be used for studying parabolic differential equations with rapidly oscillating coefficients (see Chapter VII), Theorem 7.2 may also be applied for calculating the asymptotics of the probabilities of large deviations for families of probability measures in the space $C_{0,T}(R^r)$ induced by random processes with continuous trajectories.

Consider the family μ^ε of measures in $C_{0,T}(R^r)$ corresponding to the processes $\sqrt{\varepsilon}W_t$, where W_t is a Wiener process in R^r:

$$\mu^\varepsilon(\mathfrak{A}) = P\{\sqrt{\varepsilon}W_. \in \mathfrak{A}\}.$$

Here $\mathfrak{A}$ is an element of the σ-field $\mathfrak{N}_T$ in $C_{0T}(R^r)$ generated by cylinder sets. It is clear that, in every sense, the measures μ^ε converge (as $\varepsilon\downarrow 0$) to a measure concentrated on the function from $C_{0T}(R^r)$ which is identically equal to zero. Therefore, unless the set $\mathfrak{A} \in \mathfrak{N}_T$ contains some neighborhood of the zero function, $\mu^\varepsilon(\mathfrak{A}) \to 0$ as $\varepsilon\downarrow 0$, and thereby, results on the asymptotics of $\mu^\varepsilon(\mathfrak{A})$ are limit theorems for the probabilities of large deviations.

We proceed to define the functional $S_{0,T}(\phi)$ on the space $C_{0,T}(R^r)$. To this end, we set $S_{0T}(\phi) = \frac{1}{2}\int_0^T |\dot{\phi}_s|^2 ds$ for absolutely continuous functions $\phi_.$, and $S_{0T}(\phi) = +\infty$ for the rest $\phi \in C_{0T}(R^r)$. It is possible to prove that this functional is lower semi-continuous in the uniform convergence topology, and that the set $\Phi(s) = \{\phi : S_{0T}(\phi) \leq s\}$ is compact in $C_{0T}(R^r)$. Let us denote by $\rho_{0,T}(\cdot,\cdot)$ a metric in $C_{0T}(R^r)$.

THEOREM 7.3. *For any* δ, γ, $K > 0$, *one can find* $\varepsilon_0 > 0$ *such that, for* $0 < \varepsilon < \varepsilon_0$

$$\mu^\varepsilon\{\psi \in C_{0T}(R^r) : \rho_{0T}(\psi,\phi) < \delta\} = P\{\rho_{0T}(\sqrt{\varepsilon}W_.,\phi) < \delta\} \geq$$
$$\geq \exp\{-\varepsilon^{-1}(S_{0T}(\phi)+\gamma)\},$$

where $T > 0$ *and* $\phi \in C_{0T}(R^r)$ *are such that* $\phi_0 = 0$ *and* $T + S_{0T}(\phi) \leq K$.

For any δ, γ, $s_0 > 0$, *an* $\varepsilon_0 > 0$ *exists such that, for* $0 < \varepsilon < \varepsilon_0$ *and* $s < s_0$,

$$\mu^\varepsilon\{\psi \in C_{0T}(R^r) : \rho_{0T}(\psi,\Phi(s)) \geq \delta\} = P\{\rho_{0T}(\sqrt{\varepsilon}W_.,\Phi(s)) \geq \delta\} \leq$$
$$\leq \exp\{-\varepsilon^{-1}(s-\gamma)\},$$

where $\Phi(s) = \{\psi \in C_{0T}(R^r) : S_{0T}(\psi) \leq s\}$.

This theorem may be proved by approximating the trajectories of the process $\sqrt{\varepsilon}W_t$, $t \in [0,T]$, with random broken lines having vertices at the points $\left(\frac{kT}{n}, \sqrt{\varepsilon}W_{\frac{kT}{n}}\right)$ $k = 0, 1, \cdots, n$, and by applying Theorem 7.2 to the random vector $\left(\sqrt{\varepsilon}W_{\frac{T}{n}}, \sqrt{\varepsilon}W_{\frac{2T}{n}}, \cdots, \sqrt{\varepsilon}W_T\right)$. The proof may be found in Chapter III of the above cited monograph by Wentzell and Freidlin. In §4.4 we will prove some results which imply Theorem 7.3.

Now we will introduce a general notion to be helpful when describing large deviations.

Consider a family μ^ε, $\varepsilon > 0$, of probability measures in the space $C_{0T}(R^r)$. Suppose we are given a real valued function $\lambda(\varepsilon)$, $\varepsilon > 0$, tending to $+\infty$ as $\varepsilon \downarrow 0$ and a lower semi-continuous functional $S(\phi)$ on the space $C_{0T}(R^r)$ taking non-negative values and the value $+\infty$.

The functional $\lambda(\varepsilon)S(\phi)$ is called the *action functional* for the family of measures μ^ε in $C_{0T}(R^r)$ as $\varepsilon \downarrow 0$, if for any $s > 0$, the set $\Phi(s) = \{\phi \in C_{0T}(R^r) : S(\phi) \leq s\}$ is compact and for any δ, γ, $s > 0$ and $\phi \in C_{0,T}(R^r)$, one can find $\varepsilon_0 > 0$ such that the relations

$$\mu^\varepsilon(\psi \in C_{0T}(R^r) : \rho_{0,T}(\psi,\phi)) < \delta) \geq \exp\{-\lambda(\varepsilon)(S(\phi)+\gamma)\},$$

$$\mu^\varepsilon(\psi \in C_{0T}(R^r) : \rho_{0,T}(\psi,\Phi(s)) \geq \delta\} \leq \exp\{-\lambda(\varepsilon)(s-\gamma)\}$$

are valid for all $\varepsilon \in (0, \varepsilon_0]$.

The functional $S(\phi)$ is called the *normed action functional,* the function $\lambda(\varepsilon)$ being termed the *normalizing coefficient.*

We note that since to every almost surely continuous r-dimensional random process, there corresponds some measure in $C_{0T}(R^r)$ one may talk about the action functional corresponding to a family of random processes.

Therefore, Theorem 7.3 affirms that the functional $S_{0T}(\phi)$ equal to $\frac{1}{2}\int_0^T |\dot{\phi}_s|^2\,ds$ for absolutely continuous $\phi \in C_{0T}(R^r)$ and equal to $+\infty$ for the rest $\phi \in C_{0T}(R^r)$, is the normed action functional for the family of measures μ^ε in $C_{0,T}(R^r)$ generated by the processes $\sqrt{\varepsilon}\, W_t$. Here $\lambda(\varepsilon) = \varepsilon^{-1}$.

The action functional may, certainly, be examined for a family of measures in any metric space, and not solely in $C_{0,T}(R^r)$. In particular, Theorem 7.2 may be looked upon as a theorem on the action functional for a family of measures in R^r. The notion of action functional is studied in detail in Wentzell and Freidlin [2], see also Varadhan [1], [2], [3] and Borovkov [1].

Now we will cite some properties of the action functional to be used in the sequel. Let $\lambda(h) S^\mu_{0,T}(\phi)$ be the action functional for a family of measures μ^h in $C_{0T}(R^r)$ as $h \downarrow 0$.

1. Suppose that F is a continuous mapping of the measurable space $(C_{0,T}(R^r), \mathfrak{N}_T)$ into itself, and let a new measure ν^h on the space $C_{0,T}(R^r)$ be defined by the equality $\nu^h(A) = \mu^h(F^{-1}(A))$, $A \in \mathfrak{N}_T$. Then the action functional for the family of measures ν^h, as $h \downarrow 0$, has the form $\lambda(h) S^\nu_{0T}(\phi)$, where $S^\nu_{0T}(\phi) = \min\{S^\mu_{0T}(\psi) : \psi \in F^{-1}(\phi)\}$.

2. A set $A \subseteq C_{0T}(R^r)$ is called a regular set for the functional $S^\mu_{0T}(\phi)$ on $C_{0T}(R^r)$, if the lower bound of $S^\mu_{0T}(\phi)$ over the closure $[A]$ of the set A coincides with the lower bound of this functional over the set (A) of the interior points of A:

$$\inf\{S^\mu_{0T}(\phi) : \phi \in [A]\} = \inf\{S^\mu_{0T}(\phi) : \phi \in (A)\} .$$

For any regular measurable set A

$$\lim_{h \downarrow 0} \lambda^{-1}(h) \ln \mu^h(A) = -\inf\{S^\mu_{0T}(\phi) : \phi \in A\} .$$

3. If $G(\phi)$ is a continuous bounded functional on $C_{0T}(R^r)$, then

$$\lim_{h \downarrow 0} \lambda^{-1}(h) \ln \int_{C_{0T}(R^r)} \exp\{\lambda(h) G(\phi)\} d\mu^h =$$

$$= \max_{\phi \in C_{0T}(R^r)} [G(\phi) - S^\mu_{0T}(\phi)] .$$

Property 3 shows, in particular, that estimating via the action functional is a version of the Laplace method for the asymptotic estimation of integrals depending on a parameter. Given a non-negative continuous function $g(x)$, $x \in R^1$, having its only minimum at the point $x_0 \in R^1$,

$$\lim_{h \downarrow 0} h \ln \int_{-\infty}^{\infty} e^{-h^{-1} g(x)} dx = -\min_x g(x) = -g(x_0) .$$

Property 3 is an assertion of the same kind. However, the fact that the measure itself depends on a parameter, causes the action functional to appear. If the functional G is assumed to be differentiable a sufficient number of times, then one can derive the precise asymptotic expansions for $\int_{C_{0T}(R^r)} \exp\{\lambda(h) G(\phi)\} d\mu^h$ in a number of cases (see, e.g. Schilder [1], Wentzell and Freidlin [2]).

We shall now consider some families of random processes depending on a small parameter and indicate the corresponding action functionals.

Let a Markov family $(X^{\varepsilon,x}, P)$ in R^r be determined by the stochastic differential equation

$$dX_t^{\varepsilon,x} = \sqrt{\varepsilon}\, \sigma(X_t^{\varepsilon,x})\, dW_t + b(X_t^{\varepsilon,x})\, dt , \quad X_0^{\varepsilon,x} = x .$$

The elements of the matrix $\sigma(x) = (\sigma_j^i(x))$ and of $b(x) = (b^1(x), \cdots, b^r(x))$ are assumed to be bounded and Lipschitz continuous, the matrix $(a^{ij}(x)) = \sigma(x)\sigma^*(x)$ being positive definite for all $x \in R^r$.

THEOREM 7.4. *The action functional* $\lambda(\varepsilon) S_{0,T}^x(\phi)$ *for the Markov family* $(X_t^{\varepsilon,x}, P)$ *as* $\varepsilon \downarrow 0$ *in the space* $C_{0T}(R^r)$ *is given by the formulae*:

$$\lambda(\varepsilon) = \varepsilon^{-1}, S_{0T}^x(\phi) = \frac{1}{2} \int_0^T \sum_{i,j=1}^r a_{ij}(\phi_s)(\dot{\phi}_s^i - b^i(\phi_s))(\dot{\phi}_s^j - b^j(\phi_s))\, ds ,$$

$$(a_{ij}(x)) = (a^{ij}(x))^{-1} ,$$

for absolutely continuous $\phi \in C_{0T}(R^r)$, *and by* $S^x_{0T}(\phi) = +\infty$ *for the remaining* $\phi \in C_{0T}(R^r)$.

If $(a^{ij}(x))$ is a unit matrix, then Theorem 7.4 comes immediately from Theorem 7.3 and from Property 1, since the mapping

$$F : \phi \to \psi, \quad \psi_t = x + \int_0^t b(\psi_s)\, ds + \phi_t$$

defines a continuous transformation of the space $C_{0T}(R^r)$ into itself. For $b(x) \equiv 0$, this result has been established by Varadhan in [2]. The general case was examined by Wentzell and Freidlin in [1] (see also [2]).

We will consider now the family of random processes $X_t^{\varepsilon,x}$ defined by the differential equation in R^r

$$\dot{X}_t^{\varepsilon,x} = b(X_t^{\varepsilon,x}, \xi_{t/\varepsilon}), \quad X_0^{\varepsilon,x} = x . \qquad (7.9)$$

Here $b(x, \xi) = (b^1(x, \xi), \cdots, b^r(x, \xi))$ are bounded, Lipschitz continuous functions, ε is a positive parameter, and $\xi_t = \xi_t(\omega)$ is some random process. In this book, the ξ_t will always be either a non-degenerate diffusion process in a bounded r-dimensional domain with reflection on the boundary, or a Markov process with a finite number of states. In the second case, the equations for $X_t^{\varepsilon,x}$ are fulfilled at all moments except for those when the trajectories ξ_t have jumps. At the moments of jumps, the continuity of the functions $X_t^{\varepsilon,x}$ is needed.

First, suppose that (ξ_t, P_x) is a non-degenerate diffusion process in a bounded domain $D \subset R^\ell$ with reflection on the boundary, and let L be the corresponding elliptic differential operator, $\gamma(x)$ being a vector field on ∂D along which the reflection happens. The coefficients of the operator L, the boundary ∂D, and the vector field $\gamma(x)$ are assumed to be smooth enough, the field $\gamma(x)$ being nowhere tangent to the boundary. We will denote by $m(x)$ the density function of the stationary distribution of the process (ξ_t, P_x). As it was explained in the preceding section, the function $m(y)$ is a solution of the equation $L^* m(x) = 0$ in the domain D with adjoint boundary conditions.

Let us put $\overline{b}(x) = \int_D b(x,y)\,m(y)\,dy$ and consider the averaged system of differential equations in R^r :

$$\dot{\overline{X}}{}_t^x = \overline{b}(\overline{X}_t^x), \quad \overline{X}_0^x = x .$$

It is not difficult to prove that $\lim_{\varepsilon \downarrow 0} P\{ \sup_{0 \le t \le T} |X_t^{\varepsilon,x} - \overline{X}_t^x| > \delta \} = 0$ for any $\delta > 0$ (see Chapter IV). The deviations of $X^{\varepsilon,x}$ from $\overline{X}^x$ of magnitude $0(1)$ as $\varepsilon \downarrow 0$, are described via the action functional. To define this functional, we will consider the eigenvalue problem

$$L\,u(y) + (\alpha, b(x,y))\,u(y) = \lambda \cdot u(y), y \in \mathfrak{D} ;$$

$$\left.\frac{\partial u(y)}{\partial \gamma(y)}\right|_{y \in \partial D} = 0 .$$

Here $x, \alpha \in R^r$ are parameters. For any $x, \alpha \in R^r$, this problem has a single real eigenvalue $\lambda = \lambda(x, \alpha)$ which exceeds the real parts of other eigenvalues. If the function $b(x,y)$ is differentiable, then $\lambda(x, \alpha)$ is differentiable as well (Kato [1]). The function $\lambda(x, \alpha)$ may be shown to be convex in the variable α.

Let us denote by $L(x, \beta)$ the Legendre transform of $\lambda(x, \alpha)$ with respect to variables α :

$$L(x, \beta) = \sup_{\alpha} [(\alpha, \beta) - \lambda(x, \alpha)] .$$

THEOREM 7.5. *Let* $X_t^{\varepsilon,x}$ *be a solution of equation (7.9), where* $b(x,y)$ *is a bounded differentiable function, and* (ξ_t, P_x) *is a diffusion process in the domain* $D \in R^\ell$ *governed by the operator* L *and subject to the reflection on the boundary along the field* $\gamma(x)$. *The action functional* $\lambda(\varepsilon) S_{0,T}(\phi)$ *for the family of processes* $X_t^{\varepsilon,x}$ *in the space* $C_{0T}(R^r)$ *as* $\varepsilon \downarrow 0$, *is defined by the equalities*

$$\lambda(\varepsilon) = \varepsilon^{-1}, \quad S_{0T}(\phi) = \int_0^T L(\phi_s, \dot{\phi}_s)\, ds$$

for absolutely continuous ϕ, *and by* $S_{0T}(\phi) = +\infty$ *for the rest* $\phi \in C_{0T}(R^r)$.

Now suppose that (ξ_t, P_i) is a time homogeneous Markov process with N states, $Q = (q_{ij})$ being the matrix of transition rates, i.e. $P_i(\xi_\Delta = j) = q_{ij}\Delta + o(\Delta)$ as $\Delta \downarrow 0$, $i \neq j$, $q_{ii} = 1 - \sum_{j:j\neq i} q_{ij}$. Let all the q_{ij} be not equal to zero.

Under the assumptions made above, a unique stationary distribution $q = (q_1, \cdots, q_N)$ for the process (ξ_t, P_i) exists. The vector q may be found from the conditions $qQ = 0$, $\Sigma q_i = 1$. Put $\overline{b}(x) = \sum_{i=1}^N b(x,i) q_i$ and let $\overline{X}_t^x$ be a solution of the averaged equation

$$\dot{\overline{X}}_t^x = \overline{b}(\overline{X}_t^x), \overline{X}_0^x = x .$$

Denote by $X_t^{\varepsilon,x}$ the process defined by equation (7.9) in which a Markov process with a finite number of states is taken as ξ_t. Just as in the case of non-degenerate diffusion process, the process $X_t^{\varepsilon,x}$ converges in probability as $\varepsilon \downarrow 0$ to the solution of the averaged system, uniformly on every finite time interval. To determine the action functional describing the deviations from the averaged motion, we will introduce the $N \times N$ matrix $Q^{\alpha,x}$ whose elements $q_{i,j}^{\alpha,x}$ are specified by the equalities $q_{ij}^{\alpha,x} = q_{ij} + \delta_{ij}(\alpha, b(x,i))$, where δ_{ij} is the Kronecker delta $\alpha \in R^r$. The matrix $Q^{\alpha,x}$ may be proved to have a single real eigenvalue $\lambda(x,\alpha)$ exceeding the real parts of all other eigenvalues. The function $b(x,i)$ will be assumed differentiable in x for every $i = 1, 2, \cdots, N$. Then $\lambda(x,\alpha)$ is differentiable in the parameters $x, \alpha \in R^r$ as well. One can prove that the function $\lambda(x,\alpha)$ is convex in the variables α. Let

$$L(x,\beta) = \sup_\alpha \,[(\alpha,\beta) - \lambda(x,\alpha)] . \tag{7.10}$$

THEOREM 7.6. *Let* $X_t^{\varepsilon,x}$ *be a solution of equation (7.9), where* ξ_t *is a Markov process with a finite number of states. Then the action functional* $\lambda(\varepsilon) S_{0T}(\phi)$ *for the family of processes* $X_t^{\varepsilon,x}$ *in* $C_{0T}(R^r)$ *as* $\varepsilon \downarrow 0$ *is defined by the equalities*

$$\lambda(\varepsilon) = \varepsilon^{-1}, \; S_{0T}(\phi) = \int_0^T L(\phi_s, \dot{\phi}_s)\, ds$$

for absolutely continuous ϕ, *and by* $S_{0T}(\phi) = +\infty$ *for the remaining* $\phi \in C_{0T}(R^r)$. *The function* $L(x, \beta)$ *is determined by equality (7.10), where* $\lambda(x, \alpha)$ *is the eigenvalue of the matrix* $Q^{x,\alpha}$ *possessing the maximal real part.*

The proofs of Theorems 7.5 and 7.6 are available in Freidlin [14] (see also Wentzell and Freidlin [2], Chapter VII).

Chapter II

REPRESENTATION OF SOLUTIONS OF DIFFERENTIAL EQUATIONS AS FUNCTIONAL INTEGRALS AND THE STATEMENT OF BOUNDARY VALUE PROBLEMS

§2.1 *The Feynman-Kac formula for the solution of Cauchy's problem*

Let us consider the Cauchy problem

$$\frac{\partial u(t,x)}{\partial t} = \frac{1}{2} \sum_{i,j=1}^{r} a^{ij}(x) \frac{\partial^2 u}{\partial x^i \partial x^j} + \sum_{i=1}^{r} b^i(x) \frac{\partial u}{\partial x^i} + c(x)\, u =$$

$$= L u + c(x) u, \; t > 0, \; x \in R^r, \; u(0,x) = f(x) . \qquad (1.1)$$

We suppose the coefficients of the operator L and the function $c(x)$ to be bounded and Lipschitz continuous, and let the initial function $f(x)$ be bounded and continuous. The characteristic form $\Sigma\, a^{ij}(x)\lambda_i\lambda_j$ is assumed to be non-negative definite. Thereby we allow the operator L to be degenerate. We shall always suppose that the matrix $(a^{ij}(x))$ can be represented in the form $(a^{ij}(x)) = \sigma(x)\sigma^*(x)$, where $\sigma(x)$ is a matrix whose elements satisfy the Lipschitz condition. In §3.2 we shall prove the existence of such a matrix under the condition that the matrix $(a^{ij}(x))$ with Lipschitz continuous elements does not degenerate or at least is of a constant rank.

Let us consider the stochastic differential equation

$$dX_t^x = \sigma(X_t^x)\, dW_t + b(X_t^x)\, dt, \; X_0^x = x \in R^r , \qquad (1.2)$$

where W_t is an r-dimensional Wiener process. Under the above assumptions, equation (1.2) defines a Markov family (X_t^x, P) and a corresponding Markov process (X_t, P_x).

We denote by B the Banach space of all bounded measurable functions in R^r equipped with the uniform norm. Let us introduce a family of operators $\tilde{T}_t^c$, $t \geq 0$, acting on the functions $f \in B$ according to the formula

$$(\tilde{T}_t^c f)(x) = E f(X_t^x) \exp\left\{\int_0^t c(X_s^x)\,ds\right\} = E_x f(X_t) \exp\left\{\int_0^t c(X_s)\,ds\right\}.$$

Each operator $\tilde{T}_t^c$ is linear, bounded and preserves non-negativity. The family $(\tilde{T}_t^c)$ forms a semi-group:

$$\tilde{T}_{t+s}^c f(x) = E_x \exp\left\{\int_0^t c(X_u)\,du\right\} \theta_t\left[f(X_s) \exp\left\{\int_0^s c(X_u)\,du\right\}\right] =$$

$$= E_x \exp\left\{\int_0^t c(X_u)\,du\right\} E_{X_t} f(X_s) \exp\left\{\int_0^s c(X_u)\,du\right\} = \tilde{T}_t^c(\tilde{T}_s^c f)(x).$$

Here we have made use of the Markov property (1.4.3) of the process (X_t, P_x). Let A denote the infinitesimal operator of the process (X_t, P_x), and $\tilde{A}$ designate that of the semi-group $\tilde{T}_t^c$:

$$\tilde{A} f = \lim_{t \downarrow 0} \frac{\tilde{T}_t^c f - f}{t}.$$

LEMMA 1.1. *The domains* D_A *and* $D_{\tilde{A}}$ *coincide, and*

$$\tilde{A} f(x) = A f(x) + c(x) f(x).$$

Proof. Taking into account the definition of the infinitesimal operator A and the equality

$$\exp\left\{\int_0^t c(X_s)\,ds\right\} = 1 + \int_0^t c(X_s)\,ds + o(t)$$

as $t \downarrow 0$, we get

$$\frac{1}{t}(\widetilde{T}_t^c f(x) - f(x)) = \frac{1}{t}\left(E_x f(X_t) \exp\left\{\int_0^t c(X_s)\, ds\right\} - f(x)\right) =$$

$$= \frac{1}{t}(E_x f(X_t) - f(x)] + \frac{1}{t} E_x f(X_t) \int_0^t c(X_s)\, ds + o(t) . \tag{1.3}$$

Note that if $f \in D_A$ or $f \in D_{\widetilde{A}}$, then $\lim_{t \downarrow 0} E\, f(X_t) = f(x)$. Whence, noting that the function $c(x)$ is bounded and uniformly continuous, and $f(x)$ is bounded, we deduce that the last summand in (1.3) tends to $c(x) f(x)$ as $t \downarrow 0$ uniformly in $x \in R^r$. Thus from (1.3) it follows that the limits on the right- and left-hand sides exist simultaneously and

$$\widetilde{A} f = \lim_{t \downarrow 0} \frac{\widetilde{T}_t^c f - f}{t} = \lim_{t \downarrow 0} \frac{T_t f - f}{t} + c(x) f = A f + c(x) f ,$$

where $T_t f = E_x f(X_t)$. Lemma 1.1 is proved. □

Theorem 4.1 of Chapter I and Lemma 1.1 together imply that bounded functions possessing uniformly continuous, bounded derivatives up to second order, belong to $D_{\widetilde{A}}$. As it was explained in §1.4, for every $f(x) \in D_{\widetilde{A}}$, the function

$$u_t(x) = \widetilde{T}_t^c f(x) = E_x f(X_t) \exp\left\{\int_0^t c(X_s)\, ds\right\}$$

is a solution of the abstract Cauchy problem

$$\frac{du_t}{dt} = \widetilde{A} u_t , \quad u_0(x) = f(x) . \tag{1.4}$$

This solution is unique in the class K of functions $v(t,x)$ growing at most exponentially fast:

$$K = \{v(t,x) : \sup_x |v(t,x)| < C e^{\alpha t} \quad \text{for some} \quad C, \alpha < \infty\}.$$

THEOREM 1.1. *Let* $u(t,x)$ *be a solution of problem (1.1) which, for every* $t \geq 0$, *is bounded and has first- and second-order bounded, uniformly continuous derivatives in* x *and moreover has first-order derivative in* t *which is uniformly continuous in* $t \in [0,T]$ *and* $x \in R^r$. *Then*

$$u(t,x) = E_x f(X_t) \exp\left\{\int_0^t c(X_s) ds\right\}. \tag{1.5}$$

Proof. As it follows from Theorem 1.4.1 and Lemma 1.1, the function $u(t,x)$ having for every t second-order bounded, uniformly continuous derivatives in x, belongs to $D_{\tilde{A}}$ and $\tilde{A}\, u(t,x) = L\, u(t,x) + c(x) u(t,x)$. Since the derivative $\frac{\partial u}{\partial t}(t,x)$ exists and is uniformly continuous, we have

$$\left\| \frac{u(t+h,x) - u(t,x)}{h} - \frac{\partial u}{\partial t}(t,x) \right\| \to 0 \text{ as } h \to 0,$$

and the function $u(t,x)$ is the solution of abstract Cauchy problem (1.4).

Next observe that the solution of problem (1.1) belongs to the class K of functions growing at most exponentially fast. Indeed, choose $c > \|c(x)\|$ arbitrarily. It is easy to check that the function $v(t,x) = e^{-ct} u(t,x)$ is a solution of the following Cauchy problem:

$$\frac{\partial v}{\partial t} = Lv + (c(x) - c)\, v, \; v(0,x) = f(x). \tag{1.6}$$

Since $c(x) - c < 0$, the solution of problem (1.6) satisfies the maximum principle: $\sup_x |v(t,x)| \leq \|f\|$. Hence, $|u(t,x)| < e^{ct} |v(t,x)| \leq e^{ct} \|f\|$. But as it has been noticed above, the unique solution of problem (1.4) in the class K has the form $\tilde{T}_t^c f(x)$. Therefore,

$$u(t,x) = \tilde{T}_t^c f(x) = E_x f(X_t) \exp\left\{\int_0^t c(X_s) ds\right\}. \; \square$$

This is the well-known Feynman-Kac formula (see Feynman and Hibs [1], Kac [1]). It represents the solution of problem (1.1) in terms of the diffusion process corresponding to the operator L.

If all coefficients $a^{ij}(x)$ vanish identically, the operator L turns into a first-order differential operator. In this case, equation (1.2) defining the Markov family (X_t^x, P) becomes the equation of characteristics for the operator L, and the Feynman-Kac formula turns into the well-known representation of the solution in terms of characteristics.

Let us suppose the matrix $(a^{ij}(x))$ to be uniformly positive definite in R^r. Under this assumption, problem (1.1) has a solution for any bounded continuous function $f(x)$.

We shall show that the representation (1.5) remains valid for such functions f. To that end, choose a uniformly bounded sequence $(f^{(n)}(x))$ of functions with uniformly continuous, bounded derivatives of first- and second-order such that $f^{(n)}(x)$ tends to $f(x)$ uniformly on every compact set. Let $u^{(n)}(t,x)$ be a solution of problem (1.1) with the initial function $f^{(n)}(x)$. The functions $u^{(n)}(t,x)$ satisfy all the hypotheses of Theorem 1.1, and they admit the representation

$$u^{(n)}(t,x) = E_x f^{(n)}(X_t) \exp\left\{\int_0^t c(X_s)\,ds\right\}. \tag{1.7}$$

From the fact that the coefficients of equation (1.1) are bounded and the functions $f^{(n)}(x)$ are bounded uniformly in n, we easily deduce that $\lim_{n\to\infty} u^{(n)}(t,x) = u(t,x)$, where $u(t,x)$ is the solution of problem (1.1) with the initial function $f(x)$. The right-hand side of (1.7) has the limit

$$E_x f(X_t) \exp\left\{\int_0^t c(X_s)\,ds\right\} \quad \text{as} \quad n \to \infty .$$

Therefore, formula (1.5) remains valid for the function $u(t,x)$.

The restrictions on the function $c(x)$ may also be weakened. For example, a little later we will prove that if the operator L does not degenerate, has bounded Lipschitz coefficients and the function $c(x)$ is Hölder continuous and non-positive, then, whatever the growth of $|c(x)|$ as $x \to \pm\infty$ the solution of problem (1.1) exists and can be represented in the form (1.5). Formula (1.5) is also valid in the case of unbounded initial functions, provided their growth is not too fast.

The function $u(t,x)$ defined by formula (1.5) will be referred to as the *generalized solution of Cauchy's problem* (1.1). This definition is correct. We have shown that if a bounded, sufficiently smooth solution of Cauchy's problem exists, then it can be represented in the form (1.5).

By Lemma 4.3 of Chapter I, a generalized solution is continuous, whenever $f(x)$ and $c(x)$ are continuous. One can demonstrate (Blagoveščhenskii and Freidlin [1]) that if the initial function $f(x)$ and the functions $\sigma^i_j(x)$, $b^i(x)$, and $c(x)$ have $(k+1)$ bounded derivatives, then the generalized solution has at least k derivatives with respect to x and the k-th order derivatives satisfy a Lipschitz condition. From this one can easily deduce that the function $u(t,x)$ has at least $\left[\frac{k}{2}\right]$ derivatives with respect to t. If $k \geq 2$, then the generalized solution satisfies equation (1.1).

The generalized solution is stable with respect to small pertubations of the coefficients of the operator L. In fact, let us consider, together with the operator L, the operator

$$\tilde{L} = \frac{1}{2} \sum_{i,j=1}^{r} \tilde{a}^{ij}(x) \frac{\partial^2}{\partial x^i \partial x^j} + \sum_{i=1}^{r} \tilde{b}^i(x) \frac{\partial}{\partial x^i} .$$

The coefficients of the operator $\tilde{L}$ are assumed to be bounded and Lipschitz continuous. The form $\sum_{i,j=1}^{r} \tilde{a}^{ij}(x)\lambda_i\lambda_j$, $x \in R^r$, is non-negative. In addition a matrix $\tilde{\sigma}(x)$ with Lipschitz elements $\tilde{\sigma}^i_j(x)$ is assumed to exist such that $\tilde{\sigma}(x)\tilde{\sigma}^*(x) = (a^{ij}(x))$.

We will consider the Cauchy problem

$$\frac{\partial u^{\varepsilon}(t,x)}{\partial t} = (L + \varepsilon\widetilde{L})\, u^{\varepsilon} + c(x)\, u^{\varepsilon}, \ t > 0, \ x \in R^r ,$$

$$u^{\varepsilon}(0,x) = f(x) .$$

Let the Markov random functions $X_t^{\varepsilon,x}$ be specified by the stochastic differential equation

$$dX_t^{\varepsilon,x} = \sigma(X_t^{\varepsilon,x})\, dW_t + \sqrt{\varepsilon}\, \widetilde{\sigma}(X_t^{\varepsilon,x})\, d\widetilde{W}_t + \\ + [b(X_t^{\varepsilon,x}) + \widetilde{b}(X_t^{\varepsilon,x})]\, dt , \tag{1.8}$$

$$X_0^{\varepsilon,x} = x, \ b(x) = (b^1(x), \cdots, b^r(x)), \ \widetilde{b}(x) = (\widetilde{b}^1(x), \cdots, \widetilde{b}^r(x)) ,$$

where W_t, $\widetilde{W}_t$ are independent r-dimensional Wiener processes. For $\varepsilon = 0$, equation (1.8) turns into (1.2); the random function satisfying equation (1.2) will be denoted by X_t^x.

LEMMA 1.2. *For any* T, $\delta > 0$:

$$\lim_{\varepsilon \downarrow 0} E|X_t^{\varepsilon,x} - X_t^x|^2 = 0 ,$$

$$\lim_{\varepsilon \downarrow 0} P\{ \sup_{0 \le t \le T} |X_t^{\varepsilon,x} - X_t^x| > \delta \} = 0$$

uniformly in $x \in R^r$.

To prove these assertions, we write down the relation for $X_t^{\varepsilon,x} - X_t^x$ which follows from (1.8) and (1.2):

$$X_t^{\varepsilon,x} - X_t^x = \int_0^t [\sigma(X_s^{\varepsilon,x}) - \sigma(X_s^x)]\, dW_s + \\ + \sqrt{\varepsilon} \int_0^t \widetilde{\sigma}(X_s^{\varepsilon,x})\, d\widetilde{W}_s + \int_0^t [b(X_s^{\varepsilon,x}) - b(X_s^x)]\, ds + \varepsilon \int_0^t \widetilde{b}(X_s^{\varepsilon,x})\, ds . \tag{1.9}$$

Squaring both sides of this equality and using the properties of stochastic integrals and the fact the coefficients are bounded and Lipschitz continuous, we obtain from (1.9):

$$E|X_t^{\varepsilon,x} - X_t^x|^2 \le (c_1 + c_2 t) \int_0^t E|X_s^{\varepsilon,x} - X_s^x|^2 \, ds + c_2 \varepsilon ,$$

where c_i are constants depending only on the Lipschitz constant, on the maximum of the absolute value of the coefficients, and on the space dimension. Applying Lemma 1.4.1 we get from this the first assertion of the lemma. By the first assertion and the properties of stochastic integrals we deduce the second assertion from (1.9). □

COROLLARY. *Lemma 1.2 implies that the generalized solution of problem (1.7)*

$$u^\varepsilon(t,x) = E\, f(X_t^{\varepsilon,x}) \exp\left\{ \int_0^t c(X_s^{\varepsilon,x})\, ds \right\}$$

converges to the generalized solution of problem (1.1) as $\varepsilon \downarrow 0$ *uniformly in* $x \in R^r$, $0 \le t \le T_0 < \infty$.

In particular, if $\sum_{i,j=1}^{r} \tilde{a}^{ij}(x) \lambda_i \lambda_j > c \sum_1^r \lambda_i^2$, $c > 0$, then problem (1.7) has a classical solution $u^\varepsilon(t,x)$ tending to the generalized solution of problem (1.1) as $\varepsilon \downarrow 0$.

If all coefficients $a^{ij}(x)$ in the higher derivatives of the operator L vanish identically, then this result turns into the well-known statement that the solution of Cauchy's problem for a parabolic equation with a small parameter ε in the higher derivatives converges to the solution of Cauchy's problem for the limit first-order equation as $\varepsilon \downarrow 0$. Similar questions will be considered further in Chapter IV.

The equation for $u(t,x) = E_x f(X_t)$

$$\frac{\partial u}{\partial t} = \frac{1}{2} \sum_{i,j=1}^{r} a^{ij}(x) \frac{\partial^2 u}{\partial x^i \partial x^j} + \sum_{i=1}^{r} b^i(x) \frac{\partial u}{\partial x^i} \tag{1.10}$$

which is obtained from (1.1) for $c(x) \equiv 0$ is called *Kolmogorov's backward equation.*

Let the matrix $(a^{ij}(x))$ be positive definite uniformly in R^r, $a^{ij}(x) \in C^{2,\delta}_{R^r}$, $b(x) \in C^{1,\delta}_{R^r}$. As is customary, we denote by $C^{k,\delta}_{R^r}$ the space of real-valued functions on R^r, which have k continuous bounded derivatives, the k-th order derivatives being Hölder continuous with the exponent δ. The equation (1.10) has a fundamental solution $p(t,x,y)$ (Friedman [1]). The function $p(t,x,y)$ obeys the equation

$$\frac{\partial p}{\partial t} = L\, p(t,x,y)\ t > 0,\ x \in R^r,\ y \in R^r ,$$

in the variables (t,x) and the initial condition $p(0,x,y) = \delta(x-y)$.

For continuous and bounded f, the solution of Cauchy's problem (1.10) can be represented in the form:

$$u(t,x) = \int_{R^r} f(y)\, p(t,x,y)\, dy . \tag{1.11}$$

Formula (1.11) remains true if the function $f(x)$ has discontinuities on a set of zero measure, but in this case the initial condition in problem (1.10) must be fulfilled only at the continuity points of the function $f(x)$.

If $f(x) = \chi_D(x)$ is the indicator of a set D with a smooth boundary ∂D, then

$$u(t,x) = E_x \chi_D(X_t) = P(t,x,D) = \int_D p(t,x,y)\, dy .$$

So, the fundamental solution of equation (1.10) is the transition density of the Markov process (X_t, P_x). It is possible to prove (see, e.g. Wentzell [1]) that the function $p(t,x,y)$ satisfies the Kolmogorov forward equation

$$\frac{\partial p}{\partial t} = L^* p(t,x,y) = \frac{1}{2} \sum_{i,j=1}^{r} \frac{\partial^2}{\partial y^i \partial y^j} (a^{ij}(y) p(t,x,y)) - \sum_{i=1}^{r} \frac{\partial}{\partial y^i} (b^i(y)\, p(t,x,y)) . \tag{1.12}$$

in the variables (t,y).

Suppose that equation (1.12) has a time-independent solution $m = m(y)$ satisfying the normalizing condition $\int_{R^r} m(y)\,dy = 1$. Then m is the density function of the uniquely determined stationary distribution of the process (X_t, P_x). For instance, let $(a^{ij}(y))$ be the unit matrix and suppose that the vector field $b(x) = (b^1(x), \cdots, b^r(x))$ has a potential U, i.e. $b(x) = -\nabla U(x)$, where $U(x)$ is a non-negative function going to infinity as $|x| \to \infty$. We put $m(x) = c \cdot \exp\{-2U(x)\}$. By means of direct substitution into the equation one can check that

$$\frac{1}{2}\,\Delta m(x) + \sum_i \frac{\partial}{\partial x_i}\left(m(x)\,\frac{\partial}{\partial x_i}\,U(x)\right) = 0\,.$$

Therefore, if $\int_{R^r} \exp\{-2U(x)\}\,dx < \infty$, the process in R^r generated by the operator $Lf = \frac{1}{2}\Delta f - (\nabla U, \nabla f)$ has a stationary distribution with the density $m(x) = c\,\exp\{-2U(x)\}$; where the constant c is defined by the normalization condition: $(\int_{R^r} \exp\{-2U(x)\}dx)^{-1} = c$. In the one-dimensional case, the equation for the stationary density has the form

$$\frac{1}{2}\,\frac{d^2}{dx^2}\,(a(x)\,m(x)) - \frac{d}{dx}\,(b(x)\,m(x)) = 0\,.$$

If $a(x) > 0$ and $\int_{-\infty}^{\infty} a^{-1}(x) \exp\{\int_0^x 2b(y)a^{-1}(y)dy\}dx = c_1 < \infty$, then the only probability density which is a solution of this equation has the form

$$m(x) = \frac{1}{c_1 a(x)}\,\exp\left\{\int_0^x \frac{2b(y)}{a(y)}\,dy\right\}. \tag{1.13}$$

§2.2 *Probabilistic representation of the solution of Dirichlet's problem*

Given a domain $D \subset R^r$ with boundary ∂D, consider the Dirichlet problem

$$L\,u(x) - c(x)\,u(x) = f(x),\ x \in D\,;$$

$$\lim_{x \in D,\, x \to x_0} u(x) = \psi(x_0),\ x_0 \in \partial D\,. \tag{2.1}$$

Here L is the operator introduced at the beginning of the previous section. The conditions on its coefficients are listed there too. The functions $f(x)$, $c(x)$ and $\psi(x)$ are assumed continuous and bounded, and $c(x) \geq 0$.

Let $\tau = \tau_D^x = \inf\{t : X_t^x \in \partial D\}$, where (X_t^x, P) is a Markov family which solves equation (1.2). If X_t^x never hits ∂D, then we set $\tau = +\infty$. The random variable τ is a Markov time with respect to the increasing system of σ-fields $\mathcal{N}_t$ generated by the underlying Wiener process W_t.

THEOREM 2.1. *Suppose* $E\,\tau_D^x < \infty$ *for all* $x \in D$. *If* $u(x)$, $x \in D$, *is a bounded solution of problem 2.1 which has bounded, continuous, first- and second-order partial derivatives in any interior subdomain* $\tilde{D} \subset D$, *then*

$$u(x) = -E \int_0^{\tau_D^x} f(X_t^x) \exp\left\{-\int_0^t c(X_s^x)\,ds\right\} dt + \\ + E\,\psi(X_{\tau_D^x}^x) \exp\left\{-\int_0^{\tau_D^x} c(X_s^x)\,ds\right\}. \qquad (2.2)$$

Proof. First, we assume that the function $u(x)$ can be extended to the whole space R^r so that it is bounded and has bounded continuous first- and second-order derivatives. Let us consider the function $u(X_t^x)e^{Y_t^x}$ where X_t^x is the solution of equation (1.2) and $Y_t^x = -\int_0^t c(X_s^x)\,ds$. By Ito's formula, we have:

$$u(X_t^x)\exp\{Y_t^x\} - u(x) = \int_0^t (\nabla u(X_s^x)\exp\{Y_s^x\}, \sigma(X_s^x)\,dW_s) + \\ + \int_0^t f(X_s^x)\exp\{Y_s^x\}\,ds + \qquad (2.3) \\ + \int_0^t [Lu(X_s^x) - f(X_s^x) - c(X_s^x)u(X_s^x)]\exp\{Y_s^x\}\,ds\,.$$

Here $\nabla u(x)$ is the gradient of the function $u(x)$, and (a,b) denotes the Euclidean scalar product of the vectors $a, b \in R^r$. With probability one, equation (2.3) holds for all $t \geq 0$. In particular, equation (2.3) remains valid if we replace t by the random variable $\tau_T = T \wedge \tau_D^x$. The trajectory X_t^x, $x \in D$, does not leave the domain D before the moment τ_T. Therefore, (2.1) implies that $Lu(X_s^x) = f(X_s^x) + c(X_s^x)u(X_s^x)$ for all $s < \tau_T$. Thus, for $t = \tau_T$, the last integral in (2.3) vanishes. The random variable τ_T is a Markov time and $E\tau_T \leq T < \infty$. Moreover,

$$\exp\left\{-\int_0^s c(X_v^x)\,dv\right\} \leq 1 .$$

Consequently,

$$E\int_0^{\tau_T} (\nabla u(X_s^x)\exp\{-Y_s^x\}, \sigma(X_s^x)\,dW_s) = 0 . \tag{2.4}$$

Together with (2.3), this yields:

$$E\,u(X_{\tau_T}^x)\exp\left\{-\int_0^{\tau_T} c(X_s)\,ds\right\} = u(x) + E\int_0^{\tau_T} f(X_s^x)\exp\left\{-\int_0^s c(X_v^x)\,dv\right\} ds . \tag{2.5}$$

Equation (2.5) holds for all $T \geq 0$, the trajectories of X_s^x are continuous with probability 1; $P\{\tau_D^x > T\}$ tends to zero as $T \to \infty$, and $E\tau_T \leq E\tau_D^x < \infty$. Therefore, using Lebesgue's Dominated Convergence Theorem, we can pass to the limit as $T \to \infty$ on the both sides of (2.5) to obtain (2.2).

To get rid of the assumption about the existence of a smooth extension of $u(x)$ onto the whole space R^r, we consider an expanding sequence of domains $D_n \subset D$ with smooth boundaries ∂D_n tending to the domain D. The solution of the Dirichlet problem

$$Lu_n(x) - c(x)u_n = f(x),\ x \in D_n,$$

$$u_n(x)|_{\partial D_n} = u(x),$$

can be extended smoothly to R^r. Therefore, u_n has a representation of (2.3) type. The function $u(x)$ is continuous on the set $D \cup \partial D$, the Markov times $\tau_n = \inf\{t : X_t^x \in \partial D_n\}$ increase monotonically and tend to $\tau = \tau_D^x$, $E\tau_D^x < \infty$; so the validity of the representation (2.2) for $u(x)$ follows from its validity for $u_n(x)$. □

The assertion of Theorem 2.1 underlies a great deal of the application of probabilistic methods to the theory of differential equations. Under some assumptions, this theorem has been proved by many authors (see e.g. Dynkin [3], where further references are available too). The above proof follows the work by Freidlin [7].

Note that if (X_t, P_x) is the Markov process corresponding to the family (X_t^x, P) and $\tau_D = \inf\{t : X_t \in \partial D\}$, then equality (2.2) may be written in the form

$$u(x) = -E_x \int_0^{\tau_D} f(X_t) \exp\left\{-\int_0^t c(X_s)\,ds\right\} dt +$$

$$+ E_x \psi(X_{\tau_D}) \exp\left\{-\int_0^{\tau_D} c(X_s)\,ds\right\},$$

provided $E_x\tau_D < \infty$, $x \in D$. We remind that the operation of taking expectation E_x is an integration with respect to the measure P_x in the space $C_{0,\infty}(R^r)$. Thus, the last equality is the representation of a solution of problem (2.1) in the form of functional integral.

REMARK 1. If $c(x) > \alpha > 0$ for all $x \in D$, then the assumption $E\tau_D^x < \infty$ may be dropped in Theorem 2.1. In this case the finiteness of all needed expectations and equality (2.4) are ensured by the presence of the integrand $\exp\{-\int_0^t c(X_s^x)\,ds\} < e^{-\alpha t}$.

REMARK 2. If $f(x) \equiv 0$ in place of the condition $E\tau_D^x < \infty$, then it is sufficient to suppose $P\{\tau_D^x < \infty\} = 1$ for all $x \in D$. In fact, if $f(x) \equiv 0$, then the last term on the right-hand side of equation (2.5) vanishes, and by letting $T \to \infty$, we obtain from (2.5) that

$$u(x) = E\,\psi(X^x_{\tau_D^x})\exp\left\{-\int_0^{\tau_D^x} c(X_s^x)\,ds\right\}.$$

REMARK 3. Suppose $\sup_{x \in D} E \exp\{\alpha\tau_D^x\} < \infty$ for some $\alpha > 0$. Then representation (2.2) remains true for the functions $c(x)$ with negative values if $c(x) > -\alpha$ for all x. In this case the condition $E\exp\{\alpha\tau_D^x\} < \infty$ allows one to pass to the limit in (2.5) as $T \to \infty$. In particular, if $\tau_D^x(\omega) \leq T(x) < \infty$ with probability 1, then formula (2.2) holds for any bounded function $c(x)$.

REMARK 4. Let D be a bounded domain with a smooth boundary ∂D. Assume that the operator L is non-degenerate and has sufficiently smooth coefficients. It is not hard to prove (see Lemma 3.3.1) that in this case $E\tau_D^x < c < \infty$. For every continuous function $\psi(x)$ on ∂D, the solution $u(x)$ of problem (2.1) with the properties required in Theorem 2.1, exists, provided, e.g. $c(x)$ and $f(x)$ are Hölder continuous and $c(x) \leq 0$. In particular, for $c(x) \equiv f(x) \equiv 0$, the solution of problem (2.1) can be written in the form:

$$u(x) = E\,\psi(X^x_{\tau_D^x}) = \int_{\partial D} \psi(y)\,\Pi(x, dy),$$

where $\Pi(x, dy) = P\{X^x_{\tau_D^x} \in dy\}$, $dy \subset \partial D$, $x \in D$. Alternatively, the function $u(x)$ can be represented as a Poisson integral (Miranda [1]):

$$u(x) = \int \psi(y)\,\pi(x,y)\,dy,$$

where $\pi(x,y)$, $x \in D$, $y \in \partial D$, is Poisson's kernel for the operator L in the domain D. Comparing these formulae, we conclude that the exit probabilities $\Pi(x,\gamma) = P\{X^x_{\tau^x_D} \in \gamma\}$, $\gamma \subset \partial D$, have densities: $\Pi(x,\gamma) = \int_\gamma \pi(x,y)dy$. The function $\pi(x,y)$ is strictly positive, twice continuously differentiable in $x \in D$; $\pi(x,y) \to 0$ as $x \to y_0 \in \partial D$, $y \neq y_0$. The function $\pi(x,y)$ has a singularity as $x \to y \in \partial D$.

Let us consider the operator

$$\mathcal{L}u(t,x) = -\frac{\partial u}{\partial t} + \frac{1}{2}\sum_{i,j=1}^{r} a^{ij}(t,x)\frac{\partial^2 u}{\partial x^i \partial x^j} + \sum_{1}^{r} b^i(t,x)\frac{\partial u}{\partial x^i} = -\frac{\partial u}{\partial t} + Lu. \tag{2.6}$$

Assume that $(a^{ij}(t,x)) = \sigma(t,x)\,\sigma^*(t,x)$, where the elements of the matrix $\sigma(t,x) = \{\sigma^i_j(t,x)\}$ are continuous and bounded on the set $[0,\infty) \times R^r$ and Lipschitz continuous in x with a Lipschitz constant which does not depend on t. The functions $b(t,x)$ are also assumed to be continuous, bounded uniformly in t and Lipschitz continuous in x. To avoid Markov families and processes which are inhomogeneous in time, operator (2.5) will be thought of as a degenerate one acting in the space of functions $f(t,x)$ of $(r+1)$ space variables $(t, x^1, \cdots, x^r)$. The Markov family in R^{r+1} corresponding to the operator L can be constructed using the stochastic differential equations

$$dX^{t,x}_s = \sigma(t^{t,x}_s, X^{t,x}_s)\,dW_s + b(t^{t,x}_s, X^{t,x}_s)\,ds, \quad X^{t,x}_0 = x,$$

$$t^{t,x}_s = t - s.$$

In order for the coefficients $\sigma(t,x)$ and $b(t,x)$ to be defined for all t, one can take $\sigma(-t,x) = \sigma(t,x)$, $b(-t,x) = b(t,x)$. Under the above assumptions on the coefficients, a unique solution of these equations exists and defines a Markov family $(t^{t,x}_s, X^{t,x}_s; P)$.

Let us consider the Cauchy problem

$$\frac{\partial u(t,x)}{\partial t} = L u + c(t,x) u, \ t > 0, \ x \in R^r ;$$

$$u(0,x) = f(x) , \tag{2.7}$$

where L is the operator defined by (2.6); $c(x)$, $f(x)$ are continuous bounded functions. From Theorem 2.1 and Remark 3, we obtain

THEOREM 2.2. *Let* $u(t,x)$ *be a solution of problem (2.7) which is continuous and bounded on* $[0,T] \times R^r$ *for every* $T > 0$. *Assume that the derivatives of* $u(t,x)$ *up to second order with respect to* x *and of first order with respect to* t *are bounded and continuous in the region* $(h < t < T, \ x \in R^r)$ *for every* $h \in (0,T)$.

Then $u(t,x)$ *can be represented in the form*

$$u(t,x) = E f(X_t^{t,x}) \exp\left\{ \int_0^t c(t-s, X_s^{t,x})\, ds \right\} .$$

This result extends the Feynman-Kac formula to the case of parabolic equations which are inhomogeneous in time. In particular, if the coefficients of the operator L do not depend on t and only the function $c(t,x)$ depends on t, then clearly the solution of problem (2.7) may be written as follows

$$u(t,x) = E f(X_t^x) \exp\left\{ \int_0^t c(t-s, X_s^x)\, ds \right\} ,$$

where (X_t^x, P) is a homogeneous Markov family defined by equation (1.2).

Given a domain $D \subset R^r$, consider the mixed problem

$$\frac{\partial u(t,x)}{\partial t} = L u(t,x) + c(t,x) u(t,x), \ t > 0, \ x \in D ,$$

$$u(0,x) = f(x), \ u(t,x)\big|_{x \in \partial D} = g(t,x) . \tag{2.8}$$

The functions $c(t,x)$, $f(x)$ and $g(t,x)$ are assumed bounded and continuous in their domains of definition, and such that $g(0,x) = f(x)$ for $x \in \partial D$. Let $\tau^{t,x} = \inf\{s : X_s^{t,x} \in \partial D\}$, $\tau_t = t \wedge \tau^{t,x}$.

Suppose that $G(t,x)$ is a function defined on the boundary of the cylinder $K = \{x \in D, t > 0\}$ coinciding with $f(x)$ on the base of the cylinder and with $g(t,x)$ on the lateral surface: $G(0,x) = f(x)$, $G(t,x)|_{x \in \partial D} = g(t,x)$. The function $G(t,x)$ is continuous on the boundary of the cylinder. Theorem 2.1 and Remark 3 lead to

THEOREM 2.3. *Let* $u(t,x)$ *be a solution of problem (2.8) which is continuous and bounded on* $[0,T] \times (D \cup \partial D)$ *for every* $T > 0$. *Suppose that the first-order derivatives of* $u(t,x)$ *with respect to* t *and derivatives up to second order with respect to* x *are bounded and continuous in the region* $(h < t < T,\ x \in D,\ \rho(x, \partial D) > h)$ *for every* $h \in (0,T)$ *and every* $T > 0$.

Then the following representation holds:

$$u(t,x) = E\, G(\tau_t, X_{\tau_t}^{t,x}) \exp\left\{ \int_0^{\tau_t} c(t-s, X_s^{t,x})\, ds \right\} =$$

$$= E\, f(X_t^{t,x}) \chi_{\tau_t = t} \exp\left\{ \int_0^{t} c(t-s, X_s^{t,x})\, ds \right\} +$$

$$+ E\, g(\tau^{t,x}, X_{\tau^{t,x}}^{t,x}) \chi_{\tau_t \neq t} \exp\left\{ \int_0^{\tau^{t,x}} c(t-s, X_s^{t,x})\, ds \right\},$$

where $\chi_{\tau_t = t}$ *is the indicator of the set* $\{\omega : \tau^{t,x} = t\}$, $\chi_{\tau_t \neq t} = 1 - \chi_{t = \tau_t}$.

Using Theorem 2.3 it is possible to prove that representation (1.5) for the solution of Cauchy's problem (1.1) is valid without any restrictions on the growth of $c(x)$ as $|x| \to +\infty$, provided $c(x) \leq 0$. Indeed, if $u(x)$

is the solution of problem (1.1) with such a coefficient $c(x)$, then by virtue of Theorem 2.3, for any $N > 0$ and $D = \{x \in R^r : |x| < N\}$, the formula

$$u(t,x) = E f(X_t^x) \chi_{\tau_N \geq t} \exp\left\{\int_0^t c(X_s^x)\, ds\right\} + E\, u(\tau_N, X_{\tau_N}^x) \chi_{\tau_N < t} \exp\left\{\int_0^{\tau_N} c(X_s^x)\, ds\right\} \tag{2.9}$$

holds, where $\tau_N = \inf\{t : |X_t^x| = N\}$. Letting $N \to \infty$, $\tau_N \to \infty$ with probability 1, and all the functions under the expectation sign are bounded uniformly in N. Therefore, the second summand on the right-hand side of (2.9) tends to zero, and formula (2.9) turns into (1.5). Using the above reasoning it is not difficult to deduce the existence theorem for the classical solution without assuming boundedness of $|c(x)|$, if the operator L is non-degenerate and $c(x) \leq 0$. An analogous argument is applicable to equations which are inhomogeneous in t.

We proceed to consider the time-homogeneous case. Let the operator L be the same as in (1.1). The characteristic form $\sum_1^r a^{ij}(x)\lambda_i\lambda_j$ is assumed to be non-degenerate uniformly in D and the coefficients of the operator and $c(x)$ are assumed to be bounded and smooth enough. Let the boundary functions $g(t,x)$ and $f(x)$ be bounded and continuous. Then without assuming the functions $f(x)$ and $g(t,x)$ to coincide on the set $\{t = 0, x \in \partial D\}$, we obtain that a unique bounded solution $u(t,x)$ of problem (2.8) exists which has the needed number of derivatives inside the cylinder K and takes the boundary values everywhere excluding, possibly, some points of the set $(t = 0, x \in \partial D)$. Noting that the process (X_t, P_x) corresponding to the operator L has a transition density, it is not difficult to prove that the assertion of Theorem 2.3 is valid for such a solution.

In particular, let $g(t,x) = 0$, $f(x) = 1$, $c(x) = 0$. Then the function $u(t,x) = P\{\tau_D^x > t\}$ satisfies the problem:

$$\frac{\partial u}{\partial t} = L u,\ t > 0,\ x \in D,\ u(0,x) = 1,\ u(t,x)\Big|_{x \in \partial D} = 0 .$$

Because the solution of this problem is differentiable in t, one can conclude from this that the random variable $\tau_D^x = \min\{t : X_t^x \in \partial D\}$ has a density function.

Theorems 2.1, 2.2, and 2.3 enable us to evaluate mean values of various functionals of the trajectories of the Markov family (X_t^x, P) (Markov process (X_t, P_x)) by solving the corresponding differential equations. Let us calculate the mean values of some such functionals.

As was pointed out in Remark 4, if $f(x) \equiv c(x) \equiv 0$ and the domain D is bounded and has a smooth boundary ∂D, then Theorem 2.1 implies that the function $u(x) = E\psi(X^x_{\tau_D^x})$ is the solution of the Dirichlet problem:

$$L\,u(x) = 0,\ x \in D,\ u(x)\Big|_{x \in \partial D} = \psi(x) . \tag{2.10}$$

Let us consider the one-dimensional case: $D = (\alpha, \beta) \subset R^1$, $\psi(\alpha) = 0$, $\psi(\beta) = 1$,

$$L = \frac{1}{2}\, a(x)\, \frac{d^2}{dx^2} + b(x)\frac{d}{dx}\ ,\quad a(x) > 0 .$$

Then $u(x) = P\{X^x_{\tau_D^x} = \beta\}$ is the probability that the Markov trajectory X_t^x starting from $x \in (\alpha, \beta)$ reaches the point β before α. Problem (2.10) takes the form:

$$\frac{1}{2}\, a(x)\, u''(x) + b(x)\, u'(x) = 0 \quad \text{for} \quad \alpha < x < \beta\ ;$$

$$u(\alpha) = 0,\quad u(\beta) = 1\ .$$

From this we derive:

$$u(x) = P\{X^x_{\tau_D^x} = \beta\} =$$

$$= \int_\alpha^x \exp\{-U(y)\}dy \left(\int_\alpha^\beta \exp\{-U(y)\}dy\right)^{-1} , \tag{2.11}$$

where $U(y) = \int_0^y \frac{2b(z)}{a(z)}\,dz$. In particular, if $X_t^x = W_t^x$ is a Wiener family, then $b(z) \equiv 0$, $a(z) \equiv 1$, and $P\left\{W_{\tau_D^x}^x = \beta\right\} = \frac{x-\alpha}{\beta-\alpha}$.

Let $\psi(x) \equiv 0$, $c(x) \equiv 0$, and $f(x) \equiv -1$, then by formula (2.2), $u(x) = E\tau_D^x$ is the mean exit time from the domain D (starting from the point $x \in D$). By Theorem 2.1, this function is the solution of the problem

$$L\,u(x) = -1,\ x \in D,\ u(x)\big|_{x\in\partial D} = 0 \tag{2.12}$$

at least in the case when the operator L is non-degenerate and the domain D is bounded and has smooth boundary. In the one-dimensional case, problem (2.12) may be integrated explicitly:

$$u(x) = \frac{G(x)H(\beta) - G(\beta)H(x)}{G(\beta)},\quad x \in D = (\alpha,\beta), \tag{2.13}$$

where $G(x) = \int_\alpha^x \exp\{-U(y)\}dy$, $H(x) = \int_\alpha^x \exp\{-U(y)\}\int_0^y \frac{2}{a(z)}\exp\{U(z)\}dz\,dy$.

The function $U(y)$ has been specified above.

In particular, if $X_t^x = W_t^x$ is the Wiener family, then $U(y) \equiv 0$, $G(x) = x-\alpha$, $H(x) = x^2-\alpha^2$ and $E\tau_D^x = u(x) = (x-\alpha)(\beta-x)$. If, for example, $-\alpha = \beta > 0$ and $x = 0$, then $E\tau_D^0 = \beta^2$. Thus, starting from zero, the Wiener trajectory hits the end-points of the interval $[-\beta,\beta]$ at the mean time β^2. This indicates once more that Wiener trajectories have infinite instantaneous velocity. On the contrary, the mean hitting time to the end-points of a large interval increases at the rate of the square of the length of this interval. This results from "frequent change of the motion direction."

For $\psi(x) \equiv 1$, $f(x) \equiv 0$ and $c(x) = \lambda > 0$, we obtain from (2.2), a boundary problem for the function $u_\lambda(x) = E\,e^{-\lambda\tau_D^x} = \int_0^\infty e^{-\lambda t}\,p_{\tau_D^x}(t)\,dt$ which is the Laplace transform for the density function $p_{\tau_D^x}(t)$ of the random variable $\tau_D^x = \min\{t : X_t^x \in \partial D\}$. Let us compute this function for the one-dimensional Wiener process in the interval $(-\beta,\beta)$, $\beta > 0$. By solving the boundary problem

$$\frac{1}{2} u_\lambda''(x) - \lambda u_\lambda(x) = 0, \ x \in (-\beta, \beta), \ u_\lambda(-\beta) = u_\lambda(\beta) = 1, \qquad (2.14)$$

we find that

$$u_\lambda(x) = \frac{e^{x\sqrt{2\lambda}} + e^{-x\sqrt{2\lambda}}}{e^{\beta\sqrt{2\lambda}} + e^{-\beta\sqrt{2\lambda}}},$$

The function $E e^{-\lambda \tau_D^x}$ is also finite for negative λ with modulus small enough: for $0 \geq \lambda > -\frac{\pi^2}{8\beta^2}$ we have

$$u_\lambda(x) = \frac{\cos x\sqrt{2|\lambda|}}{\cos \beta\sqrt{2|\lambda|}}$$

§2.3 *On the correct statement of Dirichlet's problem*

Throughout this and the following section, we will suppose in addition to the assumptions on the coefficients of the operator L made in §2.1, that the operator L is uniformly elliptic: $\sum_{i,j=1}^{r} a^{ij}(x)\lambda_i\lambda_j \geq a \sum_1^r \lambda_i^2$ for some $a > 0$, any real $\lambda_1, \lambda_2, \cdots, \lambda_r$, and $x \in R^r$.

First of all we recall the notion of regularity of a boundary point which is used in the theory of elliptic differential equations and introduce an equivalent probabilistic definition.

Suppose that D is a bounded domain in R^r, ∂D is its boundary, $D_n \subset D$ is an increasing sequence of domains, $\cup D_n = D$, having smooth boundaries ∂D_n. Let a continuous function $\psi(x)$ be given on ∂D. By the same symbol $\psi(x)$ we will denote a continuous extension of this function to the whole domain D. Let $u_n(x)$ denote the solution of the following Dirichlet problem in D_n:

$$L u_n(x) = 0, \ x \in D_n, \ u_n(x)\big|_{x \in \partial D_n} = \psi(x). \qquad (3.1)$$

The function $u(x) = \lim_{n\to\infty} u_n(x)$ is called a *generalized solution in the Wiener sense* of the Dirichlet problem

$$Lu(x) = 0, \ x \in D, \ u(x)\Big|_{x\in\partial D} = \psi(x). \tag{3.2}$$

Later on in this section we shall prove that this limit exists and does not depend on the choice of the domains D_n nor on the way the function $\psi(x)$ is extended from the boundary to the whole domain D.

A point $x_0 \in \partial D$ is called *L-regular*, if for any continuous function $\psi(x)$, the generalized solution $u(x)$ of problem (3.2) in the Wiener sense coincides with the boundary function at the point x_0: $\lim_{x\to x_0} u(x) = \psi(x_0)$.

The notion of L-regularity has been studied in detail in potential theory. It has been established (see, e.g. Krylov [1]) that, under weak assumptions on the smoothness of the coefficients, a boundary point x_0 is L-regular if and only if it is Δ-regular, where Δ is the Laplace operator. In potential theory, one has the so-called Wiener criterion giving necessary and sufficient conditions for Δ-regularity. A number of geometric criteria for regularity can be deduced from this one. For example, it is possible to prove (although it can be done more easily without Wiener's criterion) that the point $x_0 \in \partial D$ is L-regular, provided it can be touched by a cone with the vertex at the point x_0 which in some neighborhood of the point x does not intersect D (Fig. 1). In particular, if the boundary ∂D of the domain D is smooth enough, then the boundary points are all L-regular. We also note that the L-regularity is a local notion: changes of the boundary and the operator L outside a neighborhood of a point $x_0 \in \partial D$ do not

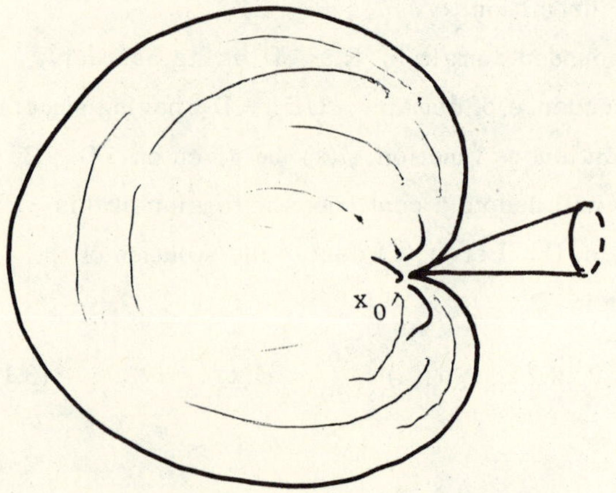

Fig. 1

affect the regularity of the point x_0. The Wiener criterion as well as basic notions of potential theory have a natural probabilistic meaning. The adequacy of the probabilistic language makes many classical results of potential theory highly transparent and intuitive. These problems have been studied in detail in several monographs (see e.g. Ito and McKean [1], Dynkin and Jushkevich [1], Meyer [1]) and we will not go into this point. We shall merely introduce the notion of regularity for Markov processes which will be widely used in the sequel and show that it is equivalent to L-regularity in the non-degenerate case.

In order to introduce the probabilistic notion of regularity, let us consider the Markov process (X_t, P_x) corresponding to the operator L. This process is defined by equations (1.2).

We will say that a point $x_0 \epsilon \partial D$ is *regular for the process* (X_t, P_x) (*for the operator* L) in the domain D, if

$$\lim_{x \to x_0} P_x\{X_{\tau_D} \epsilon U_\delta(x_0) \cap \partial D\} = 1$$

for any $\delta > 0$, where $U_\delta(x_0)$ is the δ-neighborhood of the point x_0 and $\tau_D = \inf\{t : X_t \epsilon \partial D\}$.

Note that in the case of a bounded domain, $P_x\{\tau_D < \infty\} = 1$ for $x \epsilon D$.

LEMMA 3.1. *The function* $u(x) = E_x \psi(X_{\tau_D})$ *is the generalized solution in the Wiener sense of problem (3.2). A point* $x_0 \epsilon \partial D$ *is* L*-regular, if and only if it is regular for the process* (X_t, P_x) *in the domain* D.

Proof. By virtue of Theorem 2.1 and Remark 4, the function $u_n(x)$ which is a solution of problem (3.1) can be written in the form: $u_n(x) = E_x \psi(X_{\tau_n})$, where $\tau_n = \inf\{t : X_t \epsilon \partial D_n\}$. Since the trajectories X_t are continuous with probability one and $\lim_{n\to\infty} \tau_n = \tau_D$, $\psi(X_{\tau_n}) \to \psi(X_{\tau_D})$ as $n \to \infty$ and by Lebesgue's Dominated Convergence Theorem, $\lim_{n\to\infty} u_n(x) = u(x)$ exists and $u(x) = E_x \psi(X_{\tau_D})$.

Now we shall prove the equivalence of the regularity definitions. Let $\psi(x)$ be a continuous function which is equal to 1 for $x \epsilon U_{\delta/2}(x_0) \cap \partial D$, equal to 0 outside $U_\delta(x_0)$, and which takes values between 0 and 1 on the rest of the boundary. The L-regularity of the point x_0 implies that the Wiener generalized solution $u(x) = E_x \psi(X_{\tau_D})$ of the Dirichlet problem coincides with the boundary function at the point x_0. Therefore,

$$\lim_{x \to x_0} E_x \psi(X_{\tau_D}) = 1 .$$

Since $P_x\{X_{\tau_D} \in U_\delta(x_0) \cap \partial D\} > E_x \psi(X_{\tau_D})$, we have

$$\lim_{x \to x_0} P_x\{X_{\tau_D} \in U_\delta(x_0) \cap \partial D\} = 1$$

and thus the point x_0 is regular for the process $(X_t; P_x)$.

Conversely, if $x_0 \in \partial D$ is a regular point for the process (X_t, P_x), then for any continuous boundary function $\psi(x)$, the function $u(x) = E_x \psi(X_{\tau_D})$ takes the boundary value at x_0:

$$|u(x) - \psi(x_0)| = |E_x \psi(X_{\tau_D}) - \psi(x_0)| \leq$$

$$\leq P_x\{X_{\tau_D} \notin U_\delta(x_0) \cap \partial D\} \max_{x \in \partial D} |\psi(x)| +$$

$$+ P_x\{X_{\tau_D} \in U_\delta(x_0) \cap \partial D\} \max_{x \in \partial D, |x - x_0| < \delta} |\psi(x) - \psi(x_0)| \to 0$$

provided that first we let $x \to x_0$, and then $\delta \downarrow 0$. Since $u(x) = E_x \psi(X_{\tau_D})$ is the Wiener generalized solution of problem (3.2), this implies the L-regularity of the point $x_0 \in \partial D$. □

To check the regularity of the point $x_0 \in \partial D$ for the process (X_t, P_x), we must verify, in accordance with the definition, that the trajectories starting from a point $x \in D$ close to x_0, hit ∂D near the point x_0. It turns out that it is possible to determine, whether the point $x_0 \in \partial D$ is regular or not by observing only the trajectories starting from the point x_0.

We shall denote by τ_D^+ the first exit time after zero from the domain $D: \tau_D^+ = \inf\{t > 0 : X_t \notin D\}$. It has been established (see, e.g. Dynkin and Jushkevich [1]) that a point $x_0 \in \partial D$ is regular if and only if $P_{x_0}\{\tau_D^+ = 0\} = 1$. From this, one can see that a point x_0 is irregular, if and only if the exterior to the domain D near x_0 is a very "thin" set. For instance, if D is a disk with deleted center, then this center is an irregular point: there are no other points of the complement of the set D near this point. One can provide more interesting examples of irregular points in 3-dimensional space. For example, if the boundary of the domain D near a point x_0 is a surface of revolution of the curve $y = e^{-\frac{1}{x-x_0}}$, $x > x_0$, around the x-axis (see Fig. 2), then it is possible to prove that the Wiener

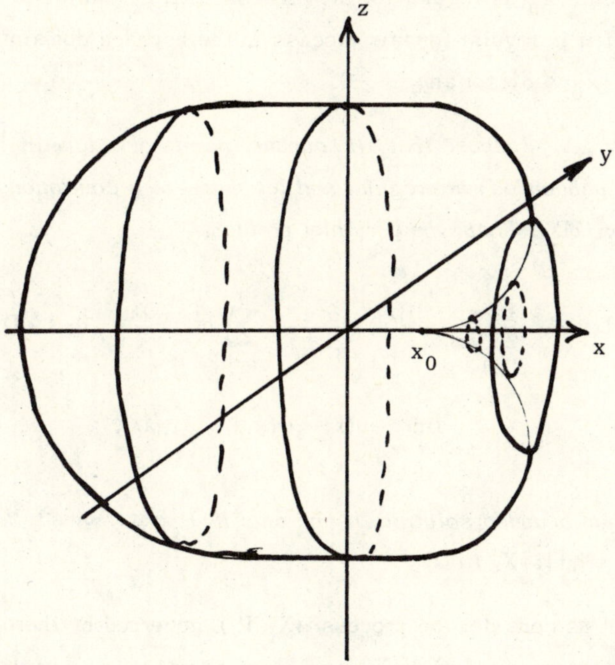

Fig. 2

trajectory in R^3 starting from x_0 does not leave D immediately, and thereby the point x_0 is irregular. We emphasize that this section concerns non-degenerate operators. The regularity of boundary points in the degenerate case is treated in Chapter 3.

In order to exclude infinite exit times from the domain, we assumed the domain to be bounded. If the domain is unbounded, then τ_D may, in general, be equal to $+\infty$ with positive probability. Therefore, in this case, we must give a slightly different definition.

A point $x_0 \in D$ is called *regular for the process* (X_t, P_x) (or *for the corresponding operator* L) in the domain D, provided

$$\lim_{x \in D, x \to x_0} P_x\{X_{\tau_D} \in U_\delta(x_0) \cap \partial D, \tau_D < \infty\} = 1$$

for any $\delta > 0$.

If one takes into account the fact that the operator L is non-degenerate, and its coefficients are bounded, then it is easily established that the point x_0 is regular for the process (X_t, P_x) in the domain D if and only if it is regular for this process in the bounded domain $D_\rho = D \cap \{x : |x - x_0| < \rho\}$ for any $\rho > 0$.

THEOREM 3.1. *Suppose that all boundary points of a domain* D *(generally speaking, unbounded) are regular and let* $\psi(x)$ *be a continuous bounded function on* ∂D. *Then, the Dirichlet problem*

$$L\,u(x) = \frac{1}{2} \sum_{i,j=1}^{r} a^{ij}(x) \frac{\partial^2 u}{\partial x^i \partial x^j} + \sum_{i=1}^{r} b^i(x) \frac{\partial u}{\partial x^i} = 0, \; x \in D,$$

$$\lim_{x \in D, x \to x_0} u(x) = \psi(x_0), \; x_0 \in \partial D\,, \tag{3.3}$$

has a unique bounded solution, if and only if $P_x\{\tau_D < \infty\} = 1$ *for all* $x \in D$, *where* $\tau_D = \inf\{t : X_t \in \partial D\}$.

Proof. Let us consider the process (X_t, P_x) governed by the operator L and the function $u(x) = E_x \psi(X_{\tau_D}) \chi_{\tau_D < \infty}$, where $\chi_{\tau_D < \infty}$ is the indicator of the set $\{\tau_D < \infty\} \subset C_{0,\infty}(R^r)$. We shall show that this function is a solution of problem (3.3). To begin with, let us check that it obeys the boundary conditions. In fact, since the point $x_0 \in \partial D$ is regular for the process (X_t, P_x), we obtain that

$$|u(x)-\psi(x_0)| = |E_x \psi(X_{\tau_D}) \chi_{\tau_D<\infty} - \psi(x_0)| \le$$

$$\le P_x\{|X_{\tau_D} - x_0| > \delta, \tau_D < \infty\} \cdot \sup_{x\epsilon\partial D} |\psi(x)| + P_x\{\tau_D = \infty\} \cdot \sup_{x\epsilon\partial D} |\psi(x)| +$$

$$+ P_x\{X_{\tau_D} \epsilon U_\delta(x_0) \cap \partial D, \tau_D < \infty\} \cdot \max_{x\epsilon\partial D, |x-x_0|<\delta} |\psi(x)-\psi(x_0)| \to 0$$

provided that first we let $x \to x_0$ and then $\delta \downarrow 0$.

Let us show that $u(x)$ is differentiable the required number of times for $x \epsilon D$ and satisfies the equation $L u(x) = 0$ for $x \epsilon D$. Suppose that G is a ball with the center at $x \epsilon D$ of such a small radius that it lies entirely in D, and let $\tau_G = \inf\{t : X_t \epsilon \partial G\}$ be the first exit time of the process (X_t, P_x) to the boundary of the ball G. The fact that the operator L is non-degenerate yields: $E_x \tau_G < \infty$ for $x \epsilon G$. Since (X_t, P_x) is a strong Markov process, we have for $x \epsilon G$

$$u(x) = E_x \psi(X_{\tau_D}) \chi_{\tau_D<\infty} = E_x E_{X_{\tau_G}} \psi(X_{\tau_D}) \chi_{\tau_D<\infty} =$$

$$= E_x u(X_{\tau_G}) . \tag{3.4}$$

It follows from Remark 4 of Theorem 2.1 that, under the assumptions of this section, a Poisson kernel $\pi(x,y)$, $x \epsilon G$, $y \epsilon \partial G$, exists such that $u(x) = \int_{\partial G} \pi(x,y) u(y) dy$. Since $\pi(x,y)$ is smooth in x and the function $u(y)$ is bounded, we conclude that $u(x)$ has continuous first- and second-order derivatives at the point x. These derivatives are bounded in any subdomain of D having a positive distance from the boundary of the domain D. On the basis of Remark 4, (3.4) implies that the function $u(x)$ satisfies the equation $L u(x) = 0$ inside the ball G, which establishes the existence of a bounded solution of problem (3.3).

Now we will prove that if $P_x\{\tau_D < \infty\} = 1$ for all $x \epsilon D$, then the solution of problem (3.3) is unique in the class of all bounded functions. Obviously, for this it is sufficient to demonstrate that every bounded solution $V(x)$ vanishing on ∂D, is identically equal to zero.

Suppose that $D_N = D \cap \{x \in R^r : |x| < N\}$ and let $\tau_N = \min\{t : X_t \in \partial D_N\}$ be the first hitting time to the boundary of the domain D_N. In accordance with Theorem 2.1 and Remark 4, for $x \in D_N$,

$$V(x) = E_x V(X_{\tau_N}) = E_x V(X_{\tau_N}) \chi_{\tau_N < \tau_D}, \tag{3.5}$$

where $\chi_{\tau_N < \tau_D}$ is the indicator of the set $\{\tau_N < \tau_D\}$. Since the coefficients of equation (1.2) are bounded and $P_x\{\tau_D > t\} \to 0$ as $t \to \infty$ for all $x \in D$, we conclude that $P_x\{\tau_N < \tau_D\} \to 0$ as $N \to \infty$. Whence, by passing in equality (3.5) to the limit as $N \to \infty$, we get: $V(x) = 0$ for all $x \in D$.

To complete the proof of Theorem 3.1, it remains only to check that the solution of problem (3.3) is non-unique, whenever $P_x\{\tau_D < \infty\} \neq 1$ for some $x \in D$. By the regularity of the boundary points, the function $W(x) = P_x\{\tau_D = \infty\}$ takes zero boundary values on ∂D. By the strong Markov property of the process (X_t, P_x) with respect to the Markov time τ_G, it follows that

$$W(x) = E_x W(X_{\tau_G}) \text{ for } x \in G .$$

Here τ_G denotes the first exit time to the boundary of the ball $G \subset D$. As was explained previously, this relation implies that $LW(x)$ makes sense and $LW(x) = 0$ for $x \in D$. Consequently, for $\psi(x) \equiv 0$, in addition to the solution $\widetilde{W}(x) \equiv 0$, we also have the solution $W(x) = P_x\{\tau_D = \infty\} \not\equiv 0$. This finishes the proof of Theorem 3.1. □

Hence, if $P_x\{\tau_D < \infty\} = 1$, $x \in D$, and all boundary points of the domain are regular, we conclude that, to every continuous bounded function $\psi(x)$ on ∂D, there corresponds a unique solution of the equation $Lu(x) = 0$, $x \in D$; namely, $u(x) = E_x \psi(X_{\tau_D})$. Of course, bounded solutions of this equation exist to which no continuous function on ∂D corresponds: for example, the function $u(x) = P_x\{X_{\tau_D} \in U_\delta(x_0)\}$, where $x_0 \in \partial D$, and $\delta > 0$ is small enough. How can the set of all bounded solutions of the equation $Lu = 0$ in D be described? A wider set, namely, the set of all non-negative solutions of the equation $Lu = 0$, proves, generally speaking, to be more suitable for describing. The set of such solutions forms a convex cone, and every point of this cone can be represented in the form of a superposition of extreme elements of the cone. It often turns out that these extreme elements which admit no non-trivial expansion into other extreme elements, are unbounded positive solutions of the equation $Lu = 0$. Because of this, it is convenient to consider from the beginning the set of all (and not only bounded) non-negative solutions of this equation.

Suppose, for example, that D is a ball in R^r, and let $\pi(x,y)$, $x \epsilon D$, $y \epsilon \partial D$, be the Poisson kernel for the operator L in the ball D. With every boundary point y of this ball, one can associate a non-negative function $\pi_y(x) = \pi(x,y)$ which satisfies the equation $L\pi_y(x) = 0$, $y \epsilon D$, vanishes on ∂D, except at the point y, and tends to $+\infty$ as $x \to y$. It is possible to prove that the $\pi_y(x)$ are indecomposible minimal functions (the minimality means that if $Lu(x) = 0$, $0 < u(x) < \pi_y(x)$, for $x \epsilon D$, then a $c > 0$ exists such that $u(x) = c\pi_y(x)$) and every non-negative solution $u(x)$ of the equation $Lu = 0$ in the ball can be represented as a convex linear combination of the functions $\pi_y(x)$, $y \epsilon \partial D$,

$$u(x) = \int_{\partial D} \pi_y(x)\,\mu(dy) . \tag{3.6}$$

Thus in the case when the domain D is a ball, every non-negative solution of the equation $Lu(x) = 0$ is defined by a measure on the boundary of the ball. If D is a domain with a smooth boundary ∂D which is at the same time the boundary of the complement of the closure, then the situation is similar: every non-negative solution of the equation $Lu(x) = 0$ in such a domain can be represented in the form (3.6), where the integral is taken over the natural boundary of the domain D. However, for domains of general type, this is not the case. Let, for instance, D be a unit disk in R^2 with a cut along some radius (Fig. 3). It is easily seen that in this case the integration must be carried out over the set $\partial^* D$

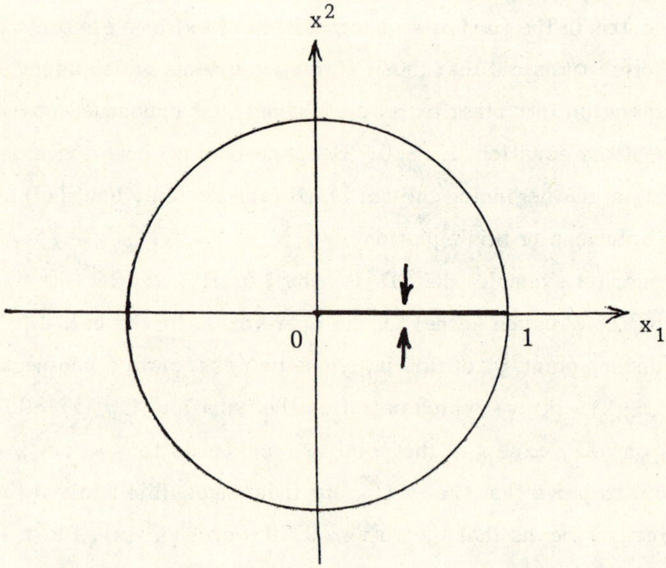

Fig. 3

consisting of the circumference and the two "banks" of the interval along which the cut has been made. Hence, here one has to split some points of the natural boundary of the domain D.

Martin [1] has suggested a construction with the help of which such a splitting for every domain D can be performed. Doob [2], [3] clarified the probabilistic meaning of Martin's boundary. After these works, generalizations to the case of Markov chains and to the case of general Markov process appeared. A different boundary construction has been proposed by Feller. We will not dwell on these general constructions (see, e.g. Ito and McKean [1], Dynkin [4]). In the following section we shall consider the problem of describing the set of all bounded solutions of the equation $L u(x) = 0$, $x \in D$, taking prescribed values on the boundary ∂D of a given domain D. By Theorem 3.1, this problem is, certainly, of interest only in case $P_x\{\tau_D < \infty\} \neq 1$. In the problem which will be considered, only a finite number of points is added to the natural boundary and so we may restrict ourselves to considering only bounded solutions.

§2.4 *Dirichlet's problem in unbounded domains*

So, for Dirichlet's problem (3.3) to have a unique solution, it is necessary that $P_x\{\tau_D < \infty\} = 1$ for $x \in D$. When is this condition fulfilled? If there is a straight line ℓ such that the projection of the domain D onto ℓ is bounded, then it will follow from Lemma 3.3.1 that $P_x\{\tau_D < \infty\} = 1$. So, in this case, there is uniqueness in the class of bounded functions.

Let us consider the question of uniqueness of the solution of the exterior Dirichlet problem, i.e. the case when $D \cup \partial D$ is the exterior to a bounded domain $D' \subset R^r$. As before, let the boundary ∂D of the domain D be regular and the operator L be uniformly elliptic.

A Markov process (X_t, P_x) in R^r is called *recurrent,* if $P_x\{\tau_G < \infty\} = 1$ for every domain $G \subset R^r$ and arbitrary $x \in R^r$, where $\tau_G = \inf\{t : X_t \in G\}$. If in addition $E_x \tau_G < \infty$ for all $x \in R^r$ and every domain $G \subset R^r$, then the process (X_t, P_x) is termed *positively recurrent.* If still $E_x \tau_G = \infty$ for some domain $G \subset R^r$ and some $x \in R^r$, then the recurrent process (X_t, P_x) is called a *null recurrent process.* A non-recurrent process is called *transient.*

Clearly, the process (X_t, P_x) is recurrent if and only if problem (3.3) has a unique bounded solution in the exterior to any bounded domain D′ which has smooth boundary $\partial D' = \partial(R^r \setminus (D' \cup \partial D'))$.

LEMMA 4.1. *Let* (X_t, P_x) *be a Markov process in* R^r *governed by a uniformly elliptic operator* L *with bounded coefficients. Then the following hypotheses are equivalent:*

1. *The process* (X_t, P_x) *is recurrent.*

2. *A ball* $G_0 \subset R^r$ *exists such that* $P_{x_0}\{\tau_{G_0} < \infty\} = 1$ *for some* $x_0 \notin G_0 \cup \partial G_0$.

3. *For any* $x \in R^r$, *any open set* G *and any* $s > 0$, $P_x\{\inf\{t > s : X_t \in G\} < \infty\} = 1$.

4. *There exist a bounded domain* $G_0 \subset R^r$, *a point* $x_0 \in R^r$, *and a number* $\alpha > 0$ *such that, for any* $s > 0$, $P_{x_0}\{\inf\{t > s : X_t \in G_0\} < \infty\} > \alpha$.

If the process (X_t, P_x) *is transient, then* $\lim_{t\to\infty} |X_t| = \infty$ P_x*-a.s. for any* $x \in R^r$.

Proof. First we demonstrate the equivalence of hypotheses 1-4. Let us show that $2 \Rightarrow 1$. Let Γ be a ball such that $G_0 \subset \Gamma$, $\tau_\Gamma = \inf\{t : X_t \in \partial\Gamma\}$. We can apply the strong Markov property to get

$$1 = P_{x_0}\{\tau_{G_0} < \infty\} =$$

$$E_{x_0}\chi_{\tau_\Gamma < \tau_{G_0} < \infty} P_{x_{\tau_\Gamma}}\{\tau_{G_0} < \infty\} + P_{x_0}\{\tau_{G_0} < \tau_\Gamma < \infty\}.$$

It is possible to show that the distribution of X_{τ_Γ} on $\partial\Gamma$ has a nowhere vanishing density. This together with the fact that $P_y\{\tau_{G_0} < \infty\}$ is continuous implies that

$$P_y\{\tau_{G_0} < \infty\} = 1, \ \forall y \in \partial\Gamma.$$

Since Γ was an arbitrary ball containing G_0 we see that

$$P_y\{\tau_{G_0} < \infty\} = 1, \ \forall y \in R^r.$$

From this it is readily deduced that the trajectory X_t will arrive at G_0 an infinite number of times P_x-a.s. for all $x \in R^r$, and what is more, the arrivals at G_0 will happen after arbitrarily large time s (P_x-a.s.).

Let us now show that $P_x\{\tau_K < \infty\} = 1$ for any ball K and $x \in R^r$. In fact, let Π be a ball containing G_0 and K. Let $\alpha = \inf_{x \in G_0} P_x\{\tau_{\partial\Pi} > \tau_K\}$. The function under the infimum is continuous and positive for $x \in G_0 \cup \partial G_0$. Therefore, $\alpha > 0$. Since the process X_t arrives at G_0 an infinite number of times, it will perform an infinite number of transitions from G_0 to the exterior to the ball Π. By the strong Markov property, during every such a transition, the trajectory of the process will arrive at the set K with a probability not less than α. This implies that $P_x\{\tau_K < \infty\} = 1$, $x \in R^r$.

Thus our argument results in $2 \Longrightarrow 1 \Longrightarrow 3$. The implications $1 \Longrightarrow 2$, $3 \Longrightarrow 4$, and $3 \Longrightarrow 1$ are straightforward. We shall show that $4 \Longrightarrow 2$.

Without any loss of generality, one can think of the domain G_0 (which is mentioned in property 4) as being a ball and $x_0 \in G_0$. Let us consider another ball Π which contains $G_0 \cup \partial G_0$. Let $\tau_{\partial\Pi} = \inf\{t : X_t \in \partial\Pi\}$, $\tau_{\partial G_0} = \inf\{t : X_t \in \partial G_0\}$. We denote by $\pi(x,y)$, $x \in \Pi$, $y \in \partial\Pi$, the Poisson kernel for the operator L in the domain Π. We put $V(x) = P_x\{$the trajectory X_t performs an infinite number of transitions from G_0 to $\partial\Pi\}$, $A = \sup_{x \in G_0 \cup \partial G_0} V(x) \geq V(x_0) > a > 0$. The strong Markov property yields

$$V(x) = \int_{\partial\Pi} \int_{\partial G_0} \pi(x,y) P_y\{X_{\tau_{\partial G_0}} \in \partial z\} V(z)\, dy\,, \quad x \in \Pi\,.$$

This shows that $A \leq A \sup_{y \in \partial\Pi} P_y\{\tau_{\partial G_0} < \infty\}$. Since $A > a > 0$, noting that the function $P_y\{\tau_{\partial G_0} < \infty\}$ is continuous, we infer that $P_y\{\tau_{\partial G_0} < \infty\} = 1$ for some $y \in \partial\Pi$. Therefore, property 2 holds.

The last assertion of Lemma 4.1 is straightforward from the fact that property 4 is equivalent to the process being recurrent. □

THEOREM 4.1. *Suppose that there exists a non-negative function* $V(x)$ *defined in* R^r *for* $|x| > r_1 > 0$ *and having continuous second-order derivatives such that* $LV(x) \leq 0$ *for* $|x| > r_1$, $\lim_{|x|\to\infty} V(x) = +\infty$. *Then the process* (X_t, P_x) *corresponding to the operator* L *is recurrent. If the function* $LV(x)$ *satisfies the stronger inequality* $LV(x) \leq -a < 0$ *for* $|x| > r_1$, *then the process* (X_t, P_x) *is positively recurrent.*

Proof. Together with the process (X_t, P_x) governed by the operator L, we shall consider the corresponding Markov family (X_t^x, P). Choose a small number $\lambda > 0$ and an $r_2 > r_1$ large enough so that $V(x) > \frac{1}{\lambda}$ for $|x| \geq r_2$. Put $\tau_i^x = \inf\{t : |X_t^x| = r_i\}$, $\bar{\tau}^x = \tau_1^x \wedge \tau_2^x$ and let x be such that $r_1 < |x| < r_2$. Since L is uniformly elliptic, it follows that $E\bar{\tau}^x < \infty$. Hence we can apply Theorem 2.1 to get

$$V(x) = -E \int_0^{\bar{\tau}^x} LV(X_t^x)\,dt + E\,V(X_{\bar{\tau}^x}^x) \tag{4.1}$$

and since $LV \leq 0$ we have

$$\begin{aligned} V(x) &\geq E\,V(X_{\bar{\tau}^x}^x) \geq P\{\bar{\tau}^x = \tau_1^x\} \min_{|x|=r_1} V(x) + \\ &+ \lambda^{-1} P\{\bar{\tau}^x = \tau_2^x\} \geq \frac{1}{\lambda} P\{\tau_1^x > \tau_2^x\}. \end{aligned} \tag{4.2}$$

Here we have used the fact that $\tau_2^x < \infty$ P-a.s., $r_1 < |x| < r_2$. From (4.2) we get:

$$P\{\tau_1^x < \infty\} \geq P\{\tau_1^x < \tau_2^x\} \geq 1 - \lambda V(x).$$

Since λ is arbitrary, this yields $P\{\tau_1^x < \infty\} = 1$ and therefore, the trajectories starting from x reach the smaller of the two spheres with probability 1. On the basis of Lemma 4.1 we conclude from this that the process (X_t, P_x) is recurrent.

If the inequality $LV(x) < -a < 0$ is valid, then (4.1) yields:

$$0 \leq V(x) - aE\bar{\tau}^x.$$

Consequently

$$E(\tau_1^x \wedge \tau_2^x) \leq a^{-1} V(x).$$

The expression under the expectation increases monotonically as $r_2 \to \infty$ tends to τ_1. Therefore

$$E\tau_1^x \leq a^{-1} V(x). \tag{4.3}$$

To complete the proof of Theorem 4.1, we must also show that $E_x \tau_D < \infty$, $\tau_0 = \inf\{t : X_t \in D\}$ for the domains D lying inside the sphere $|x| = r_1$. Let us designate $\sup_{|x|=r_1} E_x \tau_2 = A$, $\sup_{|x|=r_2} E_x \tau_1 = B$, $\inf_{|x|=r_1} P_x\{\tau_D < \tau_2\} = \beta$; $A, B < \infty$, $\beta > 0$. Using the strong Markov property, for $|x| = r_1$, we obtain

$$E_x \tau_D = E_x \tau_D \chi_{\tau_D < \tau_2} + E_x \tau_D \chi_{\tau_D > \tau_2} \leq E_x \tau_2 +$$

$$+ E_x \chi_{\tau_D > \tau_2} (\tau_2 + E_{X_{\tau_2}} (\tau_1 + E_{X_{\tau_1}} \tau_D)) \leq 2A + B +$$

$$+ (1 - \beta) \sup_{|x|=r_1} E_x \tau_D .$$

This results in $\sup_{|x|=r_1} E_x \tau_D \leq \frac{2A+B}{\beta} < \infty$. Together with the bound (4.3), this yields that $E_x \tau_D < \frac{1}{\alpha} V(x) + (2A + B)\beta^{-1}$ for any $x \in R^r$. □

THEOREM 4.2. *Given a domain* D *in* R^r, *assume that in* $R^r \setminus D$ *a continuous function* $w(x) \geq 0$ *exists possessing second-order continuous derivatives in* $R^r \setminus D$ *and such that* $Lw(x) \leq 0$, $\inf_{x \in \partial D} w(x) > \alpha > 0$ *and* $w(x_0) < \alpha$ *for some point* $x_0 \in R^r \setminus D$. *Then the process* (X_t, P_x) *corresponding to the operator* L *is transient.*

Proof. Let τ_D be the first hitting time to the domain D, $\tau_N = \inf\{t : |X_t| = N\}$, $\tau_{N,D} = \tau_D \wedge \tau_N$. Again we apply Theorem 2.1 to get that

$$E_{x_0} w(X_{\tau_{N,D}}) \leq w(x_0) < \alpha .$$

This implies that

$$\alpha > E_{x_0} w(X_{\tau_{N,D}}) \geq \inf_{y \in \partial D} w(y) \cdot P_{x_0}\{\tau_D < \tau_N\} .$$

Passing in this inequality to limit as $N \to \infty$ yields

$$\alpha \geq \inf_{y \in \partial D} w(y) P_{x_0}\{\tau_D < \infty\} > \alpha P_{x_0}\{\tau_D < \infty\} .$$

Thus $P_{x_0}\{\tau_D < \infty\} < 1$ and the process (X_t, P_x) is transient. □

Now we will apply Theorems 4.1 and 4.2 to some examples. Let (X_t, P_x) be the one-dimensional process generated by the operator $L = \frac{1}{2} a(x) \frac{d^2}{dx^2} + b(x) \frac{d}{dx}$. We set

$$U(x) = \int_0^x 2b(y) a^{-1}(y) dy , \quad V(x) = \int_0^x \exp\{-U(y)\} dy .$$

It is easily seen that $L V(x) = 0$. Suppose that $\lim_{|x| \to \infty} |V(x)| = \infty$. Then the function $|V(x)|$ satisfies the conditions of Theorem 4.1 for $|x| > 1$. Therefore, the condition $\lim_{|x| \to \infty} |V(x)| = \infty$ is sufficient for the process (X_t, P_x) to be recurrent.

If the function $|V(x)|$ has a finite limit as $x \to +\infty$ or as $x \to -\infty$ ($V(x)$ is monotone, so the limits exist), then the functions $w_+(x) = -V(x) + \lim_{x \to \infty} V(x)$ or $w_-(x) = V(x) - \lim_{x \to -\infty} V(x)$ obey the corresponding conditions of Theorem 4.2. Thus in this case, the process (X_t, P_x) is transient. Hence, the condition $\lim_{|x| \to \infty} |V(x)| = \infty$ is necessary and sufficient for a one-dimensional process to be recurrent. We note that in the one-dimensional case this assertion can be also deduced from formula (2.11) without using Theorems 4.1 and 4.2.

In the multi-dimensional case, the function $V(x)$ which is mentioned in Theorem 4.1 may sometimes be chosen in the form $V(x) = \ln (Cx,x)$, where C is a positive definite matrix. This function is positive outside some ball and the inequality $L V(x) \leq 0$ holds, provided

$$(Cx,x) [(Cx,b(x)) + \mathrm{Tr}\, a(x)C] \leq 2(a(x)Cx,Cx) , \qquad (4.4)$$

where $a(x) = (a^{ij}(x))$, $b(x) = (b^1(x), \cdots, b^r(x))$. Choosing as C the unit matrix, we get that the process (X_t, P_x) is recurrent whenever $(x, b(x)) \leq -\operatorname{Tr} a(x) + \varepsilon)$. In particular, a process is recurrent if $(x, b(x)) < -\beta|x|, \beta > 0$, outside some sphere. It is not hard to check that this process is positively recurrent.

If $L = \frac{1}{2}\Delta + \sum_{i=1}^{r} b^i(x) \frac{\partial}{\partial x^i}$, then choosing as C the unit matrix, we derive from (4.4) the following recurrence condition: $(x, b(x)) \leq 2 - r$ outside some sphere. In particular, this implies that the Wiener process is recurrent for $r = 1, 2$.

To verify the transience via Theorem 4.2, it is convenient to select the function $w(x)$ in the form $w(x) = (Cx, x)^{-\alpha}$, $\alpha > 0$. This function is positive. For the inequality $L w(x) \leq 0$ to be fulfilled, it is necessary that

$$(Cx, x)\,[(Cx, b(x)) + \operatorname{Tr} a(x) C] \geq 2(1+\alpha)\,(a(x)\,Cx, Cx)\,. \tag{4.5}$$

This shows that if, outside some ball, the projection of $b(x)$ on the radius-vector is directed away from the origin and $(x, b(x)) > \beta|x|$, then the corresponding process is transient.

In the case when $a(x)$ is the unit matrix, choosing as C the unit matrix, we obtain from (4.5) the following transience condition: $(x, b(x)) \geq 2 - r + \varepsilon$, $\varepsilon > 0$. This implies that the Wiener process is not recurrent for $r \geq 3$. In the formula for $w(x)$, one can put C to be a degenerate non-negative definite matrix. If for example, all elements c_{ij} vanish identically except for c_{ii}, then it follows from (4.5) that the process (X_t, P_x) is transient whenever $x^i b^i(x) \geq (1+\varepsilon)\, a^{ii}(x)$, $\varepsilon > 0$.

Sufficient conditions for recurrence and transience have been examined by Has'minskii [2,6]. Similar questions in the theory of differential equations have been studied by Meyers and Serrin [1].

Thus, in the case when the process (X_t, P_x) corresponding to the operator L is recurrent, the exterior Dirichlet problem

$$L u(x) = 0, x \in R^r \setminus (D \cup \partial D),\ u(x)|_{x \in \partial D} = \psi(x)\,, \tag{4.6}$$

has a unique bounded solution. Here it is assumed that ∂D is also the boundary of the domain $R^r \setminus (D \cup \partial D)$, all boundary points are regular for the process (X_t, P_x) in the domain $R^r \setminus (D \cup \partial D)$, and $\psi(x)$ is a continuous bounded function.

In the case when the process is transient, the situation is more complicated. Let for example, $L = \frac{1}{2}\Delta$ and $r \geq 3$. The corresponding process is the Wiener process and as was observed previously, it is transient. It is known from the theory of differential equations, that to single out a unique solution of problem (4.6), one can assign $\lim_{|x|\to\infty} u(x) = c$. With this condition, problem (4.6) has a solution for any c and this solution is unique in the class of bounded functions. But one cannot assign the supplementary condition in the form $\lim_{|x|\to\infty} u(x) = c$ for other operators corresponding to a transient process. The solution of problem (4.6) with such a supplementary condition may not exist. For example, let $D = \{x \in R^2 : |x^1| < 1, |x^2| < 1\}$. Consider the Dirichlet problem

$$L\,u(x) = \frac{1}{2}\Delta u(x) + \frac{\partial u}{\partial x^1} - \operatorname{arctg} x^2 \cdot \frac{\partial u}{\partial x^2} = 0\,, \quad (x^1, x^2) \notin D\,,$$

$$u(x)\big|_{x \in \partial D} = \psi(x)\,, \tag{4.7}$$

in the domain $R^2 \setminus (D \cup \partial D)$.

The two-dimensional process generated by the operator L has independent components: the process X_t^1 on the x^1-axis is governed by the operator $\frac{1}{2}\frac{d^2}{(\partial x^1)^2} + \frac{d}{dx^1}$, its trajectories starting from an arbitrary initial point tend to $+\infty$; the process X_t^2 on the x^2-axis is governed by the operator $\frac{1}{2}\frac{d^2}{(\partial x^2)^2} - \operatorname{arctg} x^2 \frac{d}{dx^2}$, it is positively recurrent. Since the process X_t^1 is transient, we conclude that the two-dimensional process is surely transient. So the solution of problem (4.7) is non-unique. We shall seek a solution of problem (4.7) satisfying the condition

$$\lim_{x^1 \to \infty} u(x^1, x^2) = c = \text{const.} \tag{4.8}$$

for any $x^2 \in (-\infty, \infty)$. We put $u(x) = E_x \psi(X_{\tau_D}) \chi_{\tau_D < \infty} + c\, P_x\{\tau_D = \infty\}$. Using the strong Markov property, one can show that $u(x)$ is the solution of problem (4.7). This may be done in the same way as in the proof of Theorem 3.1. From the fact that $\lim_{t\to\infty} X_t^1 = +\infty$ P_x a.s., it follows immediately that condition (4.8) holds.

To establish the uniqueness of a bounded solution of problem (4.7)-(4.8), we let $\tau_t = \tau_D \wedge t$. By Ito's formula, the stochastic process $u(X_{\tau_t}) = Y_t$ is a bounded martingale. So, given $\psi(x)\big|_{x \in \partial D} = c = 0$, we get

$$u(x) = \lim_{t\to\infty} E_x u(X_{\tau_t}) = E_x \lim_{t\to\infty} u(X_{\tau_t}) = 0$$

because either $\tau_D < \infty$ and $X_{\tau_D} \in \partial D$ or $\lim_{t\to\infty} X_{\tau_t}^1 = \infty$, P_x a.s., $x \notin D$. Thus, problem (4.7)-(4.8) has a bounded solution for any c and the solution is unique.

But, generally speaking, $\lim_{x^1 \to -\infty} u(x^1, x^2) \neq c$. To show this, we set $c = 0$, $\psi(x) \equiv 1$. Let $\tau_1 = \min\{t : X_t^1 = -1\}$ denote the first hitting time of the process X_t^1 to the line $x^1 = -1$. Since $\lim_{t\to\infty} X_t^1 = +\infty$, P_x a.s., we have $\tau_1 < \infty$ with probability 1 if $x = (x^1, x^2)$ and $x^1 < -1$. It is easy to see that $\lim_{x^1 \to -\infty} P_x\{\tau_1 > t\} = 1$ for any t. It follows from (1.13) that the second component has a limit distribution for large t and $\lim_{t\to\infty} P_x\{X_t^2 \in (a,b)\} = \int_a^b m(y)\,dy$, where $m(y)$ is the density of the invariant measure of the process X_t^2, $m(y) > 0$. As a consequence of the fact that τ_1 and X_t^2 are independent variables $X_{\tau_1}^2$ has a limit distribution with density $m(y)$ for the trajectories starting from the point (x^1, x^2) as

$x^1 \to -\infty$. So $P_x\{|X^2_{\tau_1}| < 1\} \to \int_{-1}^{1} m(y)\,dy = M > 0$ for $x = (x^1, x^2), x^1 \to -\infty$. On account of the inclusion $\{X^1_{\tau_D} = -1\} \supset \{|X^2_{\tau_1}| < 1\}$ for large $-x^1$,

$$u(x^1, x^2) = E_{x^1, x^2}\psi(X_{\tau_D})\chi_{\tau_D < \infty} = P_{x^1, x^2}\{\tau_D < \infty\} >$$

$$> P_{x^1, x^2}\{|X^2_{\tau_1}| < 1\} > M - \varepsilon$$

for any preassigned $\varepsilon > 0$ and, therefore, $\lim_{x^1 \to -\infty} u(x_1, x_2) > 0$.

It is not difficult to prove that the solution of problem (4.7)-(4.8) is unique, provided (4.8) is fulfilled solely for $x^2 = 0$. If equality (4.8) holds for $x^2 = 0$, then it is valid for any x^2.

We will remark that if in equation (4.7) the minus sign preceding arctg x is changed to plus, then the solution of problem (4.7) with the supplementary condition $\lim_{|x| \to \infty} u(x) = c$ exists (and is unique). That this limit exists, is a consequence of the fact that the trajectories of the process (X_t, P_x) corresponding to such a modified operator, which start far enough from the origin, will run to infinity without returning to the bounded part of the plane: $\lim_{|x| \to \infty} P_x\{\tau_D < \infty\} = 0$. It is easy to establish the following simple statement:

THEOREM 4.3. *Let* $D \cup \partial D$ *be the exterior to a bounded domain in* R^r. *The boundary* ∂D *of the domain* D *is assumed to be regular for the operator* L. *Then a solution of the exterior Dirichlet problem*

$$L u(x) = 0, \quad x \in D, \quad u(x)\big|_{\partial D} = \psi(x), \quad \lim_{|x| \to \infty} u(x) = c$$

exists for any continuous function $\psi(x)$ *and any* c, *if and only if*

$$\lim_{|x| \to \infty} P_x\{\tau_{\partial D} < \infty\} = 0, \tag{4.9}$$

where $\tau_{\partial D} = \inf\{t : X_t \in \partial D\}$. *This solution is unique in the class of bounded functions. For (4.9) to hold, it is sufficient that, outside some ball* Π, *a continuous function* $w(x)$, $x \notin \Pi$, *should exist having continuous first- and second-order derivatives such that* $w(x)|_{\partial\Pi} > 0$, $Lw(x) \leq 0$ *for* $x \notin \Pi$, $\lim_{|x|\to\infty} w(x) = 0$.

Proof of this theorem is similar to the above reasoning and so we drop it. □

As $w(x)$ is in the form $w(x) = (Cx,x)^{-\alpha}$, where C is a positive definite matrix, $\alpha > 0$. This function is non-negative and tends toward zero as $|x| \to \infty$.

For the inequality $Lw(x) \leq 0$ to hold, it is necessary that inequality (4.5) be valid. In particular, condition (4.9) is fulfilled, if, outside some sphere, the field $b(x)$ has a positive projection, bounded away from zero, on the radius-vector connecting the origin with the point x. If $b(x) = 0$ outside some sphere and $a(x)$ tends to a constant non-degenerate matrix A as $|x| \to \infty$, then it follows from (4.5) that condition (4.9) holds for $r \geq 3$.

In some cases, for instance, if $L = \Delta$, it is possible to prove that every bounded solution of the exterior Dirichlet problem has a limit as $|x| \to \infty$. Roughly speaking, this means that there is a unique boundary point at infinity. In general, there may exist a sufficiently extensive boundary at infinity. These questions are treated in the works by Freidlin [6], Cranston, Orey and Rosler [1]. We shall cite some results of the first of these works. For brevity, we restrict ourselves to the two-dimensional case and make also some simplifying assumptions.

So let D be the exterior to the closure of a bounded domain in the plane R^2. The boundary ∂D is assumed regular. In particular, the domain D may coincide with the entire plane. We will consider the differential operator

$$L_\alpha u(x) = \frac{\alpha}{2} \sum_{i,j=1}^{2} a^{ij}(x) \frac{\partial^2 u}{\partial x^i \partial x^j} + \sum_{i=1}^{2} b^i(x) \frac{\partial u}{\partial x^i} .$$

Here α is a positive parameter, L_1 is a uniformly elliptic operator with bounded, sufficiently smooth coefficients. To formulate supplementary assumptions on the coefficients $b(x) = (b^1(x), b^2(x))$, let us consider the dynamical system

$$\dot{X}_t(x) = b(X_t(x)), \quad X_0(x) = x \in R^2. \tag{4.10}$$

Let us call system (4.10) a Z-system, if for some R, h, a, $b > 0$, the following hypotheses hold (Fig. 4):

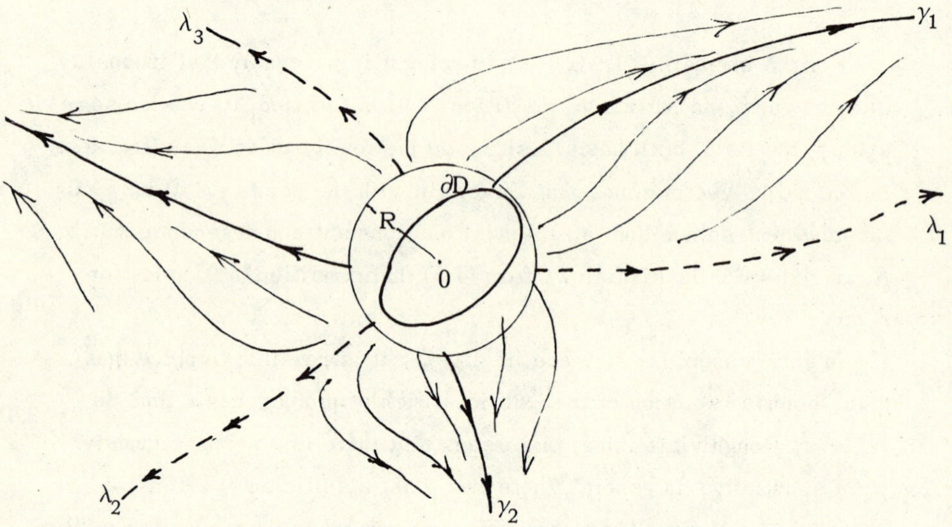

Fig. 4

1. $(\nabla\rho(x), b(x)) > a$ for $\rho(x) \geq R$, where $\rho(x) = |x|$.

2. One can determine integral curves $\gamma_1(t), \cdots, \gamma_\ell(t), \lambda_1(t), \cdots, \lambda_\ell(t)$ starting from the circle $|x| = R$ at $t = 0$ and neighborhoods[7] $\Gamma_1, \cdots, \Gamma_\ell$,

---

[7] γ_i, λ_i (without the argument t) designate the entire semi-trajectories $\gamma_i(t)$, $\lambda_i(t)$, $t \geq 0$ respectively; $\rho(x,A)$ is the distance from a point x to the set A in the plane R^2.

$\Lambda_1, \cdots, \Lambda_\ell$, $\gamma_i \subset \Gamma_i \subset \{x : \rho(x,\gamma_i) < h\}$, $\lambda_i \subset \Lambda_i \subset \{x : \rho(x,\lambda_i) < h\}$, $i = 1, \cdots, \ell$, of them such that the functions $\rho_i(x) = \rho(x,\gamma_i)$, $\overline{\rho}_i(x) = \rho(x,\lambda_i)$ have first- and second-order bounded continuous derivatives outside the closure of Γ_i and Λ_i respectively, and for every $x \notin \{|x| < R\} \cup (\bigcup_1^\ell \Lambda_i)$ one can find a number $\kappa(x) = 1, 2, \cdots, \ell$ such that $\lim_{t\to\infty} \rho_{\kappa(x)}(X_t(x)) = 0$. It is also assumed that $(\nabla\rho_k(x), b(x)) < -b < 0$ for $x \in G_k \setminus (\Gamma_k \cup (\bigcup_{i=1}^\ell \Lambda_i))$, where $G_k = \{x \in R^2 : |x| > R, \kappa(x) = k\}$. The curves λ_k separate the domains of attraction for different γ_k.

3. For any $k = 1, \cdots, \ell$, the set G_k contains a neighborhood of the curve γ_k which expands linearly as $|x|$ gets large: $\{x \in R^2 : |x| > R, \rho_k(x) < \mu\rho(x) - N\} \subset G_k$ for some μ, $N > 0$.

THEOREM 4.4. *Suppose that the dynamical system (4.10) is a Z-system. Then one can find a positive* $\overline{a}$ *such that, for* $a \in (0, \overline{a}]$, *the problem*

$$L_a u(x) = 0, \quad x \in D, \quad u(x)\big|_{\partial D} = \psi(x),$$

$$\lim_{t\to\infty} u(\gamma_k(t)) = a_k, \quad k = 1, 2, \cdots, \ell, \tag{4.11}$$

has a unique bounded solution for any continuous function $\psi(x)$ *on* ∂D *and any* $a_1, \cdots, a_\ell$.

Proof. Let (X_t^a, P_x^a) and $(X_t^{a,x}, P)$ denote the Markov process and the Markov family corresponding to the operator L_a. The trajectories of the family are specified by the stochastic differential equation:

$$dX_t^{a,x} = \sqrt{a}\,\sigma(X_t^{a,x})\,dW_t + b(X_t^{a,x})\,dt,$$

where $\sigma(x)$ is such that $\sigma(x)\,\sigma^*(x) = (a^{ij}(x))$.

For sufficiently small a, with probability 1, the trajectories either hit the boundary ∂D of the domain D or run fast enough to infinity as $t \to \infty$.

In fact, the trajectory $X_t^{\alpha,x}$ either hits ∂D or, from some time $t = t(\omega)$ on, it will no longer arrive at the R-neighborhood of the origin. Unless it arrives at the R-neighborhood of the origin, we apply Ito's formula to the function $\rho(X_t^{\alpha,x})$

$$\rho(X_t^{\alpha,x}) = \rho(x) + \int_0^t (\nabla\rho(X_s^{\alpha,x}), \sigma(X_s^{\alpha,x})\, dW_s + \int_0^t L_\alpha \rho(X_s^{\alpha,x})\, ds\ .$$

Let $\alpha_1 > 0$ be such that $L_\alpha \rho(x) > \frac{\alpha}{2}$ for $\alpha < \alpha_1$, $|x| > R$. Using a random change of time, the stochastic integral can be reduced to the Wiener process: $\int_0^t (\nabla\rho, \sigma dW_s)$ has the same distribution as $\widetilde{W}_{\zeta_\tau}$ has, where $\zeta_t = \frac{\alpha}{2} \int_0^t \sum_1^2 a^{ij}(X_s^{\alpha,x}) \frac{\partial\rho}{\partial x^i} \frac{\partial\rho}{\partial x^j}\, ds$, W_s being a one-dimensional Wiener process. In view of this,

$$\varliminf_{t\to\infty} t^{-1}\rho(X_t^{\alpha,x}) \geq \frac{\alpha}{2} \tag{4.12}$$

for those trajectories which do not hit ∂D.

Similarly, relying on Property 2 of Z-systems, one can establish that if a trajectory $X_t^{\alpha,x}$ never leaves $G_k \setminus ((\bigcup_1^\ell \Lambda_i) \cup \Gamma_k)$ then for $\alpha < \alpha_2$

$$\varlimsup_{t\to\infty} t^{-1}\rho_k(X_t^{\alpha,x}) < -\frac{b}{2}\ .$$

Since the last inequality cannot hold, the trajectories not leaving $G_k \setminus (\bigcup_1^\ell \Lambda_i)$ must enter Γ_k and (noting that the diffusion matrix is non-degenerate) they must intersect the curve γ_k. And what is more, $P\{X_t^{\alpha,x}$ intersects γ_k without leaving $G_k\} > d > 0$ for all $x \in G_k \setminus (\{|x| < R\} \cup (\bigcup_1^\ell \Lambda_i))$. From this, applying the strong Markov property we find that the trajectories $X_t^{\alpha,x}$ either hit ∂D or intersect one of the curves γ_k P-a.s., $x \in D$, for every $\alpha < \alpha_2$. Moreover, the trajectories not hitting ∂D, intersect at least one of the curves γ_k after any preassigned time.

By virtue of Properties 2 and 3 of Z-systems,

$$\lim_{\substack{|x|\to\infty \\ \rho_k(x)<c}} P\{X_t^{\alpha,x} \text{ never leaves } G_k\} = 1 \tag{4.13}$$

for any $c > 0$, provided $\alpha < \alpha_1 \wedge \alpha_2 = \overline{\alpha}$.

In fact, by Property 2, via Ito's formula we see that

$$\rho_k(X_t^{\alpha,x}) = \rho_k(x) + \int_0^t (\nabla\rho_k, \sigma dW_s) + \int_0^t L_\alpha \rho_k ds .$$

Note that the stochastic integral can be written in the form of a Wiener process at a random (but growing no faster than linearly with t) time. From this, using the law of iterated logarithm as $t \to \infty$, we obtain that $\lim_{t\to\infty} t^{-\frac{1+\delta}{2}} \int_0^t (\nabla\rho_k, \sigma dW_s) ds = 0$, $\delta > 0$. Whence, taking into account (4.12) and Properties 2 and 3, one can deduce (4.13).

Hence, for $\alpha < \overline{\alpha} = \alpha_1 \wedge \alpha_2$ with P-a.a. trajectories $X_t^{\alpha,x}$ not hitting D, one can associate a unique number $K(\omega)$ such that from some time on, the trajectories do not leave the set $G_{K(\omega)}$. By the above, these trajectories intersect $\gamma_{K(\omega)}$ infinitely many times, with the intersection times going to infinity.

We will define a random variable $\overline{\psi}(X_\infty^{\alpha,x})$ by setting

$$\overline{\psi}(X_\infty^{\alpha,x}) = \begin{cases} \psi(X_{\tau_D^x}^{\alpha,x}), & \text{if } \tau_D^x = \inf\{t : X_t^{\alpha,x} \in \partial D\} < \infty \\ \\ a_{K(\omega)}, & \text{if } \tau_D^x = \infty . \end{cases}$$

By the preceding, this definition is correct. Let $u(x) = E\overline{\psi}(X_\infty^{\alpha,x})$. We shall show the function $u(x)$ to be a solution of problem (4.11). First of all, by the strong Markov property, for any disk $G \subset D$, the equality

$$u(x) = E\, u(X^{\alpha,x}_{\tau^x_G}), \quad \tau^x_G = \inf\{t : X^{\alpha,x}_t \in \partial G\}, \quad x \in G,$$

holds. From this, just as we have done repeatedly in this chapter, we conclude that $u(x)$ has the required derivatives and $L\,u(x) = 0$. The fulfillment of the boundary conditions on ∂D is due to the boundary points being regular. That the conditions $\lim_{t\to\infty} u(\gamma_k(t)) = a_k$ hold, results from relation (4.13).

Therefore, the function $u(x) = E\bar{\psi}(X^{\alpha,x}_\infty)$ is a solution of problem (4.11).

We now proceed to prove uniqueness. Let $\psi(x) = a_1 = a_2 = \cdots = a_k = 0$. From Ito's formula, it follows that for any $t > 0$,

$$u(x) = E\, u(X^{\alpha,x}_{\tau^x_D \wedge t}),$$

where $u(x)$ is a solution of problem (4.11).

By passing in this equality to the limit as $t \to \infty$ and remembering that P-a.s. the trajectory $X^{\alpha,x}_t$ either hits ∂D or, arbitrarily far, intersects one of the curves γ_k along which $\lim_{t\to\infty} u(\gamma_k(t)) = 0$, we obtain that $u(x) = \lim_{t\to\infty} E\, u(X^{\alpha,x}_t) = 0$. Consequently, the solution of problem (4.11) is unique in the class of bounded functions. □

Under hypotheses similar to those of Theorem 4.4, one can prove that every bounded solution of the equation $L_\alpha u(x) = 0$ in D has limits along the curves $\gamma_k(t)$ as $t \to \infty$. Thereby, the set of all bounded solutions of the exterior Dirichlet problem $L_\alpha u(x) = 0$, $x \in D$, $u(x)|_{x\in\partial D} = \psi(x)$, is found to be an ℓ-dimensional space, and the above constructed boundary at infinity is (in a sense) complete. We shall illustrate the proof of such statements by a simple example. Let the curve γ_1 coincide with the positive semi-axis x^1 and suppose that equation (4.11) in the angle $\{x^1 > 0, |x^2| < \beta x^1\}$ can be written in the form:

$$\frac{1}{2}\Delta u + b^1(x^1)\frac{\partial u}{\partial x^1} + b^2(x^2)\frac{\partial u}{\partial x^2} = 0 . \qquad (4.14)$$

Moreover, let in this angle, $0 < \underline{b}^1 < b^1(x) < \overline{b}^1$, $b^2(x^2) > H > 0$ for $x^2 < -h < 0$ and $b^2(x^2) < -H$ for $x^2 > h$. We shall show that every bounded solution of equation (4.14) in R^2 has a limit along the x^1-axis. Let $\underline{\lim}_{x^1\to\infty} u(x^1,0) = u^-$, $\overline{\lim}_{x^1\to\infty} u(x^1,0) = u^+$, $0 < \varepsilon < \frac{u^+-u^-}{2}$. By a priori bounds of the modulus of continuity for bounded solutions of elliptic equation (4.14), the function $u(x)$ is uniformly continuous in R^2. Thus if a_k^+ is a sequence of points on the x^1-axis running to infinity, along which the function $u(x)$ attains its upper limit, then, for sufficiently large n_0 and δ^{-1}, we have $|u(x)-u^+| < \varepsilon$ for $x \in \bigcup_{k>n_0} \{x : |x-a_k^+| < \delta\}$. Let $x = (x^1,x^2)$. We put $\tau_n^{x^1} = \inf\{t : X_t^{1,x^1} = a_n^+\}$, where X_t^{1,x^1} is the first component of the trajectory $X_t^x = (X_t^{1,x}, X_t^{2,x})$. The pair (X_t^{1,x^1}, P) itself forms a one-dimensional Markov family, just as (X_t^{2,x^2}, P) does. Since the component X_t^{2,x^2} is a positively recurrent Markov process and the variables $\tau_n^{x^1}$ are defined by the component X_t^{1,x^1} independent of X_t^{2,x^2}, it is possible to choose a, x_0^1, $\mu > 0$ in such a way that for all $x^1 > x_0^1$ and n such that $a_n^+ > x^1$, the bound

$$P\{|X^{2,0}_{\tau_n^{x^1}}| < a\} > \mu$$

is valid. Since the operator is uniformly elliptic and the coefficients are bounded, this inequality implies that for any $\delta > 0$, one can find a $\kappa > 0$ for which

$$P\{\sigma^x < \infty\} > \kappa \qquad (4.15)$$

where σ^x is the first hitting time of the set $\{x \in R^2 : |x-a_n^+| < \delta\}$ and $a_n^+ > x^1 > x_0^1$, $x = (x^1, 0)$. From (4.15) it follows that the trajectories not leaving the angle $\{x^1 > 0, |x^2| < \beta x^1\}$ arrive at the set $\bigcup_{n>n_1} \{x : |x-a_n^+| < \delta\}$

for any n_1 and $\delta > 0$. Therefore, if the initial point $x = (x^1, 0)$ is chosen far enough from the origin, then, with a probability larger than $\frac{3}{4}$, the trajectories will arrive at the set $A^+ = \{x : |u(x) - u^+| < \varepsilon\}$ for arbitrarily large t. One can prove in a similar way that, for arbitrarily large t, the trajectories starting far enough from the origin, will arrive at the set $A^- = \{x : |u(x) - u^-| < \varepsilon\}$ with a probability larger than $\frac{3}{4}$. Consequently, with a probability larger than $\frac{1}{2}$, for arbitrarily large t, the trajectories will arrive both at A^+ and at A^-. If $u^+ \neq u^-$, this contradicts the existence of $\lim_{t\to\infty} u(X_t^x)$. The last limit has to exist, since $u(X_t^x)$ is a bounded martingale. So, $u^+ = u^- = \lim_{x^1\to\infty} u(x^1, 0)$.

Therefore, the statement of the exterior Dirichlet problem is completely determined by the behavior of the trajectories of the corresponding diffusion process. Supplementary conditions at infinity must be set so that with them one might define the value of $\lim_{t\to\infty} u(X_t^x)$ for those trajectories which do not hit ∂D. In doing so one may prescribe different values of $\lim_{t\to\infty} u(X_t^x)$ along the trajectories running to infinity in different ways.

Concluding this section we formulate some results on ergodic properties of diffusion processes.

THEOREM 4.5. *Suppose that the process* (X_t, P_x) *in the state space* $(R^r, \mathcal{B}^r)$ *corresponding to a uniformly elliptic operator* L *is positively recurrent. Let* $a^{ij}(x) \in C^{2,\delta}_{R^r}$, $b^i(x) \in C^{1,\delta}_{R^r}$. *Then a unique stationary distribution* $\mu(A)$, $A \in \mathcal{B}^r$, *of the process* (X_t, P_x) *exists. The measure* $\mu(A)$ *has a density* $m(x)$ *with respect to the Lebesgue measure in* R^r; $m(x)$ *is the unique solution of the problem*

$$L^*m(x) = \frac{1}{2} \sum_{i,j=1}^{r} \frac{\partial^2}{\partial x^i \partial x^j} (a^{ij}(x)\, m(x)) - \sum_{i=1}^{r} \frac{\partial}{\partial x^i} (b^i(x)\, m(x)) = 0 ,$$

$$\int_{R^r} m(x)\, dx = 1 .$$

If the function f(x) *is integrable with respect to the measure* $\mu(A)$, *then*

$$\lim_{T\to\infty} \frac{1}{T} \int_0^T f(X_t)\,dt = \int_{R^r} f(x)\,\mu(dx) \quad P_x \text{ a.s.}\ .$$

For any set $A \in \mathcal{B}^r$

$$\lim_{t\to\infty} P(t,x,A) = \mu(A)\ .$$

For any continuous bounded function f(x), $x \in R^r$,

$$\lim_{t\to\infty} E_x f(X_t) = \int_{R^r} f(x)\,\mu(dx)\ .$$

The proof of these assertions may be found in the monograph by Has'minskii [6] (see also [2]). From this theorem, it follows immediately that the solution u(t,x) of the Cauchy problem

$$\frac{\partial u(t,x)}{\partial t} = L\,u(t,x),\ \ u(0,x) = f(x) \tag{4.16}$$

tends to $\int_{R^r} f(x)\,\mu(dx)$ as $t \to \infty$. In the cited works there are also results concerning ergodic properties of null recurrent processes.

If the process (X_t, P_x) corresponding to the operator L is transient, then for any compact set $K \subset R^r$ and arbitrary $x \in R^r$,

$$\lim_{t\to\infty} P(t,x,K) = 0\ .$$

This assertion follows from Lemma 4.1. Whence it follows that the solution of Cauchy's problem (4.16) tends to zero as $t \to \infty$ in the case when the process (X_t, P_x) is transient, whenever the initial function has compact support.

If the hypotheses of Theorem 4.4 hold and the initial function f(x) is bounded and has limits along the curves $\gamma_k(t)$ as $t \to \infty$ in a strong enough sense, then the solution of the Cauchy problem $\frac{\partial u(t,x)}{\partial t} = L_\alpha u$, $u(0,x) = f(x)$ tends to a solution of problem (4.11) as $t \to \infty$; the constants a_k are defined as limits of the initial function along the curves $\gamma_k(t)$.

§2.5 *Probabilistic representation of solutions of boundary problems with reflection conditions*

Let D be a domain in R^r with a smooth boundary ∂D and let $\gamma(x)$, $x \in \partial D$, be a smooth[8] vector field on ∂D forming an acute angle with the inward normal to ∂D. We will suppose that either the domain D is bounded or the conditions formulated in Remark 2 of §1.6 are fulfilled. Let us denote by (X_t^x, P) a Markov family in $D \cup \partial D$ governed by the operator L inside the domain D with reflection on the boundary in the direction $\gamma(x)$ (see §1.6). The trajectories X_t^x and functions ξ_t^x are defined as solutions of the stochastic differential equation

$$dX_t^x = \sigma(X_t^x)dW_t + b(X_t^x)dt + \chi_{\partial D}(X_t^x)\gamma(X_t^x)d\xi_t^x$$

$$X_0^x = x, \quad \xi_0^x = 0.$$

(5.1)

The operator L is assumed not to degenerate on the boundary at least in the direction $\gamma(x)$: $\sum\limits_{i,j=1}^{r} a^{ij}(x)\gamma_i(x)\gamma_j(x) > a > 0$, $x \in \partial D$, where $\gamma(x) = (\gamma_1(x), \cdots, \gamma_r(x))$. We denote by (X_t, P_x) the Markov process in the state space $D \cup \partial D$ corresponding to the family (X_t^x, P).

Let us consider the mixed problem

$$\frac{\partial u(t,x)}{\partial t} = Lu(t,x) - c(t,x)u(t,x), \quad t > 0, \; x \in D$$

$$u(0,x) = f(x), \; (\nabla_x u(t,x), \gamma(x)) - \lambda(x)u(t,x)\big|_{x \in \partial D} = h(x).$$

(5.2)

---

[8]It is quite sufficient to assume that the normal to ∂D exists, and its direction cosines are defined by three times continuously differentiable functions. The field $\gamma(x)$ may be thought of as twice continuously differentiable.

The functions $f(x), c(x), \lambda(x), h(x)$ are assumed continuous and bounded.

This section gives the representation of the solutions of problem (5.2) as well as of stationary problems (5.5), (5.8), and (5.14) in the form of the expectations of appropriate functionals of the trajectories of the Markov family (X_t^x, P). These representations may be rewritten in an obvious way in terms of the corresponding Markov process (X_t, P_x), that is in the form of integrals over the space of continuous functions $C_{0,\infty}(D \cup \partial D)$.

THEOREM 5.1. *Suppose that* $u(t,x)$ *is a solution of problem (5.2) having bounded continuous derivatives* $\frac{\partial u}{\partial t}(t,x)$, $\frac{\partial u}{\partial x^i}(t,x)$, *and* $\frac{\partial^2 u}{\partial x^i \partial x^j}(t,x)$ *for* $t \in (0,T]$, $x \in D \cup \partial D$, $i,j = 1,2,\cdots,r$. *Then the representation*

$$u(t,x) = E f(X_t^x) \exp\left\{-\int_0^t c(X_s^x)ds - \int_0^t \lambda(X_s^x)\, d\xi_s^x\right\} - \tag{5.3}$$

$$- E\int_0^t h(X_s^x) \exp\left\{-\int_0^s c(X_{s_1}^x)ds_1 - \int_0^s \lambda(X_{s_1}^x)\, d\xi_{s_1}^x\right\} d\xi_s^x$$

holds true.

Proof. We introduce the notations:

$$Y_t = -\int_0^t c(X_s^x)ds\,, \quad Z_t = -\int_0^t \lambda(X_s^x)\, d\xi_s\,.$$

Let us apply Ito's formula (1.6.10) to the function $f(t, X_t^x, Y_t, Z_t) = u(T-t, X_t^x)\exp\{Y_t + Z_t\}$:

$$u(0,X_t^x)\exp\{Y_t+Z_t\} = u(t,x) + \int_0^t \left(-\frac{\partial u}{\partial t}+Lu-cu\right)(t-s,X_s^x)\times$$

$$\times \exp\{Y_s+Z_s\}ds + \int_0^t \exp\{Y_s+Z_s\}(\nabla_x u(t-s,X_s^x),\sigma(X_s^x)\,dW_s) + \tag{5.4}$$

$$+\int_0^t \exp\{Y_s+Z_s\}(\nabla_x u(t-s,X_s^x),\gamma(X_s^x))\,d\xi_s^x -$$

$$-\int_0^t \lambda(X_s^x)\,u(t-s,X_s^x)\exp\{Y_s+Z_s\}d\xi_s^x .$$

Note that the expectation of the stochastic integral is zero, $-\frac{\partial u}{\partial t}+Lu-cu=0$ for $x\in D$, $s\in(0,T]$, and that the process ξ_s^x grows only when X_s^x is on the boundary of the domain. Taking this into account, we derive from (5.4) the statement of the theorem:

$$u(t,x) = E\,f(X_t^x)\exp\{Y_t+Z_t\} - E\int_0^t h(X_s^x)\exp\{Y_s+Z_s\}ds . \quad \square$$

REMARK 1. If the coefficients of the operator L and the functions λ, c, h depend also on t, then an analogous representation can be derived for the solution of the mixed problem, provided one considers the time-space process in the cylinder $(-\infty,\infty)\times(D\cup\partial D)$, governed by the operator $-\frac{\partial u}{\partial t}+Lu$ inside this cylinder and subject to reflection in the direction of $\gamma(x)$ on its boundary. We shall utilize such representations in Chapters V and VI.

The function $u(t,x)$ defined by formula (5.3) may be understood as a generalized solution of problem (5.2). Such an approach is convenient in particular when studying problem (5.2) for degenerate equations (see §3.6). Under weak extra assumptions, this generalized solution is established to be a solution in the "small viscosity" sense as well.

Now let us consider the stationary problem

$$L\,u(x) - c(x)\,u(x) = f(x),\ x \in \partial D$$

$$(\nabla u(x), \gamma(x)) - \lambda(x)\,u(x)\Big|_{x\in\partial D} = h(x)\,. \tag{5.5}$$

THEOREM 5.2. *Suppose* $c(x) \geq c > 0$, $\lambda(x) \geq 0$. *Let* $u(x)$ *be a solution of problem (5.5) possessing first- and second-order continuous and bounded derivatives in* $D \cup \partial D$. *Then*

$$\begin{aligned} u(x) = E \int_0^\infty f(X_s^x) \exp\left\{- \int_0^s c(X_{s_1}^x)\,ds_1 - \int_0^s \lambda(X_{s_1}^x)\,d\xi_{s_1}^x\right\} ds\ - \\ - E \int_0^\infty h(X_s^x) \exp\left\{- \int_0^s c(X_{s_1}^x)\,ds_1 - \int_0^s \lambda(X_{s_1}^x)\,d\xi_{s_1}^x\right\} d\xi_s\,. \end{aligned} \tag{5.6}$$

Proof. We shall apply Ito's formula (1.6.10) to the function $u(X_t^x)\exp\{Y_t + Z_t\}$, where Y_t, Z_t have been defined when proving Theorem 5.1:

$$\begin{aligned} u(X_t^x)\exp\{Y_t + Z_t\} - u(x) &= \int_0^t (\nabla u(X_s^x), \sigma(X_s^x)\,dW_s)\exp\{Y_s + Z_s\} + \\ &+ \int_0^t L\,u(X_s^x)\exp\{Y_s + Z_s\}ds + \int_0^t (\nabla u(X_s^x), \gamma(X_s^x))\exp\{Y_s + Z_s\}d\xi_s^x - \\ &- \int_0^t c(X_s^x)\exp\{Y_s + Z_s\}u(X_s^x)\,ds - \int_0^t \lambda(X_s^x)\,u(X_s^x)\exp\{Y_s + Z_s\}d\xi_s^x = \\ &= \int_0^t (\nabla u(X_s^x), \sigma(X_s^x)\,dW_s)\exp\{Y_s + Z_s\} + \int_0^t f(X_s^x)\exp\{Y_s + Z_s\}ds + \\ &+ \int_0^t h(X_s^x)\exp\{Y_s + Z_s\}d\xi_s^x\,. \end{aligned}$$

On taking expectation of both sides of this equality, we get

$$E\, u(X_t^x) \exp\{Y_t + Z_t\} - u(x) =$$

$$= E \int_0^t f(X_s^x) \exp\{Y_s + Z_s\} ds + E \int_0^t h(X_s^x) \exp\{Y_s + Z_s\} d\xi_s^x . \tag{5.7}$$

The assumption that $\lambda(x) \geq 0$, $c(x) \geq c > 0$ implies that $\exp\{Y_t + Z_t\} < \exp\{-ct\}$ P-a.s. for all $x \in D \cup \partial D$. Whence, we conclude that the first summand on the left of equality (5.7) tends to zero as $t \to \infty$, and the integrals on the right-hand side converge, whenever t is substituted by $+\infty$ when integrating. Thus, passing in (5.7) to the limit as $t \to \infty$, we obtain the claim of Theorem 5.2. □

If $c(x) = 0$, $\lambda(x) = 0$, then a new situation arises which must be discussed separately. For brevity, let D be a bounded domain, and suppose that the operator L does not degenerate in $D \cup \partial D$ and $a^{ij}(x) \in C^{2,\delta}_{R^r}$, $b^i(x) \in C^{1,\delta}_{R^r}$, $i, j = 1, \cdots, r$. We consider the problem

$$L\, u(x) = f(x), x \in D;\ (\gamma(x), \nabla u(x))\big|_{x \in \partial D} = 0 . \tag{5.8}$$

As is known from the theory of differential equations (see, e.g. Miranda [1]), problem (5.8) is not solvable for arbitrary right-hand side f(x). In order to formulate solvability conditions, a homogeneous boundary problem adjoint to (5.8) must be considered. This adjoint problem has the form (see Miranda [1]):

$$L^* m(x) = \frac{1}{2} \sum_{i,j=1}^{r} \frac{\partial^2}{\partial x^i \partial x^j} (a^{ij}(x) m(x)) - \sum_{i=1}^{r} \frac{\partial}{\partial x^i} (b^i(x) m(x)) = 0 ,$$

$$a^{\gamma^*}(x) (\gamma^*(x), \nabla m(x)) + \beta(x) m(x)\big|_{x \in \partial D} = 0 . \tag{5.9}$$

In general, the adjoint vector field $\gamma^*(x)$ and the functions $\beta(x), a^{\gamma^*}(x)$ are expressed rather clumsily in terms of the coefficients of the operator L and of the field $\gamma(x)$ (see the above cited monograph by Miranda), and so we will give here the expressions for $\gamma^*(x)$ and $\beta(x)$ only in the case when $\gamma(x)$ is a field of co-normals.

The vector field

$$\gamma(x) = (\gamma_1(x), \cdots, \gamma_r(x)), \; x \in \partial D ,$$

$$\gamma_i(x) = \frac{1}{a} \sum_{j=1}^{r} a^{ij}(x) X_j(x), \; i = 1, \cdots, r ,$$

is called the *field of co-normals* to the boundary ∂D of the domain D for the given operator L. Here $(X_1(x), \cdots, X_j(x))$ are direction cosines of inward normals to ∂D, and $a = a(x)$ is a normalizing factor defined so that $\sum_1^r \gamma_i^2(x) = 1$. As is known, the co-normal direction has the following remarkable property: for $i \neq 1$, the coefficients $a^{1i}(x)$ vanish in the coordinate system in which the x^1-axis is directed along the co-normal, and the remaining axes lie in a hyperplane tangent to ∂D at the point $x \in \partial D$.

If $\gamma(x)$ is a field of co-normals, then $\gamma^*(x) = \gamma(x)$, $a^{\gamma^*} = a$ and in the boundary condition (5.9), the function $\beta(x)$ is given by the formula

$$\beta(x) = -\sum_{i=1}^{r} \left(b^i(x) - \frac{1}{2} \sum_{k=1}^{r} \frac{\partial a^{ik}(x)}{\partial x^k} \right) X_i(x) .$$

Under the above conditions, it is possible to prove that problem (5.9) has a solution unique up to a constant factor. If we normalize this factor by the condition $\int_D m(x)\,dx = 1$, then $m(x)$ is the density of the invariant measure of the process (X_t, P_x) in $D \cup \partial D$ governed by the operator L with the boundary conditions $(\gamma(x), \nabla) = 0$.

Green's formula implies that, if $u, v \in C_D^2(R^1) \cap C_{D \cup \partial D}^1(R^1)$, then

$$\int_D (v\,Lu - u\,L^*v)\,dx = \int_{\partial D} (a^{\gamma^*}(\gamma^*, \nabla v)\,u + \beta uv - a^\gamma(\gamma, \nabla u)v)\,ds \;, \quad (5.10)$$

where $a^\gamma = a^\gamma(x) > 0$, $a^{\gamma^*} = a^{\gamma^*}(x) > 0$. If $\gamma(x)$ are co-normals then
$a^\gamma = a^{\gamma^*} = \frac{1}{2}\,(\sum_i (\sum_j a^{ij}X_j)^2)^{\frac{1}{2}}$.

If u is a solution of problem (5.8) and $v = m(x)$ is a solution of adjoint problem (5.9), then (5.10) implies that

$$\int_D f(x)\,m(x)\,dx = 0 \;. \tag{5.11}$$

It is this relation that is the solvability condition for problem (5.8). Under minor assumptions on the function $f(x)$, condition (5.11) can be established to be not only necessary but also sufficient for the solvability of problem (5.8).

If the domain D is bounded and the operator L does not degenerate in $D \cup \partial D$, then for any bounded measurable function $f(x)$, $x \in D \cup \partial D$, the relation

$$|E\,f(X_t^x) - \int_D f(x)\,m(x)\,dx| < c_1\,e^{-\alpha t} \tag{5.12}$$

holds for some c_1, $\alpha > 0$. This follows, for example, from Lemma 3.7.3. Noting that $u(t,x) = E\,f(X_t^x)$ is a solution of problem (5.2) for $c(x) = 0$, $\lambda(x) = h(x) = 0$, relation (5.12) is readily deduced from analytical considerations; for the constant α we may choose the first eigenvalue after zero of the operator L with the boundary condition $(\gamma(x), \nabla u) = 0$ on ∂D.

For $c(x) = 0$, $\lambda(x) = h(x) = 0$, equality (5.7) takes the form:

$$E\,u(X_t^x) - u(x) = \int_0^t E\,f(X_s^x)\,ds\ . \tag{5.13}$$

If relation (5.11) is fulfilled for the function $f(x)$, then $|E\,f(X_s^x)| < c_1 e^{-as}$. In addition, we observe that the solution of problem (5.8) is defined up to an additive constant. The function $u(x)$ is assumed to be orthogonal to $m(x)$: $\int_D u(x)\,m(x)\,ds = 0$. This may be achieved by adding an appropriate constant to $u(x)$. For such a choice of $u(x)$, by passing in (5.13) to the limit as $t \to \infty$, we obtain

$$u(x) = -\int_0^\infty E\,f(X_t^x)\,dt\ .$$

Therefore, we have proved the following result:

THEOREM 5.3. *Suppose that the domain D is bounded and the operator L does not degenerate in $D \cup \partial D$. Let (5.11) be fulfilled. Then the solution of problem (5.8) can be represented in the form:*

$$u(x) = -\int_0^\infty E\,f(X_t^x)\,dt + \text{const.}$$

In Chapter III we shall return to the consideration of problem (5.8) in the case when the operator L is degenerate.

Now a few words will be said concerning the problem with inhomogeneous boundary conditions:

$$L\,u(x) = 0,\ x \in D,\ (\gamma(x), \nabla u(x))\Big|_{x \in \partial D} = h(x)\ . \tag{5.14}$$

Let $\gamma(x)$ be the field of co-normals.

Just as above, the domain D is viewed as bounded and the operator L as non-degenerate. From formula (5.10) it follows that for this problem to be solvable, it is necessary that

$$\int_{\partial D} a(x)h(x)m(x)\,dx = 0 , \tag{5.15}$$

where $m(x)$ is the density of the stationary distribution of the family (X_t^x, P) (to be more exact, a trace of this density on ∂D). If condition (5.15) is fulfilled, then problem (5.14) has a solution which is unique up to an additive constant, and this solution can be represented in the form:

$$u(x) = -\lim_{t\to\infty} E\int_0^t h(X_s^x)\,d\xi_s + \text{const.}$$

Condition (5.15) guarantees that the limit on the right-hand side of this equality exists (see, e.g. Korostelev [1]).

To clarify the probabilistic sense of condition (5.15), consider the random functions $Y_t^y = X^y((\xi^y)^{-1}(t))$, $(\xi^y)^{-1}(t) = (\xi^y)^{-1}(t+0)$, $y \in \partial D$, taking values on ∂D. These functions form a Markov family (Y_t^y, P) on ∂D. hard to verify that these functions form a Markov family (Y_t^y, P) on ∂D. This family (the Markov process $(Y_t, \tilde{P}_y)$ corresponding to the family (Y_t^y, P)) is referred to as the trace of the family (Y_t^y, P) (of the process $(Y_t, \tilde{P}_y)$) on ∂D. Under the above hypotheses on the operator L and the domain D, the process $(Y_t, \tilde{P}_y)$ has a unique stationary distribution on ∂D. Let $a(x) \equiv 1$. Then, up to a constant factor, the density of this stationary distribution coincides with the trace of the density $m(x)$ of the stationary distribution of the process (X_t, P_x) in $D \cup \partial D$ on the boundary of the domain. Therefore, condition (5.15) is a condition of orthogonality between the function $h(x)$ and the stationary density of the process $(Y_t, \tilde{P}_y)$ on ∂D.

The trace of the Markob process on the boundary of a domain has been treated by a number of authors (Molchanov [1], Sato and Veno [1], Ueno [1]).

Relying on Ito's formula and (5.10), it is possible to understand how the local time ξ_t^x behaves as $t \to \infty$. As before, let $\gamma(x)$ be the field of co-normals. Let a smooth function $u(x)$, $x \in D \cup \partial D$, be such that $(\gamma(x), \nabla u(x))|_{x \in \partial D} = 1$. We will apply Ito's formula to the function $u(X_t^x)$:

$$u(X_t^x) = u(x) + \int_0^t (\nabla u(X_s^x), \sigma(X_s^x)\, dW_s) + \int_0^t L\, u(X_s^x)\, ds + \xi_t^x\,. \quad (5.16)$$

Noting that $\lim_{t\to\infty} \frac{1}{t} \int_0^t f(s,\omega)\, dW_s = 0$, provided $|f(s,\omega)| \le c < \infty$ P-a.s. (this can be proven, for example, with random change of time), we obtain from (5.16) that

$$-\lim_{t\to\infty} t^{-1} \xi_t^x = \lim_{t\to\infty} \frac{1}{t} \int_0^t L\, u(X_s^x)\, ds, \quad \text{P-a.s.}, \ x \in D \cup \partial D\,.$$

Applying Theorem 1.7.1, it is not difficult to prove that the limit on the right-hand side of the last equality exists and equals

$$\int_D L\, u(x)\, m(x)\, dx\,.$$

Using relation (5.10) we find that

$$\lim_{t\to\infty} \frac{1}{t} \int_0^t L\, u(X_s^x)\, ds = \int_D L\, u(x)\, m(x)\, dx = -\int_{\partial D} a(x)\, m(x)\, dx\,.$$

Since the operator L does not degenerate, we have $m(x) > 0$ for $x \in D \cup \partial D$. So $\int_{\partial D} a(x)\, m(x)\, dx = \mu > 0$. Thus,

$$\xi_t^x \sim \mu t\,, \quad \text{as } t \to \infty\,.$$

Note that the invariant measure of a set in the state space is the fraction of time in the time interval $[0,T]$ spent by the process in this set as $T \to \infty$. Because of this, one may characterize the constant μ as the limit of the fraction of time spent in a narrow strip near the boundary divided by the width of this strip as the width tends to zero. In a similar way one can show that

$$\lim_{t\to\infty} \frac{1}{t} \int_0^t h(X_s^x)\, d\xi_s^x = \int_{\partial D} a(x)\, h(x)\, m(x)\, dx .$$

Now we turn to problems in unbounded regions. We shall not strive for generality and restrict our discussion to examples illustrating what problems and results may arise here.

Let us denote by $D = \{x = (x^1, x^2) : |x^2| \le 1\}$ a strip in the plane R^2 and by $\gamma(x) = (\gamma_1(x), \gamma_2(x))$ a smooth inward directed vector field on the boundary of this strip, $\gamma_2(x^1, 1) = -1$, $\gamma_2(x^1, -1) = 1$.

We will consider the boundary value problem:

$$L\,u(x) = \frac{1}{2}\,\Delta u(x) + b^1(x^1)\,\frac{\partial u}{\partial x^1} + b^2(x^2)\,\frac{\partial u}{\partial x^2} = 0\,,$$

$$(\gamma(x), \nabla u(x))\Big|_{x\in\partial D} = 0\,. \tag{5.17}$$

The functions $b^1(x)$, $b^2(x)$, $\gamma_1(x)$ are assumed to be bounded and Lipschitz continuous. Hence, the operator L can be decomposed into the sum of two operators L_1 and L_2, the former acts with respect to x^1, the latter with respect to x^2:

$$L_1 = \frac{1}{2}\,\frac{\partial^2}{(\partial x^1)^2} + b^1(x^1)\,\frac{\partial}{\partial x^1}\,, \quad L_2 = \frac{1}{2}\,\frac{\partial^2}{(\partial x^2)^2} + b^2(x^2)\,\frac{\partial}{\partial x^2}\,.$$

We will denote by $(X_t^x, P) = (X_t^{1,x}, X_t^{2,x};\, P)$ a Markov family in D governed by the operator L inside the domain with reflection on ∂D in the direction $\gamma(x)$, (X_t, P_x) being the corresponding Markov process. The second component of this family forms itself a Markov family in the state space $[-1,1]$, and its trajectories are defined by the stochastic differential equation (see §1.6)

$$dX_t^{2,x^2} = dW_t^2 + b^2(X_t^{2,x^2})\,dt + \gamma^2(X_t^{2,x^2})\,d\xi_t^{x^2}, \tag{5.18}$$

where $\xi_t^{x^2}$ is a continuous non-decreasing function which grows only when $|X_t^{2,x^2}| = 1$, $\gamma^2(1) = -\gamma^2(-1) = 1$.

The first component is a solution of the stochastic equation

$$\begin{gathered} dX_t^{1,x} = dW_t^1 + b^1(X_t^{1,x})\,dt + \gamma^1(X_t^{1,x}, X_t^{2,x^2})\,d\xi_t^{x^2}, \\ X_0^{1,x} = x^1, \quad x = (x^1, x^2). \end{gathered} \tag{5.19}$$

We shall be interested solely in the question of the uniqueness of the problem (5.17). Because of this, we will consider the equation with zero right-hand side and zero boundary conditions. If $u(x)$ is a bounded solution of problem (5.17), then the random process $Y_t^x = u(X_t^x)$ is a martingale for any $x \in D$ (with respect to the family of σ-fields N_t adapted to the two-dimensional Wiener process $W_t = (W_t^1, W_t^2)$). This follows from Ito's formula:

$$Y_t^x = u(x) + \int_0^t (\nabla u(X_s^x), dW_s)$$

with the other terms on the right-hand side vanishing. Since $u(X_t^x)$ is a bounded martingale, $\lim_{t\to\infty} u(X_t^x)$ exists for any $x \in D$ P-a.s.. Suppose that the coefficients of equation (5.17) and the boundary conditions are such that a trajectory X_t^x starting from a point $x \in D$, after arbitrarily large time t, returns into a bounded domain $G \subset D$ (in this case, the family (X_t^x, P) and the process with reflection (X_t, P_x) in the strip D are said to be recurrent). Then, just as in Lemma 4.1, one can easily prove that the trajectories X_t^x will visit any open subset of the domain D P-a.s. for every $x \in D$. In this case, $u(x) = E \lim_{t\to\infty} u(X_t^x) = u(y)$ for every $x, y \in D$. Here we made use of the fact that the trajectories starting from a point x, will arrive at any neighborhood of the point $y \in D$ with probability 1 and that the function u is continuous at the point y. Therefore, every solution of problem (5.17) is a constant, whenever the family (X_t^x, P) is recurrent.

If the Markov family (X_t^x, P) is transient, then necessarily $\lim_{t\to\infty} |X_t^{1,x}| = \infty$ P-a.s. for every $x \in D$. The proof of this assertion also repeats the corresponding argument of Lemma 4.1. If $\lim_{t\to\infty} |X_t^{1,x}| = \infty$, then the following two cases are possible: either $\lim_{t\to\infty} X_t^{1,x} = \pm\infty$ P-a.s. for all $x \in D$, or this does not hold. In the first case, from $u(X_t^x)$ being a martingale, it follows immediately that $u(x) = E \lim_{t\to\infty} u(X_t^x) = \text{const.}$ Thus, in this case, the solution is also unique up to a constant.[9]

Once the Markov family (X_t^x, P) is transient, it is possible that $P\{\lim_{t\to\infty} X_t^{1,x} = \infty\} = \alpha(x) > 0$ and $P\{\lim_{t\to\infty} X_t^{1,x} = -\infty\} = 1 - \alpha(x) > 0$. Then the general solution of problem (5.17) depends on two arbitrary constants and has the form $u_{c_1,c_2}(x) = c_1\alpha(x) + c_2(1-\alpha(x))$. That the function $u_{c_1c_2}(x)$ is a solution of problem (5.17) can be checked with the strong Markov property analogously to the way this has been done in §2.3. To prove that these functions exhaust all solutions, one should use the following two facts. The first one is the existence (with probability 1)

[9]The distinction between this case and the case of a recurrent process becomes apparent if we consider problem (5.17) not in the entire strip, but, for instance, in a strip with an eliminated small disk $G \subset D \setminus \partial D$ with Dirichlet's condition $(u(x)|_{\partial G} = 0)$ on its boundary ∂G. Then in the case when the process is recurrent, this problem has a unique solution in the class of all bounded functions. However, if $\lim_{t\to\infty} X_t^x = \pm\infty$ P-a.s., then the condition $\lim_{x\to\infty} u(x) = c$ $(\lim_{x\to-\infty} u(x) = c)$ must be given in addition.

of $\lim_{t\to\infty} u(X_t^x)$. The second fact is that the trajectories run to $+\infty$ or to $-\infty$ with probability 1, $\lim_{x^1\to\infty} \alpha(x^1,x^2)$ being equal to 1 and $\lim_{x^1\to-\infty} \alpha(x^1,x^2)$ vanishing. Whence we conclude that the limits $\lim_{x^1\to\infty} u(x) = c_1$ and $\lim_{x^1\to-\infty} u(x) = c_2$ exist uniformly in $x^2 \in [-1,1]$. By the strong Markov property, the function $u(x)$ having such limits at $\pm\infty$ can be represented in the form $u(x) = c_1\alpha(x) + c_2(1-\alpha(x))$. Therefore, if c_1, c_2 vary arbitrarily, then the functions $u_{c_1,c_2}(x)$ exhaust all bounded solutions of problem (5.17).

How can one determine the behavior of the trajectories as $t \to \infty$ from the coefficients of the operator and the field $\gamma(x)$? Let the function $f(x^2) \in C^2_{[-1,1]}$, $f'(-1) = 0$, $f'(1) = 1$. We will apply Ito's formula to the function $f(x_t^{2,x^2})$;

$$f(X_t^{2,x^2}) = f(x^2) + \int_0^t f'(X_s^{2,x^2})\,dW_s^2 + \int_0^t L_2 f(X_s^{2,x^2})\,ds - \\ - \int_0^t \chi_+(X_s^{2,x^2})\,d\xi_s^{x^2}, \tag{5.20}$$

where $\chi_+(x) = 1$ for $x = 1$ and $\chi_+(x) = 0$ at the remaining points of $[-1,1]$. From (5.20), it follows that

$$\lim_{t\to\infty} \frac{1}{t} \int_0^t \chi_+(X_s^{2,x^2})\,d\xi_s^{x^2} = \lim_{t\to\infty} \frac{1}{t} \int_0^t L_2 f(X_s^{2,x^2})\,ds = \\ = \int_{-1}^1 L_2 f(x^2)\,m(x^2)\,dx^2 = \frac{m(1)}{2},$$

where $m(x)$ is the density of the invariant distribution for the family (X_t^{2,x^2}, P) on the interval $[-1,1]$ with reflection at the end-points. In accordance with (5.9), the function $m(x)$ is a solution of the boundary value problem

$$\frac{1}{2}\frac{d^2m}{dx^2} - \frac{d}{dx}(b^2(x)m(x)) = 0, \; x \in (-1,1),$$
$$\frac{dm}{dx}(\pm 1) - 2b^2(\pm 1)m(\pm 1) = 0. \tag{5.21}$$

Solving this problem we derive that $m(x) = c\exp\{\int_0^x 2b(y)dy\}$, where $c = (\int_{-1}^1 \exp\{2\int_0^x b(y)dy\}dx)^{-1}$. Hence,

$$\lim_{t\to\infty} \frac{1}{t}\int_0^t \chi_+(X_s^{2,x^2})d\xi_s^{x^2} = \frac{c}{2}\exp\left\{\int_0^1 2b^2(y)dy\right\} = t_+.$$

Similarly, for the local time at the point -1, we have the relation

$$\lim_{t\to+\infty} \frac{1}{t}\int_0^t \chi_-(X_s^{2,x^2})d\xi_s^{x^2} = \frac{c}{2}\exp\left\{-\int_{-1}^0 2b^2(y)dy\right\} = t_-.$$

We will assume $\lim_{x^1\to\pm\infty} b^1(x^1) = b_\pm^1$, $\lim_{x^1\to\pm\infty} \gamma^1(x^1,\pm 1) = \gamma_\pm(\pm 1)$. Equation (5.19) and the above bounds imply that the family (X_t^x, P) is recurrent, provided

$$A_+ = b_+^1 + t_+\gamma_+(1) + t_-\gamma_+(-1) < 0,$$

$$A_- = b_-^1 + t_+\gamma_-(1) + t_-\gamma_-(-1) > 0.$$

If the inequalities $A_+ < 0$, $A_- < 0$ are fulfilled, then the trajectories go to $-\infty$ along the x^1-axis with probability 1. If $A_+ > 0$, $A_- > 0$, then they tend to $+\infty$. In the case when $A_+ > 0$, $A_- < 0$, the trajectories run to infinity with probability 1, and what is more, with positive probability, the trajectories run both to $+\infty$ and to $-\infty$.

Completing this section, we touch on the problem with singularities on the boundary. For brevity, we restrict ourselves to the two-dimensional case. Let D be a bounded domain in the plane with a smooth boundary. The operator L or the field $\gamma(x)$ defined on ∂D are assumed to have singularities at a finite number of points $a_1, \cdots, a_n \in \partial D$. For example, these singularities may be the degeneracy of the operator, unboundedness of its coefficients, discontinuities of the field $\gamma(x)$, or tangency of $\gamma(x)$ to the boundary. For $x \in D \cup \partial D \setminus (\bigcup_1^n a_i)$, the operator L is assumed to be non-degenerate with smooth enough coefficients, and, for $x \in \partial D \setminus (\bigcup_1^n a_i)$, the field $\gamma(x)$ is supposed to be smooth and to form an acute angle with the inward normal to ∂D.

To study the problem

$$L u(x) = f(x), x \in D; \; (\gamma(x), \nabla u(x))\Big|_{x \in \partial D \setminus (\bigcup_1^n a_i)} = 0,$$

we shall construct a process with reflection along the field $\gamma(x)$ with absorption on the set $\{a_1, \cdots, a_n\}$. To this end, we shall change the operator and the field, if necessary, on the set $\bigcup_{i=1}^n U_\delta(a_i)$, where $U_\delta(a_i) = \{x \in R^2 : |x - a_i| < \delta\}$, so that the new operator L^δ does not degenerate on $D \cup \partial D$ and so that the new field $\gamma^\delta(x)$ is smooth and not be tangent to the boundary. The Markov family corresponding to this changed operator with reflection along the changed field, will be denoted by $(X_t^{\delta,x}, P)$. Let $(\overline{X}_t^{\delta,x}, P)$ be a family derived from $(X_t^{\delta,x}, P)$ by stopping at the first hitting time to $\bigcup_{i=1}^n U_\delta(a^i)$:

$$\overline{X}_t^{\delta,x} = \begin{cases} X_t^{\delta,x} & \text{for } \; t \leq \tau_\delta^x = \inf\{t : X_t^{\delta,x} \in \bigcup_{i=1}^n U_\delta(a_i) \\ X_{\tau_\delta^x}^{\delta,x} & \text{for } \; t > \tau_\delta^x . \end{cases}$$

The Markov time τ_δ^x grows as $\delta \downarrow 0$. We denote $\tau^x = \lim_{\delta \downarrow 0} \tau_\delta^x$, $\tau^x \leq \infty$; $\overline{X}_t^x = \lim_{\delta \downarrow 0} X_t^{\delta,x}$. The last limit exists with probability 1, and the functions $\overline{X}_t^x$ together with the measure P form a Markov family.

The point a_i is called positive if $\lim_{x \to a_i, x \in D \cup \partial D} P\{\lim_{t\to\infty} X_t^x = a_i\} = 1$, and negative if $P\{\lim_{t\to\infty} X_t^x = a_i\} = 0$ for $x \in D \cup \partial D$.

Generally speaking, the singular points a_i do not necessarily belong to one of these two classes: namely, there may be also points a_i such that, when approaching them in one way, the probability that X_t^x goes to a_i tends to 1 as $t \to \infty$, and when the initial point x approaches a_i in another way, the probability is less than 1.

Conditions for a point to be positive or negative can be given in terms of barriers (test functions). For example, for a point $a \in \{a_1, \cdots, a_n\}$ to be positive, it is sufficient that in a neighborhood $\mathcal{E}$ of the point a, a continuous function $v(x) \in C^2_{\mathcal{E}\setminus a}$ exist, positive in $\mathcal{E}\setminus a$ and equal to zero at the point a, for which $Lv(x) \leq 0$ for $x \in \mathcal{E} \cap D$, and $(\gamma(x), \nabla v(x)) \leq 0$ for $x \in \partial D \cap (\mathcal{E}\ a)$. The proof of this assertion is analogous to that of Theorem 4.1: using Ito's formula, $v(X_t^x)$ is established to be a supermartingale, which implies that the point a is positive.

We will assume that the domain D is simply connected, and that all singular points are either positive or negative. Furthermore, suppose that at least one point is positive. Then, for arbitrary $c_1, c_2, \cdots, c_k$, the following problem has a unique bounded solution:

$$L u(x) = 0 \quad \text{for} \quad x \in D, (\nabla u(x), \gamma(x))\Big|_{x \in \partial D \setminus \{a_1, \cdots, a_n\}} = 0,$$

$$\lim_{x \to a_i} u(x) = c_i \quad \text{for} \quad i = 1, 2, \cdots, k,$$

where $a_1, a_2, \cdots, a_k$ are the positive points and $a_{k+1}, \cdots, a_n$ are the negative ones.

Problems of this kind arise, for example, in the study of the Poincaré problem

$$L\,u(x) = 0\,, \quad \frac{\partial u}{\partial \ell(x)}\bigg|_{x \in \partial D} = 0\,,$$

where $\ell(x)$ is a smooth vector field on ∂D which for some x is directed inwards and for other x outwards, and which is tangent to the boundary at isolated points. In the general situation, these tangent points are divided into positive and negative (see Maljutov [1], Dynkin [2], McKean [1]). We remark that, for arbitrary choice of the constants c_i at positive points, the solution may have discontinuities at negative points.

Chapter III

BOUNDARY VALUE PROBLEMS FOR EQUATIONS WITH NON-NEGATIVE CHARACTERISTIC FORM

§3.1 *On peculiarities in the statement of boundary value problems for degenerate equations*

Given the differential operator

$$Lu = \frac{1}{2} \sum_{i,j=1}^{r} a^{ij}(x) \frac{\partial^2 u(x)}{\partial x^i \partial x^j} + \sum_{i=1}^{r} b^i(x) \frac{\partial u(x)}{\partial x^i}, \quad x \in R^r, \tag{1.1}$$

We suppose that the coefficients of this operator are bounded and the characteristic form $\sum_{i,j=1} a^{ij}(x)\lambda_i\lambda_j$ is non-negative for every $x \in R^r$. Assumptions on the smoothness of the coefficients will be refined later on. This chapter will be concerned with boundary value problems for the operators L and $L + c(x)$ and with the corresponding diffusion processes.

If the operator L were elliptic, i.e. if the form $\Sigma a^{ij}(x)\lambda_i\lambda_j$ were strongly positive definite, then, as is known, the Dirichlet problem for the equation $Lu(x) = 0$ in a bounded domain $D \subset R^r$ would have a unique solution for any continuous bounded function $\psi(x)$. Of course, one must make minor assumptions on the smoothness of the coefficients of the operator and require the boundary Γ of the domain D to be regular. Under these hypotheses, the solution of the Dirichlet problem is smooth at all interior points of the domain D.

If the characteristic form of the operator L is only non-negative, then the situation changes essentially. We shall point out the basic effects which occur when the operator L degenerates.

First, boundary conditions should be assigned at not all boundary points. The regularity of boundary points can no longer be ensured by requiring smoothness of the coefficients and the boundary, as in the elliptic case. This effect is also displayed in first-order equations, which may be looked upon as a special case of degenerate equations. It is known that, for the first-order equations, the boundary function should be assigned only on that part of the boundary across which characteristics leave. The question about the regularity of boundary points, i.e. points where Dirichlet boundary conditions can be satisfied, arises also when studying equations degenerating only on the boundary of the domain. Such equations were discussed in a number of works (see, e.g. Keldysh [1], Has'minskii [1]), where sufficient regularity conditions were suggested for equations degenerating on the boundary. In the case of degenerations of general form, a simple regularity condition for the points of a smooth boundary for the operator $L - c(x)$ has been given by Fichera [1], [2] for $c(x) \geq c > 0$. If $c(x) = 0$, then the situation becomes more complicated. For example, it may happen that, on a part of the boundary, one may assign only the function equal to a constant: other Dirichlet boundary conditions will never be satisfied. On the other hand, sometimes it is possible, at one point of the geometrical boundary, to assign different limits along different directions of the entrance at this point; and these conditions can be satisfied. Still more "extensive unglueing" of boundary points is also possible.

When studying boundary value problems for degenerate equations, it is necessary to take into account the points where the boundary value is satisfied only while approaching these points over some special set (partly regular points). In the usual sense, the boundary conditions are not satisfied at these points. If no conditions are laid down at the partly regular points, then uniqueness will not hold.

Another effect to be kept in mind when studying degenerate equations, is the non-uniqueness of the solution of the first boundary value problem. For example, consider, in the ring $K = \{x \in R^2, r_0 < |x| < r_1\}$, an operator having the form $\ell = \frac{\partial}{\partial \phi} + a(r) \frac{\partial^2}{\partial r^2}$ in the polar coordinates (r, ϕ). Here $a(r)$ is positive in a neighborhood of the points r_0 and r_1 and is zero on the interval $[r_0+h, r_1-h]$. Then all boundary points for the equation $\ell u(x) = 0$ in the domain K are regular. However, assigning $u(x)$ on the boundary of the ring does not define $u(x)$ in a unique way: if $u(x)$ is a solution, and $v(r)$ is zero for $r \notin [r_0+h, r_1-h]$, then $u(x) + v(|x|)$ is also a solution satisfying the same boundary conditions.

Therefore, some requirements should be imposed on the operator to ensure the uniqueness of the solution of the first boundary value problem. On the other hand, it is of interest to examine additional conditions which single out one solution when the uniqueness conditions are not fulfilled. Such a question arises, for example, when studying the stabilization of a solution of the mixed problem for degenerate parabolic equation in case the solution of the limit stationary problem is not unique.

If the non-homogeneous equation $\ell u(x) = f(x)$ is considered in the ring K, then, as is easy to see, a solution fails to exist for some function f.

Here solvability conditions are of interest, in particular, the conditions on the operator under which the equation $Lu = f$ is solvable for any, say, sufficiently smooth function $f(x)$.

Similar problems arise when considering the second boundary value problem for degenerate equations.

When examining equations with non-negative characteristic form, one should also bear in mind that the solution of degenerate equations will not be, generally speaking, smooth or even continuous. For example, if no special conditions are imposed on the boundary function, then the solution of the equation $\frac{\partial u}{\partial y} = 0$ in the ring $r_0^2 < x^2+y^2 < r_1^2$ on the plane has a discontinuity along the vertical segments tangent to the circle $x^2+y^2 = r_0^2$ in the lower half-plane. One can give still more convincing examples (see §3.6) illustrating that the solution may be not smooth even if the equation does not degenerate on the boundary and has a unique solution only. And moreover, smoothness of the solution cannot be generated by increasing the smoothness of the coefficients or boundary function.

Since in the general case one cannot expect the existence of a smooth solution, it is necessary to introduce the notion of a generalized solution just as is always done in the theory of differential equations. To define this generalized solution, one can consider equation $Au = 0$ rather than equation $Lu = 0$, where A is some extension of the operator L. On the one hand, this extension should be wide enough to ensure existence of a solution to the equation $Au = 0$ with the corresponding boundary conditions. On the other hand, under proper extra assumptions, this generalized solution must be unique. It turns out that, as such an extension of the operator L, it is convenient to take the infinitesimal operator A of the Markov process governed by the operator L inside the domain the problem is considered in, and by the corresponding boundary conditions on its boundary. In particular, if the first boundary value problem is dealt with, then one should consider processes absorbed at the boundary. All the peculiarities of the statement of boundary value problems for degenerate equations are a display of the properties of the corresponding Markov process. Degenerate equations of the general form were examined in this way in the works of Freidlin [1, 2, 4, 5, 11]. Local properties of the generalized solutions were studied there as well. Close results were obtained via analytical methods by Oleinik (see Oleinik and Radkevich [1], there are other references there too). The works of Pinsky [1, 2] and Sarafian [1] are also devoted to analyzing degenerate equations with the help of probabilistic methods. See also the book of Friedman [2].

Put $a(x) = (a^{ij}(x))$, $b(x) = (b^1(x), \cdots, b^r(x))$. Then the Markov family (and the Markov process) in the space R^r corresponding to the operator L may be constructed with the help of the stochastic differential equations

$$dX_t^x = \sigma(X_t^x)\,dW_t + b(X_t^x)\,dt, \quad X_0^x = x\,, \tag{1.2}$$

where $\sigma(x) = (\sigma^i_j(x))$ is a matrix (not necessarily square) such that $\sigma(x)\,\sigma^*(x) = a(x)$; W_t is a Wiener process whose dimension is equal to the number of columns of the matrix $\sigma(x)$. For the solution of this equation to exist and be unique, it is necessary to impose some smoothness conditions on its coefficients. We shall assume that a matrix $\sigma(x)$ may be found such that the coefficients of equation (1.2) are bounded in modulus and satisfy the Lipschitz condition:

$$\sum_{i,j=1}^{r} |\sigma_j^i(x) - \sigma_j^i(y)|^2 + \sum_{i=1}^{r} |b^i(x) - b^i(y)|^2 \leq K^2 |x-y|^2 . \quad (1.3)$$

When constructing the process (X_t, P_x) corresponding to the operator L, it is natural to formulate all necessary conditions in terms of the coefficients of the operator L. So, in the next section, we will investigate when a factorization $a(x) = \sigma(x)\sigma^*(x)$, can be found with the functions $\sigma_j^i(x)$ satisfying the Lipschitz condition.

§3.2 *On the factorization of non-negative definite matrices*

Let $a(x) = (a^{ij}(x))$ be a non-negative definite matrix of order r whose elements depend on the point $x \in R^r$. In this section we will give conditions ensuring the representability of the matrix $a(x)$ in the form $a(x) = \sigma(x)\sigma^*(x)$ with the Lipschitz continuous functions $\sigma_j^i(x)$ and obtain a bound for the Lipschitz constant of these functions. Here we follow the note of Freidlin [10]. Close results are in the work of Philips and Sarason [1].

Unless the matrix $a(x)$ degenerates, it is not hard to prove that, for the Lipschitz factorization to exist, it suffices that the functions $a^{ij}(x)$ be Lipschitz continuous. As may be shown by simple examples, this is insufficient in the case when the matrix $a(x)$ is degenerate. However, it is in the degenerate case that such a representation is especially desired, because the stochastic equations may be the only known way of constructing a Markov process with a given degenerate diffusion matrix. Many properties of the process and solutions of boundary value problems depend on the magnitude of the Lipschitz constant.

THEOREM 2.1. *If* $\sum_1^r a^{ij}(x)\lambda_i\lambda_j \geq 0$, $a^{ij}(x) = a^{ji}(x)$, *for* $x \in R^r$ *and* $a^{ij}(x) \in C^2_{R^r}(R^1)$ *for* $i,j = 1,2,\cdots,r$, *then a symmetric matrix* $\sigma(x)$ *exists such that* $\sigma^2(x) = a(x)$, *the elements of the matrix being Lipschitz continuous*: $|\sigma^i_j(x) - \sigma^i_j(y)| < c\sqrt{H}|x-y|$, *where the constant* c *depends only on the dimension of the space, and* $H = \sup_{x\in R^r, i,j,k,\ell=1,\cdots,r} \left|\frac{\partial^2 a^{ij}(x)}{\partial x^k \partial x^\ell}\right|$.

The proof of this theorem relies on the following lemma.

LEMMA 2.1. *If* $\sum_1^r a^{ij}(x)\lambda_i\lambda_j \geq 0$ *and* $\sup_{x,i,j,k,\ell} \left|\frac{\partial a^{ij}(x)}{\partial x^k \partial x^\ell}\right| = H < \infty$, *then for any* $\xi \in R^r$ *and* $k = 1,\cdots,r$ *the inequality holds*:

$$\left(\frac{\partial a(x)}{\partial x^k}\xi,\xi\right) \leq 2rH\cdot(\xi,\xi)\cdot(a(x)\xi,\xi).$$

To prove the lemma we shall expand the function $g(x) = (a(x)\xi,\xi)$ by the Taylor formula in a neighborhood of the point x:

$$g_h(x) = g(x^1,\cdots,x^{k-1},x^k+h,x^{k+1},\cdots,x^r) =$$

$$= (a(x)\xi,\xi) + h\left(\frac{\partial a(x)}{\partial x^k}\xi,\xi\right) + \frac{h^2}{2}\left(\frac{\partial^2 a(\tilde{x})}{(\partial x^k)^2}\xi,\xi\right),$$

where $\tilde{x}$ denotes some point in a neighborhood of the point x. It is easily checked that $\left|\frac{\partial^2 a(x)}{(\partial x^k)^2}\xi,\xi\right| < rH\cdot|\xi|^2$ for any $x \in R^r$. Hence

$$0 \leq g_h(x) \leq (a(x)\xi,\xi) + h\left(\frac{\partial a(x)}{\partial x^k}\xi,\xi\right) + \frac{h^2}{2}rH\cdot(\xi,\xi) \tag{2.1}$$

for all $h(-\infty,\infty)$. Inequality (2.1) says that the quadratic form in h is non-negative on the entire axis. This implies that the discriminant is non-positive:

$$\left(\frac{\partial a(x)}{\partial x^k}\xi, \xi\right)^2 \leq 2r\,H \cdot (\xi, \xi)(a(x)\xi, \xi)$$

which was to be proved. □

We turn now to proving Theorem 2.1. Let $a^\varepsilon(x) = a(x) + \varepsilon E$, where E is the unit matrix, ε is a small real number. Of the positive definite matrix $a^\varepsilon(x)$ one can extract in a unique way the symmetric positive definite square root $\sigma^\varepsilon(x)$ whose elements $\sigma_j^{i,\varepsilon}(x)$ possess the same smoothness in $x \in R^r$, as do the elements of the matrix $a^\varepsilon(x)$ (see, e.g. Theorem 2.2).

Clearly, $\sum_{i,j=1}^{r} [\sigma_j^{i,\varepsilon}(x)]^2 = \operatorname{Tr} a^\varepsilon(x) = r\varepsilon + \operatorname{Tr} a(x)$, hence, the elements of the matrix $\sigma^\varepsilon(x)$ are bounded uniformly in $\varepsilon \in [0,1]$. If we now can show that the first-order derivatives of the elements of the matrix $\sigma^\varepsilon(x)$ are bounded uniformly in ε, then by the Arzelà theorem we will be able to choose a subsequence $\sigma^{\varepsilon^1}(x)$ converging to the limit $\sigma(x)$ as $\varepsilon^1 \downarrow 0$. The matrix $\sigma(x)$ is symmetric, has Lipschitz continuous elements and $\sigma^2(x) = a(x)$.

So let us demonstrate that the elements of the matrix $\sigma^\varepsilon(x)$ have first-order derivatives, bounded uniformly in $\varepsilon \in [0,1]$. By differentiating the equality $[\sigma^\varepsilon(x)]^2 = a^\varepsilon(x)$ in x^k and denoting by $\sigma'(x)$ and $a'(x)$ first-order derivatives in x of $\sigma^\varepsilon(x)$ and $a^\varepsilon(x)$, we derive:

$$a'(x) = \sigma^\varepsilon(x)\,\sigma'(x) + \sigma'(x)\,\sigma^\varepsilon(x) . \tag{2.2}$$

Denote by $\theta^\varepsilon(x) = \theta(x)$ an orthogonal matrix transforming $\sigma^\varepsilon(x)$ to the diagonal form $\Lambda^\varepsilon = \Lambda^\varepsilon(x)$. Multiplying (2.2) by $\theta(x)$ from the left and by $\theta^{-1}(x)$ from the right, we arrive at the equality:

$$\bar{a}(x) = \theta(x)\,a'(x)\,\theta^{-1}(x) = Y\Lambda + \Lambda Y , \tag{2.3}$$

where $Y = Y(x) = (y_{ij}(x)) = \theta(x)\,\sigma'(x)\,\theta^{-1}(x)$. From (2.3), we find that $y_{ij}(x) = \bar{a}^{ij}(x)\,[\lambda_i(x) + \lambda_j(x)]^{-1}$. Next, applying Lemma 2.3 we get:

$$(\bar{a}(x)\xi,\xi)^2 = (\theta(x)a^1(x)\theta^{-1}(x)\xi,\xi)^2 = (a^1(x)\theta^{-1}(x)\xi,\theta^{-1}(x)\xi)^2 \le$$
$$\le 2rH\cdot(\theta^{-1}\xi,\theta^{-1}\xi)(a(x)\theta^{-1}(x)\xi,\theta^{-1}(x)\xi) = 2rH\cdot(\xi,\xi)(\Lambda^2\xi,\xi)\,. \tag{2.4}$$

Substituting in (2.4) the vector $\xi = (\xi_1,\cdots,\xi_r)$, having 1 on the i-th place and zero as the remaining coordinates, we have: $(\bar{a}^{ii}(x))^2 \le 2rH\lambda_i^2$. This implies that

$$|y_{ii}(x)| = \left|\frac{\bar{a}^{ii}(x)}{2\lambda_i(x)}\right| \le \sqrt{\frac{r}{2}}\,H\,. \tag{2.5}$$

Now substitute in (2.4) the vector ξ which has $\xi_i = \xi_j = 1$ and zero as the remaining coordinates:

$$(\bar{a}^{ii}(x)+\bar{a}^{jj}(x)+2\bar{a}^{ij}(x))^2 \le 4rH\cdot(\lambda_i^2+\lambda_j^2)\,. \tag{2.6}$$

Divide the last equality by $(\lambda_i+\lambda_j)^2$. Since $\lambda_i,\lambda_j > 0$, we get $\lambda_i^2+\lambda_j^2 <$ $(\lambda_i+\lambda_j)^2$, and thereby (2.6) yields:

$$\left(\frac{\bar{a}^{ii}(x)}{\lambda_i+\lambda_j}+\frac{\bar{a}^{jj}(x)}{\lambda_i+\lambda_j}+\frac{2\bar{a}^{ij}(x)}{\lambda_i+\lambda_j}\right)^2 < 4r\cdot H\,. \tag{2.7}$$

From (2.5) and (2.7) it follows that $2|a^{ij}(x)|(\lambda_i(x)+\lambda_j(x))^{-1} < 2\sqrt{rH}+\sqrt{2rH}$, whence

$$|y_{ij}(x)| = |\bar{a}^{ij}(x)|\,[\lambda_i(x)+\lambda_j(x)]^{-1} < 2\sqrt{rH}\,. \tag{2.8}$$

Since $(\sigma_j^i(x))' = \theta^{-1}(x)Y(x)\theta(x)$, where $\theta(x) = (q_{ij}(x))$ is an orthogonal matrix, (2.8) yields the required bound:

$$\left|\frac{\partial\sigma_j^i(x)}{\partial x^k}\right| = \left|\sum_{m,\ell=1}^{r} q_{im}(x)\,y_{m\ell}(x)\,q_{\ell j}(x)\right| \le \max_{x;m,\ell}|y_{m\ell}(x)|\cdot\left|\sum_{m,\ell}q_{im}(x)\,q_{\ell j}(x)\right| < 2r\sqrt{rH}\,,$$

completing the proof of Theorem 2.1. □

If one removes the condition $a^{ij}(x) \in C^2_{R^r}(R^1)$ and makes no assumptions on the nature of degeneration, then the Lipschitz continuous root of

$a(x)$ may not exist. For example, the function $a(x) = |x|^{1+\alpha}, x \in R^1$, cannot be represented in the form of the square of a function satisfying a Lipschitz condition. Moreover, such a function cannot be represented in the form of a sum of the squares of a finite number of the Lipschitz functions.

However, when there are certain special kinds of degeneration, the root of the matrix may be chosen in such a way that its elements have the same smoothness as do the elements of the original matrix.

THEOREM 2.2. *Suppose that* D *is a bounded domain in* R^r, $\Sigma a^{ij}(x)\lambda_i\lambda_j \geq 0$ *for* $x \in R^r$, *and* $a^{ij}(x) \in C_D^{k,\lambda}(R^1)$. *Moreover, let the matrix* $a(x) = (a^{ij}(x))$ *have a constant rank everywhere in the closure* $[D]$ *of the domain* D. *Then a symmetric non-negative definite matrix* $\sigma(x) = (\sigma_j^i(x))$ *exists such that*

$$\sigma^2(x) = a(x), \ \sigma_j^i(x) \in C_D^{k,\lambda}(R^1).$$

Proof. All the eigenvalues of the non-negative definite matrix $a(x)$ are real and non-negative. Denote by $\rho(x)$ the minimal positive eigenvalue of the matrix $a(x)$. The function $\rho(x)$ is continuous in $[D]$ and not equal to zero. This follows from the fact that the rank of the matrix $a(x)$ is constant. Hence, $\rho_0 = \min_{x \in [D]} \rho(x) > 0$. In the complex plane Z, consider a closed loop Γ that lies in the right half-plane and contains inside itself all positive eigenvalues of the matrices $a(x)$. For this, the loop Γ must intersect the real axis to the left of the point ρ_0.

We set

$$\sigma(x) = \frac{1}{2\pi i} \int_\Gamma \sqrt{z}\,(a(x) - zE)^{-1}\,dz\,. \tag{2.9}$$

Since on the loop Γ the matrix $a(x) - zE$ does not degenerate, we conclude that the elements of the matrix $\sigma(x)$ defined by (2.9) have the same smoothness in the parameter $x \in D$, as do the elements of the matrix $a(x)$.

To be sure, note that unless the matrix $B(x) = a(x) - zE$ degenerates in a neighborhood of some point x_0, $B(x) = B(x_0)[B^{-1}(x_0)B(x)]$ and $B^{-1}(x) = [B^{-1}(x_0)B(x)]^{-1}B^{-1}(x_0)$. For sufficiently small $|x - x_0|$, the matrix $B^{-1}(x_0)B(x)$ is close to the identity matrix, and therefore, the reciprocal matrix may be written in the form of the series

$$[B^{-1}(x_0)B(x)]^{-1} = \sum_{k=0}^{\infty} ([B^{-1}(x_0)B(x) - E)^k ,$$

which converges in the norm for $|x - x_0|$ small enough. Whence, $B^{-1}(x) = (a(x) - zE)^{-1} = \Sigma (B^{-1}(x_0)B(x) - E)^k B^{-1}(x_0)$. The last formula implies that $(a(x) - zE)^{-1}$ has the same smoothness in the parameter x as $a(x)$ does. Consequently, the elements of the matrix $\sigma(x)$ defined by equality (2.9) belong to $C_D^{k,\lambda}(R^1)$, whenever $a^{ij}(x) \in C_D^{k,\lambda}(R^1)$. To complete the proof it remains to note that, by the Cauchy integral formula, for the matrix defined by equality (2.9), we have the following equality $\sigma^2(x) = a(x)$. □

In the conclusion of this section we make some remarks of a general nature concerning the factorization of non-negative definite matrices.

First of all, we observe that, for our goals, it is sufficient to represent the matrix $a(x)$ in the form $a(x) = \sigma(x)\sigma^*(x)$, where the matrix $\sigma(x)$ may also be rectangular. If the matrices $a_1(x)$ and $a_2(x)$ of order r admit the factorization

$$a_1(x) = \sigma^{(1)}(x)[\sigma^{(1)}(x)]^*, \quad a_2(x) = \sigma^{(2)}(x)[\sigma^{(2)}(x)]^* ,$$

and $\sigma_j^{i,(1)}, \sigma_j^{i,(2)} \in C_D^{k,\lambda}(R^1)$, then the matrix $a_1(x) + a_2(x)$ also admits a factorization of class $C_D^{k,\lambda}(R^1)$. To see this, it suffices to consider the rectangular matrix $\sigma(x) = (\sigma^{(1)}(x), \sigma^{(2)}(x))$. It is clear that the elements of the matrix $\sigma(x)$ belong to class $C_D^{k,\lambda}(R^1)$ and $\sigma(x)\sigma^*(x) = \sigma^{(1)}(x)[\sigma^{(1)}(x)]^* + \sigma^{(2)}(x)[\sigma^{(2)}(x)]^* = a_1(x) + a_2(x)$. This remark, in particular, implies that if the matrices $a_1(x)$ and $a_2(x)$ admit the

Lipschitz factorization, then the matrix $a^{\varepsilon}(x) = a_1(x) + \varepsilon a_2(x)$ admits the Lipschitz factorization $a^{\varepsilon}(x) = \sigma^{\varepsilon}(x)[\sigma^{\varepsilon}(x)]^*$, the elements of the matrix $\sigma^{\varepsilon}(x)$ satisfying the Lipschitz condition with a constant independent of ε.

Finally, note that if, for every point x belonging to the closure $[D]$ of a bounded domain D, one can find a neighborhood $U(x_0)$, where the matrix $a(x)$ admits the factorization $a(x) = \sigma(x)\sigma^*(x)$ and $\sigma_j^i(x) \in C^{k,\lambda}_{U(x_0)}(R^1)$, then there exists, generally speaking, a rectangular matrix $\tilde{\sigma}(x)$, $x \in D$, with elements of class $C^{k,\lambda}_D(R^1)$ such that $\tilde{\sigma}(x)\tilde{\sigma}^*(x) = a(x)$ for $x \in D$. This is proved by using a partition of the identity argument.

§3.3 *The exit of a process from a domain*

If relations (1.3) hold, then equation (1.2) defines a Markov family (X_t^x, P) and the corresponding Markov process (X_t, P_x) in the state space R^r. Lemma 1.4.4 implies that this process is a Feller strong Markov process.

Given, in the space R^r, a domain D with the boundary ∂D, denote by τ_D the first exit time of the process (X_t, P_x) from the domain $D : \tau_D = \inf\{t : X_t \notin D\}$. It is clear that, due to the continuity of the trajectories X_t, $\tau_D = \inf\{t : X_t \in \partial D\}$ P_x-a.s. for $x \in D \cup \partial D$.

Together with τ_D, we shall consider the random variable $\tau_D^x = \inf\{t : X_t^x \notin D\}$ the first exit time of the Markov random function X_t^x from D. We remind that the random variables τ_D and τ_D^x are defined on different spaces of elementary events: τ_D is defined on the space $C_{0,\infty}(R^r)$ equipped with the σ-field of the cylindrical sets $\mathfrak{N}$, and τ_D^x is defined on the space of elementary events that the Wiener process is defined on. We denote by $\mathfrak{N}_t$ the σ-field in $C_{0,\infty}(R^r)$ generated by the sets $\{\phi \in C_{0,\infty}(R^r) : \phi_s \in B\}$ for $s \leq t$, $B \in \mathcal{B}^r$; N_t, $t \geq 0$, being an increasing family of σ-fields adapted to the underlying Wiener process W_t. The random variables τ_D and τ_D^x are Markov times with respect to the families of σ-fields $\mathfrak{N}_t$ and N_t respectively.

We shall (usually without special explanations) often switch between the notations connected with a process and the notations connected with the corresponding Markov family. If both probabilities and expectations have the index x, then the process is considered. If the trajectories have the index x, then the corresponding family is considered.

Recall that if $F[\phi]$ is a functional on $C_{0,\infty}(R^r)$, then $E_x F[X_\cdot] = E F[X_\cdot^x]$.

Let us denote by γ_D the set of limit points of the trajectory X_t as $t \to \tau_D$. Generally speaking, the set γ_D may be empty (provided the domain D is unbounded), may consist of more than one point, or may include points of the domain D. It is not difficult to provide examples when γ_D coincides with $D \cup \partial D$ with probability 1 for $x \in D$. However, in the "good" cases, $\gamma_D \subset \partial D$.

We shall say that *Condition* $\mathfrak{A}_1$ is fulfilled, if, for any $x \in D$,

$$P_x\{\gamma_D \neq \emptyset, \gamma_D \subset \partial D\} = 1 .$$

As will be explained later on, this condition is closely related to the uniqueness of the solution of the first boundary value problem for the equation $Lu = 0$ in the domain D. For verifying the condition $\mathfrak{A}_1$ one can use the following lemmas.

LEMMA 3.1. *Suppose that at least one of the following conditions is fulfilled:*

1. *The domain* D *is bounded and a function* $f(x) \in C^2_{R^r}(R^1)$ *exists such that* $f(x) \geq 0$ *and* $Lf(x) \leq -1$ *for* $x \in D$;

2. *There are* $T, \delta > 0$ *such that* $P_x\{\tau_D < T\} > \delta$ *for all* $x \in D$;

3. *One can find* $i \in \{1, 2, \cdots, r\}$ *and a positive constant* a *such that* $D \subset \{x \in R^r : r_0 \leq x^i \leq R_0\}$ *for some* $r_0, R_0, -\infty < r_0 < R_0 < \infty$, *and moreover, either* $a^{ii}(x) \geq a > 0$ *for all* $x \in D$, *or* $b^i(x)$ *preserves its sign in* D *and* $|b^i(x)| \geq a > 0$ *for* $x \in D$.

Then the condition $\mathfrak{A}_1$ *holds, the set* γ_D *consists of one point* P_x a.s. *for* $x \in D$, $\lim_{t\to\infty} P_x\{\tau_D > t\} = 0$ *uniformly in* $x \in D$, *and* $E_x \tau_D \leq c < \infty$.

Proof. Let Condition 1 hold. Since the domain D is bounded, we conclude that, without loss of generality, one can assume the function $f(x)$ to be bounded together with its derivatives everywhere in R^r. Otherwise, it may be changed outside a sufficiently large ball. Let us apply Ito's formula to $f(X_t^x)$:

$$f(X_t^x) = f(x) + \int_0^t (\nabla f(X_s^x), \sigma(X_s^x)\, dW_s) + \int_0^t L f(X_s^x)\, ds\,. \tag{3.1}$$

We shall substitute in (3.1) the random variable $t \wedge \tau_D^x$ for t and take the expectation of both sides of the equality obtained. Noting that the mean value of stochastic integral is zero and $Lf(x) \le -1$ for $x \in D$, we derive

$$E f(X^x_{\tau_D^x \wedge t}) \le f(x) - E(\tau_D^x \wedge t)\,.$$

On account of the fact that the function $f(x)$ is non-negative, this equality yields that $E(\tau_D^x \wedge t) \le f(x)$ for any $t \ge 0$. Letting $t \to \infty$, we have

$$E_x \tau_D = E\tau_D^x \le \max_{x \in D \cup \partial D} f(x) = c < \infty\,. \tag{3.2}$$

It appears clear from (3.2) that $P_x\{\tau_D < \infty\} = 1$ for $x \in D$, and that γ_D consists of one point X_{τ_D} P_x-a.s., Chebyshev's inequality and (3.2) together imply that, uniformly in $x \in D$,

$$P_x\{\tau_D > t\} \le t^{-1} E_x \tau_D \le \frac{c}{t} \to 0,\, t \to \infty\,.$$

Note that the assumption that the domain D is bounded, was needed for us only to guarantee that the function f and its derivatives are bounded. If one knows beforehand that f and its derivatives are bounded, then the assumption on the boundedness of the domain D may be dropped.

Now let Condition 2 be fulfilled. By the strong Markov property, for any integer n and $x \in D$.

$$P_x\{\tau_D > nT\} = E_x \chi_{\{\tau_D > (n-1)T\}} P_{X_{(n-1)T}}\{\tau_D > t\} <$$
$$< (1-\delta) E_x \chi_{\{\tau_D > (n-1)T\}} = (1-\delta) P_x\{\tau_D > (n-1)T\}, \tag{3.3}$$

where $\chi_{\{\tau_D > (n-1)T\}}$ is the indicator of the set $\{\tau_D > (n-1)T\}$. Relation (3.3) implies that

$$E_x \tau_D \le \sum_{k=1}^{\infty} kT P_x\{\tau_D > (k-1)T\} \le T \sum_{k=1}^{\infty} k(1-\delta)^{k-1} = c < \infty .$$

At last, if Condition 3 holds, then one can find a non-negative function $f(x)$, $x \in R^r$, bounded together with its first- and second-order derivatives for which $Lf(x) \le -1$ for $x \in D$. By virtue of the above reasoning, this implies the claim of the lemma.

If $a^{ii}(x) \ge a$, then one can put $f(x) = A \cdot (e^{\lambda R_0} - e^{\lambda x^i})$ for $r_0 \le x^i \le R_0$, where $\lambda = 2(B+1)a^{-1}$, $B = \sup_{x \in D} |b^i(x)|$, $A = \lambda^{-1} \exp\{-\lambda r_0\}$. Outside the set $\{x \in R^r, r_0 \le x^i \le R_0\}$, the function $f(x)$ may be defined in an arbitrary fashion if only it is bounded together with its first- and second-order derivatives. One trivially checks that $Lf(x) \le -1$, $f(x) \ge 0$ for $x \in D$.

If $b^i(x) \ge a$ for $x \in D$, then one can take $f(x) = a^{-1} e^{-r_0}(e^{R_0} - e^{x^i})$ for $r_0 \le x^i \le R_0$. In a similar way, it is possible to choose the function $f(x)$ also in the case when $b(x) \le -a < 0$. □

REMARK. Let D be an arbitrary (not necessarily bounded) domain in R^r. Suppose that one can find a function $f(x) \in C^2_{R^r}(R^1)$ such that $f(x) \ge 0$, $Lf(x) \le -1$ for $x \in D$ (the function $f(x)$ or its derivatives may be unbounded). Then for any $x \in D$

$$E_x \tau_D \le f(x), \ x \in D .$$

Indeed, put $D_N = D \cap \{x \in R^r : |x| < N\}$, $\tau_N = \inf\{t : X_t \notin D_N\}$. Then, in view of (3.2), $E_x \tau_N \le f(x)$. Since τ_N converges monotonically to τ_D as $N \to \infty$, the last bound leads to the needed claim.

If the conditions of Lemma 3.1 are valid, then $P_x\{\tau_D < \infty\} = 1$, $x \in D$. Condition $\mathcal{C}_1$ may also hold when the boundary cannot be reached in a finite time starting from some (or from all) points of the domain. In this case the set γ_D may consist of more than one point.

EXAMPLE 3.1. Let $D = \{(x,y) \in R^2 : |x| < 1, y \in (0,1)\}$, $L = \frac{a(x)}{2}\frac{\partial^2}{\partial x^2} - \beta x \frac{\partial}{\partial x} - y\frac{\partial}{\partial y}$. Suppose that $a(1) = a(-1) = 0$, $a(x) > 0$ for $-1 < x < 1$, $\beta > 0$. Let $(X_t^{x,y}, Y_t^{x,y}; P)$ be the Markov family corresponding to the operator L. The corresponding stochastic equation yields that

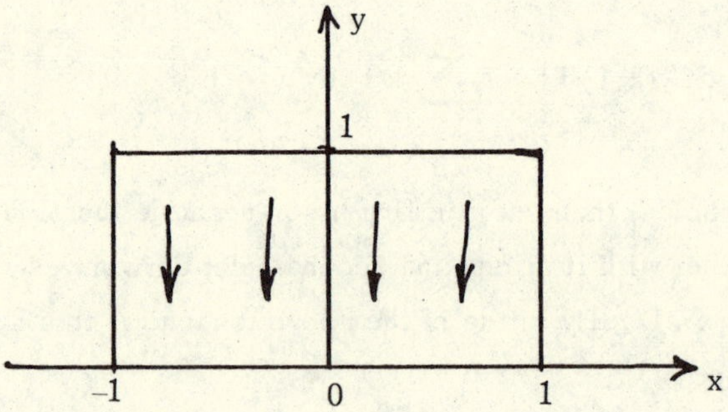

Fig. 1

$$Y_t^{x,y} = y e^{-t}, X_t^{x,y} = x + \int_0^t \sqrt{a(X_s^{x,y})}\, dW_s - \beta \int_0^t X_s^{x,y} ds .$$

This implies that, with probability 1, starting from any point of the domain D, trajectories will tend to the line segment $[-1,1]$ on the x-axis as $t \to \infty$. And the entire segment will consist of the limit points of the trajectory $(X_t^{x,y}, Y_t^{x,y})$ as $t \to \infty$. Therefore, in this case, with probability 1, starting from $(x,y) \in D$ the set γ_D coincides with $[-1,1]$ on the x-axis.

If $a(x) \equiv 0$, $\beta > 0$, then the exit time from D is also infinite. It is readily seen that in this case the set γ_D consists of one point $(0,0)$.

If $a(x) > 0$ for $x \in (-1,1)$ and $\beta < 0$, then one easily checks that Condition 1 of Lemma 3.1 holds. In this case $P_{x,y}\{\tau_D < \infty\} = 1$ for $(x,y) \in D$.

If, for $a(-1) = a(1) = 0$, $\beta > 0$, the operator L is considered in the domain $D' = \{(x,y) \in R^2 : |x| < 1, |y| < 1\}$, then, as it is easily verified, Condition $\mathcal{A}_1$ does not hold. In this case $\gamma_D = \{(x,y) : |x| \le 1, y = 0\}$ $P_{x,y}$ a.s., $(x,y) \in D$, provided $a(x) > 0$ for $x \in (-1,1)$.

The following lemma enables Condition $\mathcal{A}_1$ to be checked when $P_x\{\tau_D < \infty\} \not\equiv 1$.

LEMMA 3.2. *Suppose that a compact set* $\Gamma \subset \partial D$ *is such that* $P_x\{\gamma_D \cap \Gamma \neq \emptyset\} = 1$ *for any* $x \in D$. *Suppose further that in some neighborhood* $\mathcal{E}$ *of the set* Γ *a continuous function* $f(x)$ *is defined with the following properties*: 1) $f(x) > 0$ *for* $x \in (\mathcal{E} \cap (D \cup \partial D)) \setminus \Gamma$ *and* $f(x) = 0$ *for* $x \in \Gamma$; 2) $f(x)$ *has continuous first- and second-order derivatives for* $x \in \mathcal{E} \cap (D \cup \partial D)$ *and* $Lf(x) \le 0$.

Then $P_x\{\gamma_D \subset \Gamma\} = 1$ *for any* $x \in D$.

Proof. Let $\tau^x_{\mathcal{E}\cap D} = \inf\{t : X^x_t \notin \mathcal{E} \cap D\}$, $\tau^x_t = \tau^x_{\mathcal{E}\cap D} \wedge t$. By Ito's formula

$$f(X^x_{\tau^x_t}) = f(x) + \int_0^{\tau^x_t} (\nabla f(X^x_s), \sigma(X^x_s)\, dW_s) + \int_0^{\tau^x_t} Lf(X^x_s)\, ds .$$

Noting that $Lf(x) \le 0$ for $x \in \mathcal{E} \cap D$, we imply from this that the process $Z^x_t = f(X^x_{\tau^x_t})$ is a continuous non-negative supermartingale with respect to the family of σ-fields N_t. By Doob's theorem, with probability 1, the limit $Z^x_\infty = \lim_{t\to\infty} Z^x_t$ exists, and

$$\lim_{t\to\infty} E f(X^x_{\tau^x_t}) = \lim_{t\to\infty} E Z^x_t = E Z^x_\infty \le f(x) .$$

Whence, remembering that the function $f(x)$ is continuous, non-negative and vanishes only on Γ, we conclude that

$$\lim_{\rho(x,\Gamma)\to 0} P_x \left\{ \begin{array}{l} \text{starting from some time, the trajectory } X_t \\ \text{does not leave the set } \mathcal{E}_1 \end{array} \right\} = 1 , \quad (3.4)$$

where $\mathcal{E}_1$ is an arbitrary neighborhood of the set Γ, $\rho(x,\Gamma)$ being the distance from x to Γ. Hence, if the initial point is sufficiently close to Γ, then the set γ_D lies inside $\mathcal{E}_1$ with probability arbitrarily close to 1. Now, if we use that $P_x\{\gamma_D \cap \Gamma = \emptyset\} = 1$ for any $x \in D$, then (3.4) implies that $P_x\{\gamma_D \subset \mathcal{E}_1\} = 1$. Since $\mathcal{E}_1$ is an arbitrary neighborhood of the set Γ, this implies the claim of Lemma 3.2. □

For examining the smoothness of the generalized solutions of degenerate differential equations we shall need more refined characteristics of how fast the process leaves the domain. Let

$$\gamma(t) = \sup_{x \in D} P_x\{\tau_D > t\} .$$

Using the Markov property in the form (14.3), we get

$$\gamma(t+s) = \sup_{x \in D} P_x\{\tau_D > t+s\} = \sup_{x \in D} E_x \chi_{\tau_D > t}\, \theta_t \chi_{\tau_D > s} =$$

$$= \sup_{x \in D} E_x \chi_{\tau_D > t} P_{X_t}\{\tau_D > s\} \leq \gamma(t)\gamma(s) .$$

Thus the function $\ln \gamma(t)$ is subadditive

$$\ln \gamma(t+s) \leq \ln \gamma(t) + \ln \gamma(s), \; -\infty \leq \ln \gamma(t) \leq 0 .$$

For such functions the limit

$$\lim_{t\to\infty} t^{-1} \ln \gamma(t) = \inf_{t>0} t^{-1} \ln \gamma(t) = -\alpha_{L,D} \qquad (3.5)$$

exists (see, e.g. Kingman [1]).

We shall say that a process (X_t, P_x) leaves a domain D *uniformly exponentially fast* if $\alpha_{L,D} > 0$.

LEMMA 3.3. *If at least one of Conditions 1, 2, or 3 of Lemma 3.1 is fulfilled, then the process* (X_t, P_x) *leaves the domain* D *uniformly exponentially fast and*

$$\alpha_{L,D} \geq (\sup_{x \in D} E_x \tau_D)^{-1} .$$

Proof. First, we note that according to Lemma 3.1, if at least one of Conditions 1, 2, 3 is valid, then $E_x \tau_D \leq c < \infty$ for $x \in D$. It is immediate from this that every moment of the random variable τ_D is bounded uniformly in $x \in D$. Actually, Chebyshev's inequality yields that $P_x\{\tau_D > T\} \leq c\, T^{-1}$. Just as it was done when proving Lemma 3.1, one can deduce from this inequality that there are $\tilde{c}, \tilde{\alpha} > 0$ such that for any integer n

$$P_x\{\tau_D > nT\} < \left(\frac{c}{T}\right)^n < c\, e^{-\tilde{\alpha} \cdot n} .$$

Hence it appears clear that the moments are bounded uniformly in x, and moreover, $E_x \exp\{\lambda \tau_D\}$ is finite for $\lambda < \tilde{\alpha}$.

We will derive a more delicate bound of the k-th order moment of the random variable τ_D. To this end, let us utilize the identity

$$\frac{1}{k} \tau_D^k = \int_0^{\tau_D} (\tau_D - t)^{k-1} dt .$$

Whence, using (1.4.3) and Fubini's theorem, we get

$$\frac{1}{k} E_x \tau_D^k = E_x \int_0^{\tau_D} (\tau_D - t)^{k-1} dt = \int_0^\infty E_x \chi_{\tau_D > t} \cdot (\tau_D - t)^{k-1} dt =$$

$$= \int_0^\infty E_x \chi_{\tau_D > t}\, \theta_t \tau_D^{k-1} dt = \int_0^\infty E_x \chi_{\tau_D > t} E_{X_t} \tau_D^{k-1} dt \leq$$

$$\leq \sup_{x \in D} E_x \tau_D^{k-1} E_x \int_0^{\tau_D} dt \leq c \cdot \sup_{x \in D} E_x \tau_D^{k-1} .$$

Therefore,

$$\frac{1}{k} \sup_{x \in D} E_x \tau_D^k \le c \cdot \sup_{x \in D} E_x \tau_D^{k-1} .$$

By induction, this implies that

$$E_x \tau_D^k \le k!\, c^k , \quad x \in D . \tag{3.6}$$

From (3.6) it follows that, for any $\lambda < c^{-1}$

$$E_x e^{\lambda \tau_D} = E_x \sum_{k=0}^{\infty} \frac{\lambda^k \tau_D^k}{k!} \le \sum_{k=0}^{\infty} (\lambda c)^k = \frac{1}{1-\lambda c} = K < \infty .$$

This inequality with the aid of the exponential Chebyshev's inequality leads to the bound

$$P_x\{\tau_D > t\} \le e^{-\lambda t} E_x e^{\lambda \tau_D} = K e^{-\lambda t}$$

which implies that $\alpha_{L,D} > \lambda$ for any $\lambda < c^{-1}$. Consequently, $\alpha_{L,D} \ge c^{-1}$, which proves Lemma 3.1. □

We observe that the bounds obtained in Lemma 3.1, together with the last result, allow $\alpha_{L,D}$ to be bounded from below via the coefficients of the operator and the size of the domain.

To clarify the meaning of the constant $\alpha_{L,D}$, let us assume for a while that L is a strongly elliptic, self-adjoint operator with smooth enough coefficients, D being a bounded domain with a good boundary. Then the problem

$$L\, u(x) = \lambda\, u(x), \; x \in D, \; u(x)\big|_{\partial D} = 0 , \tag{3.7}$$

has a complete set of eigenfunctions. Let $e_1(x), e_2(x), \cdots$ be the orthonormal system of eigenfunctions, $\lambda_1, \lambda_2, \cdots$ being the corresponding eigenvalues. One can assume that all λ_k are negative and the sequence λ_k monotonically tends to $-\infty$. As it follows from the results of §2.2, the function $u(t,x) = P_x\{\tau_D > t\}$ is a solution of the problem

$$\frac{\partial u}{\partial t} = Lu,\ t > 0,\ x \in D,\ u(0,x) = 1,\ u(t,x)\Big|_{x \in \partial D} = 0 .$$

One can solve this problem with the Fourier method and obtain for the solution the expression

$$u(t,x) = \sum c_k e_k(x) \exp\{\lambda_k t\},\ c_k = \int_D e_k(x)\,dx .$$

Noting that $\lambda_1 > \lambda_k$ for $k > 1$, and $e_1(x) > 0$ for $x \in D$, the above formula implies:

$$\lim t^{-1} \ln P_x\{\tau_D > t\} = \lambda_1 = -a_{L,D} .$$

Thus, $a_{L,D}$ is the first eigenvalue of problem (3.7).

If L is not a self-adjoint non-degenerate operator, then $a_{L,D}$ also has the meaning of the eigenvalue corresponding to the non-negative eigenfunction. It is possible to prove that such an eigenfunction exists and the corresponding eigenvalue is real and has multiplicity one (see, e.g. Wentzel and Freidlin [2], Chapter V).

If there are degenerations, then, of course the constant $a_{L,D}$ is still the first eigenvalue of problem (3.7), but now in a generalized sense. In particular, the corresponding "eigenfunction" may be nonzero only on a set of a dimension smaller than r. In this case the eigenvalue may have higher multiplicity. Formula (3.5) may be used for studying the first eigenvalue of problem (3.7).

Finally, note that $a_{L,D}$ may be equal to $+\infty$. In particular, such is the case when the projection of the domain D on some x^i axis is bounded, and in the direction x^i everywhere in $D \cup \partial D$ there is only a separated from zero drift $b^i(x)$, and also $a^{ii}(x) = 0$ for $x \in D$. We faced such a situation when considering parabolic equations.

If at some point $x \in D$ the operator does not degenerate, then, as can be easily shown, necessarily $a_{L,D} < \infty$. Hence, adding a non-degenerate term into the operator does not always accelerate the exit of the process from the domain.

§3.4 *Classification of boundary points*

We turn now to studying the behavior of the process (X_t, P_x) near the boundary ∂D of the domain D. As will be seen later on, this question is closely related to how one should prescribe boundary value problems for the operator L corresponding to the process (X_t, P_x). First of all we shall introduce the notion of the regularity of a boundary point for the operator L (for the process (X_t, P_x)). Roughly speaking as regular points one should consider those points on the boundary where Dirichlet boundary conditions may be satisfied. In Section 2.3 we discussed the regularity of boundary points for elliptic operators. In the study of degenerate equations, the notion of regularity gets, in a sense, more interesting and significant. Basically this may be explained as follows. As is known, in the elliptic case a boundary point is regular or non-regular simultaneously for all elliptic operators whose coefficients meet minor smoothness requirements. In this case the regularity is in essence a condition on the geometry of the domain. In particular, if the boundary is smooth enough in the vicinity of a point $x_0 \in \partial D$, then the point x_0 is regular. However, if there are degenerations, then generally speaking, regularity depends on the behavior of all the coefficients of the operator. The requirements of the smoothness of the coefficients and the boundary of the domain are not sufficient for ensuring the regularity of a boundary point.

In the case of degenerate equations, the notion of regularity grows more interesting also because, unlike elliptic equations, the smoothness properties of solutions of degenerate equations are not, generally speaking, of local nature. These properties depend on the properties of the boundary function and on the availability of boundary bounds. The latter, as we shall see below, may be deduced under the assumption of the regularity of the boundary in a fairly strong sense. The notion of regularity of a boundary point falls into several types.

A point $x_0 \in \partial D$ is said to be *regular* for the operator L (for the process (X_t, P_x) corresponding to L) in the domain D if, for any $\delta > 0$

$$\lim_{x \in D, x \to x_0} P_x\{\gamma_D \subset U_\delta(x_0) \cap \partial D\} = 1 , \tag{4.1}$$

where $U_\delta(x_0)$ is the δ-neighborhood of the point x_0, γ_D being the set of the limit points of the trajectory X_t as $t \to \tau_D = \inf\{t : X_t \notin D\}$.

It follows from Lemma 2.3.1 that, in the case when the operator L is non-degenerate, this notion of regularity coincides with that which is dealt with in the theory of elliptic differential equations. As will be seen later on, in the case of degenerate equations, regularity in this sense is also sufficient for there to exist a solution (generalized) of the homogeneous equation $Lu = 0$ satisfying given boundary conditions. When considering non-homogeneous equations or equations of the form $Lu - c(x)u = 0$, one has to introduce a stronger notion of regularity: relation (4.1) no longer ensures the existence of a solution taking on given boundary values.

We observe that in the case of degenerate equations, the regularity of a boundary point (in the sense of (4.1)) is not a local property. For example, let D be a domain in the plane (x,y) lying between two circles $x^2 + (y-1)^2 = 1$ and $x^2 + (y-2)^2 = 4$. The process $(X_t, Y_t; P_{x,y})$ is a motion clockwise along the family of the circles $x^2 + (y-c)^2 = c^2$ with velocity vanishing only at the point (0,0). In this case the point (0,0) is regular. However, changing the operator or the domain near the points (0.2) or (0,4) may destroy the regularity of the point (0,0).

Certainly, the definition of regularity may easily be modified by making it local: a point $x_0 \in \partial D$ will be termed *locally regular* for the operator L (for the process (X_t, P_x)) in the domain D, if for any neighborhood U of the point x and for arbitrary $\delta > 0$

$$\lim_{x \in D, x \to x_0} P_x\{\gamma_{D \cap U} \subset U_\delta(x_0) \cap \partial D\} = 1 ,$$

where $\gamma_{D\cap U}$ is the totality of the limit points of the trajectory X_t as $t \to \tau_{D\cap U} = \inf\{t : X_t \notin D \cap U\}$.

Obviously, local regularity implies the usual regularity. For verifying local regularity, one can make use of the following

LEMMA 4.1. *Suppose that the condition* $\mathfrak{A}_1$ *is fulfilled for any subdomain* $\tilde{D}$, $\tilde{D} \cup \partial\tilde{D} \subset D : P_x\{\gamma_{\tilde{D}} \neq \emptyset, \gamma_D \subset \partial\tilde{D}\} = 1$. *For a point* $x_0 \in \partial D$ *to be locally regular, it is sufficient that a neighborhood* U *of the point* x_0 *exist such that, on the set* $U \cap (D \cup \partial D)$ *a continuous function* $f(x)$ *(barrier) be defined, positive everywhere but the point* $x_0, f(x_0) = 0$, *having continuous first- and second-order derivatives in* $U \cap D$, *and such that* $Lf(x) \leq 0$ *for* $x \in U \cap D$.

Proof. With Ito's formula one can check that the process $Z_t = f(X_{\tau_{U\cap D}\wedge t})$ is a continuous non-negative supermartingale with respect to the family of σ-fields $\mathfrak{N}_t$ and the measure P_x, $x \in D$. From this, as was done when proving Lemma 3.2, one can deduce that, for any neighborhood $\mathcal{E}$ of the point $x_0 \in \partial D$, there is a $t_0 = t_0[X_.] > 0$ such that

$$\lim_{x \in D, x \to x_0} P_x\{X_t \in \mathcal{E} \cap D \text{ for } t_0 < t < \tau_{U\cap D}\} = 1 .$$

Noting that in every subdomain $\tilde{D} \subset D$ Condition $\mathfrak{A}_1$ holds, this equality yields that

$$\lim_{x \in D, x \to x_0} P_x\{X_t \in \mathcal{E} \cap D \text{ for } t_0 < t \leq \tau_{U\cap D}, \gamma_{D\cap U} \subset \partial D \cap \mathcal{E}\} = 1 ,$$

which implies the local regularity of the point x_0. □

Before citing sufficient conditions for regularity in terms of the coefficients of L, we will also introduce the notion of t-regularity and that of normal regularity.

A point $x_0 \in \partial D$ will be referred to as *t-regular* for the operator L (for the process (X_t, P_x)) in the domain D, if for any $t > 0$

$$\lim_{x \in D, x \to x_0} P_x\{\tau_D > t\} = 0 .$$

The t-regularity ensures the existence of a solution of equations of the form $Lu - c(x)u = f(x)$ satisfying given boundary conditions. It also guarantees the existence of solutions of the first boundary value problem for the corresponding parabolic equations.

It is easy to provide examples showing that a locally regular point may not be t-regular. However, t-regularity implies local regularity.

Finally, we will introduce the notion of the normal regularity of a boundary point. This notion will be needed when studying local properties of generalized solutions.

A point $x_0 \in \partial D$ will be called *normally regular* for the operator L in the domain D, if one can find a neighborhood $U = U_h(x_0)$ such that

$$\overline{\lim_{x \in D, x \to x_0}} \frac{E_x \tau_{U \cap D}}{|x - x_0|} < \infty ,$$

where $\tau_{U \cap D} = \inf\{t : X_t \notin U \cap D\}$.

If $E_x \tau_D < c < \infty$ for all $x \in D$, then for every normally regular point $x_0 \in \partial D$

$$\overline{\lim_{x \in D, x \to x_0}} \frac{E_x \tau_D}{|x - x_0|} < \infty .$$

In fact, by the strong Markov property

$$\begin{aligned} E_x \tau_D &= E_x \tau_{U \cap D} \chi_{\tau_{U \cap D} = \tau_D} + E_x \chi_{\tau_D > \tau_{U \cap D}} (E_{X_{\tau_{U \cap D}}} \tau_D + \tau_{U \cap D}) \le \\ &\le E_x \tau_{U \cap D} + c P_x\{X_{\tau_{U \cap D}} \notin \partial D\} . \end{aligned} \tag{4.2}$$

Using Ito's formula, one can check that $E_x |X_{\tau_{U \cap D}} - x|^2 \le c_1 E_x \tau_{U \cap D}$, where $x \in D$ and c_1 is some constant depending on h and on the maximum of the moduli of the coefficients of the operator. The last inequality yields:

$$P_x\{|X_{\tau_{U\cap D}} - x| > h/2\} \leq \frac{4c_1}{h^2} E_x \tau_{U\cap D} . \tag{4.3}$$

From (4.2) and (4.3) it follows that

$$\overline{\lim_{x\epsilon D, x\to x_0}} \frac{E_x \tau_D}{|x-x_0|} < \left(1 + \frac{4cc_1}{h^2}\right) \overline{\lim_{x\epsilon D, x\to x_0}} \frac{E_x \tau_{U\cap D}}{|x-x_0|} .$$

It is readily checked that the normal regularity of a boundary point implies its t-regularity. It is possible to provide an example showing that the converse is not true.

LEMMA 4.2. *For a point* $x_0 \epsilon \partial D$ *to be normally regular (and thus, t-regular) it suffices that at least one of the following conditions be fulfilled*:

1. *In a neighborhood* U *of the point* x_0 *a continuous function* f(x) *(a barrier) is defined such that* $0 \leq f(x) \leq k|x-x_0|$, $k < \infty$, *for* $x \epsilon D \cap U$, f(x) *having continuous first- and second-order derivatives for* $x \epsilon D \cap U$, *and such that* $L f(x) \leq -c < 0$.

2. *In a neighborhood of the point* $x_0 \epsilon \partial D$, *the direction cosines of the outward normal* $n(x) = (n_1(x), \cdots, n_r(x))$ *are defined and three times continuously differentiable, and* $\sum_{i,j=1}^{r} a^{ij}(x_0) n_i(x_0) n_j(x_0) > 0$.

3. *In a neighborhood of the point* $x_0 \epsilon \partial D$, *the direction cosines of the outward normal* n(x) *are defined and three times continuously differentiable. The point* x_0 *belongs to the closure of some subset of* ∂D *open with respect to* ∂D, *on which* $\sum_{i,j=1}^{r} a^{ij}(x) n_i(x) n_j(x) = 0$ *and* $\sum_{i=1}^{r} b^i(x_0) n_i(x_0) > 0$.

4. *The point* $x_0 \epsilon \partial D$ *may be touched by a half-space* Π *lying outside the set* $D \cap U_\delta(x_0)$ *for sufficiently small* $\delta > 0$, *where* $U_\delta(x_0) = \{x \epsilon R^r : |x-x_0| < \delta\}$. *At least one of the relations*

$$\sum_{i=1}^{r} b^i(x_0)\tilde{n}_i(x_0) > 0, \quad \sum_{i,j=1}^{r} a^{ij}(x_0)\tilde{n}_i(x_0)\tilde{n}_j(x_0) > 0$$

holds, where $\tilde{n}(x) = (\tilde{n}_1(x), \cdots, \tilde{n}_r(x))$ *is the normal vector to the boundary of the half-space* Π *directed toward the interior of the half-space.*

Proof. Suppose Condition 1 to be fulfilled. Together with the process (X_t, P_x), we shall consider the corresponding Markov family (X_t^x, P). Put $U^{(n)} = \{x \in D \cap U, \rho(x, \partial D) > \frac{1}{n}\}$, $\tau_t = \tau_{U \cap D} \wedge t$, $\tau_{U^{(n)}} = \inf\{t : X_t \notin U^{(n)}\}$, $\tau_t^{(n)} = \tau_{U^{(n)}} \wedge t$, $\tau^x_{U^{(n)}} = \inf\{t : X_t^x \notin U^{(n)}\}$, $\tau_t^{x,(n)} = \tau^x_{U^{(n)}} \wedge t$. Without loss of generality one can assume that in the domain $U^{(n)}$ the function $f(x)$ has bounded first- and second-order derivatives. Applying Ito's formula we obtain

$$f(X^x_{\tau_t^{x,(n)}}) = f(x) + \int_0^{\tau_t^{x,(n)}} (\nabla f(X_s^x), \sigma(X_s^x)\,dW_s) +$$

$$+ \int_0^{\tau_t^{x,(n)}} Lf(X_s^x)\,ds \le -c\,\tau_t^{x,(n)} + \int_0^{\tau_t^{x,(n)}} (\nabla f(X_s^x), \sigma(X_s^x)\,dW_s)\,.$$

Whence we conclude that $E_x \tau_t^{(n)} \le \frac{f(x)}{c}$. Letting $n \to \infty$, $t \to \infty$ we derive from this inequality that

$$E_x \tau_{U \cap D} \le \frac{f(x)}{c} < \frac{k \cdot |x - x_0|}{c}\,,$$

i.e. the point x_0 is normally regular.

Now, let Condition 2 hold. Then in a neighborhood of the point x_0 one can introduce new coordinates $(\tilde{x}^1, \cdots, \tilde{x}^r)$ so that the boundary ∂D is given by the equation $\tilde{x}^1 = 0$, the point x_0 is the origin and the domain D lies in the half-space $\tilde{x}^1 < 0$. By Condition 2, in the new coordinates the coefficient $\tilde{a}^{11}(x_0)$ is positive. As the function $f(x)$, we shall pick the function

$$f(x) = -\tilde{x}^1 - b\cdot(\tilde{x}^1)^2 + c\sum_{i=2}^{r}(\tilde{x}^i)^2 .$$

It is not hard to check that the function $f(x)$ obeys the hypotheses of Condition 1 whenever $b = \text{const.}$ is fairly large, c is small, and the neighborhood U is chosen small enough. By the preceding, this implies the claim of the lemma.

Now, let Condition 3 hold. In terms of the coordinates $(\tilde{x}^1, \cdots, \tilde{x}^r)$ the diffusion coefficient in the direction $\tilde{x}^1$ vanishes together with the derivatives in the directions tangent to ∂D. This follows from the fact that the diffusion along the normal to ∂D vanishes on a set, open with respect to ∂D and having x_0 as its limit point. This implies that $\tilde{b}^1(x_0) > 0$. In this case as the barrier one can choose the function

$f(x) = -\tilde{x}^1 + c\sum_{i=2}^{r}(\tilde{x}^i)^2$ for sufficiently small $c > 0$.

Finally, if Condition 4 is satisfied, then we shall make a linear change of variables so that the boundary of the half-space is given by the equation $\tilde{x}^1 = 0$, the point x_0 is the origin, and in a neighborhood of the point x_0 the domain D lies in the half-space $\tilde{x}^1 < 0$. It is clear that $\tilde{b}^1(x_0) > 0$. From the preceding condition it follows that the point x_0 is normally regular for the domain $\tilde{D} = \{x \in R^r, \tilde{x}^1 < 0\}$. Since $\tau_{D\cap U} \le \tau_{\tilde{D}\cap U}$ for a sufficiently small neighborhood U of the point x_0, the point x_0 is regular also for the domain D. □

A set $\Gamma \subseteq \partial D$ is said to be *uniformly normally regular,* if one can find $h > 0$ and $H < 0$, such that

$$\overline{\lim_{x\in D, x\to x_0}} \frac{E_x \tau_{U_h(x_0)\cap D}}{|x - x_0|} = H(x_0) \le H$$

uniformly in $x_0 \in \Gamma$.

From Lemma 4.2, one can deduce sufficient conditions for uniform normal regularity of a set $\Gamma \subset \partial D$. It suffices, for example, that the sizes of the neighborhoods U and the constants c and k in Condition 1 be chosen the same for all the point $x_0 \in \Gamma$. One can readily provide some sufficient conditions for the uniform normal regularity of a smooth part Γ of the boundary ∂D in terms of the coefficients of L.

As has already been said, in the elliptic case a boundary point is regular or irregular simultaneously for all operators with sufficiently smooth coefficients. In particular, adding the terms with first-order derivatives to an elliptic operator does not destroy regularity (Oleinik [1]). It is easy to provide an example showing that if the operator degenerates at a point x_0, then adding terms with first-order derivatives already can violate the regularity of the point x_0, or, on the contrary, can make an irreguiar point regular. The following theorem gives sufficient conditions for the addition of a first-order operator not to destroy t-regularity.

THEOREM 4.1. *Suppose that one can find a neighborhood* $\mathcal{E}$ *of a point* $x_0 \in \partial D$ *such that for* $x \in \mathcal{E}$ *there exists a bounded solution* $\phi(x) = (\phi_1(x), \cdots, \phi_r(x))$ *of the following system of linear algebraic equations*

$$(a^{ij}(x))^{1/2}\phi(x) = \tilde{b}(x), \tilde{b}(x) = (\tilde{b}^1(x), \cdots, \tilde{b}^r(x)) . \qquad (4.4)$$

Then the point x_0 *is* t*-regular for the operator* $L_1 = \frac{1}{2} \sum_{i,j=1}^{r} a^{ij}(x) \frac{\partial^2}{\partial x^i \partial x^j} + \sum_{i=1}^{r} b^i(x) \frac{\partial}{\partial x^i}$ *if and only if it is* t*-regular for the operator*

$$L_2 = L_1 + \sum_{i=1}^{r} \tilde{b}^i(x) \frac{\partial}{\partial x^i} .$$

Proof. Let (X_t^1, P_x^1) and (X_t^2, P_x^2) be Markov processes governed by the operators L_1 and L_2 respectively. By the Cameron-Martin-Girsanov formula, in the case when a bounded solution of system (4.4) exists, the measures P_x^1 and P_x^2 are absolutely continuous with respect to each other, provided they are considered in the space $C_{0,t}(R^r)$, $t < \infty$. From this, using the fact that the vector $\phi(x)$ is bounded, it is not difficult to deduce the claim of the theorem. □

It is easy to provide an example illustrating the importance of the assumption concerning the solvability of system (4.4). Note that the local regularity of a point is not preserved when a first-order operator is added even if the corresponding system has a bounded solution. This is due to the fact that the local regularity may be caused by the behavior of the process for infinite time. The existence of a bounded solution of system (4.4) is not sufficient for the absolute continuity of the measures on the infinite time interval.

A set $\Gamma \subset \partial D$ open with respect to ∂D, is called *inaccessible* for the operator L (for the process (X_t, P_x)) in the domain D, if for any $x \in D$,

$$P_x\{\gamma_D \cap \Gamma = \emptyset\} = 1 .$$

In what follows we shall see that, under minor extra conditions, boundary conditions should not be assigned on the inaccessible part of the boundary.

LEMMA 4.3. *Let $\Gamma \subset \partial D$ be an open set with respect to ∂D. Suppose that a neighborhood $\mathcal{E}$ of the set Γ exists such that $\inf_{x \in \Pi} P_x\{\gamma_D$ contains at least one regular point of the boundary $\partial D\}$ is positive, where Π is the part of the boundary of the set $\mathcal{E}$ belonging to D. Moreover, let at least one of the following conditions hold:*

1. *On the set $\mathcal{E} \cap D$, a twice continuously differentiable non-negative function $f(x)$ is defined for which $\lim_{x \in D, \rho(x,\Gamma) \to 0} f(x) = \infty$, $L f(x) \leq 0$ for $x \in \mathcal{E} \cap D$.*

2. *Suppose that the set Γ is such that the coefficients of the operator and the direction cosines $(n_i(x))$ of the outward normal to ∂D at the points $x \in \Gamma$ are three times continuously differentiable and*

$$\sum_{i,j=1}^{r} a^{ij}(x) n_i(x) n_j(x) = 0, \quad \sum_{i=1}^{r} b^i(x) n_i(x) < 0 \quad \textit{for } x \in \Gamma .$$

Then the set Γ is inaccessible.

Proof. Let Condition 1 be valid. Using Ito's formula one can verify that the process $Z_t = f(X_{t \wedge \tau_{\mathcal{E} \cap D}})$ is a continuous non-negative supermartingale with respect to the family of σ-fields $\mathcal{N}_t$ and to the measure P_x, $x \in \mathcal{E} \cap D$. Consequently, $E_x Z_t \leq f(x), t > 0$. Noting that $\lim_{t \to \infty} Zt$ exists P_x-a.s., this inequality implies that with probability arbitrarily close to 1, starting from some $t_0 = t_0[X_\cdot]$ and up to $\tau_{\mathcal{E} \cap D}$, the trajectories X_t will not enter a neighborhood $\mathcal{E}_1$ of the set Γ, whenever this neighborhood is small enough. On account of the strong Markov property of the process (X_t, P_x) one can deduce from this that the set Γ is a limit set only for those trajectories which enter the set Π at most a countable number of times. The probability of such an event is 0. This follows from the process (X_t, P_x) being strong Markovian and from the assumption that $\lim_{x \in \Pi} P_x\{\gamma_D$ contains regular points of the boundary$\}$ is positive. Thus $P_x\{\gamma_D \cap \Gamma = \emptyset\} = 1$.

Now, suppose that Condition 2 is fulfilled. It suffices to carry out the proof for the case when one can introduce new coordinates $(\tilde{x}^1, \cdots, \tilde{x}^r)$ so that $\Gamma \subset \{\tilde{x} \in R^1 : \tilde{x}^1 = 0\}$, $D \in \{\tilde{x} \in R^r : \tilde{x}^1 > 0\}$. Let us show that in a neighborhood of the set Γ one can define a function $f(\tilde{x})$ meeting the requirements listed in Condition 1. Put $f(\tilde{x}) = \frac{1}{\tilde{x}^1}$.

This function is non-negative for $\tilde{x} \in D$ and is infinite on Γ,

$$Lf(\tilde{x}) = \tilde{a}^{11}(\tilde{x}) \frac{1}{(\tilde{x}^1)^3} - \tilde{b}^1(\tilde{x}) \frac{1}{(\tilde{x}^1)^2} .$$

It is easily checked that on Γ the sign of the scalar product of $b(x)$ and the normal vector does not change when going over to the new coordinates. Hence, $\tilde{b}(\tilde{x})\big|_{\tilde{x}^1=0} > 0$. Since the function $\tilde{a}^{11}(x)$ is non-negative, smooth, and $\tilde{a}^{11}(\tilde{x})\big|_{\tilde{x}^1=0} = 0$, we conclude that it has a zero of at least second order as $\tilde{x}^1 \to 0$. This implies that in a sufficiently small neighborhood of the set Γ the inequality $Lf(\tilde{x}) \leq 0$ holds. Thereby, in view of the above proved, the set Γ is inaccessible. □

Note that the inaccessibility of the set $\Gamma \subset \partial D$ is not always determined completely by the behavior of the diffusion and coefficients in the direction normal to the boundary. For example, let $D = \{(x,y) \in R^2 : |x| < 1, 0 < y < 1\}$, $L = \frac{a}{2} \frac{\partial^2}{\partial x^2} - y \frac{\partial}{\partial y}$ (Fig. 1). It is easy to see that for $a > 0$ only the vertical segments $\{x = \pm 1, 0 \leq y \leq 1\}$ are regular for the operator L in the domain D. The remaining part of the boundary is inaccessible. However, if $a = 0$, then the part of the boundary $\{(x,y) : -1 \leq x \leq 1, y = 0\}$ consists of regular points. The other sides of the rectangle D are inaccessible.

Regular points and inaccessible sets do not exhaust all points of the boundary. Let $K = \{(x,y) \in R^2 : 1 < x^2 + y^2 < 4\}$ (Fig. 2), $L = \frac{\partial^2}{\partial y^2}$. We shall denote by A and B the points of the smaller circumference lying on the x-axis. The trajectories of the process corresponding to the operator L which start from a point $(x,y) \in K$ close to A, with large probability leave the domain K across a small neighborhood of the point A, provided $x > -1$. If a trajectory starts at a point $(x,y) \in K$, $x < -1$, then the first exit from K will be across the large circumference. The trajectories starting from the points close to B behave in a similar way. Hence, the points A and B are not regular: arbitrarily close to

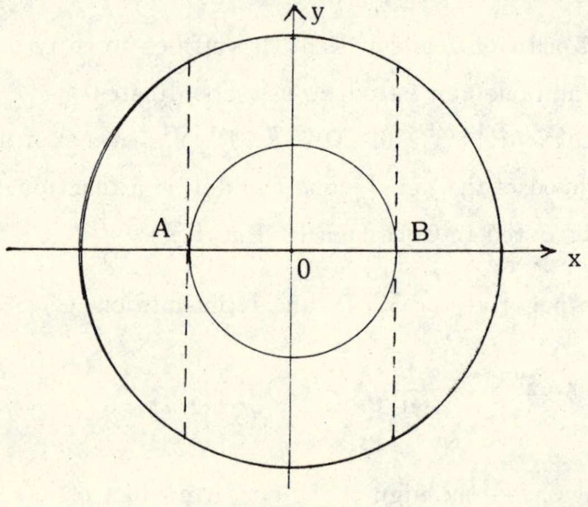

Fig. 2

these points, one can find points such that starting from them the process, with probability 1, leaves K far from these points. On the other hand, if the initial point approaches A (or B) over the set $\{(x,y) \in K : |x|<1\} = G$, then the probability that the exit from K will occur near A (near B) tends to 1. As it will be seen in the next section, such a behavior of the trajectories means that the solution (generalized) of the Dirichlet problem $Lu = 0$ in K takes boundary values at the points A and B, whenever approaching the points A and B over the set G. When approached other ways, the boundary conditions, generally speaking, are not satisfied. We shall also see that unless the boundary conditions at the points A and B are set, the generalized solution of the first boundary value problem is no longer unique: the value of the solution on the set $\{(x,y) \in K : |x| = 1\}$ is not defined.

Such points of the boundary will be called *partly regular*, or *regular over a set*. Notice that in the above example the exit from the domain K occurs in a finite time with probability $P_{x,y} = 1$, $(x,y) \in K$. We can reach the points A or B only from a subset of the domain K which has Lebesgue measure zero. One can provide examples in which entering one partly regular point has a positive probability starting from the points of some open set. In such cases it will apparently take infinite time for a trajectory to reach a partly regular point.

Suppose that a set $G \subset D$ has a point $x_0 \subset \partial D$ as a limit point, and let $G \cup \partial D$ be closed. The point $x_0 \in \partial D$ is said to be *regular over the set* G, whenever for any $\delta > 0$

$$\lim_{x \in G, x \to x_0} P_x\{\gamma_D \subset U_\delta(x_0) \cap \partial D\} = 1 .$$

In particular, if a point $x_0 \in \partial D$ is regular, then it is regular over the set G, consisting of the intersection of the domain D with an entire neighborhood of the point x_0. In the foregoing example the points A and B are regular over the set G.

It is not difficult to demonstrate that the set of partly regular (including also regular) points of a boundary is closed. Under minor extra assumptions (for example, if the domain D is bounded and at least one coefficient of the operator is nonzero in $D \cup \partial D$), the partly regular (including also regular) points and the inaccessible set exhaust all the boundary. As it will be seen below, in the general situation such is not the case.

If a point $x_0 \in \partial D$ is regular over a set $G_{x_0} \subset D$, then it is also regular over any subset $\tilde{G}_{x_0} \subset G_{x_0}$ having x_0 as its limit point. It is essential for what follows, that the sets G_x corresponding to partly regular points $x \in \partial D$ be chosen as large as possible. This requirement is connected with the uniqueness of the generalized solution of the corresponding boundary value problem. To formulate this requirement accurately, we shall denote by Γ_1 the set of the points of the boundary ∂D of the domain D which are regular for the operator L, Γ_2 denoting the set of partly regular (but not regular) points of the boundary. To every point $x \in \Gamma_2$, a set $G_x \subset D$ corresponds such that $G_x \cup \partial D$ is closed and

$$U_\delta(x_0) \cap G_x \neq \emptyset, \quad \lim_{y \in G_x, y \to x} P_y\{\gamma_D \subset U_\delta(x)\} = 1$$

for any $\delta > 0$. We shall assume that the sets G_x for partly regular points are picked so that the following condition be fulfilled: Let Q_1 be the set of trajectories $X_.$ of the process (X_t, P_x) for which $\gamma_D(X_.) \cap \Gamma_2 \neq \emptyset$; and let Q_2 be the set of the trajectories $X_.$ for which one can find $t = t(X_.) \in (0, \tau_D(X))$ and $x = x(X_.) \in \Gamma_2$ such that $X_s \in G_{x(X_.)}$ for $s \in (t(X_.), \tau_D(X_.))$. Then for any $y \in D$

$$P_y(Q_1) = P_y(Q_1 Q_2) \,. \tag{4.5}$$

Condition (4.5) means that a trajectory cannot enter a partly regular point $x \in D$ otherwise than over the set G. In particular, in our example $\Gamma_2 = \{A, B\}$ and condition (4.5) holds, provided $G_A = G_B = G$. Of course, condition (4.5) is somewhat bulky. But in what follows we shall mainly deal with the problems where there are no partly regular points. So, here we shall not go into the detailed analysis of condition (4.5).

As has been said, under minor supplementary conditions, the boundary ∂D of a domain D consists of the closed set $\Gamma_1 \cup \Gamma_2$ (the union of the set of regular and that of partly regular points) and of the inaccessible set $\Gamma_0 = \partial D \setminus (\Gamma_1 \cup \Gamma_2)$. However, in the general situation such is not the case. Let us return to example 3.1. One trivially checks that if $a(x) > 0$ for $x \in (-1,1)$, $a(-1) = a(1) = 0$ and $\beta > 0$, then with probability 1, starting from any point $(x,y) \in D$, γ_D is the segment $[-1,1]$ of the x-axis. The upper base of the rectangle D and its lateral sides are inaccessible. Of course the points of $[-1,1]$ are neither regular nor partly regular.

A set $\Gamma \subset \partial D$ is said to be an *attracting* set if

$$\lim_{x \in D, \rho(x,\Gamma) \to 0} P_x\{\gamma_D = \Gamma\} = 1 .$$

For equations degenerating only on the boundary, the attracting sets have been studied by Has'minskii [1]. First-order equations have attracting sets as well. For example, if the boundary of a domain D contains a stable limit cycle Γ of the dynamical system $\dot{X} = b(X)$, $b(x) = (b^1(x), \cdots, b^r(x))$, then the set Γ is an attracting one for the operator $\sum_{i=1}^{r} b^i(x) \frac{\partial}{\partial x^i}$.

Some sufficient conditions for a set $\Gamma \subset \partial D$ to be attracting may be found in the work of Has'minskii [1].

As will be seen, when setting a boundary value problem in a domain D, one can assign only a constant on the attracting set. Boundary conditions other than a constant will not be satisfied. If no conditions are assigned on such a set, then there is no uniqueness. The whole set Γ should be looked upon as one point of the boundary. On the other hand, at some boundary points, on the contrary, several boundary conditions may be assigned: when approaching such a point in different ways, the limits of the solution may be different. Such a point should be divided into an entire segment or a more extensive set. This effect is due to the fact that the trajectories of the process may enter such a point in essentially different ways. For example, consider the equation

$$Lu(x,y) = a(x,y)\Delta u - x\frac{\partial u}{\partial x} - y\frac{\partial u}{\partial y} = 0$$

in the disk $x^2+y^2<1$ with deleted center, and let $a(x,y) = 0$ in a neighborhood U of the point $(0,0)$, $a(x,y) > 0$ for $x^2+y^2 \notin U$. The boundary conditions should be assigned on the circumference $x^2+y^2 = 1$ and at the center of the disk. Note that if $f(\phi)$ is a continuous function on $[-\pi,\pi]$, $f(-\pi) = f(\pi)$, then at the point $(0,0)$ one can assign the condition: $\lim_{xy^{-1}=\mathrm{tg}\phi,(x,y)\to(0,0)} u(x,y) = f(\phi)$, $\phi \in [-\pi,\pi]$. Such a statement of the boundary value problem for the operator L is connected with the fact that with positive probability, the trajectories of the process corresponding to the operator L, can approach the origin remaining inside any non-zero angle with the vertex at the point $(0,0)$. We have faced a similar situation when considering the exterior Dirichlet problem.

Notice that "dividing" the boundary points may be the implication of the fairly complicated arrangement of the boundary of a domain. However, if $P_x\{\tau_D = \infty\} > 0$ for some $x \in D$, then the points of a smooth boundary may also get divided.

Generally speaking, similar to the points regular over some set, one can consider subsets of the boundary which are attracting only when approaching them in a special fashion. Considering such partly attracting sets involves no special difficulties. For brevity, we shall assume that there are no such sets. For the sake of brevity, we also exclude from consideration the cases when the sets γ_D are of a more complicated structure.

The union of all attracting sets of the boundary ∂D of the domain D will be denoted by Γ_3. Individual attracting sets will be denoted by $\Gamma_3^\alpha : \Gamma_3 = \bigcup_\alpha \Gamma_3^\alpha, \Gamma_3^\alpha \cap \Gamma_3^\beta = \emptyset$ for $\alpha \neq \beta$.

We shall say that Condition $\mathfrak{A}_2$ is fulfilled if the boundary ∂D of a domain D is a union of the inaccessible set Γ_0, the set of regular points Γ_1, the set of partly regular points Γ_2 for which condition (4.5) holds, and attracting sets Γ_3^α, $\Gamma_3 = \bigcup \Gamma_3^\alpha$.

§3.5 *First boundary value problem. Existence and uniqueness theorems for generalized solutions*

Given a bounded domain $D \subset R^r$ with a boundary ∂D, consider the operator

$$L = \frac{1}{2} \sum_{i,j=1}^{r} a^{ij}(x) \frac{\partial^2}{\partial x^i \partial x^j} + \sum_{i=1}^{r} b^i(x) \frac{\partial}{\partial x^i}, \quad x \in R^r,$$

with the non-negative characteristic form whose coefficients obey condition (1.3).

Suppose that (X_t^x, P) is a Markov family governed by the operator L in R^r, and let (X_t, P_x) be the corresponding Markov process. Denote by $(X_t^{D,x}, P)$ the Markov family derived from (X_t^x, P) by means of stopping at the first exit time from the domain D. This family is defined by the stochastic equation

$$X_t^{D,x} = x + \int_0^{t\wedge\tau_D^x} \sigma(X_s^{D,x})\, dW_s + \int_0^{t\wedge\tau_D^x} b(X_s^{D,x})\, ds,$$

$$\tau_D^x = \inf\{t : X_t^{D,x} \notin D\}, \quad x \in D \cup \partial D.$$

Here $\sigma(x)$ is a matrix with Lipschitz continuous elements such that $\sigma(x)\sigma^*(x) = (a^{ij}(x))$, W_s is a Wiener process, $b(x) = (b^1(x), \cdots, b^r(x))$. Clearly, τ_D^x coincides with the first exit time of the trajectory X_t^x from D. Together with the family $(X_t^{D,x}, P)$, we shall consider the corresponding Markov process (X_t^D, P_x^D). This process is obtained from (X_t, P_x) by stopping at the time $\tau_D = \inf\{t : X_t \notin D\}$.

Denote by T_t^D the semi-group of operators corresponding to the process (X_t^D, P_x^D) (to the Markov family $(X_t^{D,x}, P)$). These operators act in the space of bounded measurable functions on $D \cup \partial D$ by the formula

$$(T_t^D f)(x) = E_x^D f(X_t^D) = E\, f(X_t^{D,x}) .$$

Let A^D be the infinitesimal operator of this semi-group. The operator A^D is just that extension of the operator L which is useful for studying the first boundary value problem for the equation $Lu = 0$.

Suppose that Condition $\mathfrak{A}_2$ is fulfilled, and let $\psi(x)$ be a bounded function on $\Gamma_1 \cup \Gamma_2 \cup \Gamma_3 \subset \partial D$, constant on each component Γ_3^α of the set $\Gamma_3 : \psi(x) = \psi_\alpha$ for $x \in \Gamma_3^\alpha \subset \Gamma_3$.

By a *generalized solution of Dirichlet's problem* for the equation $Lu(x) = 0$ in the domain D with the boundary function $\psi(x)$, we mean a function $u(x)$, $x \in D$, which satisfies the boundary conditions

$$\lim_{x \to x_0} u(x) = \psi(x_0), \quad x_0 \in \Gamma_1$$

$$\lim_{x \in G_{x_0}, x \to x_0} u(x) = \psi(x_0), \quad x_0 \in \Gamma_2$$

$$\lim_{\rho(x, \Gamma_3^\alpha) \to 0} u(x) = \psi_\alpha, \quad \Gamma_3^\alpha \subset \Gamma_3$$

and obeys the equation

$$A^D u(x) = 0, \quad x \in D, \tag{5.1}$$

if on ∂D this function is defined as equal to $\psi(x)$ for $x \in \Gamma_1 \cup \Gamma_2 \cup \Gamma_3$, and for $x \in \Gamma_0$ we define it in an arbitrary fashion (but so that it remains bounded).

Let us show that such a definition of the generalized solution is correct. Indeed, let a bounded function $u(x)$ satisfy the above listed boundary conditions, be twice continuously differentiable, and obey the equation $Lu(x) = 0$ in D. We shall check that, in this case, $u(x) \in D_{A^D}$ and $A^D u(x) = 0$. Consider the expanding sequence of domains D_n such that $(D_n \cup \partial D_n) \subset D_{n+1} \subset D$ for all $n = 1, 2, \cdots$. Let $\tau^{(n),x} = \inf\{t, X_t^x \notin D_n\}$, $x \in D_n$. By Ito's formula

$$u(X^x_{t\wedge\tau^{(n),x}}) = u(x) + \int_0^{t\wedge\tau^{(n),x}} (\nabla u(X_s^x), \sigma(X_s^x)\, dW_s), \quad x \in D_n,$$

(the non-stochastic integral vanished because $Lu(X_s^x) = 0$ for $0 \le s \le \tau^{(n),x}$). This implies that $E\, u(X^x_{t\wedge\tau^{(n),x}}) = u(x)$. Next, letting n approach infinity and noting that $\tau^{(n),x} \to \tau_D^x$, $X^x_{\tau^{(n),x}\wedge t} \to X_t^{D,x}$, with probability 1, we obtain by the Lebesgue dominated convergence theorem:

$$E\, u(X^x_{t\wedge\tau_D^x}) = T_t^D u(x) = u(x), \quad t \ge 0. \tag{5.2}$$

The same equality holds at all the points $x \in \partial D$, since the process (X_t^D, P_x^D) stops on the boundary. From (5.2) it follows immediately that $A^D u(x) = 0$.

Suppose now that a bounded function $u(x)$ is twice continuously differentiable in D and $A^D u(x) = 0$. We will show that then $Lu(x) = 0$ for $x \in D$. Denote $Lu(x) = g(x)$ and apply Ito's formula:

$$u(X^x_{\bar\tau^x\wedge t}) = u(x) + \int_0^{t\wedge\bar\tau^x} (\nabla u(X_s^x), \sigma(X_s^x)\, dW_s) + \int_0^{t\wedge\bar\tau^x} g(X_s^x)\, ds, \tag{5.3}$$

where $\bar\tau^x$ is the first exit time from some domain D^1 such that $D^1 \cup \partial D^1 \subset D$, $x \in D^1$. Remembering that the coefficients of the operator L are bounded and using properties of stochastic integrals, it is not difficult to verify that for some $c_1 > 0$, for any $\delta > 0$ and sufficiently small t

$$P_x\{ \sup_{0 \le s \le t} |X_s^x - x| > \delta\} \le \exp\left\{-\frac{c_1\delta}{t}\right\}. \tag{5.4}$$

From (5.3) and (5.4) it follows that

$$A^D u(x) = \lim_{t \downarrow 0} \frac{E\, u(X_t^{D,x}) - u(x)}{t} = g(x)$$

for $x \in D$. Since $A^D u(x) = 0$, we infer from this that $L u(x) = g(x) = 0$.

Notice that if L is a first-order operator whose characteristics hit the boundary, then the solution of the first boundary value problem for the equation $L u(x) = 0$ may be obtained by carrying the boundary conditions along the characteristics towards the interior of the domain. In this case, it is the equation $A^D u = 0$ which is the condition of the constancy of the solution on the characteristics, and as regular points we have just those of the boundary across which the characteristics leave the domain. Unless the characteristics of a first-order equation hit the boundary, it is not sufficient to assign the function $u(x)$ on the regular part of the boundary in order to single out a unique solution. Therefore, in the case of first-order equations, the above introduced notion of the generalized solution is quite natural as well.

Finally, as will be seen in the next chapter, under some natural extra assumptions, the generalized solution introduced above is also the generalized solution in the sense of small parameter. This means that the solution of problem (5.1) coincides with $\lim_{\varepsilon \downarrow 0} u^\varepsilon(x)$, where $u^\varepsilon(x)$ is a solution (classical) of the non-degenerate problem: $(L + \varepsilon L_1)\, u^\varepsilon(x) = 0$ for $x \in D$, $u^\varepsilon(x)|_{\partial D} = \tilde{\psi}(x)$. Here $\tilde{\psi}(x)$ is a continuous function on ∂D coinciding with $\psi(x)$ on $\Gamma_1 \cup \Gamma_2 \cup \Gamma_3$; L_1 is a second-order operator, elliptic in $D \cup \partial D$, with Lipschitz continuous coefficients.

Hence, the above introduced definition of the solution for the first boundary value problem is correct.

Let Condition $\mathfrak{A}_2$ be fulfilled. In this case, one of the following three possibilities describes the behavior of the trajectories of the process (X_t, P_x) for $x \in D$, as $t \to \tau_D$:

1. $\gamma_D \cap (\Gamma_1 \cup \Gamma_2 \cup \Gamma_3) = \emptyset$;

2. $\gamma_D \cap (\Gamma_1 \cup \Gamma_2) \neq \emptyset$ and then the set γ_D consists of the only point $X_{\tau_D} \in \Gamma_1 \cup \Gamma_2$, i.e. $P_z\{\gamma_D \cap (\Gamma_1 \cup \Gamma_2) \neq \emptyset$, γ_D contains more than one point$\} = 0$, $z \in D$;

3. $\gamma_D \cap \Gamma_3 \neq \emptyset$ and then γ_D consists of some set $\Gamma_3^\alpha \subset \Gamma_3$, i.e. $P_z\{\gamma_D \cap \Gamma_3 \neq \emptyset$, γ_D does not coincide with some $\Gamma_3^\alpha \subset \Gamma_3\} = 0$, $z \in D$.

In fact, let us verify Item 2. At first, we put $\Gamma_2 = \emptyset$. This case will be basic for us in what follows. Since in this case the set Γ_1 is compact, we conclude that for any $\varepsilon > 0$ one can find $\delta > 0$ such that for every $x \in \Gamma_1$, $P_z(\gamma_D \subset U_\varepsilon(x)) > 1 - \varepsilon$ for $z \in U_\delta(x)$. This easily implies that

$$P_y \left\{ \begin{array}{l} \gamma_D \cap \Gamma_1 \neq \emptyset, \text{ one can find in } \gamma_D \\ \text{two points with the distance between them} \\ \text{more than } \varepsilon \end{array} \right\} \leq \varepsilon, \ y \in D .$$

Since ε is arbitrary, the last inequality implies the claim of Item 2 for the case when $\Gamma_2 = \emptyset$. The case when $\Gamma_2 \neq \emptyset$ may be considered in a similar way, provided one takes in account condition (4.5).

Now we dwell on Item 3. From the definition of attracting set it follows that there may be at most a countable number of different elements Γ_3^α in Γ_3, because each Γ_3^α has a neighborhood inside which there are no other attracting sets. Therefore, if $P_z\{\gamma_D \cap \Gamma_3 \neq \emptyset\} > 0$ for some $z \in D$, then one can find $\Gamma_3^\alpha \subset \Gamma_3$ such that $P_z\{\gamma_D \cap \Gamma_3^\alpha \neq \emptyset\} > 0$. Since for every $\varepsilon > 0$ one can find a neighborhood $\mathcal{E} \supset \Gamma_3^\alpha$ such that $P_y\{\gamma_D = \Gamma_3^\alpha\} > 1 - \varepsilon$ for $y \in \mathcal{E}$, we conclude that $P_z\{\gamma_D \cap \Gamma_3^\alpha \neq \emptyset\} = P_z\{\gamma_D = \Gamma_3^\alpha\}$ for any $z \in D$. This leads to the statement of Item 3.

If Condition $\mathfrak{A}_2$ holds, then the following equality defines the random variable $\hat{\psi}$ on the space $(C_{0,\infty}(R^r), \mathfrak{N})$ almost everywhere with respect to the measure P_x, $x \in D$:

$$\hat{\psi} = \begin{cases} \psi(X_{\tau_D}), & \text{if } X_{\tau_D} \in \Gamma_1 \cup \Gamma_2 , \\ \psi_\alpha , & \text{if } \gamma_D = \Gamma_3^\alpha \subset \Gamma_3 , \\ 0 , & \text{if } \gamma_D \cap (\Gamma_1 \cup \Gamma_2 \cup \Gamma_3) = \emptyset . \end{cases}$$

THEOREM 5.1. *Suppose that Condition $\mathfrak{A}_2$ holds, and let the function $\psi(x)$, $x \in \Gamma_1 \cup \Gamma_2 \cup \Gamma_3$, be bounded, continuous and take constant values $\psi(x) = \psi_\alpha$ for $x \in \Gamma_3^\alpha \subset \Gamma_3$. Then the solution of problem (5.1) exists and is given by the formula $u(x) = E_x^D \hat{\psi} = E_x \hat{\psi}$, $x \in D$.*

Proof. We shall check that the function $u(x) = E_x^D \hat{\psi}$ satisfies the equation $A^D u(x) = 0$. To this end, we shall make use of the strong Markov property of the process (X_t, P_x). Let θ_t be the shift operators connected with the process (X_t, P_x) (see §1.4). Using the strong Markov property we derive

$$T_t^D u(x) = E_x^D u(X_t^D) = E_x^D E_{X_t^D}^D \hat{\psi} =$$

$$= E_x E_{X_{t \wedge \tau_D}} \hat{\psi} = E_x \theta_{t \wedge \tau_D} \hat{\psi} .$$

(5.5)

We now show that $\theta_{\tau_D \wedge t} \hat{\psi} = \hat{\psi}$. For this, it suffices to verify the validity of the following equalities: 1) $\theta_{\tau_D \wedge t} \{\gamma_D = \Gamma_3^\alpha\} = \{\gamma_D = \Gamma_3^\alpha\}$; 2) $\theta_{\tau_D \wedge t} \{\gamma_D \cap (\Gamma_1 \cup \Gamma_2 \cup \Gamma_3) = \emptyset\} = \{\gamma_D \cap (\Gamma_1 \cup \Gamma_2 \cup \Gamma_3) = \emptyset\}$; 3) $\theta_{\tau_D \wedge t} \{\lim_{t \uparrow \tau_D} X_t = a\} = \{\lim_{t \uparrow \tau_D} X_t = a\}$.

Relations 1 and 2 are straightforward from the fact that if the sequence X_{t_i} converges to some point $a \in D \cup \partial D$ as $t_i \to \infty$, then the sequence $\theta_t X_{t_i - t}$ also converges to this point. In order to check equality 3, we observe that $\{\lim_{t \uparrow \tau_D} X_t = a\} = \{\lim_{t \to \infty} X_t = a, \tau_D = \infty\} \cup \{\lim_{t \uparrow \tau_D} X_t = a, \tau_D < \infty\}$. The invariance of the first of these events with respect to the operators $\theta_{\tau_D \wedge t}$ follows from the argument just cited. The invariance of the second event results from the fact that the relations

$$\theta_{\tau_D \wedge t}\{X_{\tau_D} = a\} = \{X_{\tau_D} = a\}$$

are fulfilled for any $t > 0$.

Therefore, $\theta_{\tau_D \wedge t}\psi = \psi$, and from (5.5) we conclude that $T_t^D u(x) = u(x)$, $t > 0$. Hence it appears clear that $u(x) \in D_{A^D}$ and $A^D u(x) = 0$.

Now we turn to verifying that the boundary conditions are fulfilled. Let $x_0 \in \Gamma_1$. Since the boundary function is continuous at the point x_0, we conclude that for a preassigned $\varepsilon > 0$ one can find $\delta_1 > 0$ such that $|\psi(x) - \psi(x_0)| < \varepsilon$ for $|x - x_0| < \delta_1$, $x \in \Gamma_1 \cup \Gamma_2 \cup \Gamma_3$. By using the regularity of the point x_0, one can choose $\delta_2 > 0$ so that for $x \in U_{\delta_2}(x_0)$

$$P_x\{\gamma_D \subset (\partial D \cap U_{\delta_1}(x_0))\} > 1 - \varepsilon .$$

Let $\Lambda = \{\gamma_D \subset (\partial D \cap U_{\delta_1}(x_0))\}$, χ_Λ being the indicator of the set $\Lambda \subset C_{0,\infty}(R^r)$. Then, for $|u(x) - \psi(x_0)|$, we arrive at the bound:

$$|u(x) - \psi(x_0)| \le E_x^D |\hat{\psi} - \psi(x_0)| \chi_\Lambda + E_x^D |\hat{\psi} - \psi(x_0)|(1 - \chi_\Lambda) \le$$

$$\le \varepsilon + 2 \sup_x |\psi(x)| \cdot \varepsilon$$

for $|x - x_0| < \delta_2$, $x \in D$. Noting that ε is arbitrary, we conclude that $\lim_{x \to x_0} u(x) = \psi(x_0)$, $x_0 \in \Gamma_1$. If the point x_0 is regular only over some set G_{x_0}, then the boundary conditions are satisfied when approaching x_0 over the set G_{x_0}. The proof is the same as in the case of a regular point, but $P_x(\Lambda) > 1 - \varepsilon$ for the points $x \in G_{x_0} \cap U_{\delta_2}(x_0)$. The definition of attracting set implies immediately that the function $u(x)$ tends to ψ_a as x approaches Γ_3^a. □

THEOREM 5.2. *Suppose that the hypotheses of Theorem 5.1 hold. For a solution of problem (5.1) to be unique in the class of bounded functions, it is necessary and sufficient that Condition* $\mathfrak{A}_1$ *be fulfilled.*

Proof. Let Condition $\mathfrak{A}_1$ be fulfilled:

$$P_x\{\gamma_D \neq \emptyset,\ \gamma_D \subset \partial D\} = 1\ ,\ x \in D\ .$$

We shall show that in this case the solution is unique. If $A^D v(x) = 0$, then $T_t^D v(x) \equiv v(x)$. In fact, if $v(x)$ belonged to D_{A^D}, then the equalities

$$\frac{d\,T_t^D v}{dt} = T_t^D A^D v\ ,\ \lim_{t \downarrow 0} T_t^D v = v \tag{5.6}$$

would be valid (see, e.g. Dynkin [3]). If $A^D v(x) = 0$, then (5.6) yields that $T_t^D v = v$ for $t \geq 0$.

Noting that (X_t^D, P_x^D) is a Markov process, we conclude that the process $v(X_t^D) = Z_t$ is a continuous, bounded martingale. In order to prove uniqueness, clearly it is sufficient to show that the solution of problem (5.1) with zero boundary condition is zero. By Doob's theorem, the limit $v_\infty = \lim_{t\to\infty} v(X_t^D)$ exists. Conditions $\mathfrak{A}_1$ and $\mathfrak{A}_2$ imply that, with probability 1, starting from any $x \in D$, the trajectory X_t^D has the following opportunities. First, as its limit point, the trajectory X_t^D has a point from Γ_1 or from Γ_2, and in the last case, by virtue of condition (4.5), the trajectories approach this point along the corresponding set G_x. The other opportunity is that the set γ_D for the trajectory coincides with some $\Gamma_3^\alpha \subset \Gamma_3$. Whence we conclude that if $v(x)$ is a solution of problem (5.1) with zero boundary conditions, then

$$P_x^D\{\lim_{t\to\infty} v(X_t^D) = 0\} = 1\ ,\ x \in D\ .$$

Remembering that $v(x)$ is bounded, on the basis of Lebesgue's dominated convergence theorem we obtain from the equality $v(x) = E_x^D v(X_t^D)$, that

$$v(x) = \lim_{t\to\infty} E_x^D v(X_t^D) = E_x^D \lim_{t\to\infty} v(X_t^D) = 0\ .$$

Thus, if Condition $\mathfrak{A}_1$ and $\mathfrak{A}_2$ hold, then the solution of problem (5.1) is unique.

To prove the necessity of Condition $\mathfrak{A}_1$, we shall introduce the function $w(x) = P_x\{\gamma_D \neq \emptyset, \gamma_D \subset \partial D\}$. Just as when proving the existence theorem, one can verify that the function $w(x)$ takes the boundary value $\psi(x) = 1$ for $x \in \Gamma_1 \cup \Gamma_2 \cup \Gamma_3$ and obeys the equation $A^D w = 0$. Obviously, the function $v(x) \equiv 1$ also satisfies the equation $A^D v(x) = 0$ and the boundary conditions with $\psi(x) \equiv 1$. If Condition $\mathfrak{A}_1$ is not fulfilled, then $w(x) \not\equiv v(x)$, and hence, the solution of problem (5.1) is not unique. □

REMARK 1. If Conditions $\mathfrak{A}_1$ and $\mathfrak{A}_2$ hold, then the maximum principle is fulfilled for the bounded solution of problem (5.1): $u(x) \leq \sup_{y \in \Gamma_1 \cup \Gamma_2 \cup \Gamma_3} \psi(y)$, $x \in D$. This follows from the equality

$$u(x) = E_x^D \lim_{t \to \infty} u(X_t^D) . \tag{5.7}$$

If Condition $\mathfrak{A}_1$ does not hold, then the maximum of the function $u(x)$ may be attained inside the domain.

REMARK 2. Each solution of problem (5.1), continuous on the set $\Gamma_3^a \subset \Gamma_3$, takes a constant value on Γ_3^a. This results from (5.7), if one takes into account that $\lim_{\rho(x,\Gamma_3^a) \to 0} P_x\{\gamma_D = \Gamma_3^a\} = 1$.

REMARK 3. Let us consider the peculiarities which emerge when considering the first boundary value problem for the equation

$$Lu(x) - c(x)\,u(x) = f(x), \quad c(x) \geq c_0 > 0 . \tag{5.8}$$

Here L is the same operator as before, $f(x)$ and $c(x)$ being bounded continuous functions. The statement of the boundary value problem for such an equation is simplified. The boundary conditions should be assigned only at boundary points which are t-regular for the operator L. Boundary conditions should not be defined at the boundary points which are regular, but not t-regular, as well as on attracting sets. Let Γ_1 be the set of t-regular points of the boundary ∂D of the domain D. Suppose that the set $\partial D \setminus \Gamma_1$ is inaccessible for the process (X_t, P_x) corresponding to the operator L. To define the generalized solution of the first boundary value problem for equation (5.8) in a domain $D \subset R^r$ it is suitable to use the characteristic operator $\mathfrak{A}$ of the process (X_t, P_x) introduced by Dynkin:

$$\mathfrak{A} f(x) = \lim_{U \downarrow x} \frac{E_x f(X_{\tau_U}) - f(x)}{E_x \tau_U},$$

where U is an arbitrary neighborhood of the point x, $\tau_U = \inf\{t : X_t \notin U\}$ (see Dynkin [3]). This operator is an extension of the operator L.

By a generalized solution of the first boundary value problem for equation (5.8) in the domain D, we mean the function $u(x)$ for which

$$\mathfrak{A} u(x) - c(x) u(x) = f(x), \quad \lim_{x \to x_0 \in \Gamma_1} u(x) = \psi(x_0).$$

It is not difficult to prove that, under the above assumptions, the generalized solution is unique and defined by the formula

$$u(x) = E_x \psi(X_{\tau_D}) \exp\left\{-\int_0^{\tau_D} c(X_s)\, ds\right\} -$$

$$- E_x \int_0^{\tau_D} f(X_s) \exp\left\{-\int_0^s c(X_{s_1})\, ds_1\right\} ds .$$

The correct statement of the first boundary value problem for equation (5.8) was first put forward by Ficera [1, 2]. This problem is dealt with in the monograph of Friedman [2].

REMARK 4. Consider now the statement of the first boundary value problem for the equation $Lu = 0$, when Condition $\mathfrak{A}_1$ does not hold. We shall discuss only the case of bounded domain D. In this case, starting from $x \in D$, the set γ_D is non-empty with probability 1, and Condition

$\mathfrak{A}_1$ may be formulated as follows: $P_x\{\gamma_D \subset \partial D\} = 1$, $x \in D$. It follows from Theorem 5.2 that, if Condition $\mathfrak{A}_1$ is not fulfilled, then the generalized solution is non-unique. In order to understand how a unique solution should be singled out, let us consider some boundary problems which do not satisfy Condition $\mathfrak{A}_1$. First, as was said in Chapter II, the generalized solution of the mixed problem in the cylinder $\Pi = (0, \infty) \times D$ for the equation $\frac{\partial u}{\partial t} = L\,u(t,x)$ is unique. Generally speaking, it is possible that the limit $\lim_{t\to\infty} u(t,x)$ does not exist. However, if such a limit exists, then $u(x) = \lim_{t\to\infty} u(t,x)$ must be a solution (generalized) of the problem $L\,u(x) = 0$, $x \in D$, $u(x)\big|_{\Gamma_1} = \psi(x)$, where Γ_1 is the part of the boundary of the domain D which is t-regular for the operator L. The set $\partial D \setminus \Gamma_1$ is assumed to be inaccessible. If Condition $\mathfrak{A}_1$ is fulfilled, then the function $u(x)$ is defined in a unique way. Unless Condition $\mathfrak{A}_1$ is fulfilled, the equality $u(x) = \lim_{t\to\infty} u(t,x)$ may be looked upon as the condition singling out a unique generalized solution of the problem $L\,u = 0$, $u(x)\big|_{\Gamma_1} = \psi(x)$. In the case when Condition $\mathfrak{A}_1$ does not hold, the solution, singled out in such a way, depends on the initial function $f(x)$. Therefore the problem $L\,u = 0$, $u\big|_{\Gamma_1} = \psi(x)$, may be regularized by considering the problem of stabilizing the solution of the corresponding evolution equation.

Another way of regularization consists in considering the equation $L\,u^\varepsilon(x) - \varepsilon c(x) u^\varepsilon(x) = 0$, $c(x) \ge c_0 > 0$, $u^\varepsilon(x)\big|_{\Gamma_1} = \psi(x)$. Such a problem already has a unique solution. If ε tends to zero, we obtain a solution of the equation $L\,u = 0$. Notice, that such a solution is sure to attain its maximum on the boundary. The regularization with the help of the corresponding evolutionary equation may lead to a solution with the maximum inside the domain.

Finally, a third way of regularizing is to single out the generalized solution as the limit of solutions of perturbed equations for which uniqueness holds.

Consider the simplest example. Let $D = (-1,1) \subset R^1$, $\ell = a(x) \frac{d^2}{dx^2} - x \frac{d}{dx}$, where $a(x) = 0$ for $x \in [-0.9, 0.9]$ and is positive outside this segment. The end-points of the segment $[-1,1]$ are regular for the operator L. In the domain $D = (-1,1)$ Condition $\mathfrak{A}_1$ does not hold: with a positive probability, the set γ_D consists of the point 0. To single out a unique solution of the first boundary value problem

$$\ell u(x) = 0, \; -1 < x < 1, \; u(-1) = \psi_-, \, u(+1) = \psi_+ , \tag{5.9}$$

the condition $u(0) = c$ may be assigned in addition to the conditions at the end-points of the segment $[-1,1]$. It is easy to make sure that the solution of problem (5.9) exists and is unique in the class of bounded functions for any c. If one considers the limit of the solution of the problem $\frac{\partial u}{\partial t} = \ell u(t,x), u(t, \pm 1) = \psi_\pm, u(0,x) = f(x)$, as $t \to \infty$, then one should set $c = f(0)$. When regularizing with the aid of killing (i.e. with the aid of introducing the term $-\varepsilon c(x) u(x)$ into the equation) one should, obviously, assume that $c = 0$. Finally, when regularizing with a perturbed equation, in the case of general position, c should be assumed to equal one of the boundary values: either ψ_+ or ψ_-. In the next chapter, this point is considered more accurately.

Therefore, in this example, when Condition $\mathfrak{A}_1$ does not hold, singling out a unique solution may be performed by means of indicating some set (in the present case—the point 0) of interior points at which additional conditions should be assigned. Generally speaking, this set should be picked out so that one may describe with its help all possible behaviors of the trajectories as $t \to \infty$.

Some questions of singling out such a set are treated in Freidlin [5]. Note that a problem of that kind has been considered in Chapter II, when studying the outer Dirichlet problem.

§3.6 *The Hölder continuity of generalized solutions. Existence conditions for derivatives*

It is easy to provide examples showing that the generalized solution of the first boundary value problem may be discontinuous even if the coefficients of the equation, the boundary of the domain, and the boundary function are infinitely differentiable. However, a more thorough analysis shows that these discontinuities arise due to "unsufficiently regular" behavior of the trajectories of the corresponding process near the boundary. If all points of the accessible part of the boundary are regular, the boundary function is continuous, and, say, some coefficient of the operator vanishes nowhere, then the generalized solution of the first boundary value problem is already continuous. On the other hand, as it will be seen below, even if the operator does not degenerate near the boundary, the generalized solution may have no derivatives inside the domain.

In order to understand what smoothness properties may be expected, let us consider the following example (Freidlin [4]) which we shall repeatedly return to.

EXAMPLE 6.1. Suppose that the domain D is a square: $D = \{(x,y) \in R^2 : |x| < 1, |y| < 1\}$. Denote by $\phi(x,y)$ a non-negative, infinitely differentiable function on R^2, even in y and vanishing outside the ε-neighborhood of the boundary of the square D, $\phi(x,y) \le 1$. Consider in D the Dirichlet problem

$$\ell u(x,y) = \frac{a}{2} \frac{\partial^2 u}{\partial x^2} + \beta y \frac{\partial u}{\partial y} + \frac{1}{2} \phi^2(x,y) \Delta u = 0 , \qquad (6.1)$$

$u(x,y)\big|_{\partial D} = y$; a, β are positive constants.

The operator ℓ does not degenerate in a neighborhood of the boundary, $a > 0$; hence, all boundary points are regular in the strongest sense. The trajectories of the corresponding random process leave the domain uniformly exponentially fast. By Theorems 5.1 and 5.2, a unique generalized solution of problem (6.1) exists. By symmetry properties we conclude that $u(x,y) = -u(x,-y)$ and $u(x,0) = u(0,0) = 0$. Since Condition $\mathfrak{A}_1$ is fulfilled in the domain $D \cap \{y > 0\}$, we deduce that the maximum principle holds in this domain, which implies that $u(x,y) > 0$ for $y > 0$. In a similar way we deduce that $u(x,y) < 0$ for $y < 0$. If ε is small enough, then $u(x,y) > 0.99$ for $1 - y \le \varepsilon$ and $u(x,y) < -0.99$ for $1 + y \le \varepsilon$.

Denote by $(X_t^{x,y}, Y_t^{x,y}; P)$ the Markov family in R^2 governed by the operator ℓ:

$$X_t^{x,y} - x = \int_0^t \sqrt{a+\phi^2(X_s^{x,y}, Y_s^{x,y})}\, dW_s^1 \,,$$
$$Y_t^{x,y} - y = \int_0^t \phi(X_s^{x,y}, Y_s^{x,y})\, dW_s^2 + \int_0^t \beta Y_s^{x,y}\, ds \,. \tag{6.2}$$

Denote by $(X_t, Y_t; P_{x,y})$ the Markov process corresponding to $(X_t^{x,y}, Y_t^{x,y}; P)$. Let D^ε be the domain obtained from D after eliminating the ε-neighborhood of the boundary. Denote $\tau_{D^\varepsilon} = \tau^\varepsilon = \inf\{t : (X_t, Y_t) \notin D^\varepsilon\}$. The strong Markov property of the process $(X_t, Y_t; P_{x,y})$ yields the relation $u(x,y) = E_{x,y} u(X_{\tau^\varepsilon}, Y_{\tau^\varepsilon})$.

Let $y > 0$. Since $P_{0,y}\{Y_{\tau^\varepsilon} > 0\} = 1$,

$$u(0,y) = E_{0,y} u(X_{\tau^\varepsilon}, Y_{\tau^\varepsilon}) \geq {}^{99}/_{100} P_{0,y}\{Y_{\tau^\varepsilon} = 1-\varepsilon\}\,. \tag{6.3}$$

We shall estimate the right-hand side of this inequality. The motion along the y-axis in the domain D^ε is deterministic. Integrating the equation $\dot{Y} = \beta Y$ with the initial condition $Y_0 = y$, we derive that it takes the time $t(y) = \frac{1}{\beta} \ln \frac{1-\varepsilon}{y}$ to reach the point $y = 1-\varepsilon$.

Observe that

$$P_{0,y}\{Y_{\tau^\varepsilon} = 1-\varepsilon\} = P_{0,y}\{\sup_{0 \leq s \leq t(y)} |X_s| < 1-\varepsilon\}\,.$$

Since the component X_s in the domain D^ε is a one-dimensional Markov process with the generator $\frac{a}{2}\frac{d^2}{dx^2}$, the function

$$v_\varepsilon(t,x) = P_x\{\sup_{0 \leq s \leq t} |X_s| < 1-\varepsilon\}$$

is a solution of the problem

$$\frac{\partial v_\varepsilon}{\partial t} = \frac{\alpha}{2}\frac{\partial^2 v_\varepsilon}{\partial x^2},\ t > 0,\ |x| < 1-\varepsilon,\ v_\varepsilon(t, \pm(1-\varepsilon)) = 0,\ v_\varepsilon(0,x) = 1\ .$$

By solving this problem with the Fourier method we obtain for $v_\varepsilon(t,0)$ the expression

$$v_\varepsilon(t,0) = \sum_{n=1}^{\infty} c_n \exp\left\{-\frac{\alpha\pi^2 t\cdot(2n-1)^2}{8(1-\varepsilon)^2}\right\}, \tag{6.4}$$

where c_n are the Fourier coefficients of the initial function $f(x) \equiv 1$. From (6.4) it follows that

$$v_\varepsilon(t,0) > K_1 \exp\left\{-\frac{\alpha\pi^2 t}{8(1-\varepsilon)^2}\right\}$$

for some $K_1 > 0$. Using this bound we obtain

$$P_{0,y}\{Y_{\tau^\varepsilon} = 1-\varepsilon\} = P_{0,y}\{\sup_{0\le s\le t(y)} |X_s| < 1-\varepsilon\} >$$

$$> K_1 \exp\left\{-\frac{\alpha\pi^2}{8\beta(1-\varepsilon)^2}\ln\frac{1-\varepsilon}{y}\right\},$$

and, thereby, in view of (6.3),

$$u(0,y) \ge 0.99\cdot P_{0,y}\{Y_{\tau^\varepsilon} = 1-\varepsilon\} > K_2\, y^{\frac{\alpha\pi^2}{8\beta(1-\varepsilon)^2}}\ .$$

Whence, taking into account that $u(0,0) = 0$, we infer that the function $u(x,y)$ may have no derivatives at the point $(0,0)$, unless supplementary assumptions on the magnitude of α/β are made. And what is more, for any $\gamma \in (0,1]$, one can find α/β so small that the function $u(x,y)$ will not satisfy the Hölder condition with the exponent γ. If, at last, we put $\alpha = 0$, then, generally the solution will be discontinuous on the segment $[-1+\varepsilon, 1-\varepsilon]$ of the x-axis (then Condition $\mathfrak{A}_1$ does not hold; the generalized solution is non-unique, but among the solutions there are no continuous ones).

The term $\frac{1}{2}\phi^2(x,y)\Delta$ has been introduced into the operator L solely to eliminate the possibility of "rough" solutions due to degeneration on the boundary. If $\phi(x,y) \equiv 0$, then all the bounds of the function $|u(0,y) - u(0,0)|$ are preserved, only $\varepsilon = 0$ should be substituted into them.

Now we are going to obtain an upper bound for $u(0,y) - u(0,0)$ assuming for brevity $\phi(x,y) \equiv 0$. Since in this case the motion along the y-axis is deterministic everywhere in R^2, we have

$$|u(0,y) - u(0,0)| = E_{0,y}|Y_{\tau_D}| = \int_0^\infty Y_t^y P_\tau(t)\,dt\,, \tag{6.5}$$

where Y_t^y is a solution of the equation $\dot{Y}_t^y = \beta Y_t^y, Y_0^y = y$; $P_\tau(t)$ is the density function of the random variable $\tau = \inf\{t : |X_t| = 1\}$ starting from the point $X_0 = 0$.

From the above cited formulae it follows that

$$Y_t^y = y\exp\{\beta t\},\ P_\tau(t) = -\frac{\partial v_0}{\partial t}(0,t) =$$

$$= \sum_{n=1}^\infty \frac{c_n \alpha\pi^2(2n-1)^2}{8}\exp\left\{-\frac{\alpha\pi^2(2n-1)^2 t}{8}\right\} < K_3\exp\left\{-\frac{\alpha\pi^2 t}{8}\right\}$$

for a certain $K_3 > 0$ independent of t. Substituting this bound into (6.5) we derive

$$|u(0,y) - u(0,0)| < K_3 y \int_0^\infty e^{\beta t - \frac{\alpha\pi^2}{8}t}\,dt\,. \tag{6.6}$$

If $\beta < \frac{\alpha\pi^2}{8}$, then the integral on the right-hand side of (6.6) converges, and the function $u(x,y)$ is Lipschitz continuous at the point $(0,0)$. However, if $\beta > \frac{\alpha\pi^2}{8}$, then the foregoing bound for $|u(0,y) - u(0,0)|$ implies that the function $u(x,y)$ is not Lipschitz continuous.

Let us show that if $\alpha \neq 0$, then the generalized solution of problem (6.1) is sure to satisfy some Hölder condition. For brevity, we assume

$\phi \equiv 0$. Let $\kappa = \frac{\alpha\pi^2}{8\beta} \wedge 1$. Since $|Y_{\tau_D}| \le 1$ and $\kappa \le 1$, we have

$$|u(0,y) - u(0,0)| = E_{0,y}|Y_{\tau_D}| \le E_{0,y}|Y_{\tau_D}|^\kappa \le$$

$$\le \int_0^\infty (Y_t^y)^\kappa p_\tau(t)\,dt \le K_3 y^\kappa \int_0^\infty \exp\left\{\left(\kappa\beta - \frac{\alpha\pi^2}{8}\right)t\right\} dt \le K_3 K_4 y^\kappa ,$$

where $K_4 = \int_0^\infty \exp\left\{\left(\kappa\beta - \frac{\alpha\pi^2}{8}\right)t\right\} dt < \infty$ for $\kappa < \frac{\alpha\pi^2}{8\beta}$.

Thus, we have obtained rather precise upper and lower bounds

$$K_5 y^{\frac{\alpha\pi^2}{8\beta}} \le |u(0,y) - u(0,0)| \le K_6 y^\kappa$$

for any $\kappa < 1 \wedge \frac{\alpha\pi^2}{8\beta}$.

Therefore, in this example, one can ensure only Hölder continuity of the generalized solution, unless special assumptions are made on the coefficients. To have more smoothness, for example, Lipschitz continuity, one already needs to assume that the exit from the domain is sufficiently fast (a sufficiently large coefficient α). Another possible assumption may be like this: the trajectories of the corresponding Markov family, starting from neighboring points, diverge slowly (sufficiently small $\beta > 0$).

The degree of smoothness is determined by the relation between the divergence velocity of the trajectories, starting from close points, and the distribution of the exit time from the domain. In this section, following the works of Freidlin [4], [11], we shall show that the same characteristics determine the degree of smoothness of the generalized solution in the general case as well. It is worthwhile noting that here the class of all degenerate equations is dealt with. If at some point $y \in D$ the operator does not degenerate, then it is easy to show that near this point the generalized solution is smooth. For example, if the coefficients of the operator are Lipschitz continuous near the point y, then the generalized solution has two continuous derivatives. Here the smoothness is of a local nature. It is possible to select certain classes of degenerate equations for which smoothness is also defined by local properties (see Hörmander [1]). An elegant probabilistic method of examining hypoellipticity was developed by Malliavin [1,2]. His results are presented in the monograph of Ikeda and Watanabe [2]. As follows from the above cited example, in the class of all equations with non-negative characteristic form such is not the case.

First of all, let us estimate the divergence velocity for the solutions of a stochastic differential equation starting from different points. In this section, it will be convenient to denote by $\overline{X}_t$ the solution of the stochastic equation $d\overline{X}_t = \sigma(\overline{X}_t)dW_t + b(\overline{X}_t)dt$, $\overline{X}_0 = x$ and by $\overline{Y}_t$ the solution of the same equation with initial condition $\overline{Y}_0 = y$. As always, the coefficients are assumed to be bounded and Lipschitz continuous:

$$|\sigma_j^i(x) - \sigma_j^i(y)| < K|x-y|, \ |b^i(x) - b^i(y)| < K|x-y|$$

for all $i,j = 1,2,\cdots,r$ and $x,y \in R^r$. We remind that the Lipschitz constant K is bounded via the corresponding norms of the coefficients of the operator L (see §3.2).

LEMMA 6.1. *For any integer* m

$$E|\overline{X}_t - \overline{Y}_t|^{2m} \le |x-y|^{2m} e^{\beta_m t} ,$$

where

$$\beta_m = \beta_m(K) = 2m(m-1)r^2K^2 + mr^2K^2 + 2mrK .$$

Proof. Using the fact that the coefficients of the stochastic equation are bounded, one can readily check that constants $c_{m,T} < \infty$ exist such that

$$E|\overline{X}_t|^{2m} < E \sup_{0 \le s \le t} |\overline{X}_s|^{2m} = c_m(t,x) < c_{m,T} \tag{6.7}$$

for any $m > 0$ and $0 \le t \le T$. Now, we shall apply Ito's formula to the function $\rho^{2m}(\overline{X}_t, \overline{Y}_t) = |\overline{X}_t - \overline{Y}_t|^{2m}$:

$$\rho^{2m}(\overline{X}_t,\overline{Y}_t) = \rho^{2m}(x,y) + \int_0^t 2m\rho^{2m-2}(\overline{X}_s,\overline{Y}_s)\left(\sum_{i=1}^r (\overline{X}_s^i - \overline{Y}_s^i)(b^i(\overline{X}_s) - b^i(\overline{Y}_s))\right) ds +$$

$$+ \int_0^t 2m\rho^{2m-2}(\overline{X}_s,\overline{Y}_s) \sum_{i=1}^r \left[(\overline{X}_s^i - \overline{Y}_s^i) \sum_{j=1}^r ((\sigma_j^i(\overline{X}_s) - \sigma_j^i(\overline{Y}_s))dW_s^j)\right] +$$

$$+ \int_0^t 2m(m-1)\rho^{2m-4}(\overline{X}_s,\overline{Y}_s)\,|(\sigma(\overline{X}_s) - \sigma(\overline{Y}_s))(\overline{X}_s - \overline{Y}_s)|^2 ds + \tag{6.8}$$

$$+ \int_0^t m\rho^{2m-2}(\overline{X}_s,\overline{Y}_s) \sum_{i,k=1}^r [\sigma_k^i(\overline{X}_s) - \sigma_k^i(\overline{Y}_s)]^2 ds .$$

By virtue of (6.7) the expectation of the stochastic integral in (6.8) is zero. For any $u,v \in R^r$, the inequalities hold:

$$\left|\sum_{i=1}^r (u^i - v^i)(b^i(u) - b^i(v))\right| \le rK|u-v|^2 ,$$

$$|(\sigma(u) - \sigma(v))(u-v)|^2 \le r^2K^2|u-v|^4 ,$$

$$\sum_{i,k=1}^r |\sigma_k^i(u) - \sigma_k^i(v)|^2 \le r^2K^2|u-v|^2 .$$

Utilizing these inequalities, (6.8) yields:

$$E\rho^{2m}(\overline{X}_t,\overline{Y}_t) \le \rho^{2m}(x,y) + \beta_m \int_0^t E\rho^{2m}(\overline{X}_s,\overline{Y}_s)ds .$$

Whence we conclude by Lemma 1.4.1 that

$$E\rho^{2m}(\bar{X}_t - \bar{Y}_t) \le \rho^{2m}(x,y)\exp\{\beta_m t\}. \quad \square$$

COROLLARY. *For any* $T > 0$, *the inequality holds*

$$E\sup_{0\le t\le T}|\bar{X}_t - \bar{Y}_t|^2 \le 3|x-y|^2 + 3E\sup_{0\le t\le T}\left|\int_0^t (\sigma(\bar{X}_s) - \sigma(\bar{Y}_s))\,dW_s\right|^2 +$$

$$+ 3E\left(\int_0^T |b(\bar{X}_s) - b(\bar{Y}_s)|ds\right)^2 \le 3|x-y|^2 + 12\int_0^T \sum_{j=1}^r \left[\sum_{i=1}^r (\sigma_j^i(\bar{X}_s) - \sigma_j^i(\bar{Y}_s))\right]^2 ds +$$

$$+ 3TE\int_0^T |b(\bar{X}_s) - b(\bar{Y}_s)|^2\,ds \le c(T)\,|x-y|^2\exp\{\beta_1 T\},$$

where $c(T)$ *grows in* T *at most linearly.*

When bounding the stochastic integral, we made use of the bound (see Doob [1]):

$$E\sup_{0\le t\le T}\left|\int_0^t f(s,\omega)\,dW_s\right|^2 \le 4\int_0^T Ef^2(s,\omega)\,ds.$$

THEOREM 6.1. *Suppose that* D *is a bounded domain in* R^r *and let* $\psi(x)$ *be a function on* ∂D, *satisfying a Hölder condition with exponent* μ. *Suppose that a set* $\Gamma_0 \subset \partial D$ *is inaccessible, and the set* $\Gamma_1 = \partial D \setminus \Gamma_0$ *is uniformly normally regular. Moreover, suppose that the process* (X_t, P_x) *leaves the domain* D *uniformly exponentially fast. Then the generalized solution of the problem* $Lu(x) = 0$, $x \in D$, $u(x)|_{\Gamma_1} = \psi(x)$ *satisfies the Hölder condition:* $|u(x) - u(y)| < \text{const}\cdot|x-y|^\gamma$ *for some* $\gamma > 0$; $x,y \in D$.

Proof. We put $\tau^x = \inf\{t : \overline{X}_t \notin D\}$, $\tau^y = \inf\{t : \overline{Y}_t \notin D\}$, $\overline{\tau} = \tau^x \wedge \tau^y$. The strong Markov property implies that

$$u(x) = E\, u(\overline{X}_{\overline{\tau}}),\; u(y) = E\, u(\overline{Y}_{\overline{\tau}}) .$$

Using these equalities we obtain:

$$|u(x) - u(y)| \le E\,\chi_{\overline{\tau}=\tau^x} |\psi(\overline{X}_{\overline{\tau}}) - u(\overline{Y}_{\overline{\tau}})| + \\ + E\,\chi_{\overline{\tau}\ne\tau^x} |\psi(\overline{Y}_{\overline{\tau}}) - u(\overline{X}_{\overline{\tau}})| , \tag{6.9}$$

where $\chi_{\overline{\tau}=\tau^x}$ and $\chi_{\overline{\tau}\ne\tau^x}$ are the indicators of the sets $\{\tau = \tau^x\}$ and $\{\overline{\tau} \ne \tau^x\}$ respectively.

If $a \in D$, $b \in \Gamma_1$, then

$$|u(a) - u(b)| \le E|\psi(X^a_{\tau^a_D}) - \psi(b)| < \\ < c_1 E|X^a_{\tau^a_D} - b|^\mu < c_1 (E|X^a_{\tau^a_D} - b|^2)^{\mu/2} . \tag{6.10}$$

Here and henceforth in this section, c_i are some constants. When deducing (6.10), we made use of the fact that $(E|\xi|^\alpha)^{1/\alpha}$ is an increasing function of the parameter $\alpha > 0$.

By applying Ito's formula to the function $f(X^a_t) = |X^a_t - b|^2$ for $t < \tau^a_D = \inf\{t : X^a_t \notin D\}$, we derive:

$$|X^a_{\tau^a_D} - b|^2 = |a-b|^2 + \int_0^{\tau^a_D} (\nabla f(X^a_s), \sigma(X^a_s)\, dW_s) + \int_0^{\tau^a_D} L f(X^a_s) ds .$$

whence we deduce that $E|X^a_{\tau^a_D} - b|^2 \le |a-b|^2 + c_2 E \tau^a_D$. Taking into account that all points of the set Γ_1 are uniformly normally regular, we obtain

$$E|X^a_{\tau^a_D} - b|^2 < c_3 |a-b| .$$

This inequality along with (6.10) yields that

$$|u(a)-u(b)| < c_4 |a-b|^{\mu/2} . \tag{6.11}$$

Notice that in view of the uniform normal regularity of the set Γ_1, the constant c_4 may be chosen independent of the point $b \in \Gamma_1$. Next, from (6.9) and (6.11) it follows that

$$|u(x)-u(y)| \le c_5 E|\bar{X}_{\bar{\tau}}-\bar{Y}_{\bar{\tau}}|^{\mu/2}; \ x,y \in D . \tag{6.12}$$

Therefore, the problem has reduced to bounding the magnitude of divergence of the trajectories by the time $\bar{\tau}$.

Suppose that $\kappa \in (0,1)$, χ_n is the indicator of the set $\{\omega : n \le \bar{\tau}(\omega) < n+1\}$ and let d be the diameter of the domain D. Noting that $|\bar{X}_{\bar{\tau}}-\bar{Y}_{\bar{\tau}}| d^{-1} \le 1$, we arrive at

$$E|\bar{X}_{\bar{\tau}}-\bar{Y}_{\bar{\tau}}|^{\mu/2} = d^{\mu/2} E[d^{-1}|\bar{X}_{\bar{\tau}}-\bar{Y}_{\bar{\tau}}|]^{\mu/2} \le d^{(1-\kappa)\mu/2} E|\bar{X}_{\bar{\tau}}-\bar{Y}_{\bar{\tau}}|^{\frac{\kappa\mu}{2}} \le$$

$$\le d^{\frac{(1-\kappa)\mu}{2}} \sum_{n=0}^{\infty} E|\bar{X}_{\tau}-\bar{Y}_{\tau}|^{\frac{\kappa\mu}{2}} \chi_n \le$$

$$\le c_6 \sum_{n=0}^{\infty} E\{ \sup_{n\le s\le n+1} |\bar{X}_s-\bar{Y}_s|^{\frac{\kappa\mu}{2}} \cdot \chi_n\} \le c_6 \sum_{n=0}^{\infty} (P\{\bar{\tau}>n\})^{1/2}(E\{ \sup_{s\le n+1} |\bar{X}_s-\bar{Y}_s|^2\})^{\frac{\kappa\mu}{4}} .$$

By the corollary of Lemma 6.1

$$E\{ \sup_{s\le n+1} |\bar{X}_s-\bar{Y}_s|^2\} \le c(n+1)\, |x-y|^2 \exp\{(n+1)\beta_1\} ;$$

the function $c(n+1)$ growing at most linearly.

Since the trajectories of the process (X_t, P_x) leave the domain D uniformly exponentially fast, one can find $a_{L,D} > 0$ such that

$$P\{\bar{\tau}>n\} < c_7 \exp\{-a_{L,D} \cdot n\} .$$

Relying on these bounds we have

$$E|\bar{X}_\tau - \bar{Y}_\tau|^{\mu/2} \le$$

$$\le c_8|x-y|^{\frac{\kappa\mu}{2}} \sum_{n=0}^{\infty} \sqrt{c(n+1)}\, \exp\left\{(n+1)\left(\frac{1}{4}\beta_1\kappa\mu - \frac{1}{2}\alpha\right)\right\}. \tag{6.13}$$

Now, let κ be picked in such a way that $\frac{1}{2}\beta_1\kappa\mu - \alpha < 0$. Then the series on the right-hand side of (6.13) converges and

$$E|\bar{X}_\tau - \bar{Y}_\tau|^{\mu/2} \le c_9|x-y|^{\frac{\kappa\mu}{2}}. \tag{6.14}$$

From (6.12) and (6.14) we conclude that, for $\kappa < \left(1 \wedge \frac{2\alpha_{L,D}}{\beta_1\mu}\right)$

$$|u(x) - u(y)| \le c_{10}|x-y|^{\frac{\kappa\mu}{2}}.$$

Therefore, everywhere in D, the function u(x) satisfies a Hölder condition with exponent $\gamma < \mu/2 \wedge \frac{\alpha_{L,D}}{\beta_1}$.

As follows from Example 4.1, in the general case one cannot expect more than Hölder continuity of the generalized solution. To obtain Lipschitz continuity, special assumptions concerning the magnitude of $\alpha_{L,D}$ and β_1 are already needed.

THEOREM 6.2. *Suppose that* D *is a bounded domain in* R^r, *and let a set* $\Gamma_0 \subset \partial D$, *be open with respect to* ∂D, *inaccessible, and such that the set* $\Gamma_1 = \partial D \setminus \Gamma_0$ *is uniformly normally regular. Suppose that* $\psi(x)$ *is the restriction to* ∂D *of some twice continuously differentiable function* $v(x)$, $x \in R^r$. *Then the generalized solution of the problem* $L\,u(x) = 0$, $x \in D$, $u(x)|_{\Gamma_1} = \psi(x)$ *is Lipschitz continuous, provided* $\alpha_{L,D} > \beta_1$.

Proof. First we shall deduce the bound near the boundary. To this end, we shall apply Ito's formula to the function $v(X_t^a)$:

$$v(X_t^a) = v(a) + \int_0^t (\nabla v(X_s^a), \sigma(X_s^a)dW_s) + \int_0^t Lv(X_s^a)ds\,, \quad a \in D\,.$$

This equality leads to

$$|E[v(X^a_{\tau^a_D}) - v(a)]| = |E \int_0^{\tau^a_D} Lv(X_s^a)ds| \le \sup_{x \in D \cup \partial D} |Lv(x)| \cdot E\tau^a_D . \tag{6.15}$$

Since $u(a) = E\,v(X^a_{\tau^a_D})$, (6.15) yields that for $b \in \Gamma_1$

$$|u(a)-u(b)| \le |u(a)-v(a)| + |v(a)-v(b)| \le$$

$$\le \max_{x \in D \cup \partial D} |Lv(x)| \cdot E\tau^a_D + |a-b| \cdot \max_{x \in D \cup \partial D} |\nabla v(x)|\,.$$

On account of the uniform normal regularity of Γ_1, this inequality results in

$$|u(a)-u(b)| \le c_{11}\,|a-b|\,, \tag{6.16}$$

where c_{11} is some constant, one and the same for all the points $a \in D$, $b \in \Gamma_1$.

From (6.9) and (6.16) we conclude that

$$|u(x)-u(y)| \le c_{12} E|\overline{X}_{\overline{\tau}} - \overline{Y}_{\overline{\tau}}|\,. \tag{6.17}$$

Further, relying on the corollary of Lemma 6.1 and using the notations introduced in the proof of Theorem 6.1, we obtain

$$\begin{aligned} E|\overline{X}_{\overline{\tau}} - \overline{Y}_{\overline{\tau}}| &\le \sum_{n=0}^{\infty} E\{ \sup_{0 \le s \le n+1} |\overline{X}_s - \overline{Y}_s|\,\chi_n\} \le \\ &\le c_{13} \sum_{n=1}^{\infty} (E \sup_{0 \le s \le n} |\overline{X}_s - \overline{Y}_s|^2 \cdot P\{\overline{\tau} > n\})^{1/2} \le \\ &\le c_{14}|x-y| \cdot \sum_{n=1}^{\infty} c(n) \exp\left\{\frac{1}{2}(\beta_1 - a_{L,D})n\right\} \le c_{15}|x-y|\,, \end{aligned} \tag{6.18}$$

whenever $\beta_1 < \alpha_{L,D}$. From (6.17) and (6.18) follows the claim of Theorem 6.2. □

We observe that §3.2 and Lemmas 3.1 and 3.3 present simple bounds for $\alpha_{L,D}$ and β_1 through the coefficients of the operator L and the domain D. These bounds enable one to estimate the best Hölder constant in Theorem 6.1 and verify the condition $\alpha_{L,D} > \beta_1$ in Theorem 6.2.

REMARK. If the boundary of the domain is smooth, then one of our assumptions is that the boundary function is twice continuously differentiable. In the case of a less smooth boundary function, the generalized solution may not be Lipschitz continuous, even if the remaining assumptions of the theorem are fulfilled. The corresponding example may be obtained if we consider the heat equation

$$-\frac{\partial u}{\partial x} + \frac{\partial^2 u}{\partial y^2} = 0 \text{ in the domain } D = \{(x,y) \in R^2 : 0 < x < 1, |y| < 1\}$$

with the boundary function $\psi = |y|^{1+\alpha}$, $0 < \alpha < 1$. It is readily checked that, for small x, the generalized solution $u(x,0)$ is equivalent to $c \cdot x^{\frac{1+a}{2}}$

Now we shall discuss conditions which ensure the existence of higher derivatives of the generalized solution; in particular, conditions under which the generalized solution is a classical one. The existence proof of the k-th order derivatives of the generalized solution is arranged according to the following scheme. The Dirichlet problem is considered for the non-degenerate equation $Lu^\varepsilon + \varepsilon\Delta u^\varepsilon = 0$, $x \in D$, $u^\varepsilon|_{\partial D} = \psi$. Then we prove that, under some minor assumptions, $u^\varepsilon(x)$ converges, uniformly on every compact set lying in D, to the generalized solution $u(x)$ of the degenerate problem as $\varepsilon \downarrow 0$. Next we give the conditions under which (k+1)st-order derivatives of $u^\varepsilon(x)$ admit an a priori estimate uniform in ε. Hence it appears clear that, under these conditions, $u(x)$ has k derivatives, and its kth-order derivatives are Lipschitz continuous.

As in Theorems 6.1 and 6.2, obtaining the necessary bounds consists of two parts: first, we obtain a bound near the boundary; secondly, this bound is carried to the interior of the domain. The boundary bounds may be deduced by using sufficiently high smoothness of the boundary function and the regularity, in a fairly strong sense, of the boundary points. It is in this way that we acted when proving Theorems 6.1 and 6.2. Now, for brevity, we will assume that the operator does not degenerate on the regular part of the boundary of the domain. This permits us to use the usual boundary bounds, known in the theory of elliptic partial differential equations. The primary attention will be payed to the extension of these bounds inside the domain. It is this carrying over which requires assumptions of inequalities type.

First, we introduce the notations. Let $f(x)$ be a function, defined for $x \in R^r$ with values in R^r. We shall define the m-th difference of the function $f(x)$ by the equalities:

$$\delta_1^h f(x) = f(x+h) - f(x), \quad h \in R^r,$$

$$\delta_m f(x) = \delta_m^{h_1,\cdots,h_m} f(x) = \delta_{m-1}^{h_1,\cdots,h_{m-1}} f(x+h_m) - $$

$$- \delta_{m-1}^{h_1,\cdots,h_{m-1}} f(x); \; h_1, \cdots, h_m \in R^r .$$

If the function $f(x)$ is smooth enough, then by selecting the increments $h_1, \cdots, h_m \in R^r$, $|h_i| = h$, one can ensure that

$$\lim_{h \downarrow 0} \frac{1}{h^m} \delta_m^{h_1,\cdots,h_m} f(x) = \frac{\partial^m f(x)}{(\partial x^1)^{\ell_1} \cdots (\partial x^r)^{\ell_r}}$$

for any natural $\ell_1, \ell_2, \cdots, \ell_r$; $\sum_1^r \ell_k = m$. Denote by $H = H[x; h_1, \cdots, h_m]$ a set in R^r consisting of the point x and of the points of the form $x + h_{i_1} + h_{i_2} + \cdots + h_{i_k}$ for $i_1, \cdots, i_k < m$, $k \le m$; H_f denoting the convex hull of the values of the function $f(y)$ for $y \in H$. Taylor's formula yields the following inequality

$$|\delta_m u(f(x))| \leq |((\nabla u)(f(x)), \delta_m f(x))| + \\ + \|D_m u\|_{H_f} \cdot \sum |\delta_{\ell_1} f(x)|^{m_1} \cdots |\delta_{\ell_k} f(x))|^{m_k}, \tag{6.19}$$

where $\|D_m u\|_A = \sup_{x \epsilon A, \Sigma m_i = k \leq m, m_i \geq 0} \left| \frac{\partial^k u(x)}{(\partial x^1)^{m_1} \cdots (\partial x^r)^{m_r}} \right|$, and the sum is taken over all possible products

$$|\delta_{\ell_1} f(x)|^{m_1} \cdots |\delta_{\ell_k} f(x)|^{m_k} \text{ with } \ell_i < m, \ \Sigma m_i \ell_i = m .$$

If X_t^x is a random Markov function, $x \epsilon R^r$, then $\delta_k^{h_1, \cdots, h_k} X_t^x = \delta_k X_t^x$ means that the difference operator δ_k is applied to X_t^x as to the function of the initial point $x = X_0^x$.

LEMMA 6.2. *Suppose that the functions* $\sigma_j^i(x)$, $b^i(x)$, $(i, j = 1, \cdots, r)$, $x \epsilon R^r$ *have* k *bounded continuous derivatives. Then one can find* $\beta_{k,m}$ *depending only on the Lipschitz constant of the functions* $\sigma_j^i(x)$ *and* $b^i(x)$, *on the dimension of the space and on the numbers* k *and* m, *such that for any* $h_1, h_2, \cdots, h_k \epsilon R^r$, $|h_i| = h$,

$$E|\delta_k X_t^x|^{2m} < c_{k,m} h^{2km} \exp\{t \beta_{k,m}\},$$

where the constants $c_{k,m}$ *are determined by the maxima of the moduli of the functions* $\sigma_j^i(x)$, $b^i(x)$ *and their derivatives up to* k-th *order inclusively. The sequence* $\beta_{k,m}$ *may be thought of as monotonically increasing in each index.*

The proof of this lemma is arranged by induction in k. For $k = 1$, it turns into Lemma 6.1. To verify the validity of Lemma 6.2 for some $k > 1$ (provided it is known to be valid for all smaller values of k), we apply Ito's formula to the function $|\delta_k X_t^x|^{2m}$. Then the claim of Lemma 6.2 will be derived with the help of bound (6.19) and elementary inequalities analogous to the way it was done when proving Lemma 6.1. The detailed proof of this lemma is available in Freidlin [4, 11]. In particular, a recurrent relation for determining the numbers $\beta_{k,m}$ is written down there. For $k=1$, the sequence coincides with the sequence β_m introduced in Lemma 6.1.

Note that stochastic differential equations with smooth coefficients may be differentiated with respect to the initial conditions (for the precise formulations and references, see Freidlin and Wentzell [2]). In particular, under the conditions of Lemma 6.2, $\dfrac{\partial^k X_t^x}{(\partial x^1)^{\ell_1}\cdots(\partial x^r)^{\ell_r}}$, $\ell_i \geq 0$, $\sum_1^r \ell_i = k$, exists. Lemma 6.2 implies the bound

$$E\left|\frac{\partial^k X_t^x}{(\partial x^1)^{\ell_1}\cdots(\partial x^r)^{\ell_r}}\right|^{2m} \leq c_{k,m} \exp\{t\beta_{k,m}\}. \tag{6.20}$$

LEMMA 6.3. *Let* $f(x,\omega)$ *be a bounded, progressively measurable function. Then, for any integer* m,

$$E\left(\int_0^T f(s,\omega)\,dW_s\right)^{2m} \leq c_m T^{m-1}\int_0^T E|f(s,\omega)|^{2m}\,ds\,,$$

where $c_m = (m(2m-1))^m$.

Proof. Applying Ito's formula to the function $(\int_0^t f(s,\omega)\,dW_s)^{2m}$ we obtain

$$g_m(T) = E\left(\int_0^T f(s,\omega)\,dW_s\right)^{2m} =$$

$$= m(2m-1)\int_0^T E\left(\int_0^t f(s,\omega)\,dW_s\right)^{2m-2} f^2(t,\omega)\,dt\,.$$

This formula implies that the function $g_m(T)$ increases monotonically in T. In view of the monotonicity, the Hölder inequality yields:

$$E\left(\int_0^T f(s,\omega)\,dW_s\right)^{2m} \le m(2m-1)\left[\int_0^T E\left(\int_0^t f(s,\omega)\,dW_s\right)^{2m} dt\right]^{\frac{m-1}{m}} \times$$

$$\times \left[\int_0^T E f^{2m}(t,\omega)\,dt\right]^{1/m} \le m(2m-1)\,T^{\frac{m-1}{m}}\left[E\left(\int_0^T f(s,\omega)\,dWs\right)^{2m}\right]^{\frac{m-1}{m}} \times$$

$$\times \left[\int_0^T E f^{2m}(t,\omega)\,dt\right]^{1/m}.$$

Which implies the assertion of the lemma. □

LEMMA 6.4. *For any integer positive* k *and* m, *the inequality*

$$E \sup_{0\le s\le t} \left|\delta_k^{h_1,\cdots,h_k} X_s^x\right|^{2m} \le \tilde{c}_{k,m}(t)\exp\{\beta_{k,m} t\} h^{2km}$$

holds, where $\tilde{c}_{k,m}$ *grows as* $t \to \infty$ *not faster than some* $(2m-1)$*-degree polynomial,* $|h_i| = h$.

This lemma can be established just as the corollary of Lemma 6.1 was. One should use the fact that the 2m-th power of a stochastic integral is a semi-martingale for which Doob's inequality holds (Doob [1]). To bound the 2m-th power of the stochastic integral, one can make use of Lemma 6.3. □

Let $\tau_D^y = \inf\{t : X_t^y \notin D\}$, $\bar{\tau} = \min\{\tau_D^y : y \in H(x;\, h_1, \cdots, h_k)\}$. Obviously, $P\{\bar{\tau} > t\} \le P_x\{\tau_D > t\} < c\cdot\exp\{-\alpha_{L,D}\cdot t\}$.

LEMMA 6.5. *Suppose that the coefficients of the stochastic differential equation have* k *bounded continuous derivatives. If* $\alpha_{L,D} > \beta_{k,m}$, *then*

$$E\left|\delta_k^{h_1,\cdots,h_k} X_{\bar\tau}^x\right|^m \le \tilde{\tilde{c}}_{k,m} h^{km}$$

where $|h_i| = h$, *the constants* $\beta_{k,m}$ *being defined in Lemma 6.2.*

Proof. Let χ_n be the indicator of the set $\{\omega : n \le \bar\tau < n+1\}$. Using the previous lemma we get

$$E|\delta_k X_{\bar\tau}^x|^m \le \sum_{n=0}^{\infty} E\{\sup_{n\le s\le n+1} |\delta_k X_s^x|^m \chi_n \le$$

$$\le \sum_{n=0}^{\infty} (E \sup_{0\le s\le n+1} |\delta_k X_s^x|^{2m} E\,\chi_n)^{1/2} \le$$

$$\le \sum_{n=0}^{\infty} \tilde{c}_{k,m}(n+1) \exp\left\{\frac{n}{2}(\beta_{k,m} - a_{L,D})\right\} h^{km} \le \tilde{\tilde{c}}\, h^{km} .$$

The series on the right-hand side converges, because $\tilde{c}_{k,m}(n)$ grows polynomially with n, and $a_{L,D} > \beta_{k,m}$, by the condition of the lemma. □

LEMMA 6.6. *Suppose that the coefficients of the stochastic equation are* k *times continuously differentiable, and let* $\beta_{2k,2k} < a_{L,D}$. *Then, if the generalized solution* $u(x)$ *is* k *times continuously differentiable up to the boundary* ∂D *of the domain* D, *then the bound*

$$\|D_k u(x)\|_D < c\|D_k u\|_{\Gamma_1}$$

holds, where Γ_1 *is the regular part of the boundary* ∂D *and the set* $\partial D \setminus \Gamma_1$ *is inaccessible.*

Proof. We shall choose $h_1, \cdots, h_m$, $|h_i| = h$, $m \le k$, so that

$$\lim_{h\downarrow 0} h^{-m} \delta_m^{h_1,\cdots,h_m} u(x) = \frac{\partial^m u(x)}{(\partial x^1)^{\ell_1} \cdots (\partial x^r)^{\ell_r}} .$$

Let $\bar\tau = \min\{\tau_D^y : y \in H(x; h_1, \cdots, h_m)\}$. Since $\bar\tau$ does not depend on future, by the strong Markov property of the family (X_t^x, P) we have: $u(x) = E\,u(X_{\bar\tau}^x)$.

We note that at the time $\bar{\tau}$ at least one point of $X_{\bar{\tau}}^{y}$, $y \in H(x; h_1, \cdots, h_m) = H$, belongs to Γ_1. Hence, for sufficiently small h with a probability larger than $1-\varepsilon$, all the points $X_{\bar{\tau}}^{y}$, $y \in H$, belong to some neighborhood $U_\delta(\Gamma_1)$ of the set Γ_1. Taking this into account, we have

$$\left|\delta_m^{h_1,\cdots,h_m} u(x)\right|^2 \leq (E|\delta_m u(X_{\bar{\tau}}^{x})|)^2 \leq$$

$$\leq c_1 \Bigg[\|D_m u\|^2_{U_\delta(\Gamma_1)} (1-\varepsilon) E\Bigg(|\delta_m X_{\bar{\tau}}^{x}|^2 + \sum_{0<\ell_i<m,\ \sum_1^p \ell_i m_i = m,\ p\geq 1} |\delta_{\ell_1} X_{\bar{\tau}}^{x}|^{2m_1} \cdots |\delta_{\ell_p} X_{\bar{\tau}}^{x}|^{2m_p}\Bigg) +$$

$$+ \varepsilon \|D_m u\|^2_D E\Bigg(|\delta_m X_{\bar{\tau}}^{x}|^2 + \sum_{0<\ell_i<m,\ \sum_1^p \ell_i m_i = m,\ p\geq 1} |\delta_{\ell_1} X_{\bar{\tau}}^{x}|^{2m_1} \cdots |\delta_{\ell} X_{\bar{\tau}}^{x}|^{2m_p}\Bigg)\Bigg]. \quad (6.21)$$

Relying on Lemma 6.5 and noting the condition $\alpha_{L,D} > \beta_{2m,2m}$ and the monotonicity of $\beta_{k,m}$ in k and m, we deduce the bounds

$$E|\delta_m X_{\bar{\tau}}^{x}| \leq c_2 h^{2m},$$

$$E|\delta_{\ell_1} X_{\bar{\tau}}^{x}|^{2m_1} \cdots |\delta_{\ell_p} X_{\bar{\tau}}^{x}|^{2m_p} \leq \sum_{\ell<m} (E|\delta_\ell X_{\bar{\tau}}^{x}|^{2m})^{1/\ell} \leq c_3 h^{2m}. \quad (6.22)$$

From (6.21) and (6.22) it follows that

$$|h^{-m} \delta_m u(x)|^2 \leq c(\|D_m u\|^2_{U_\delta(\Gamma_1)} (1-\varepsilon) + \varepsilon \|D_m u\|_D)$$

which implies the claim of Lemma 6.6. □

So, let

$$L = \frac{1}{2} \sum_{i,j=1}^{r} a^{ij}(x) \frac{\partial^2}{\partial x^i \partial x^j} + \sum_{i=1}^{r} b^i(x) \frac{\partial}{\partial x^i}.$$

We suppose that a factorization $(a^{ij}(x)) = \sigma(x)\sigma^*(x)$ exists, where the matrix $\sigma(x)$ (not necessarily square) has k times continuously differentiable elements. The functions $b^i(x)$ are also assumed to be k times continuously differentiable. Denote by $\beta_{k,m}$ the sequence introduced in Lemma 6.2. Given a bounded domain $D \subset R^r$ with a boundary ∂D, we shall say that *Condition* $\mathcal{A}_3$ holds whenever $\partial D = \Gamma_0 \cup \Gamma_1$, where the sets Γ_0 and Γ_1 are closed with respect to ∂D. The set Γ_0 is inaccessible for the process corresponding to the operator L, and the set Γ_1 consists of the points regular for the Laplace operator, and $\det\{a^{ij}(x)\} \neq 0$ for $x \in \Gamma_1$.

Consider the Dirichlet problem

$$L\,u(x) = 0,\ x \in D,\ u(x)\Big|_{\Gamma_1} = \psi(x)\,, \tag{6.23}$$

where $\psi(x)$ is a continuous function on Γ_1. Suppose that $\alpha_{L,D} > 0$. Then, as follows from Theorems 5.1 and 5.2, the generalized solution of problem (6.23) exists and is unique. By Theorem 6.1, it is Hölder continuous.

THEOREM 6.3. *Suppose that Condition* $\mathcal{A}_3$ *holds and* $\alpha_{L,D} > \beta_{2k,2k}$. *Then, for* $x \in D$, *the generalized solution of problem (6.23) has Lipschitz continuous* (k–1)-*order derivatives. In particular, if* $\alpha_{L,D} > \beta_{6,6}$ *then the generalized solution has Lipschitz continuous second-order derivatives and* $L\,u(x) = 0$.

Proof. Let $\mathcal{E} = D \cap U_\delta(\Gamma_1)$; the $\delta > 0$ being so small that the operator L does not degenerate in $\mathcal{E} \cup \partial\mathcal{E}$. Let a domain $\tilde{D}$ be such that $\partial\tilde{D} = \Gamma_0 \cup \tilde{\Gamma}_1$, $\tilde{\Gamma}_1$ will be a smooth manifold lying in $\mathcal{E}$, $\{D \setminus U_{\delta/4}(\Gamma_1)\} \supset \tilde{D} \supset \{D \setminus U_{\delta/2}(\Gamma_1)\}$ (see Fig. 3, where the shaded domain is the domain $\mathcal{E}$).

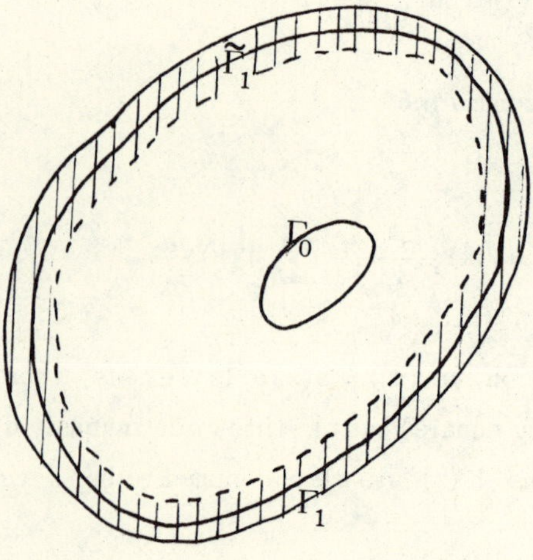

Fig. 3

Consider the Dirichlet problem $L^\varepsilon u^\varepsilon = L\,u^\varepsilon + \frac{\varepsilon^2}{2}\Delta u^\varepsilon = 0$, $x \in D$,

$u^\varepsilon(x)|_{\partial D} = \tilde{\psi}(x)$, where $\tilde{\psi}(x)$ is a continuous function on ∂D such that $\tilde{\psi}(x) = \psi(x)$ on Γ_1 and $|\tilde{\psi}(x)| \leq \max_{x \in \Gamma_1} |\psi(x)|$. As will be shown in §4.2, $u(x) = \lim_{\varepsilon \downarrow 0} u^\varepsilon(x)$, the convergence being uniform on every compact set lying in D. If we show that, under the condition of the theorem, there are bounds, uniform in ε, of the first k derivatives of the function $u^\varepsilon(x)$, then by the Arzelà theorem, this will imply the claim of Theorem 6.3.

Denote by $\sigma^\varepsilon(x)$ the rectangular matrix $(\sigma(x), \varepsilon E)$ where E is the unit matrix of order r. Let $\tilde{W}_t$ be a Wiener process whose dimension is equal to the number of columns in the matrix $\sigma^\varepsilon(x)$. It is clear that $\sigma^\varepsilon(x)[\sigma^\varepsilon(x)]^* = (a^{ij}(x)) + \varepsilon E$. The trajectories of the Markov family corresponding to the operator L^ε may be constructed with the stochastic equation

$$dX_t^{\varepsilon,x} = \sigma^\varepsilon(X_t^{\varepsilon,x})d\tilde{W}_t + b(X_t^{\varepsilon,x})dt,\ X_0^{\varepsilon,x} = x \in R^r. \tag{6.24}$$

The sequence $\beta_{2k,2k}$ for equation (6.24) may be chosen the same for all ε because the Lipschitz constant does not depend on ε.

Next, denote $\tau_{\tilde{D}}^{\varepsilon,x} = \inf\{t : X_t^{\varepsilon,x} \notin \tilde{D}\}$. For any t, $\mu > 0$, for sufficiently small ε, the bound holds

$$\begin{aligned} P\{\tau_{\tilde{D}}^{\varepsilon,x} > t\} &\leq P\{\tau_D^x > t\} + \\ &+ P\left\{\sup_{0 \leq s \leq t} |X_s^{\varepsilon,x} - X_s^x| > \frac{\delta}{4}\right\} < c_1 \exp\{-\alpha_{L,D} t\} + \mu. \end{aligned} \tag{6.25}$$

For any $\lambda > 0$, one can select t and μ so that the inequality

$$\left|\frac{1}{t} \ln (c_1 e^{-\alpha_{L,D} t} + \mu) + \alpha_{L,D}\right| < \lambda$$

can be fulfilled. This inequality together with (6.25) yields that, for ε small enough, $\alpha_{L^\varepsilon,\tilde{D}} > \alpha_{L,D} - \lambda$. Since by the condition $\alpha_{L,D} > \beta_{2k,2k}$, we have that $\alpha_{L^\varepsilon,\tilde{D}} > \beta_{2k,2k}$ for ε small enough.

Since the coefficients are smooth and the operator L is non-degenerate in $\mathcal{E}$, one can conclude that the bounds, uniform in ε, of the first k

derivatives of the function $u^{\varepsilon}(x)$ exist for $x \in \widetilde{\Gamma}_1$. Taking into account that $\alpha_{L^{\varepsilon}\widetilde{D}} > \beta_{2k,2k}$ and relying on Lemma 6.6, we obtain that the bounds, uniform in ε, of the derivatives of $u^{\varepsilon}(x)$ exist everywhere in $\widetilde{D}$ up to the k-order inclusively. This implies that derivatives of the generalized solution of problem (6.23) exist. □

REMARK 1. Consider the (possibly degenerate) parabolic equation

$$\mathfrak{L}u(t,x) = -\frac{\partial u}{\partial t} + Lu = 0,\ t > 0,\ x \in R^r .$$

The process corresponding to the operator $\mathfrak{L}$ performs deterministic motion along the coordinate t. Therefore, if the equation $\mathfrak{L}u = 0$ is considered in the domain $\mathfrak{D} = \{(t,x) : -\infty < t_0 < t < t_1 < \infty,\ x \in D \subset R^r\}$, then it will take a finite time to hit $\partial\mathfrak{D}$, and thus one can assume that $\alpha_{\mathfrak{L},D} = \infty$. This implies that in considering the equation $\mathfrak{L}u = 0$, the smoothness of the generalized solution is defined only by the existence of boundary bounds. For sufficiently smooth coefficients the carrying over of the bounds inside the domain is possible without any extra conditions. In particular, in the case of Cauchy's problem, derivatives of the generalized solution exist, whenever the initial function and the coefficients are sufficiently smooth. The corresponding exact result can be easily deduced via bound (6.20) (see, e.g. Blagoveščhenskii and Freidlin [1]).

REMARK 2. Let us consider the Dirichlet problem for the equation $L u(x) - c(x) u(x) = 0$ in the domain D, where $c(x)$ is a strictly positive, fairly smooth function. This killing term permits us to carry over onto the entire domain bounds of derivatives having higher order than for the equation $L u(x) = 0$. We shall clarify this assuming that $c(x) = c =$ const > 0; otherwise the equation should be divided by $c(x)$. By using the same notions as in Lemma 6.5, one can write down the following equality: $u(x) = E\, u(X^x_\tau) \exp\{-c\tau\}$. Let us trace, for instance, how first-order derivatives are bounded inside the domain, assuming that, near the

the boundary, the derivatives have been bounded:

$$|u(x)-u(y)| = |E(u(\overline{X}_\tau)-u(\overline{Y}_\tau))e^{-c\tau}| \le$$

$$\le c_1 E|\overline{X}_\tau - \overline{Y}_\tau| e^{-c\tau} \le c_1 \sum_{n=0}^{\infty} E\{ \sup_{n \le s < n+1} |\overline{X}_s - \overline{Y}_s| e^{-cn} \chi_n\} \le$$

$$\le c_1|x-y| \sum_{n=0}^{\infty} \exp\left\{\frac{n}{2}[(\beta_1 - \alpha_{L,D}) - 2c]\right\} < c_2|x-y|$$

provided $c > \frac{1}{2}(\beta_1 - \alpha_{L,D})$. Therefore, in the presence of killing, the first-order derivative may be bounded also when $\alpha_{L,D} < \beta_1$. Notice that if the equation involves the term $-cu$, then the boundary bounds at normally regular points are preserved. Similar is the case for higher derivatives.

If the coefficients of the equation are sufficiently smooth and bounds near the boundary are available, then for every i, one can find $c_i > 0$ such that, for $c > c_i$, the bounds of k-order derivatives may be carried over from the boundary to the inside of the domain. The numbers c_i depend on the Lipschitz constant of the coefficients of the corresponding stochastic equation.

§3.7 *Second boundary value problem*

Here we shall briefly discuss boundary value problems of the form

$$Lu(x) - c(x)u(x) = f(x),\ x \in D \subset R^r,\ \left.\frac{\partial u(x)}{\partial \ell(x)}\right|_{\Gamma_1} = 0, \tag{7.1}$$

where, just as everywhere in this chapter, L is an operator with a non-negative characteristic form, $c(x) \ge 0$, $\ell(x)$ is a smooth vector field on the boundary of the domain D which is nowhere tangent to the boundary; $\Gamma_1 \subset \partial D$ is the part of this boundary on which the boundary conditions are assigned.

We shall assume that the domain D is bounded, and its boundary ∂D has the outward normal $n(x) = (n_1(x), \cdots, n_r(x))$, $x \in \partial D$, whose direction cosines are three times continuously differentiable.

Let ∂D consist of two closed $(r-1)$-dimensional manifolds Γ_0 and Γ_1. Suppose that on Γ_1 the diffusion in the direction of the normal to Γ_1 is distinct from zero:

$$\sum_{i,j=1}^{r} a^{ij}(x) n_i(x) n_j(x) \neq 0, \; x \in \Gamma_1 ;$$

the condition

$$\sum_{i,j=1}^{r} a^{ij}(x) n_i(x) n_j(x) = 0, \; \sum_{i=1}^{r} b^i(x) n_i(x) < 0$$

being fulfilled on Γ_0.

We assume that the coefficients $a^{ij}(x)$ are twice continuously differentiable in R^r, $b^i(x)$ are once continuously differentiable, the functions $c(x)$ and $f(x)$ are continuous in R^r. The vector field $\ell(x) = (\ell_1(x), \cdots, \ell_r(x))$, $x \in \Gamma_1$, is assumed to be three times continuously differentiable.

Denote by (X_t, P_x) a Markov process in the state space $D \cup \Gamma_1$ which, inside the domain D, is governed by the operator L, and on the boundary Γ_1 is subject to reflection in the direction of the field $\ell(x)$. No conditions should be assigned on Γ_0, since, with probability 1 starting from any $x \in D \cup \Gamma_1$, the trajectories X_t will never hit Γ_0. The construction of the process (X_t, P_x) may be performed with the help of stochastic differential equations. The corresponding construction is described in §1.6. Some properties of the process (X_t, P_x) are considered there as well. Denote by A the infinitesimal operator of the process (X_t, P_x). As follows from §1.6, the domain of the operator A contains functions $f(x)$, twice continuously differentiable in $D \cup \partial D$, for which $\frac{\partial f(x)}{\partial \ell(x)} = 0$ for $x \in \Gamma_1$.

By a *generalized solution of problem (7.1)*, we mean a function $u(x)$, $x \in D \cup \Gamma_1$, such that

$$A\,u(x) - c(x)\,u(x) = f(x)\ .$$

This definition is correct. If a function $v(x)$ is twice continuously differentiable in the domain D up to the boundary, then $v \in D_A$ if and only if $\frac{\partial v(x)}{\partial \ell(x)}\Big|_{\Gamma_1} = 0$, and in this case $Av = Lv$. This assertion comes from the results of §1.6.

THEOREM 7.1. *Suppose that* $c(x) \geq c_0 > 0$. *Then for any function* $f(x)$, *continuous in* $D \cup \partial D$, *a unique generalized solution of problem (7.1) exists. This solution is representable in the form*

$$u(x) = -E_x \int_0^\infty f(X_t) \exp\left\{-\int_0^t c(X_s)\,ds\right\} dt\ .$$

Proof. Since $c(x) \geq c_0 > 0$, we have that $\sup_{x \in D} |u(x)| \leq \sup_{x \in D} |f(x)| \cdot \int_0^\infty \exp\{-c_0 t\} dt = c_0^{-1} \sup_{x \in D} |f(x)|$. Let us check that $u(x) \in D_A$ and $A\,u(x) - c(x)\,u(x) = f(x)$:

$$T_h u(x) - u(x) = E_x \left[\int_0^\infty \exp\left\{-\int_0^t c(X_s)\,ds\right\} f(X_t)\,dt - \right.$$

$$\left. - \int_h^\infty f(X_t) \exp\left\{-\int_h^t c(X_s)\,ds\right\} dt \right] = E_x \int_0^h f(X_t) \exp\left\{-\int_0^t c(X_s)\,ds\right\} dt +$$

$$+ E_x \int_h^\infty f(X_t) \exp\left\{-\int_h^t c(X_s)\,ds\right\} \left[\exp\left\{-\int_0^h c(X_s)\,ds\right\} - 1 \right] dt\ .$$

Whence, noting the uniform continuity of the functions $f(x)$ and $c(x)$ on $D \cup \partial D$, we conclude that $h^{-1}[T_h f(x) - f(x)]$ converges to $f(x) + c(x)u(x)$ uniformly in $x \in D \cup \Gamma_1$ as $h \downarrow 0$. Therefore, $u(x) \in D_A$ and $Au(x) = f(x) + c(x)u(x)$.

Now, we shall show that the generalized solution is unique. We introduce the family of the operators $\widetilde{T}_t f(x) = E_x f(X_t) \times \exp\{-\int_0^t c(X_s)ds\}$, acting on measurable functions $f(x)$, $x \in D \cup \Gamma_1$. As is known, the operators $\widetilde{T}_t$ form a contraction semi-group. Theorem 9.7 of the monograph by Dynkin [3] implies that if $u \in D_A$, then u belongs to the domain of the definition of the generator $\widetilde{A}$ of the semi-group $\widetilde{T}_t$ and $\widetilde{A}u = Au - c(x)u$. Let u be a generalized solution of problem 7.1 with $f(x) \equiv 0$. Then $\widetilde{A}u = Au - c(x)u = 0$. It is known from the theory of semi-groups that if $u \in D_{\widetilde{A}}$, then

$$\frac{d\widetilde{T}_t u}{dt} = \widetilde{T}_t \widetilde{A} u .$$

Since $\widetilde{A}u = 0$, this implies that $\dfrac{d\widetilde{T}_t u}{dt} = 0$, and thereby, the equality

$$u(x) = \widetilde{T}_t u(x) = E_x u(X_t) \exp\left\{-\int_0^t c(X_s)\,ds\right\}$$

holds for all $t \geq 0$.

Since $u(x)$ is bounded and $c(x) > 0$, one can pass to the limit under the expectation sign on the right-hand side:

$$u(x) = E_x \lim_{t \to \infty} u(X_t) \exp\left\{-\int_0^t c(X_s)\,ds\right\} = 0 ,$$

which completes the uniqueness proof. □

Now we proceed to consider problem (7.1) for $c(x) \equiv 0$. Just as in the classical case, for such a problem to be solvable, one should impose on the right-hand side some conditions of orthogonality to the space of the solutions of the adjoint homogeneous problem. In doing so, the solution will be defined up to some subspace. Unlike the classical case, degenerations may cause this subspace (even in the case of a connected domain D) to have a larger or even infinite dimension. We shall provide conditions under which the solution of problem (7.1) with $c(x) \equiv 0$ is defined up to an additive constant.

Let B^* be the space of countably additive functions of the set $\mu(G)$, $G \subset D \cup \Gamma_1$, with finite total variation, $\|\mu\|$ being the total variation of μ. Denote by U_t the semi-group of operators in B^* acting by the formula

$$U_t \mu(G) = \int_{D \cup \Gamma_1} p(t,x,G)\, \mu(dx) .$$

Here $p(t,x,G)$ is the transition function of the process (X_t, P_x). It is known that the space B^* is adjoint to the space B of bounded measurable functions $f(x)$ on $D \cup \Gamma_1$, $\|f\|_B = \sup_{x \in D \cup \Gamma_1} |f(x)|$. The semi-group U_t is adjoint to the semi-group $T_t : T_t f(x) = E_x f(X_t)$ (see §1.4).

Let μ be the invariant measure for the process $(X_t, P_x) : U_t \mu(G) = \mu(G)$ for $t \geq 0$ and any Borel $G \subset D \cup \Gamma_1$. If A^* designates the infinitesimal operator of the semi-group U_t, then the definition of the invariant measure may be written down in the form $A^* \mu = 0$.

If a function $u(x)$ satisfies the equation $A\, u(x) = f(x)$, then, for any invariant measure μ, the following chain of equalities holds:

$$\int_{D \cup \Gamma_1} f(x)\, \mu(dx) = \int_{D \cup \Gamma_1} A\, u(x)\, \mu(dx) = \int_{D \cup \Gamma_1} \lim_{t \downarrow 0} \frac{1}{t} (T_t u(x) - u(x))\, \mu(dx) =$$

$$= \lim_{t \downarrow 0} \frac{1}{t} \int_{D \cup \Gamma_1} (T_t u(x) - u(x))\, \mu(dx) = \lim_{t \downarrow 0} \frac{1}{t} \left[\int_{D \cup \Gamma_1} u(x)\, (U_t \mu)(dx) - \int_{D \cup \Gamma_1} u(x)\, \mu(dx) \right] = 0$$

Therefore, for the solvability of the equation $Au = f$ it is necessary that $\int_{D \cup \Gamma_1} f(x)\,\mu(dx) = 0$ for any invariant measure μ.

We shall say that *Condition* $\mathcal{B}$ holds, if one can find in the domain D an open set K such that $\inf_{x \in K} \det(a^{ij}(x)) > 0$ and $p(t_0, x, K) > h$ for some t_0, $h > 0$ for any $x \in D \cup \Gamma_1$.

LEMMA 7.1. *Let the process* (X_t, P_x) *satisfy Condition* $\mathcal{B}$ *and suppose that one can select in* D ℓ *non-intersecting Borel sets* $\mathcal{E}_1, \cdots, \mathcal{E}_\ell$ *satisfying the following requirements:*

1. *with probability* 1, *the trajectories, starting in* $\mathcal{E}_i$, *never leave* $\mathcal{E}_i$;
2. *the set* $\mathcal{E}_i \cap K = K_i$ *is non-empty and homeomorphic to the interior of a ball in* R^r;
3. $K = \bigcup_{i=1}^{\ell} K_i$ (*the set* K *was introduced in Condition* $\mathcal{B}$).

Then, for every $i = 1, \cdots, \ell$, *there exist a unique normed invariant measure* μ_i, *concentrated on* $\mathcal{E}_i$, *and a constant* $a_i > 0$ *such that*

$$|P(t,x,G) - \mu_i(G)| \le e^{-a_i t}$$

for any Borel $G \in \mathcal{E}_i$ *and* $x \in \mathcal{E}_i$.

Every invariant measure μ *of the process* (X_t, P_x) *is a linear combination of measures* μ_i *with positive coefficients. If the set* K *may be chosen homeomorphic to the interior of a ball, then the normed invariant measure is unique.*

Proof. Consider the process (X_t, P_x) on the set $\mathcal{E}_i$. We shall check that Doeblin's condition is fulfilled for this process on $\mathcal{E}_i$ (see §1.7): there exist a finite measure $\phi(\cdot)$, $\phi(\mathcal{E}_i) > 0$, defined on the Borel subsets of the space $\mathcal{E}_i$, and numbers ε, $T > 0$ such that $p(T,x,G) < 1 - \varepsilon$ for an arbitrary Borel $G \subset \mathcal{E}_i$, $\phi(G) < \varepsilon$ and $x \in \mathcal{E}_i$.

Condition $\mathcal{B}$ implies that $P(t_0,x,K) > h$ for $x \in \mathcal{E}_i$. Let $\phi(G) = \lambda(G \cap K_i)$ for any Borel $G \subset \mathcal{E}_i$, where λ is the Lebesgue measure in R^r.

Denote by $q(t,x,y)$ the transition density function with respect to the measure ϕ for the process obtained from (X_t, P_x) by means of stopping at the first exit time from K_i. Such a density exists and is continuous for $t > 0$, because the process does not degenerate on $K_i \cup \partial K_i$. Let $\mu = \inf\{q(t,x,y) : x,y \in K_i \cup \partial K_i, t \in [t_0, 2t_0]\}$. By the maximum principle we conclude that $\mu > 0$. We put $\overline{\tau} = \inf\{t : X_t \in K_i \cup \partial K_i\}$. Let $\widetilde{T} = 2t_0$, $x \in \mathcal{E}_i$ and $G \subset K_i$ be a Borel set. Then, relying on the strong Markov property of the process (X_t, P_x) we see that:

$$P\{\widetilde{T},x,G\} \geq E_x \chi_{\overline{\tau}\leq t_0} P_{X_{\overline{\tau}}}\{X_{\widetilde{T}-\overline{\tau}} \in G\} >$$

$$> P_x\{\overline{\tau} < t_0\} \inf_{\substack{t_0 \leq t \leq \widetilde{T} \\ x \in K_i}} \int_G q(t,x,y)\,dy > h \cdot \mu \cdot \phi(G)\,, \tag{7.2}$$

where $\chi_{\overline{\tau}\leq t_0}$ is the indicator of the set $\{\overline{\tau} < t_0\}$. From (7.2) we conclude that Doeblin's condition is fulfilled for $T = 2t_0$, for the above constructed measure ϕ and $\varepsilon = \frac{1}{2}(\phi(\mathcal{E}_i) \wedge h\mu\phi(\mathcal{E}_i))$.

If Doeblin's condition is satisfied, there exists a normed invariant measure $\mu_i(\cdot)$ with support $\mathcal{E}_i$ such that, for any $x \in \mathcal{E}_i$ and Borel set $G \subset \mathcal{E}_i$, the relation holds:

$$|P(t,x,G) - \mu_i(G)| < [1 - h\mu\phi(\mathcal{E}_i)]^{t/T} = e^{-\alpha_i t}\,. \tag{7.3}$$

From (7.3) one can readily deduce the uniqueness of the normed invariant measure concentrated on $\mathcal{E}_i$, and all remaining claims of Lemma 7.1. □

LEMMA 7.2. *Let the conditions of Lemma 7.1 be fulfilled. Then* $C_1, \beta_1 > 0$ *exist such that, for* $x \in \mathcal{E}_i$, *and for any bounded measurable function* $f(x)$ *and any* $j \in \{1, 2, \cdots, \ell\}$ the inequality

$$\left|E_x f(X_t) - \int_{\mathcal{E}_j} f(x)\,\mu_j(dx)\right| \leq C_1 \exp\{-\beta_1 t\} \cdot \sup_{x \in \mathcal{E}_i} |f(x)|$$

will be valid.

Proof. It suffices to consider the case $|f(x)| \le 1$. We put $\gamma_i = \left\{y \in \mathcal{E}_j : \frac{i}{n} > f(y) \ge \frac{i-1}{n}\right\}$. Then, taking into account that $|f(x)| \le 1$, we have

$$\left| \int_{\mathcal{E}_j} f(y)\, P(t,x,dy) - \sum_{i=-n}^{n} \frac{i}{n}\, P(t,x,\gamma_i) \right| < \frac{1}{n},$$

$$\left| \int_{\mathcal{E}_j} f(y)\, \mu_j(dy) - \sum_{i=-n}^{n} \frac{i}{n} \mu_j(\gamma_i) \right| < \frac{1}{n}.$$

From these inequalities, on the basis of Lemma 7.1 we conclude that

$$\left| E_x f(X_t) - \int_{\mathcal{E}_j} f(y)\, \mu_j(dy) \right| < \frac{2}{n} + 2n e^{-\alpha_i t}.$$

Choosing $n \sim \exp\left\{\frac{\alpha_i t}{2}\right\}$ we obtain

$$\left| E_x f(X_t) - \int_{\mathcal{E}_j} f(y)\, \mu_j(dy) \right| < 5e^{-\frac{\alpha_i t}{2}},$$

which implies the claim of Lemma 7.2. □

Let us set $\tau = \inf\{t : X_t \in K \cup \partial K\}$, $\pi_i(x) = P_x\{X_\tau \in \mathcal{E}_i\}$. Condition $\mathcal{B}$ yields that, for some $\beta_2, C_2 > 0$ and any $x \in D$,

$$P_x\{\tau > t\} < C_2 e^{-\beta_2 t}. \tag{7.4}$$

LEMMA 7.3. *Suppose the conditions of Lemma 7.1 to be valid. Then one can find* $C, \beta > 0$ *such that*

$$\left| E_x f(X_t) - \sum_{i=1}^{\ell} \pi_i(x) \int_{\mathcal{E}_i} f(x)\,\mu_i(dx) \right| \leq$$

$$\leq C \sup_{x \in D \cup \partial D} |f(x)| \cdot \exp\{-\beta t\}$$

for any bounded measurable $f(x)$ *and arbitrary* $x \in D$.

Proof of this lemma can easily be arranged with the aid of Lemma 7.2 and bound 7.4, so we drop it. □

THEOREM 7.2. *Suppose that the conditions of Lemma 7.1 are fulfilled and let a function* $f(x)$ *be continuous in* $D \cup \partial D$. *Then the generalized solution of the problem* $L\,u(x) = f(x)$, $x \in D$, $\left.\frac{\partial u(x)}{\partial \ell(x)}\right|_{\Gamma_1} = 0$, *exists if and only if* $\int_D f(x)\,\mu_i(dx) = 0$ *for* $i = 1, \cdots, \ell$.

Suppose that points $x_1, \cdots, x_\ell$ *are picked so that* $x_1 \in K_1, \cdots, x_\ell \in K_\ell$; *and let* $c_1, \cdots, c_\ell$ *be arbitrary constants. Then there exists a unique bounded generalized solution, continuous at the points* $x_1, \cdots, x_\ell$ *and taking at these points the values* $c_1, \cdots, c_\ell$ *respectively.*

Proof. We shall put

$$v(x) = -\int_0^\infty E_x f(X_t)\,dt\,.$$

By Lemma 7.2, noting the orthogonality conditions, we conclude that this integral converges uniformly in $x \in D \cup \Gamma_1$. The calculations involved in the proof of Theorem 7.1 imply that $A\,v(x) = f(x)$.

Let us choose an open ball Π_i with center at the point x_i and lying entirely in K_i. We write $\tau_i = \inf\{t : X_t \notin \Pi_i\}$. Using the strong Markov property one can verify that for $x \in \Pi_i$

$$v(x) = -E_x \int_0^{\tau_i} f(X_s)\,ds + E_x v(X_{\tau_i})\,. \tag{7.5}$$

Since the operator L does not degenerate in Π_i and has sufficiently smooth coefficients, we conclude that it has a smooth Green's function $G(x,y)$, $(x,y \in \Pi_i)$, and Poisson's kernel $g(x,y)$, $(x \in \Pi_i,\ y \in \partial\Pi_i)$; and equality (7.5) may be rewritten in the form

$$v(x) = \int_{\Pi_i} G(x,y) f(y)\,dy + \int_{\partial\Pi_i} v(y)\, g(x,y)\,dy, \quad x \in \Pi_i \,.$$

From this formula comes the continuity of the function $v(x)$ for $x \in \Pi_i$. Consequently, $\lim_{x \to x_i} v(x) = v(x_i)$.

Denote $d_i = v(x_i)$ and consider the function

$$u(x) = v(x) + \sum_{i=1}^{\ell} \pi_i(x)(c_i - d_i) \,.$$

With the Markov property of the process (X_t, P_x), it is readily checked that $T_t \pi_i(x) = \pi_i(x)$, $x \in D \cup \Gamma_1$, $i \in \{1, \cdots, \ell\}$, and thus $A\,\pi_i(x) = 0$. Therefore, $u(x) \in D_A$ and $A\,u = A\,v + \Sigma(c_i - d_i) A\,\pi_i(x) = A\,v = f$.

From the definition of the function $\pi_i(x)$ it follows that $\pi_i(x) = 1$ for $x \in K_i$. On account of the continuity of the function $v(x)$ for $x \in \Pi_i$, this implies that $\lim_{x \to x_i} u(x) = c_i$, $i \in \{1, \cdots, \ell\}$. Hence the function $u(x)$ is a generalized solution taking given values $c_1, \cdots, c_\ell$ at the points $x_1, \cdots, x_\ell$.

We proceed to prove the uniqueness of such a solution. Let $f(x) = c_1 = c_2 = \cdots = c_\ell = 0$. The equality $A\,u(x) = 0$ yields that $T_t\,u(x) = u(x)$ for $t \geq 0$. Since $u(X_t)$ is a bounded martingale, $\lim_{t \to \infty} u(X_t)$ exists P_x-a.s. From the conditions of the theorem it follows that, for almost all trajectories $X_.$, starting from $x \in D$, one can find a number $i = i[X_.]$ such that X_t will visit any neighborhood of the point x_i for arbitrarily large t. By the condition, the function $u(x)$ is continuous at the points

x_i and $u(x_i) = 0$. Therefore, $\lim_{t\to\infty} u(X_t) = 0$, P_x-a.s., $x \in D$, which implies that $u(x) = E_x \lim_{t\to\infty} u(X_t) = 0$. □

EXAMPLE 7.1. On the segment $[-1,1]$, consider the problem

$$a(x)\frac{d^2u}{dx^2} + x\frac{du}{dx} = f, \; x \in (-1,1), \; \frac{du}{dx}(\pm 1) = 0, \tag{7.6}$$

where $a(x)$ is a smooth function, positive for $x \in [-1,-0.9) \cup (-0.1, 0.1) \cup (0.9,1)$ and vanishing on the rest of the segment $[-1,1]$. In this case Condition $\mathcal{B}$ holds: $\ell = 2$, $\mathcal{E}_1 = [-1,-0.9)$, $\mathcal{E}_2 = (0.9,1]$. The invariant measures of the corresponding process make up a two-dimensional cone: each invariant measure μ may be obtained as a linear combination of measures μ_1 and μ_2 concentrated on $\mathcal{E}_1$ and $\mathcal{E}_2$ respectively. The measures μ_1 and μ_2 are defined by their density functions m_1 and m_2 which are positive normed solutions of the problems

$$\frac{d^2}{dx^2}(a(x)m_1(x)) - \frac{d}{dx}(xm_1(x)) = 0, \; x \in (-1,-0.9), \frac{dm_1}{dx}(-1) = 0;$$

$$\frac{d^2}{dx^2}(a(x)m_2(x)) - \frac{d}{dx}(xm_2(x)) = 0, \; x \in (0.9,1), \frac{dm_2}{dx}(1) = 0.$$

From this, one can find m_1 and m_2. The conditions for the solvability of problem (7.6) have the form $\int_{-1}^{-0.9} f(x)\,m_1(x)\,dx = \int_{0.9}^{1} f(x)\,m_2(x)\,dx = 0$. Problem (7.6) has a two-dimensional space of solutions. To single out a unique solution, one can assign $u(-1) = c_1$, $u(1) = c_2$.

Chapter IV

SMALL PARAMETER IN SECOND-ORDER ELLIPTIC DIFFERENTIAL EQUATIONS

§4.1 *Classical case. Problem statement*

A great number of works dealing with special cases were followed in 1950 by the article of Levinson [1] which is concerned with the behavior as $\varepsilon \downarrow 0$ of the solution of Dirichlet's problem for the second-order elliptic equation of the general form with a small parameter in the higher order derivatives:

$$L^{\varepsilon}u^{\varepsilon}(x) = \frac{\varepsilon}{2}\sum_{i,j=1}^{r} a^{ij}(x)\frac{\partial^2 u^{\varepsilon}}{\partial x^i \partial x^j} + \sum_{i=1}^{r} B^i_{(x)}\frac{\partial u^{\varepsilon}}{\partial x^i} + C(x)u^{\varepsilon} = (\varepsilon L_1 + L_0)u^{\varepsilon} = 0\,,$$

$$x \in D \subset R^r\,,\quad u^{\varepsilon}(x)\big|_{\partial D} = \psi(x)\,. \tag{1.1}$$

Here $(a^{ij}(x))$ is a positive definite matrix, D is a bounded domain in R^r with a fairly good boundary, $\psi(x)$ being a continuous function on ∂D. The coefficients of the equation are assumed bounded and Lipschitz continuous.

Levinson singled out a relatively simple class of problems when the functions $u^{\varepsilon}(x)$ converge as $\varepsilon \downarrow 0$ to the solution $u_0(x)$ of the first-order equation obtained from (1.1) by setting $\varepsilon = 0$:

$$L_0 u_0(x) = \sum_{i=1}^{r} B^i(x)\frac{\partial u_0}{\partial x} + C(x)u_0 = 0\,,\quad x \in D\,,$$

$$u_0(x)\big|_{\widetilde{\partial D}} = \psi(x)\,. \tag{1.2}$$

Here $\widetilde{\partial D}$ is the part of the boundary of the domain D, regular for the operator L_0 (see §3.4). Under the Levinson conditions, problem (1.2) has a unique solution.

To formulate the Levinson conditions, consider the characteristics of the operator L_0. These characteristics are the solutions of the ordinary differential equation

$$\frac{dX_t^x}{dt} = B(X_t^x), \quad X_0^x = x \in R^r, \quad B(x) = (B^1(x), \cdots, B^r(x)) . \tag{1.3}$$

For brevity, we shall assume that, at every point of the boundary of the domain D, an outward normal $n(x) = (n_1(x), \cdots, n_r(x))$, $x \in D$, is defined, where $n_i(x)$ are sufficiently smooth functions. The coefficients $B^i(x)$ are also assumed fairly smooth.

We shall say that for the operator L in the domain D the *Levinson condition* holds if, starting from any $x \in D$, the characteristic X_t^x leaves D in a finite time crossing the boundary ∂D at non-zero angle (Fig. 1).

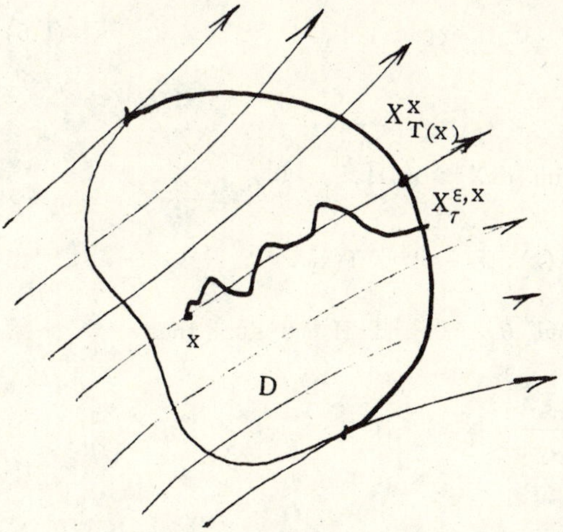

Fig. 1

If one denotes $T(x) = \inf\{t : X_t^x \in D\}$, then the Levinson condition is as follows: for $x \in D$,

1) $T(x) < \infty$;

2) $(n(X^x_{T(x)}), B(X^x_{T(x)})) > 0$.

If the Levinson condition is fulfilled, the set $\widetilde{\partial D}$ consists of those points of the boundary ∂D where $(n(y), B(y)) > 0$.

The Levinson result has a simple probabilistic meaning. Let for brevity $C(x) \equiv 0$. Along with the family of characteristics of the operator L_0, let us consider the Markov family $(X_t^{\varepsilon,x}, P)$ in the space R^r corresponding to the operator L^ε. The trajectories of this family are the solutions of the stochastic differential equation

$$dX_t^{\varepsilon,x} = \sqrt{\varepsilon}\,\sigma(X_t^{\varepsilon,x})\,dW_t + B(X_t^{\varepsilon,x})\,dt\,, \quad X_0^{\varepsilon,x} = x\,, \tag{1.4}$$

where $\sigma(x)$ is a matrix with Lipschitz continuous elements, W_t being a Wiener process.

Put

$$m^\varepsilon(t) = E \max_{0\le s\le t} |X_s^{\varepsilon,x} - X_s^x|^2 .$$

From (1.3) and (1.4), by Doob's inequality for stochastic integrals we obtain

$$m^\varepsilon(t) \le c_1 t \int_0^t m^\varepsilon(s)\,ds + \varepsilon t\, c_2 .$$

Here and henceforth, c_i are positive constants. The last inequality implies that

$$m^\varepsilon(t) \le c_2 t\varepsilon \exp\{c_1 t^2\} . \tag{1.5}$$

Put

$$\tau^{\varepsilon,x} = \tau = \inf\{t : X_t^{\varepsilon,x} \notin D\}, \quad \varepsilon \ge 0 ;$$

$$\tau^{0,x} = T(x)\,; \quad \overline{\tau} = \tau^{\varepsilon,x} \wedge T(x) .$$

For any $x \in D$, one can find $\theta_1 > 0$ and $H > 0$ such that

$$\lim_{\substack{z\to y\\ z\in D}} \frac{E\,\tau^{\varepsilon,z}}{|z-y|} = h^\varepsilon(y) < H < \infty \tag{1.6}$$

uniformly in $y \in \Gamma_{\theta_1}^x = \{y \in \partial D : |y - X_{T(x)}^x| < \theta_1\}$ and in $\varepsilon \in [0,1]$.

This assertion is established just as Lemma 3.4.2 was. For sufficiently small $\varepsilon > 0$, the role of a barrier may be played by the function $T(z) : L^\varepsilon T(z) = L_0 T(z) + \varepsilon \times$ (a bounded function) $< -1 + c_3\varepsilon < -c < 0$.

Since the coefficients of the stochastic equation are bounded, (1.6) implies that

$$E|X^{\varepsilon,z}_{\tau^{\varepsilon,z}} - y| < |z-y| + E|X^{\varepsilon,z}_{\tau^{\varepsilon,z}} - z| \le$$

$$\le |z-y| + E\left| \int_0^{\tau^{\varepsilon,z}} B(X_s^{\varepsilon,z})\,ds + \int_0^{\tau^{\varepsilon,z}} \sqrt{\varepsilon}\,\sigma(X_s^{\varepsilon,z})\,dW_s \right| \le \tag{1.7}$$

$$\le |z-y| + c_4 E\tau^{\varepsilon,z} + c_5\sqrt{\varepsilon E\tau^{\varepsilon,z}} \le c_6\sqrt{\varepsilon} + c_7|z-y| .$$

Suppose that the boundary function $\psi(x)$ satisfies a Lipschitz condition with a constant K_ψ. We will show that in this case $|u^\varepsilon(x) - u_0(x)| < c\sqrt{\varepsilon}$. From (1.7) it follows that

$$|u^\varepsilon(z) - \psi(y)| \le E|\psi(X^{\varepsilon,z}_{\tau^{\varepsilon,z}}) - \psi(y)| \le$$

$$\le K_\psi E|X^{\varepsilon,z}_{\tau^{\varepsilon,z}} - y| \le K_\psi c_6\sqrt{\varepsilon} + K_\psi c_7|z-y| . \tag{1.8}$$

for $z \in D$, $y \in \Gamma^x_{\theta_1}$, $\varepsilon \in [0,1]$.

We put $\widetilde{T} = \widetilde{T}(x) = \sup\{t < T(x) : |X_t^x - X_{T(x)}^x| = \theta_1\}$, $\theta = \inf_{0 \le t \le \widetilde{T}} \rho(X_t^x, \partial D)$. Clearly, $0 < \theta \le \theta_1$.

The strong Markov property of the family $(X_t^{\varepsilon,x}, P)$ yields

$$|u^\varepsilon(x) - u(x)| = E|u^\varepsilon(X^{\varepsilon,x}_{\widetilde{T}}) - u_0(X^x_{\widetilde{T}})| \le$$

$$\le E|u^\varepsilon(X^{\varepsilon,x}_{T(x)}) - \psi(X^x_{T(x)})|\chi_1 +$$

$$+ E|\psi(X^{\varepsilon,x}_{\tau^{\varepsilon,x}}) - u_0(X^x_{\tau^{\varepsilon,x}})|\chi_2 + \tag{1.9}$$

$$+ \sup_{x\in\partial D} |\psi(x)| \cdot P\{ \sup_{0\le t\le T(x)} |X_t^{\varepsilon,x} - X_t^x| > \theta\} ,$$

where

$$\chi_1 = \{\omega : \tau^{\varepsilon,x} \geq T(x), \sup_{0\leq t\leq T(x)} |X_t^{\varepsilon,x} - X_t^x| < \theta\},$$

$$\chi_2 = \{\omega : \tau^{\varepsilon,x} < T(x), \sup_{0\leq t\leq T(x)} |X_t^{\varepsilon,x} - X_t^x| < \theta\}.$$

From (1.8) and (1.9) we conclude that

$$|u^\varepsilon(x) - u_0(x)| < c_8 \sqrt{\varepsilon} + c_9 E \sup_{0\leq t\leq T(x)} |X_t^{\varepsilon,x} - X_t^x| +$$

$$+ \sup_{x\epsilon\partial D} |\psi(x)| P\{ \sup_{0\leq t\leq T(x)} |X_t^{\varepsilon,x} - X_t^x| > \theta\}.$$

The last inequality along with (1.5) implies that $|u^\varepsilon(x) - u_0(x)| < c_{10}\sqrt{\varepsilon}$. Therefore, our reasoning leads to the following result

THEOREM 1.1. *Suppose that the coefficients of the operator and the direction cosines* $n_i(x)$ *of the normal to the boundary of the domain are twice continuously differentiable. Suppose that the Levinson condition holds and let the boundary function* $\psi(x)$ *be Lipschitz continuous. Then for any compact set* $K \subset D$ *one can find a constant* c *such that for all* $x \in K$

$$|u^\varepsilon(x) - u_0(x)| < c\sqrt{\varepsilon},$$

where $u^\varepsilon(x)$ *and* $u_0(x)$ *are the solutions of problems (1.1) and (1.2) respectively.*

If the boundary function is merely Lipschitz continuous, then the bound given by Theorem 1.1 cannot be improved. Indeed, let $D = \{(x,y) \subset R^r, |x| < 1, 0 < y < 1\}$, $L^\varepsilon = -\frac{\partial}{\partial x} + \frac{\varepsilon}{2}\Delta$, $\psi(x,y) = |x|$. Then $u(0,y)$ is of the order of $\sqrt{\varepsilon}$ as $\varepsilon \downarrow 0$.

If stronger assumptions are made about $\psi(x)$, for example, that $\psi(x)$ is the trace of a twice continuously differentiable function $v(x)$, $x \in R^r$, onto ∂D, then the difference $u^\varepsilon(x) - u_0(x)$ may be bounded more accurately. In this case, it will be of order of ε. If the boundary of the domain, the coefficients of the operator, and the boundary function are sufficiently smooth, then, under the Levinson condition, one can write down the following terms of the asymptotic expansion of $u^\varepsilon(x)$ into integral powers of the parameter ε:

$$u^\varepsilon(x) = u_0(x) + \varepsilon u_1(x) + \varepsilon^2 u_2(x) + \cdots .$$

The functions $u_k(x)$, $k \geq 1$, are determined successively by the following chain of equations

$$L_0 u_{k+1}(x) = -L_1 u_k(x),\ x \in D;\quad u_{k+1}(x)|_{\widetilde{\partial D}} = 0 .$$

We shall return to this question later on.

If in equation (1.1) we have $C(x) \leq -c_0 < 0$ for $x \in D$, then the condition $T(x) < \infty$ may be dropped. The corresponding bounds may be easily obtained, provided one uses the representation of the solution of the corresponding problem in the form of a functional integral.

Notice that the convergence of $u^\varepsilon(x)$ to $u_0(x)$ deteriorates while x approaches $\partial D \setminus \widetilde{\partial D}$. To calculate $u^\varepsilon(x)$ for x belonging to the ε-neighborhood of the set $\partial D \setminus \widetilde{\partial D}$, one must construct the so-called boundary layer. Although such a construction has a quite transparent probabilistic meaning, we shall not go into this here.

So, if the Levinson condition holds, $\lim u^\varepsilon(x) = u_0(x)$ as $\varepsilon \downarrow 0$. The limit function $u_0(x)$ is the unique solution of problem (1.2). The function $u_0(x)$ does not depend on perturbations; that is, on the operator L_1. Our reasoning relies on the fact that, under the Levinson conditions, the trajectories $X_t^{\varepsilon,x}$ leave D in time of the order 1 as $\varepsilon \downarrow 0$, and this time is not enough for the small perturbations to deflect essentially the trajectory $X_t^{\varepsilon,x}$ from X_t^x. Hence it already appears clear that the behavior of the exit time $\tau^{\varepsilon,x}$ as $\varepsilon \downarrow 0$ is the most important characteristic of the problem. Roughly speaking, the more time as $\varepsilon \downarrow 0$, that is needed for the trajectories $X_t^{\varepsilon,x}$ to hit the boundary of a domain, the more difficult is the analysis of the behavior of $u^\varepsilon(x)$ as $\varepsilon \downarrow 0$. Non-fulfillment of the Levinson conditions (later on we assume $c(x) \equiv 0$) causes an essential complication of the problem. With the exception of some special examples, progress here did not appear until the 1960's and it was due to applying probabilistic methods. Notice that so far one has failed to describe completely all possible cases. This may partly be explained by the variety of behaviors of the trajectories of the dynamical system (1.3), especially when dealing with large dimensions. Here one has to limit oneself to considering "extreme" cases. The general case is a combination of these "extreme" ones and some degenerate situations.

So, the Levinson condition causes $\tau^{\varepsilon,x}$ to be of order 1 as $\varepsilon \downarrow 0$. Further cases are obtained when $\tau^{\varepsilon,x}$ grows as $\varepsilon \downarrow 0$, but very slowly. We face such a situation when the Levinson condition holds for some expanding sequence of domains D_n, $\cup D_n = D$. In this case, leaving any of the domains D_n happens at the expense of the drift in time of order 1 as $\varepsilon \downarrow 0$ and merely the last part near the boundary of the domain D is overcome at the expense of the perturbations. Under minor extra assumptions on the behavior of the characteristics near the boundary ensuring, in particular, the existence of $\lim\limits_{t\to\infty} X_t^x = X_\infty^x$, $\lim\limits_{\varepsilon\downarrow 0} u^\varepsilon(x) = u_0(x)$ also exists and is a unique solution of problem (1.2).

A close case will be available if system (1.3) has hyperbolic points in D. Such an example is represented in Fig. 2.

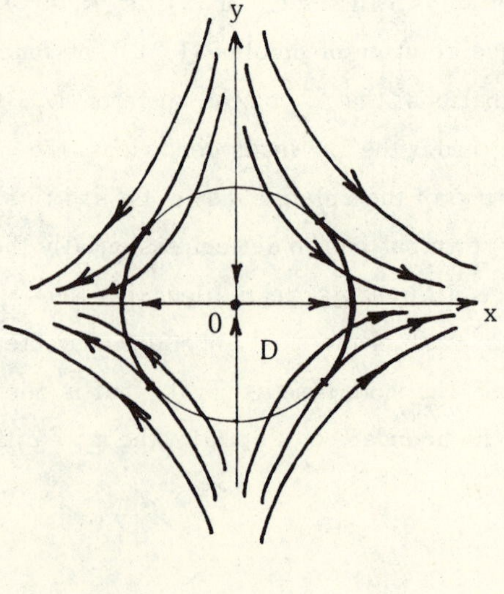

Fig. 2

This figure shows the characteristics of the operator L_0. The domain D is the circle, the heavy line marking the part of the boundary regular for the operator L_0. If the matrix $(a^{ij}(x))$ in the operator L_1 does not degenerate, then the trajectories of the Markov family $(X_t^{\varepsilon,x}, P_x)$ corresponding to the operator L^ε leave D in a finite time for $\varepsilon > 0$. If $\varepsilon \downarrow 0$, then for the trajectories starting on the y-axis, this time goes, though slowly, to $+\infty$. It is not difficult to show that in this example $\lim_{\varepsilon \downarrow 0} u^{\varepsilon}(x) = u_0(x)$ exists. The function $u_0(x)$ has, generally speaking, a discontinuity on the y-axis.

In the Levinson case, leaving a domain occurs at the expense of the terms of the operator L_0. In the above example, leaving the domain is mainly determined by the operator L_0 as well. As a further case (relative to the order of growing the exit time $\tau^{\varepsilon,x}$ from the domain) one can consider the case when the operator L_0 does not contribute to leaving the domain, as it was in the above cases, although it does not hinder this leaving. Such is the case, for example, if the domain $D \subset R^2$ is a ring with the center at the origin, the characteristics of the operator L_0 being circles with the same center. It will be convenient for us to deal with a 3-dimensional domain.

So, let $\tilde{D}$ be a bounded domain in the $(x^1; x^2)$-plane of the space R^3 lying at a positive distance from the x^2-axis (see Fig. 3). Consider the domain $D \subset R^3$ which is derived when D revolves round the x^2-axis. As the coordinates of a point $x \in R^3$, we will take the angle ϕ, $0 \leq \phi \leq 2\pi$, of the rotation which is needed to turn the $(x^1; x^2)$-plane about the x^2-axis so that it can pass through the point $x \in R^3$, and the coordinates x^1, x^2 of the point of the $(x^1; x^2)$-plane which moves to the point x at this rotation. Let the operator L^{ε} have in these coordinates the form

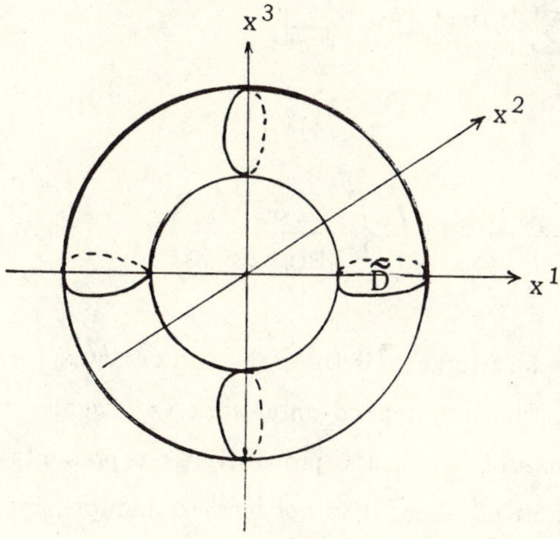

Fig. 3

$$L^{\varepsilon} = B(x)\frac{\partial}{\partial \phi} + \frac{\varepsilon}{2}\left(\sum_{i,j=1}^{2} a^{ij}(x)\frac{\partial^2}{\partial x^i \partial x^j} + 2a^{13}(x)\frac{\partial^2}{\partial x^1 \partial \phi} + 2a^{23}(x)\frac{\partial^2}{\partial x^2 \partial \phi} + a^{33}(x)\frac{\partial^2}{\partial \phi^2}\right) = L_0 + \varepsilon L_1, \quad B(x) > 0 \text{ for } x \in D \cup \partial D .$$

The coefficients of the operator and the boundary of the domain are assumed to be sufficiently smooth and the form $\Sigma a^{ij}(x)\lambda_i\lambda_j$ to be positive definite. The characteristics of the operator $L_0 = B(x^1,x^2,\phi)\frac{\partial}{\partial\phi}$ are circles lying in the plane perpendicular to the x^2-axis with centers on this axis. These characteristics neither approach nor recede from the boundary of the domain D. Therefore, leaving the domain occurs solely at the expense of perturbations. Note that the small parameter precedes the coefficients controlling the movement in the directions x^1, x^2, transversal to the characteristics. Thus, for small ε, a diffusing particle whose motion is governed by the operator L^ε, makes many revolutions round the x^2-axis along the characteristics before it makes a small movement in the $(x^1; x^2)$-plane. Hence, this time is enough for the coefficients of the diffusion along x^1, x^2 to become averaged in ϕ in accordance with the time spent by the particle near different parts of the characteristic.

According to these heuristic reasonings, for small ε movement in the variables x^1, x^2 is governed by the averaged operator

$$\frac{\varepsilon}{2}\bar{L}_1 = \frac{\varepsilon}{2}\sum_{i,j=1}^{2} \bar{a}^{ij}(x^1,x^2)\frac{\partial^2}{\partial x^i \partial x^j},$$

where

$$\bar{a}^{ij}(x^1,x^2) = \int_0^{2\pi} \frac{a^{ij}(x^1,x^2,\phi)}{B(x^1,x^2,\phi)}\,d\phi \left(\int_0^{2\pi} \frac{d\phi}{B(x^1,x^2,\phi)}\right)^{-1}.$$

Consider the Dirichlet problem: $L^\varepsilon u^\varepsilon(x) = 0$ for $x \in D$, $u^\varepsilon(x) = \psi(x)$ for $x \in \partial D$. Let, for simplicity, the boundary function depend only on x^1, x^2: $\psi(x) = \psi(x^1,x^2)$. From the above discussion, using the probabilistic representation of the solution to the Dirichlet problem, it is not hard to deduce that $\lim_{\varepsilon \downarrow 0} u^\varepsilon(x) = u_0(x)$ exists, the limit function does not depend on ϕ and is a solution of the Dirichlet problem in $\tilde{D}$: $\bar{L}_1 u_0(x^1,x^2) = 0$ for $(x^1,x^2) \in D$, $u_0(x^1,x^2)|_{\partial D} = \psi(x^1,x^2)$. Section 4.3 will present the proof of the so-called averaging principle which implies the above assertion.

It is easily checked that in the foregoing example, the exit time $\tau^{\varepsilon,x}$ from the domain D is of order ε^{-1} as $\varepsilon \downarrow 0$ (more precisely, $\lim_{\varepsilon \downarrow 0} \varepsilon E \tau^{\varepsilon,x} = m^0(x)$, where $0 < m^0(x) < \infty$ for $x \in D$). This time is already so large that the effects of the perturbations may not be neglected, and the limit function $u_0(x)$ already depends essentially on the operator L_1. For different perturbation εL_1, the limits are, generally, different.

Finally, situations are possible when the terms of the operator L_0 hinder leaving the domain. In this case, the exit time is very large, of order of $\exp\{\varepsilon^{-1} \cdot \text{const}\}$, $\varepsilon \downarrow 0$. Let system (1.3) have the asymptotically stable equilibrium point 0 in D, and the characteristics, starting from any point $x \in D$, tend to 0 without hitting ∂D. It is easy to understand (this comes from the bounds given in proving Theorem 1.1.) that, for small ε, with probability approaching 1, the trajectory $X_t^{\varepsilon,x}$, $x \in D$, will first reach any preassigned neighborhood $\mathcal{E}$ of the point 0. The presence of a non-zero, even though small diffusion causes the trajectory $X_t^{\varepsilon,x}$ to deviate from the point 0, to leave $\mathcal{E}$, and to return. The majority of such excursions will be short. A trajectory will make a lot of excursions before it hits the boundary, but at some time (for $\varepsilon \neq 0$) an excursion will occur which hits the boundary. It is this first boundary-hitting excursion that determines $X_{\tau^{\varepsilon,x}}^{\varepsilon,x}$ and thereby $u^{\varepsilon}(x) = E\psi(X_{\tau^{\varepsilon,x}}^{\varepsilon,x})$. Thus, here the behavior of $u^{\varepsilon}(x)$ as $\varepsilon \downarrow 0$ is defined by large deviations of the trajectories from their typical behavior. It turns out that in the case of the general situation for small ε, among all possible ways of hitting ∂D, there is one which is more probable than all the others taken together.

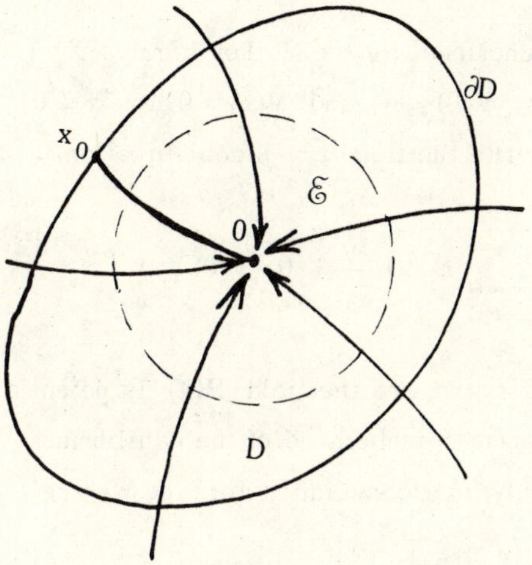

Fig. 4

In order to formulate the explicit result (Wentzell and Freidlin [1], [2]), we will consider the function $V(x)$ —the quasi-potential of the vector field $B(x) = (B^1(x), \cdots, B^r(x))$ with respect to the perturbations determined by the operator $\varepsilon L_1 = \frac{\varepsilon}{2} \sum a^{ij}(x) \frac{\partial^2}{\partial x^i \partial x^j}$:

$$V(x) = \inf\{S_{T_1,T_2}(\phi) : \phi \in C_{T_1,T_2}(R^r) : \phi_{T_1} = 0, \phi_{T_2} = x, -\infty \le T_1 < T_2 < \infty\} ,$$

$$S_{T_1,T_2}(\phi) = \frac{1}{2} \int_{T_1}^{T_2} \sum_{i,j=1}^{r} a_{ij}(\phi_s)(\dot\phi_s^i - B^i(\phi_s))(\dot\phi_s^j - B^j(\phi_s))\, ds ,$$

$$(a_{ij}(x)) = (a^{ij}(x))^{-1} .$$

Here $\varepsilon^{-1} S_{0,T}(\phi)$ is the action functional for the Markov family $(X_t^{\varepsilon,x}, P)$. The function $V(x)$ is continuous, $V(0) = 0$, and $V(x) > 0$ for $x \neq 0$. For the function $V(x)$, one can write the Hamilton-Jacobi equation

$$\sum_{i,j=1}^{r} a^{ij}(x) \frac{\partial V}{\partial x^i} \frac{\partial V}{\partial x^j} + 2 \sum_{i=1}^{r} B^i(x) \frac{\partial V}{\partial x^i} = 0, \quad V(0) = 0 .$$

In particular, if $(a_{ij}(x))$ is a unit matrix and the field $B(x)$ is potential, that is $B(x) = -\nabla U(x)$, then in some neighborhood of the equilibrium point, $V(x)$ differs from $U(x)$ only by a constant factor (see Wentzell and Freidlin [1,2]).

THEOREM 1.2. *Suppose that the vector field* $B(x)$ *has in* D *a unique, asymptotically stable equilibrium point* 0, *so that the trajectory of system (1.3), starting from any point* $x \in D$, *is attracted to* 0 *without leaving* D *as* $t \to \infty$, *and let* $(n(x), B(x)) < 0$ *for* $x \in \partial D$, *where* $n(x)$ *is the vector of the outward normal to* ∂D.

Let x_0 *be a point on* ∂D *such that* $V(x_0) = \inf\{V(x) : x \in \partial D\}$ *and suppose that* $V(x) > V(x_0)$ *for* $x \in \partial D \setminus x_0$. *Then the solution* $u^\varepsilon(x)$ *of problem (1.1) with* $C(x) \equiv 0$ *tends to the constant equal to* $\psi(x_0)$ *as* $\varepsilon \downarrow 0$.

It turns out that, for small ε, with probability approaching 1, the first hitting of the boundary of the domain D will occur near the point x_0: $X^{\varepsilon,x}_{\tau^{\varepsilon,x}} \to x_0$ in probability as $\varepsilon \downarrow 0$, for $x \in D$. Noting that $u^{\varepsilon}(x) = E\psi(X^{\varepsilon,x}_{\tau^{\varepsilon,x}})$, this implies the claim of Theorem 1.2. We shall return to this problem and some of its generalizations in §4.4. Theorem 1.2 and some of its applications will be proved in §§4.4 and 4.5.

If the vector field $B(x)$ has several stable equilibrium points or asymptotically stable limit sets in D, then extra complications appear. These complications are due to the fact that before hitting the boundary, the trajectory may travel between these ω-limit sets. Here the limit function may already be distinct from a constant. Such problems, just as the case of one equilibrium point, are studied in detail in the works of Wentzell and Freidlin [1], [2].

As was already said, the above listed situations are, in a sense, "extreme cases." In the general case, their various combinations are possible. Starting from some points of the domain, the trajectories hit the boundary in a finite time or in a time growing slowly as $\varepsilon \downarrow 0$. Starting from other points, they do this in a time of order ε^{-1}; starting from some other points — in the time $\exp\{c\varepsilon^{-1}\}$. To the general situation, one can apply not only results concerning the extreme cases, but also the ideology arising in considering them.

When treating problem (1.1), in addition to questions connected with the behavior of $u^{\varepsilon}(x)$ as $\varepsilon \downarrow 0$, a number of other questions arise as well. For example, the behavior, as $\varepsilon \downarrow 0$, of eigenvalues and eigenfunctions of the operator L^{ε} is of interest. This problem involves the question of the behavior, as $t \to \infty$, $\varepsilon \downarrow 0$, of the solution of the parabolic equation $L^{\varepsilon}u^{\varepsilon} = \dfrac{\partial u^{\varepsilon}(t,x)}{\partial t}$. Some results on these questions are available in Wentzell and Freidlin [2], Freidman [2].

Up to now we have been dealing with problems where the small parameter precedes all the second-order derivatives. If $\varepsilon = 0$, then only the first-order operator remains. This chapter will also be concerned with the problem when the operator L_0, remaining for $\varepsilon = 0$, involves second-order derivatives:

$$(\varepsilon L_1 + L_0)\, u^\varepsilon(x) = \frac{\varepsilon}{2} \sum_{i,j=1}^{r} a^{ij}(x) \frac{\partial^2 u^\varepsilon}{\partial x^i \partial x^j} + \frac{1}{2} \sum_{i,j=1}^{r} A^{ij}(x) \frac{\partial^2 u^\varepsilon}{\partial x^i \partial x^j} + \sum_{i=1}^{r} B^i(x) \frac{\partial u^\varepsilon}{\partial x^i} = 0, \quad x \in D, \quad u^\varepsilon(x)\big|_{\partial D} = \psi(x) . \tag{1.10}$$

In those cases when it may lead to some new effects, we shall consider perturbations involving first-order derivatives. Sometimes problems are of interest in which the operator L_1 contains solely first-order derivatives.

If L_0 is an elliptic non-degenerate operator, then the problem of calculating $u_0(x) = \lim_{\varepsilon \downarrow 0} u^\varepsilon(x)$ or even next terms of the expansion involves no serious difficulties. All the constructions are easy to carry out, whenever the coefficients, the boundary of the domain, and the boundary function are assumed fairly smooth. To this end, one can make use of the a priori bounds provided by the elliptic theory. The situation becomes complicated if one assumes that the operator L_0 is merely non-negative. This includes the case when L_0 is a first-order operator. We call the latter case a classical one. The results of Chapter III will help us to examine problem (1.10) for an arbitrary operator L_0 with non-negative characteristic form. Here the role of characteristics will be played by the trajectories of a Markov process governed by the operator L_0. Just as in the classical case, the nature of the results depends, in essence, on the behavior of the exit time from the domain as $\varepsilon \downarrow 0$. Many results in the general situation are similar to that in the classical case, however, there are distinctions too. We shall pay special attention to the distinctions involved.

Usually, we assume the form $\sum_{i,j=1}^{r} a^{ij}(x)\lambda_i\lambda_j$ to be positive definite, but many of our results may readily be carried over to the case of degenerate perturbations as well. However, one should bear in mind some peculiarities which arise when considering degenerate perturbations. These peculiarities may be seen in the following example.

EXAMPLE 1.1. Let $D = \{(x,y) \in R^2 : |x| < 1, |y| < 1\}$,

$$\mathcal{L}^\varepsilon u^\varepsilon(x,y) = -y\frac{\partial u^\varepsilon}{\partial y} + \alpha\varepsilon\frac{\partial^2 u^\varepsilon}{\partial y^2} - \varepsilon^2 x\frac{\partial u^\varepsilon}{\partial x} + \beta\varepsilon^3\frac{\partial^2 u^\varepsilon}{\partial x^2} =$$

$$= (L_0 + \varepsilon L_1 + \varepsilon^2 L_2 + \varepsilon^3 L_3)u^\varepsilon .$$

Consider the first boundary value problem:

$$\mathcal{L}^\varepsilon u^\varepsilon(x,y) = 0, \quad (x,y) \in D, \quad u^\varepsilon(x,y)\Big|_{\partial D} = y^2(x^2+1) . \tag{1.11}$$

The characteristics of the operator $L_0 = -y\frac{\partial}{\partial y}$ do not leave D. For the operator $L_0 + \varepsilon L_1$ in the domain D, the segments of the boundary, parallel to the x-axis, are regular. With the boundary conditions on these segments, the equation $(L_0 + \varepsilon L_1)u_1^\varepsilon(x,y) = 0$ has a unique solution $u_1^\varepsilon(x,y)$. It is easily seen that $\lim_{\varepsilon \downarrow 0} u_1^\varepsilon(x,y) = x^2+1$. For the equation $(L_0 + \varepsilon L_1 + \varepsilon^2 L_2)u_2^\varepsilon(x,y) = 0$ in D, the same segments of the boundary are regular, as for the equation $(L_0 + \varepsilon L_1)u_1^\varepsilon = 0$. However, as it is not difficult to check, $\lim u_2^\varepsilon(x,y) = 1$ as $\varepsilon \downarrow 0$. Finally, for the operator $\mathcal{L}^\varepsilon = L_0 + \varepsilon L_1 + \varepsilon^2 L_2 + \varepsilon^3 L_3$, the whole boundary of the square D is regular, and problem (1.11) has a unique solution $u^\varepsilon(x,y)$. For $\beta > \alpha$, it turns out that, for small ε, with probability approaching 1, the first hitting of the boundary will happen near the x-axis. In this case, $u^\varepsilon(x,y) \to 0$ as $\varepsilon \downarrow 0$. If $\beta < \alpha$, then hitting the boundary will occur near the y-axis, and $u^\varepsilon(x,y) \to 1$ as $\varepsilon \downarrow 0$.

Hence, though the first boundary value problem for the operator $L_0 + \varepsilon L_1$ has a unique solution, the addition of terms with the factor ε to a higher power may affect the limit of the solution as $\varepsilon \downarrow 0$. Unless the operator L_1 degenerates, no such effects take place. This example emphasizes once more that the exit time from the domain for the corresponding process is the most important characteristic of a problem.

§4.2 *The generalized Levinson condition*

So, we turn to studying problem (1.10). First, we will separate a comparatively simple case corresponding to the Levinson case (see Freidlin [16], Sarafian [2], Sarafian and Safarian [3].

The classical Levinson condition consists of two parts: first, the characteristics must leave the domain D in a finite time, and secondly, they must behave in a regular way near the boundary. To formulate similar conditions in the general case, we shall consider the Markov process (X_t, P_x) in R^r corresponding to the operator L_0; (X_t^x, P) being the corresponding Markov family. The trajectories X_t^x are the solutions of the equation

$$dX_t^x = \sigma_0(X_t^x)dW_t + B(X_t^x)dt, \quad X_0^x = x , \tag{2.1}$$

where $\sigma_0(x)$ is a matrix with the Lipschitz continuous elements such that $\sigma_0(x)\sigma_0^*(x) = (A^{ij}(x))$. We shall not strive for more generality and, in order to avoid an awkward formulation, we will assume the coefficients of the operator L^ε to be twice continuously differentiable, and, moreover, we shall make assumptions excluding the contact of the parts of the boundary near which the operator L_0 behaves in different ways. The general case requires unessential changes. Some results on the general case may be found in the above cited works.

We shall say that, for the operator L_0 in the domain D, the *generalized Levinson condition* is fulfilled if

1. the trajectories of the process (X_t, P_x) leave D uniformly exponentially fast;

2. the boundary ∂D of the domain D is the union of the three manifolds $\Gamma_0, \Gamma_1', \Gamma_1''$. Each of these manifolds is either the empty set or a closed smooth manifold of dimension (r–1) having three times continuously differentiable direction cosines $(n_1(x), \cdots, n_r(x))$ of the outward normal $n(x)$. On $\Gamma_0, \Gamma_1', \Gamma_1''$, the following conditions hold:

a) $\sum_{i,j=1}^{r} A^{ij}(x)\, n_i(x)\, n_j(x) = 0, \quad \sum_{i=1}^{r} B^i(x)\, n_i(x) \le -\beta < 0, \quad \text{for } x \in \Gamma_0,$

b) $\sum_{i,j=1}^{r} A^{ij}(x)\, n_i(x)\, n_j(x) = 0, \quad \sum_{i=1}^{r} B^i(x)\, n_i(x) \ge \beta > 0 \quad \text{for } x \in \Gamma_1',$

c) $\sum_{i,j=1}^{r} A^{ij}(x)\, n_i(x)\, n_j(x) \ge \beta > 0 \quad \text{for } x \in \Gamma_1''.$

Note that for Condition 1 to be fulfilled, one can give simple sufficient conditions in terms of the coefficients (see Lemma 3.3.3). If L_0 is a first-order operator, then the generalized Levinson condition is actually a slightly stronger condition than the classical Levinson condition.

THEOREM 2.1. *Suppose that, for the operator* L_0, *in a bounded domain* D, *the generalized Levinson condition is fulfilled. Let the boundary function be Hölder continuous. Then, for any compact set* $K \subset D \cup \partial D$ *lying at a positive distance from* Γ_0, *one can find* $c, \gamma > 0$ *such that*

$$|u^\varepsilon(x) - u_0(x)| < c\,\varepsilon^\gamma, \quad x \in K, \tag{2.2}$$

where $u^\varepsilon(x)$ *is the solution of problem (1.10), and* $u_0(x)$ *is a generalized solution of the problem*

$$L_0 u_0(x) = 0, \quad x \in D, \quad u_0(x)\Big|_{\Gamma_1' \cup \Gamma_1''} = \psi(x). \tag{2.3}$$

The generalized solution of problem (2.3) exists and is unique.

Proof will be divided into several steps.

1. First of all, notice that, under the generalized Levinson condition, problem (2.3) has a unique solution. This comes from Theorems 3.5.1 and 3.5.2. Theorem 3.6.1 implies that this solution is Hölder continuous.

2. Let $(X_t^{\varepsilon,x}, P)$ be a Markov family corresponding to the operator L^ε. Its trajectories are solutions of the equation

$$dX_t^{\varepsilon,x} = \sigma_0(X_t^{\varepsilon,x})\,dW_t + B(X_t^{\varepsilon,x})\,dt + \sqrt{\varepsilon}\,\tilde{\sigma}(X_t^{\varepsilon,x})\,d\tilde{W}_t\,,$$
$$X_0^{\varepsilon,x} = x\,, \tag{2.4}$$

where $\tilde{\sigma}(x)$ is a matrix with the Lipschitz continuous elements, $\tilde{\sigma}(x)\tilde{\sigma}^*(x) = (a^{ij}(x))$, $\tilde{W}_t$ is a Wiener process independent of the process W_t. Let us apply Ito's formula to the function $|X_t^{\varepsilon,x} - X_t^x|^2$, where $X_t^{\varepsilon,x}$ and X_t^x are the solutions of equations (2.4) and (2.1) respectively:

$$|X_t^{\varepsilon,x} - X_t^x|^2 = 2\int_0^t (X_s^{\varepsilon,x} - X_s^x, \sigma_0(X_s^{\varepsilon,x})\,dW_s - \sigma_0(X_s^x)\,dW_s) +$$
$$+ 2\int_0^t (X_s^{\varepsilon,x} - X_s^x, B(X_s^{\varepsilon,x}) - B(X_s^x))\,ds + \int_0^t \sum_{i,j=1}^r (\sigma_{0j}^i(X_s^{\varepsilon,x}) - \sigma_{0j}^i(X_s^x))^2 ds +$$
$$+ 2\sqrt{\varepsilon}\int_0^t (X_s^{\varepsilon,x} - X_s^x, \tilde{\sigma}(X_s^{\varepsilon,x})\,d\tilde{W}_s) + \varepsilon\int_0^t \sum_{i=1}^r a^{ii}(X_s^{\varepsilon,x})\,ds\,.$$

This equality implies that the bound

$$n^\varepsilon(t) = E|X_t^{\varepsilon,x} - X_t^x|^2 \le c_1 \int_0^t n^\varepsilon(s)\,ds + c_2\,\varepsilon t$$

holds for proper constants c_i. Whence we conclude that

$$n^\varepsilon(t) \le c_2\,\varepsilon t \exp\{c_1 t\}\,. \tag{2.5}$$

Subtracting the equation for X_t^x from the equation for $X_t^{\varepsilon,x}$ we obtain

$$X_t^{\varepsilon,x} - X_t^x = \int_0^t (B(X_s^{\varepsilon,x}) - B(X_s^x))\,ds + \int_0^t (\sigma_0(X_s^{\varepsilon,x}) - \sigma_0(X_s^x))\,dW_s +$$
$$+ \sqrt{\varepsilon}\int_0^t \tilde{\sigma}(X_s^{\varepsilon,x})\,d\tilde{W}_s\,. \tag{2.6}$$

Put $\xi_t^\varepsilon = X_t^{\varepsilon,x} - X_t^x$. From (2.6) we deduce that

$$\sup_{0\le s\le t} |\xi_s^\varepsilon| \le c_3 \int_0^t |\xi_s^\varepsilon| ds + \sup_{0\le s_1\le t} \left| \int_0^{s_1} [\sigma_0(X_s^{\varepsilon,x}) - \sigma_0(X_s^x)] dW_s \right| +$$

$$+ \sqrt{\varepsilon} \sup_{0\le s_1\le t} \left| \int_0^{s_1} \tilde{\sigma}(X_s^{\varepsilon,x}) d\tilde{W}_s \right| .$$

Next, using Doob's inequality for estimating the upper bound of the stochastic integrals, we have:

$$m^\varepsilon(t) = E \sup_{0\le s\le t} |\xi_s^\varepsilon|^2 \le c_4 t \int_0^t n^\varepsilon(s) ds + c_5 \int_0^t n^\varepsilon(s) ds + \varepsilon c_6 t .$$

This, along with (2.5) yields

$$m^\varepsilon(t) = E \sup_{0\le s\le t} |X_s^{\varepsilon,x} - X_s^x|^2 \le c_6 \varepsilon (t^2 + t) \exp\{c_1 t\} . \tag{2.7}$$

Note that the constant c_1 can easily be estimated using the Lipschitz constants of the coefficients of the stochastic equation.

3. One can indicate $H < \infty$ such that for $\varepsilon \in [0,1]$, $x \in D$,

$$E\tau^{\varepsilon,x} \le H\rho(x, \Gamma_1' \cup \Gamma_1'') , \tag{2.8}$$

where $\tau^{\varepsilon,x} = \inf\{t : X_t^{\varepsilon,x} \notin D\}$, $\rho(\cdot,\cdot)$ being the Euclidean distance. It is sufficient to establish this inequality for small $\varepsilon > 0$. By the second part of the generalized Levinson condition, inequality (2.8) is valid for $\varepsilon = 0$. This comes from Lemma 3.4.2. In this lemma, a barrier is constructed which bounds $E\tau^{0,x}$ from above. It is easily seen that the same barrier also fits for bounding $E\tau^{\varepsilon,x}$, whenever ε is small enough.

Equations (2.4) and (2.8) imply that

$$E|X^{\varepsilon,x}_{\tau^{\varepsilon,x}}-x|^2 \le c_7\left(E|\int_0^{\tau^{\varepsilon,x}} \sigma_0(X_s^{\varepsilon,x})\,dW_s|^2+E|\int_0^{\tau^{\varepsilon,x}} B(X_s^{\varepsilon,x})\,ds|^2+2\varepsilon E|\int_0^{\tau^{\varepsilon,x}} \sigma(X_s^{\varepsilon,x})d\widetilde{W}_s|\right)^2$$

$$\le c_8 E\tau^{\varepsilon,x} \le Hc_8\,\rho(x,\Gamma_1'\cup\Gamma_1'') \tag{2.9}$$

4. If the generalized Levinson condition holds, then one can find c_9, $\alpha > 0$ such that

$$P\{\tau^{\varepsilon,x} > t\} < c_9 \exp\{-\alpha t\} \tag{2.10}$$

for all $x \in D$ and $\varepsilon \in [0,1]$. The proof of this bound is similar to the arguments involved in the proof of Theorem 3.6.3. The first part of the generalized Levinson condition implies that $P\{\tau^{0,x} < T\} > \alpha_1 > 0$ for some T, $\alpha > 0$. From this, using bounds (2.7) and (2.8), it is not difficult to deduce, that

$$P\{\tau^{\varepsilon,x} < 2T\} > \frac{1}{2}\,\alpha_1 \tag{2.11}$$

for ε small enough. Obviously, for $\varepsilon > \varepsilon_0 > 0$ bound (2.11) is also valid (maybe for different $\alpha_1 > 0$). Therefore, (2.11) holds for all $x \in D$, $\varepsilon \in [0,1]$. From (2.11), using the Markov property, (2.10) is deduced in a standard way.

5. Let us show that the relation

$$\lim_{\varepsilon \downarrow 0} \varepsilon^n P\{X^{\varepsilon,x}_{\tau^{\varepsilon,x}} \in \Gamma_0\} = 0 \tag{2.12}$$

holds for any integer n and $x \in D$. To prove (2.12), we shall consider the set $\Gamma^d = \{x \in D : \rho(x,\Gamma_0) = d\}$, $d > 0$, (Fig. 5). We put $\tau_d^{\varepsilon,x} = \tau_d = \inf\{t : X_t^{\varepsilon,x} \in \Gamma^d \cup \Gamma_0\}$, $w^\varepsilon(x) = P\{X^{\varepsilon,x}_{\tau_d^{\varepsilon,x}} \in \Gamma_0\}$. The function $w^\varepsilon(x)$ is the solution of the boundary value problem

$$L^\varepsilon w^\varepsilon(x) = 0,\quad x \in D \cap \{x : \rho(x,\Gamma_0) < d\},\quad w^\varepsilon(x)\Big|_{\Gamma_0} = 1\ ,$$
$$w^\varepsilon\Big|_{\Gamma^d} = 0\ . \tag{2.13}$$

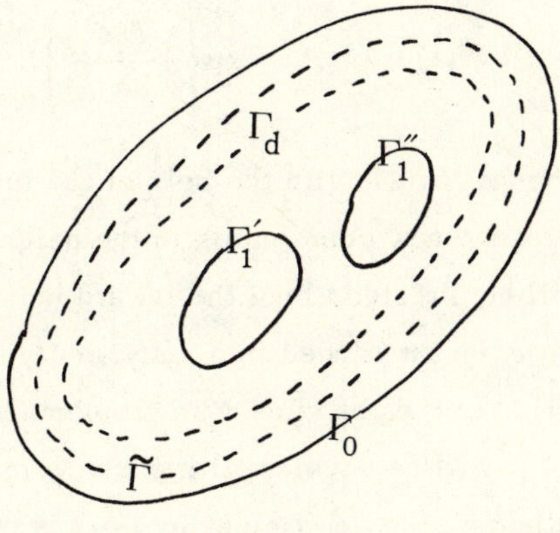

Fig. 5

Consider the one-dimensional problem

$$A(z^2+\varepsilon)\frac{d^2\bar{w}^\varepsilon}{dz^2} + \beta\frac{d\bar{w}^\varepsilon}{dz} = 0\ ,\quad z \in (0,d)$$
$$\bar{w}^\varepsilon(0) = 1\ ,\qquad \bar{w}^\varepsilon(d) = 0\ . \tag{2.14}$$

Here d, A and β are positive numbers to be chosen later on. Problem (2.14) may be integrated:

$$\bar{w}^\varepsilon(z) = \frac{\displaystyle\int_z^d \exp\left\{-\frac{\beta}{A\sqrt{\varepsilon}}\operatorname{arctg}\frac{y}{\sqrt{\varepsilon}}\right\}dy}{\displaystyle\int_0^d \exp\left\{-\frac{\beta}{A\sqrt{\varepsilon}}\operatorname{arctg}\frac{y}{\sqrt{\varepsilon}}\right\}dy}\ ,\quad z \in [0,d]\ . \tag{2.15}$$

For every $z > 0$, one can find $\varepsilon_0 = \varepsilon_0(z)$ such that, for $\varepsilon \in (0, \varepsilon_0)$, the numerator in (2.15) can be bounded from above by the quantity $(d-z)\exp\left\{-\frac{\beta}{2A\sqrt{\varepsilon}}\right\}$. For small ε, the denominator is larger that $c_{10}\varepsilon$, where c_{10} is some positive constant. Therefore,

$$\overline{w}^{\varepsilon}(z) < c_{10}\varepsilon^{-1}\exp\left\{-\frac{\beta}{2A\sqrt{\varepsilon}}\right\}. \tag{2.16}$$

We proceed to bound $w^{\varepsilon}(x)$ with the help of the function $\overline{w}^{\varepsilon}(z)$. To this end, let us introduce new coordinates in the neighborhood of Γ_0 so that the x^1-axis will be directed along the inward normal to Γ_0. Such a coordinate system may be introduced in a sufficiently small neighborhood of every point $y \in \Gamma_0$. Let d be chosen in problem (2.14) so that the d-neighborhood of Γ_0 will be covered with such coordinate systems. Note that, in these coordinates, the coefficient in $\frac{\partial}{\partial x^1}$ is strictly positive, and the coefficient $A^{11}(x)$ in $\frac{\partial^2}{(\partial x^1)^2}$ obeys the inequality $0 \leq A^{11}(x^1,\cdots,x^r) \leq c_{11}\cdot(x^1)^2$, because $A^{11}(0,x^2,\cdots,x^r) = 0$. Hence, it is possible to find coefficients A and β in equation (2.14) such that

$$L^{\varepsilon}\overline{w}^{\varepsilon}(x^1) \leq 0, \quad x^1 \in (0,d). \tag{2.17}$$

On the basis of the maximum principle, (2.13) and (2.17) together imply that $w^{\varepsilon}(x^1,\cdots,x^r) \leq \overline{w}^{\varepsilon}(x^1)$, $0 \leq x^1 \leq d$, and thus, by virtue of (2.16),

$$w^{\varepsilon}(x) < c_{10}\varepsilon^{-1}\exp\left\{-\frac{\beta}{2A\sqrt{\varepsilon}}\right\}. \tag{2.18}$$

for $\varepsilon < \varepsilon_0(x^1)$.

We shall introduce into consideration the set $\widetilde{\Gamma} = \{x \in D : \rho(x,\Gamma_0) = \widetilde{d}\}, 0<\widetilde{d}<d$ and put $\widetilde{\tau} = \inf\{t : X_t^{\varepsilon,x} \in \Gamma_1' \cup \Gamma_1'' \cup \widetilde{\Gamma}\}$. It follows from (2.18) that

$$p^{\varepsilon} = \sup_{x \in \widetilde{\Gamma}} w^{\varepsilon}(x) \leq \overline{w}^{\varepsilon}(\widetilde{d}) \leq c_{10}\varepsilon^{-1}\exp\left\{-\frac{\beta}{2A\sqrt{\varepsilon}}\right\}. \tag{2.19}$$

From bounds (2.7) and (2.9), using the first part of the generalized Levinson condition, one can easily deduce that an $\alpha > 0$ exists such that

$$\inf_{x \in \Gamma^d} P\{X^{x,\varepsilon}_{\tilde{\tau}} \in \Gamma_1' \cup \Gamma_1''\} > \alpha . \tag{2.20}$$

Starting from $x \in \tilde{\Gamma}$, the trajectory $X_t^{\varepsilon,x}$ may reach Γ_0 straightaway, without visiting Γ^d and then, without visiting $\Gamma_1' \cup \Gamma_1''$, return to $\tilde{\Gamma}$ and from here, without touching Γ^d, reach Γ. Generally, the trajectory $X_t^{\varepsilon,x}$ may perform any number $k = 0, 1, 2, \cdots$ of crossings between $\tilde{\Gamma}$ and Γ^d, before it reaches Γ_0. By the strong Markov property and by (2.19), (2.20), this reasoning yields the following bound

$$P\{X^{\varepsilon,x}_{\tau^{\varepsilon,x}} \in \Gamma_0\} < p^\varepsilon + (1-p^\varepsilon)(1-\alpha)p^\varepsilon + \cdots + (1-p^\varepsilon)^k(1-\alpha)^k p^\varepsilon + \cdots <$$

$$< p^\varepsilon \sum_{k=0}^{\infty} (1-\alpha)^k = \frac{p^\varepsilon}{\alpha} < \frac{c_{10}}{\alpha} \exp\left\{-\frac{\beta}{2A\sqrt{\varepsilon}}\right\}.$$

Which, obviously, implies equality (2.12).

6. It is possible to turn now to proving the theorem. Our reasonings are similar to those given when proving Theorem 3.6.1. Put $\bar{\tau} = \tau^{0,x} \wedge \tau^{\varepsilon,x}$, and let χ_1 be the indicator of the set $\{\omega : \bar{\tau} = \tau^{\varepsilon,x}\}$, $\chi_2 = 1 - \chi_1$. Using the strong Markov property of the family $(X_t^{\varepsilon,x}, P)$ for $\varepsilon \geq 0$, we get

$$\begin{aligned} |u^\varepsilon(x) - u_0(x)| \leq E|\psi(X^{\varepsilon,x}_{\tau^{\varepsilon,x}}) - u_0(X^x_{\tau^{\varepsilon,x}})|\chi_1 + \\ + E|\psi(X^x_{\tau^{0,x}}) - u^\varepsilon(X^{\varepsilon,x}_{\tau^{0,x}})|\chi_2 . \end{aligned} \tag{2.21}$$

Let $b \in \Gamma_1' \cup \Gamma_1''$, $a \in D$. Noting that the boundary function is Hölder continuous with some exponent $\mu > 0$, we have:

$$|\psi(b)-u^{\varepsilon}(a)| = |\psi(b)-E\,\psi(X^{\varepsilon,a}_{\tau^{\varepsilon,a}})| \le E|\psi(b)-\psi(X^{\varepsilon,a}_{\tau^{\varepsilon,a}})| <$$

$$< c_{12}E|b-X^{\varepsilon,a}_{\tau^{\varepsilon,a}}|^{\mu} \le c_{12}(E|b-X^{\varepsilon,a}_{\tau^{\varepsilon,a}}|^2)^{\mu/2} \le \tag{2.22}$$

$$\le c_{12}(|b-a|^2+E|a-X^{\varepsilon,a}_{\tau^{\varepsilon,a}}|^2)^{\mu/2} \le c_{13}|b-a|^{\mu/2}, \quad \varepsilon \ge 0 .$$

Here we have made use of (2.9) and of the fact that $(E|\xi|^{\theta})^{1/\theta}$ is an increasing function of θ. From (2.21) and (2.22) we conclude that

$$|u^{\varepsilon}(x)-u_0(x)| < c_{14}E|X^{\varepsilon,x}_{\tau}-X^{x}_{\tau}|^{\mu/2} . \tag{2.23}$$

Denote by d the diameter of the domain D. Notice that if $0 < z \le 1$ and $0 < \alpha \le 1$, then $z^{\alpha} \ge z$. From this, relying on (2.23) for $\kappa \in (0,1)$, we obtain

$$|u^{\varepsilon}(x)-u_0(x)| \le c_{14}d^{\mu}E\left|\frac{X^{\varepsilon,x}_{\tau}-X^{x}_{\tau}}{d}\right|^{\mu/2} \le$$

$$\le c_{14}d^{\frac{1}{2}(\mu-\kappa\mu)}\,E|X^{\varepsilon,x}_{\tau}-X^{x}_{\tau}|^{\frac{\kappa\mu}{2}} \le$$

$$\le c_{15}\sum_{n=1}^{\infty}(E\{\sup_{0\le s\le n}|X^{\varepsilon,x}_s-X^{x}_s|^2\})^{\frac{\kappa\mu}{2}}\,\chi_{n-1<\tau\le n} \le \tag{2.24}$$

$$\le c_{15}\sum_{1}^{\infty}(E\{\sup_{0\le s\le n}|X^{\varepsilon,x}_s-X^{x}_s|^2\})^{\frac{1}{4}\kappa\mu}\cdot(P\{\tau > n\})^{1/2} ,$$

where $\chi_{n-1<\tau\le n}$ is the indicator of the set $\{n-1 < \tau \le n\}$. Now, to bound the right-hand side of the last inequality, we shall use relations (2.7) and (2.10):

$$\sum_{n=1}^{\infty} (E \sup_{0 \le s \le n} |X_s^{\varepsilon,x} - X_s^x|^2)^{\frac{1}{4}\kappa\mu} (P\{\tau > n\})^{1/2} \le$$

$$\le \sum_{n=1}^{\infty} c_6 \varepsilon^{\frac{1}{4}\kappa\mu} (n^2+n)^{\frac{1}{4}\kappa\mu} \cdot c_9 \exp \frac{1}{4}(nc_1\kappa\mu - 2an) \le \tag{2.25}$$

$$\le c_{16}\varepsilon^{\frac{1}{4}\kappa\mu}$$

for $a > \frac{1}{4} c_1 \kappa\mu$. From (2.24) and (2.25), we conclude that, for $\gamma < \frac{a}{c_1} \wedge \frac{1}{2}\mu$ and $c = c_{15} \cdot c_{16}$, the claim of Theorem 2.1 is fulfilled. □

As was said in §4.1, in the classical case, if the coefficients, the boundary of the domain, and the function $\psi(x)$ are sufficiently smooth, then, under the Levinson condition, the difference $u^\varepsilon(x) - u_0(x)$ is of order ε. If the function $\psi(x)$ satisfies merely a Lipschitz condition, then this difference is bounded from above by a magnitude of order $\sqrt{\varepsilon}$ (Theorem 1.1). Theorem 2.1 provides for $|u^\varepsilon(x) - u_0(x)|$ only a bound not better than const. $\varepsilon^{\frac{a}{c_1}}$. Even for infinitely differentiable coefficients and boundary function, the exponent $\frac{a}{c_1}$ may be arbitrarily small. Let us show that in the case of problem (1.10), such a bound cannot, generally speaking, be improved essentially. Namely: for any $\gamma \in (0,1)$, one can find an example where problem (1.10) has infinitely differentiable coefficients, boundary, and boundary function for which the generalized Levinson condition holds, and for which $|u^\varepsilon(x) - u_0(x)|$ tends to zero as $\varepsilon \downarrow 0$ more slowly than ε^γ does.

EXAMPLE 2.1. This example is, in essence, a continuation of Example 3.6.1. Let $\tilde{D} = \{(x,y) \in R^2 : |x| < 1, |y| < 1\}$. The domain D is derived out of $\tilde{D}$ by adding the shaded parts in Fig. 6. These parts are chosen so that the domain D can have infinitely differentiable boundary. Let us consider in D the boundary value problem

$$L^{\varepsilon}u^{\varepsilon}(x,y) = \frac{\alpha}{2}\frac{\partial^2 u^{\varepsilon}}{\partial x^2} + \beta y \frac{\partial u^{\varepsilon}}{\partial y} + \frac{1}{2}\phi^2(x,y)\Delta u^{\varepsilon} + \frac{\varepsilon}{2}\Delta u^{\varepsilon} = 0\,,$$

$$(x,y) \in D\,;\quad u^{\varepsilon}(x,y)\big|_{\partial D} = y^2\,. \tag{2.26}$$

Here ϕ is an infinitely differentiable function vanishing in $\tilde{D}$ and positive for $|y| > 1$; α and β being positive parameters. At all points of the boundary of the domain D, the diffusion along the normal in the operator L_0 is distinct from zero. In other words, $\partial D = \Gamma_1''$, $\Gamma_1' = \emptyset$, $\Gamma_0 = \emptyset$. Since the coefficient in $\frac{\partial^2}{\partial x^2}$ in the operator L_0 is non-zero everywhere, we deduce that the trajectories of the corresponding process leave D uniformly exponentially fast (Lemma 3.3.3).

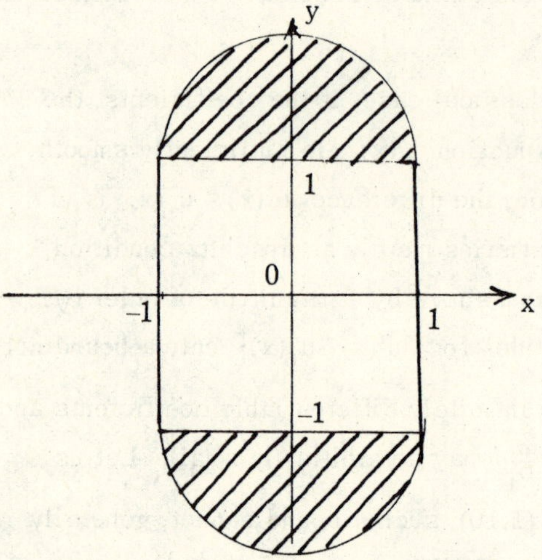

Fig. 6

Therefore, in the domain D, the operator L_0 satisfies the generalized Levinson condition.

It is readily seen that $u_0(x,0) = 0$. Let us bound $u^{\varepsilon}(x,0)$ from below. Consider the Markov family $(X_t^{\varepsilon;x,y}, Y_t^{\varepsilon;x,y}; P)$ corresponding to the operator L^{ε}. We put $\tilde{\tau}^{\varepsilon;x,y} = \inf\{t : (X_t^{\varepsilon;x,y}, Y^{\varepsilon;x,y}) \notin \tilde{D}\}$. Up to time $\tilde{\tau}^{\varepsilon;x,y}$, the trajectories $X_t^{\varepsilon;x,y}, Y_t^{\varepsilon;x,y}$ are described by the equations

$$\dot{X}_t^{\varepsilon;x,y} = \sqrt{\alpha+\varepsilon}\,\dot{W}_t^1, \quad \dot{Y}_t^{\varepsilon;x,y} = \beta Y^{\varepsilon;x,y} + \sqrt{\varepsilon}\dot{W}_t^2,$$

$$X_0^{\varepsilon;x,y} = x, \quad Y_0^{\varepsilon;x,y} = y, \quad (x,y) \in \widetilde{D}.$$

Henceforth, we shall denote by X_t, Y_t without indices the trajectories starting at the point $(0,0)$; $\tilde{\tau}^\varepsilon = \tau^{\varepsilon;0,0}$. We set $\tau_1^\varepsilon = \inf\{t : |X_t| = 1\}$, $\tau_2^{\varepsilon,y} = \inf\{t : |Y_t^{\varepsilon;x,y}| = 1\}$. It is not hard to compute (see Example 3.6.1), that

$$P\{\tau_1^\varepsilon > t\} > c_1 \exp\left\{-\frac{(\alpha+\varepsilon)\pi^2}{2}\, t\right\}. \tag{2.27}$$

Here and in what follows, c_i denote positive constants.

For the function $v_\varepsilon(y) = E\tau_2^{\varepsilon,y}$, we have the one-dimensional boundary value problem

$$\frac{\varepsilon}{2} v_\varepsilon''(y) + \beta y v_\varepsilon'(y) = -1, \quad y \in (-1,1), \quad v_\varepsilon(\pm 1) = 0.$$

Solving this boundary value problem we obtain:

$$v_\varepsilon(y) = \frac{2}{\varepsilon} \int_y^1 \int_0^z \exp\left\{-\frac{1}{\varepsilon}(z^2 - x^2)\beta\right\} dx\,dz.$$

It is not difficult to deduce from this formula (for instance, with L'Hopital's rule) that

$$E\tau_2^{\varepsilon,0} = v_\varepsilon(0) \sim \frac{1}{\beta} \ln \varepsilon^{-\frac{1}{2}}, \quad \varepsilon \downarrow 0. \tag{2.28}$$

Further, using Chebyshev's inequality and (2.28) for small ε leads to

$$P\left\{\tau_2^{\varepsilon,0} < \frac{2}{\beta} \ln \varepsilon^{-\frac{1}{2}}\right\} > 1 - \frac{E\tau_2^{\varepsilon,0}}{\frac{2}{\beta} \ln \varepsilon^{-\frac{1}{2}}} > \frac{1}{2}. \tag{2.29}$$

Denote by A^ε —the event that the trajectory (X_t, Y_t) will reach the horizontal sides of the square $\tilde{D}$ before it leaves $\tilde{D}$. From (2.27) and (2.29) it follows that

$$P(A^\varepsilon) > \frac{1}{2} P\left\{\tau_1^\varepsilon > \frac{2}{\beta} \ln \varepsilon^{-\frac{1}{2}}\right\} >$$

$$> \frac{c_1}{2} \exp\left\{-\frac{(\varepsilon+a)\pi^2 \ln \varepsilon^{-\frac{1}{2}}}{\beta}\right\} = c_2 \cdot \varepsilon^{\frac{\pi^2(\varepsilon+a)}{\beta}} > c_2\, \varepsilon^{\frac{2a\pi^2}{\beta}}$$

for ε small enough. It is readily seen that, for small ε, with probability approaching 1, the trajectories which have reached the horizontal sides of the square, will hit the boundary of the domain D for $|y| > 0.99$. Whence, noting that $u^\varepsilon(0,0) = E\, Y^2_{\tau^{\varepsilon;00}}$, we conclude that

$$u^\varepsilon(0,0) > c_3 P(A^\varepsilon) > c_4 \cdot \varepsilon^{\frac{2a\pi^2}{\beta}} .$$

Taking into account that $u_0(0,0) = 0$ we derive:

$$|u^\varepsilon(0,0) - u_0(0,0)| > c_4 \cdot \varepsilon^{\frac{2a\pi^2}{\beta}} .$$

If $\frac{a}{\beta}$ is small enough, then the exponent on the right-hand side of the last inequality may be made smaller than any preassigned number $\gamma > 0$. Observe, that in this example the difference $u^\varepsilon(x,y) - u_0(x,y)$ tends to zero so slowly only for $y = 0$, $|x| < 1$. Outside a neighborhood of the x-axis (independent of ε) this difference is of order of ε.

So, in the case of problem (1.10), without making essential additional assumptions, one cannot expect the difference $u^\varepsilon(x) - u_0(x)$ to be of order ε at the interior points of the domain D. All the more, one cannot write down the following terms of the expansion into integer powers of the small parameter. The indication on what kind of additional assumptions might be made is given by the results of §3.6. In this section, we have introduced the sequence $\beta_{2k,2k}$, $k = 0, 1, 2, \cdots$, which was determined by the Lipschitz constant of the coefficients of the corresponding stochastic equation. Let $\beta_{2k,2k}$ be such a sequence corresponding to the operator L_0, $\alpha_{L_0,D}$ being the generalized first eigenvalue of the operator L_0 in the domain D (see equality (3.3.5)).

Suppose that the coefficients of the operators L_0, L_1 are infinitely differentiable, the domain D is bounded and has infinitely differentiable boundary, and let the function $\psi(x)$ be infinitely differentiable on ∂D. Moreover, we assume that the whole boundary ∂D consists solely of the components Γ_0 and Γ_1'', the operator L_0 being non-degenerate near Γ_1''. Then, as follows from the results of §3.6, if $\alpha_{L,D} > \beta_{2k,2k}$, then the first k derivatives of the function $u^{\varepsilon}(x)$ are bounded uniformly in $\varepsilon \in [0, \varepsilon_0)$ for some $\varepsilon_0 > 0$.

We shall seek the solution of problem (1.10) in the form of the series

$$u^{\varepsilon}(x) = u_0(x) + \varepsilon u_1(x) + \varepsilon^2 u_2(x) + \cdots .$$

Substituting this equality into equation (1.10) and equating the coefficients in the same powers of ε, we obtain the following chain of equations for determining the functions $u_i(x)$: the function $u_0(x)$ is the solution of problem (2.3); for $i \geq 1$, the functions $u_i(x)$ are determined successively as the solutions of the boundary value problems

$$L_0 u_i(x) = -L_1 u_{i-1}(x), \quad x \in D, \qquad u_i(x)\big|_{\Gamma_1''} = 0. \tag{2.30}$$

THEOREM 2.2. *Suppose that the above hypotheses are fulfilled concerning the coefficients of the operators* L_0, L_1, *the boundary of the domain and the function* $\psi(x)$. *We assume that* $\alpha_{L_0,D} > \beta_{4k,4k}$. *Then problem (2.3) and problems (2.30) for* $i = 1, 2, \cdots, k-1$ *are solvable in a unique way and*

$$u^\varepsilon(x) = u_0(x) + \varepsilon u_1(x) + \cdots + \varepsilon^{k-1} u_{k-1}(x) + o(\varepsilon^{k-1}), \quad \varepsilon \downarrow 0 .$$

Proof of this theorem will be outlined only for $k = 2$. If $\alpha_{L_0,D} > \beta_{8,8}$, then there exist a priori bounds of the first four derivatives of the function $u^\varepsilon(x)$, which are uniform in $\varepsilon \in [0, \varepsilon_0]$. It is easily checked that the function $v^\varepsilon(x) = \frac{1}{\varepsilon}(u^\varepsilon(x) - u_0(x))$ obeys the conditions

$$L_0 v^\varepsilon(x) = -L_1 u^\varepsilon(x), \quad v^\varepsilon(x)|_{\Gamma_1''} = 0 . \tag{2.31}$$

First- and second-order derivatives of the function $u^\varepsilon(x)$ are bounded uniformly in $\varepsilon \in [0, \varepsilon_0]$ and tend to the corresponding derivatives of the function $u_0(x)$ as $\varepsilon \downarrow 0$. This convergence is uniform on every compact set lying inside D. From this, taking into account that

$$v^\varepsilon(x) = E \int_0^{\tau^{0,x}} L_1 u^\varepsilon(X_s^x)\, ds$$

one can readily deduce that

$$\lim_{\varepsilon \downarrow 0} v^\varepsilon(x) = E \int_0^{\tau^{0,x}} L_1 u_0(X_s^x)\, ds = u_1(x) ,$$

Since u_0 has four bounded derivatives, $L_1 u_0(x)$ has at least two bounded derivatives. Whence, noting that $\alpha_{L_0,D} > \beta_{8,8}$ we conclude that the function $u_1(x)$ defined by the last equality is a classical solution of problem (2.30) for $i = 1$ and $u^\varepsilon(x) = u_0(x) + \varepsilon u_1(x) + o(\varepsilon)$. □

We will indicate another distinction between the general case and the classical one of problem (1.10). Let λ_0^ε be the eigenvalue of the operator L^ε with zero boundary conditions on ∂D which corresponds to the non-negative eigenfunction. For $\varepsilon > 0$, such an eigenvalue exists, is unique, real, single, strictly negative, and $|\lambda^{\varepsilon}_0|$ exceeds the real part of any other eigenvalue. This eigenvalue coincides with $-\alpha_{L^{\varepsilon},D}$ (see (formula 3.3.5)). For $\varepsilon = 0$, it is natural to consider $-\alpha_{L_0,D}$ as the generalized first eigenvalue. In the classical case, when L_0 is a first-order operator, $\alpha_{L_0,D} = +\infty$, if the Levinson condition holds. It is not difficult to prove that in this case $\lambda^{\varepsilon}_0 = -\alpha_{L^{\varepsilon},D} \to -\alpha_{L_0,D} = \infty$ for $\varepsilon \downarrow 0$. This means that in the classical case, under the Levinson condition, λ^{ε}_0 possesses some continuity as $\varepsilon \downarrow 0$. It would be natural to expect that in the general case, under the generalized Levinson condition, $\lambda^{\varepsilon}_0 \to \lambda^0_0 = -\alpha_{L_0,D}$. It turns out that, generally speaking, it is not the case; $\lim_{\varepsilon \downarrow 0} \lambda^{\varepsilon}_0$ may be smaller than $-\alpha_{L_0,D}$. The corresponding example may be constructed by considering the boundary value problem of Example 3.6.1. This point is examined in detail in the works of Sarafian [2, 4], that also gives the necessary conditions under which the first eigenvalue depends continuously on ε.

Under the generalized Levinson condition, the exit time of the trajectories from the domain is of order 1 as $\varepsilon \downarrow 0$. As was explained in §4.1, another case is close to the Levinson one; namely, the case when this time grows as $\varepsilon \downarrow 0$, but rather slowly. In this last case, the limit function $u_0(x)$ also did not depend on the perturbating operator and was determined in a unique way by the equation $L_0 u = 0$ and by the corresponding boundary conditions. The analogous question for equation (1.10) was studied by Sarafian [3].

The next section will deal with the case when the exit time is of order ε^{-1}. For such time intervals, the effect of perturbations may no longer be neglected; $\lim_{\varepsilon \downarrow 0} u^{\varepsilon}(x)$ will already depend not only on L_0, but on the operator L_1 as well.

§4.3 *Averaging principle*

Consider the following stochastic differential equation

$$dX_t^{\varepsilon;x,y} = \sigma(X_t^{\varepsilon;x,y}, Y_t^{\varepsilon;x,y})dW_t + b(X_t^{\varepsilon;x,y}, Y^{\varepsilon;x,y})dt\,,$$

$$dY_t^{\varepsilon;x,y} = \frac{1}{\sqrt{\varepsilon}}\tilde{\sigma}(X_t^{\varepsilon;x,y}, Y_t^{\varepsilon;x,y})d\tilde{W}_t + \frac{1}{\varepsilon}B(X_t^{\varepsilon;x,y}, Y^{\varepsilon;x,y})dt\,,$$

$$X_0^{\varepsilon;x,y} = x,\quad Y_0^{\varepsilon;x,y} = y\,. \tag{3.1}$$

Here $X^{\varepsilon;x,y}$ and x vary in R^{ℓ_1}, $Y^{\varepsilon;x,y}$ and y vary in R^{ℓ_2}, $\ell_1+\ell_2=r$, W_t and $\tilde{W}_t$ are independent Wiener processes, and $\varepsilon > 0$. Put $(A^{ij}(x,y)) = \tilde{\sigma}(x,y)\cdot\tilde{\sigma}^*(x,y)$, $(a^{ij}(x,y)) = \sigma(x,y)\cdot\sigma^*(x,y)$. The differential operator corresponding to the Markov family $(X_t^{\varepsilon;x,y}, Y^{\varepsilon;x,y}; P)$ has the following form

$$\frac{1}{\varepsilon}L_0 + L_1 = \frac{1}{2}\sum_{i,j=1}^{\ell_1} a^{ij}(x,y)\frac{\partial^2}{\partial x^i \partial x^j} + \sum_{i=1}^{\ell_1} b^i(x,y)\frac{\partial}{\partial x^i} +$$

$$+\frac{1}{\varepsilon}\left(\frac{1}{2}\sum_{i,j=1}^{\ell_2} A^{ij}(x,y)\frac{\partial^2}{\partial y^i \partial y^j} + \sum_{i=1}^{\ell_2} B^i(x,y)\frac{\partial}{\partial y^i}\right).$$

We suppose that the coefficients of the operators L_0 and L_1 are twice continuously differentiable and bounded, and that the elements of the matrices $\sigma(x,y)$ and $\tilde{\sigma}(x,y)$ are Lipschitz continuous. The space R^{ℓ_2}, where the variables y vary, is called the space of fast movements because the velocity of the variation of $Y_t^{x,y}$ tends to infinity as $\varepsilon \downarrow 0$. The variables x are the slow ones. It is intuitively clear that, for small ε, a diffusing particle, whose motion is described by equation (3.1), will cover a large distance along the variables y before it covers a marked distance along the slow variables. So, one can expect that, for $\varepsilon \downarrow 0$, the variation of slow variables is described by the equation in which the diffusion and drift coefficients are averaged over the fast variables.

To formulate the explicit assertion, we shall consider an auxiliary process in the space of fast movements which is described by the second equation in (3.1), if the slow variables in it are "frozen":

$$dY_t^y(x) = \tilde{\sigma}(x, Y_t^y(x))\, d\tilde{\tilde{W}}_t + B(x, Y_t^y(x))\, dt\,, \qquad Y_0^y(x) = y \tag{3.2}$$

where $\tilde{\tilde{W}}_t$ is some Wiener process. The small parameter is dropped, since the solution of the equation

$$d\tilde{Y}_t(x) = \varepsilon^{-\frac{1}{2}} \tilde{\sigma}(x, \tilde{Y}_t(x))\, d\tilde{W}_t + \varepsilon^{-1} B(x, \tilde{Y}_t(x))\, dt, \quad \tilde{Y}_0(x) = y\,,$$

is obtained from $Y_t^y(x)$ by means of the time change: $\tilde{Y}_t(x) = Y^y_{t/\varepsilon}(x)$ for a proper choice of the process $\tilde{\tilde{W}}_t$.

Suppose that a vector $\bar{b}(x) = (\bar{b}^1(x), \cdots, \bar{b}^{\ell_1}(x))$ and a matrix $(\bar{a}^{ij}(x))$ exist such that for any $\delta > 0$

$$\lim_{T\to\infty} P\left\{ \left| \frac{1}{T} \int_t^{t+T} b(x, Y_s^y(x))\, ds - \bar{b}(x) \right| > \delta \right\} = 0\,,$$
$$\lim_{T\to\infty} P\left\{ \left| \frac{1}{T} \int_t^{t+T} a^{ij}(x, Y_s^y(x))\, ds - \bar{a}^{ij}(x) \right| > \delta \right\} = 0 \tag{3.3}$$

THEOREM 3.1. *Suppose that* $(X_t^{\varepsilon;x,y}, Y_t^{\varepsilon;x,y}; P)$ *is a Markov family governed by equations (3.1) and let conditions (3.3) be fulfilled. Then the measures* μ^ε *in the space* $C_{0,T}(R^{\ell_1})$, $0 < T < \infty$, *corresponding to the process* $X_t^{\varepsilon;x,y}$ *converge weakly as* $\varepsilon \downarrow 0$ *to the measure* $\bar{\mu}$ *induced by the random process* $\bar{X}_t^x$ *which is defined as the solution of the stochastic differential equation*

$$d\bar{X}_t^x = \bar{\sigma}(X_t^x)\, dW_t + \bar{b}(X_t^x)\, dt, \quad \bar{X}_0^x = x\,, \tag{3.4}$$

where $\bar{\sigma}(x) = (\bar{a}^{ij}(x))^{1/2}$.

Proof. In order to prove that μ^ε converge to μ, it suffices to verify that: first, the family $\{\mu^\varepsilon\}$ is weakly compact; secondly, this family has a unique limit point.

That $\{\mu^\varepsilon\}$ is weakly compact, follows from the fact that the coefficients of the stochastic equation are bounded (see Lemma 1.5.2).

To prove the convergence of μ^ε to μ, now it is sufficient to check that the limit points of μ^ε are the solutions of the martingale problem for the operator

$$\bar{L} = \frac{1}{2} \sum_{i,j=1}^{r} \bar{a}^{ij}(x) \frac{\partial^2}{\partial x^i \partial x^j} + \sum_{i=1}^{r} b^i(x) \frac{\partial}{\partial x^i} . \tag{3.5}$$

It is in this way that Theorem 3.1 was proved by Has'minskii in [4]. For brevity, we shall restrict ourselves by the special case when the coefficients of the second of equations (3.1) do not depend on the variables x. Under these conditions, equation (3.2) takes the form

$$dY_t^y = \tilde{\sigma}(Y_t^y) d\tilde{\tilde{W}}_t + B(Y_t^y) dt, \quad Y_0^y = y . \tag{3.6}$$

In this case it is simpler to prove that the finite-dimensional distributions of the process $X_t^{x,y}$ converge to the distributions of the process $\bar{X}_t^x$ as $\varepsilon \downarrow 0$. In order to prove the convergence of the distribution of the vector $(X_{t_1}^{\varepsilon;x,y}, X_{t_2}^{\varepsilon;x,y}, \dots, X_{t_n}^{\varepsilon;x,y})$ to the distribution of the vector $(\bar{X}_{t_1}^x, \dots, \bar{X}_{t_n}^x)$, it is sufficient to prove the convergence of the corresponding characteristic functions:

$$f^\varepsilon(\lambda) = f^\varepsilon_{t_1,\dots,t_n}(\lambda_1, \dots, \lambda_n) = E \exp\left\{ i \sum_{k=1}^{n} (\lambda_k, X_{t_k}^{\varepsilon;x,y}) \right\} \to$$
$$\to \bar{f}(\lambda) = \bar{f}_{t_1,\dots,t_n}(\lambda_1, \dots, \lambda_n) = E \exp\left\{ i \sum_{k=1}^{n} (\lambda_k, \bar{X}_{t_k}^x) \right\}, \quad \varepsilon \downarrow 0 . \tag{3.7}$$

Here $\lambda_1, \dots, \lambda_n$ are ℓ_1-dimensional real vectors.

Let us choose a small positive number Δ and, along with the processes $Y_t^{\varepsilon;x,y}$ and $\bar{X}_t^x$, consider the processes $X_t^{\varepsilon,\Delta}$ and $\bar{X}_t^\Delta$ which are defined by the following equalities

$$dX_t^{\varepsilon,\Delta} = \sigma(X^{\varepsilon,\Delta}_{\left[\frac{t}{\Delta}\right]\Delta}, Y^y_{t/\varepsilon})dW_t + b(X^{\varepsilon,\Delta}_{\left[\frac{t}{\Delta}\right]\Delta}, Y^y_{t/\varepsilon})dt, \quad X_0^{\varepsilon,\Delta} = x,$$
$$d\bar{X}_t^\Delta = \bar{\sigma}(\bar{X}^\Delta_{\left[\frac{t}{\Delta}\right]\Delta})dW_t + \bar{b}(\bar{X}^\Delta_{\left[\frac{t}{\Delta}\right]\Delta})dt, \quad \bar{X}_0^\Delta = x. \tag{3.8}$$

Here Y_t^y is the solution of equation (3.6).

We put

$$f^{\varepsilon,\Delta}(\lambda) = f^{\varepsilon,\Delta}(\lambda_1, \cdots, \lambda_n) = E \exp\left\{ i \sum_{k=1}^n (\lambda_k, X^{\varepsilon,\Delta}_{t_k}) \right\},$$

$$\bar{f}^\Delta(\lambda) = \bar{f}^\Delta(\lambda_1, \cdots, \lambda_n) = E \exp\left\{ i \sum_{k=1}^n (\lambda_k, \bar{X}^\Delta_{t_k}) \right\}.$$

Next we have

$$|f^\varepsilon(\lambda) - \bar{f}(\lambda)| \le |f^\varepsilon(\lambda) - f^{\varepsilon,\Delta}(\lambda)| + |f^{\varepsilon,\Delta}(\lambda) - \bar{f}^\Delta(\lambda)| + + |\bar{f}^\Delta(\lambda) - \bar{f}(\lambda)|. \tag{3.9}$$

We shall show that the first summand on the right-hand side of inequality (3.9) tends to zero as $\Delta \downarrow 0$ uniformly in $\varepsilon \in (0,1]$. For this, it suffices to show that $\sup_{0 \le t \le T} |X_t^{\varepsilon;\Delta} - X_t^{\varepsilon;x,y}| \to 0$ as $\Delta \downarrow 0$ uniformly in ε. Equations (3.1) and (3.8) yield:

$$\xi_t = X_t^{\varepsilon;x,y} - X_t^{\varepsilon,\Delta} = \int_0^t \left[\sigma(X_s^{\varepsilon;x,y}, Y^y_{s/\varepsilon}) - \sigma(X^{\varepsilon,\Delta}_{\left[\frac{s}{\Delta}\right]\Delta}, Y^y_{s/\varepsilon}) \right] dW_s + + \int_0^t \left[b(X_s^{\varepsilon;x,y}, Y^y_{s/\varepsilon}) - b(X^{\varepsilon,\Delta}_{\left[\frac{s}{\Delta}\right]\Delta}, Y^y_{s/\varepsilon}) \right] ds. \tag{3.10}$$

Note that the coefficients of this equation are Lipschitz continuous. Hence squaring this equality and using the properties of stochastic integrals and elementary inequalities, we have:

$$E|\xi_t|^2 \leq (c_1 + c_2 t) \int_0^t E|\xi_s|^2 ds + (c_3 + c_4 t) \int_0^t E|X_s^{\varepsilon,\Delta} - X^{\varepsilon,\Delta}_{\left[\frac{s}{\Delta}\right]\Delta}|^2 ds \,. \quad (3.11)$$

Here c_i are some positive constants. Equation (3.8) for $X_t^{\varepsilon,\Delta}$ implies that

$$E|X_s^{\varepsilon,\Delta} - X^{\varepsilon,\Delta}_{\left[\frac{s}{\Delta}\right]\Delta}|^2 < c_5 \Delta \,.$$

This, together with inequality (3.11), yields

$$E|X_t^{\varepsilon;x,y} - X_t^{\varepsilon,\Delta}| < c_6(1+T)\Delta \exp\{c_7(1+T^2)\} \quad (3.12)$$

for $t \in [0,T]$. From (3.10) and (3.12), using Doob's inequality for the mean value of the upper bound of a stochastic integral, we deduce that

$$E\{\sup_{0 \leq t \leq T} |X_t^{\varepsilon;x,y} - X_t^{\varepsilon,\Delta}|^2\} < c(T)\cdot\Delta \,.$$

This implies that the first summand on the right-hand side of (3.9) tends to zero uniformly in $\varepsilon \in (0,1]$ as $\Delta \downarrow 0$.

One can establish in the same way that the third summand in (3.9) tends to zero together with Δ.

Now we shall show that if Δ is fixed and $\varepsilon \downarrow 0$, then the second summand on the right-hand side of (3.9) also tends to zero. We consider for brevity only the one-dimensional distributions $(n=1)$, and let $t_1 \in (0,\Delta]$, $\lambda_1 = (\lambda_{1,1}, \cdots, \lambda_{1,\ell_1})$:

$$f_{t_1}^{\varepsilon,\Delta}(\lambda_1) = E \exp\left\{ i \left[\int_0^{t_1} (\lambda_1, \sigma(x, Y^y_{s/\varepsilon})\, dW_s + \int_0^{t_1} (\lambda_1, b(x, Y^y_{s/\varepsilon}))\, ds \right] \right\} =$$

$$= E \exp\left\{ i \hat{W}_{\int_0^{t_1} \sum_{k,j=1}^{\ell_1} a^{kj}(x, Y^y_{s/\varepsilon}) \lambda_{1,k} \lambda_{1,j} ds} + \int_0^t \sum_{k=1}^{\ell_1} \lambda_{1,k} b^k(x, Y^y_{s/\varepsilon})\, ds \right\} \to$$

$$\to \exp\left\{ -\frac{t_1}{2} \sum_{k,j=1}^{\ell_1} \bar{a}^{kj}(x) \lambda_{1,k} \lambda_{1,j} + t \sum_{k=1}^{\ell_1} \bar{b}^k(x) \lambda_{1,k} \right\} = \bar{f}_{t_1}^{\Delta}(\lambda_1), \quad \varepsilon \downarrow 0 . \tag{3.13}$$

Here $\hat{W}_t$ is a one-dimensional Wiener process independent of the process Y^y_s which is the solution of equation (3.6); the stochastic integral $\int_0^t f(Y^y_{s/\varepsilon})\, dW_s$ has the same distribution as $W_{\varepsilon \int_0^{t/\varepsilon} f(Y^y_s) ds}$ does. When going to the limit in (3.13), we have made use of condition (3.3) and of the fact that $E \exp\{i \hat{W}_t\} = \exp\left\{-\frac{t}{2}\right\}$.

Relation (3.13), in particular, implies that $f^{\varepsilon,\Delta}_{\Delta}(\lambda_1) \to \bar{f}^{\Delta}_{\Delta}(\lambda_1)$, i.e. $X^{\varepsilon,\Delta}_{\Delta}$ converges in distribution to $\bar{X}^{\Delta}_{\Delta}$ as $\varepsilon \downarrow 0$. On account of this fact, it is not difficult to prove the convergence of $f^{\varepsilon,\Delta}_{t_1}$ to $\bar{f}^{\Delta}_t$ for $t_1 \in (\Delta, 2\Delta)$, then for $t \in (2\Delta, 3\Delta)$, and so on. The case $n > 1$ is dealt with in a similar fashion.

Thus, choosing a sufficiently small Δ, we can make the first and third summands in (3.9) small uniformly in $\varepsilon \in (0,1]$. Then, choosing ε small enough, we can ensure the smallness of the second summand. Therefore, $\lim_{\varepsilon \downarrow 0} f^{\varepsilon}(\lambda) = \bar{f}(\lambda)$ and the finite-dimensional distributions of the processes $X^{\varepsilon;x,y}_t$ converge to the corresponding distributions of the process $\bar{X}^x_t$ as $\varepsilon \downarrow 0$. Which, on account of the weak compactness of the corresponding family of measures in $C_{0,T}(R^{\ell_1})$, implies the claim of Theorem 3.1. □

The assertion on the convergence of the slow component of the family $(X_t^{\varepsilon;x,y}, Y_t^{\varepsilon;x,y}; P)$ to the averaged process $\overline{X}_t^x$ is one of the versions of the averaging principle.

We shall dwell on some consequences of this theorem.

THEOREM 3.2. *Consider the Cauchy problem*

$$\frac{\partial u^\varepsilon(t,x,y)}{\partial t} = \left(\frac{1}{\varepsilon} L_0 + L_1\right) u^\varepsilon, \quad u^\varepsilon(0,x,y) = f(x),$$

where $f(x)$ *is a continuous bounded function. Suppose that, for the family* $(X_t^{\varepsilon;x,y}, Y_t^{\varepsilon;x,y}; P)$ *governed by the operator* $\varepsilon^{-1}L_0 + L_1$, *the conditions of Theorem 3.1 are fulfilled. Then* $\lim_{\varepsilon \downarrow 0} u^\varepsilon(t,x,y) = u(t,x)$ *exists and the limit function* $u(t,x)$ *is the solution of the Cauchy problem*

$$\frac{\partial u}{\partial t} = \overline{L}\, u(t,x), \quad u(0,x) = f(x).$$

The *proof* of this theorem follows from the equality $u^\varepsilon(t,x,y) = Ef(X_t^{\varepsilon;x,y})$ and from Theorem 3.1. □

We turn now to considering the Dirichlet problem for the operator $L_0 + \varepsilon L_1$. Let D be a ring in the plane. In polar coordinates (r, ϕ), this ring has the form $D = \{(r,\phi) : r_1 < r < r_2, 0 \le \phi \le 2\pi\}$, $0 < r_1 < r_2 < \infty$ (see Fig. 7). We put

$$L_0 + \varepsilon L_1 = \frac{1}{2} A(r,\phi) \frac{\partial^2}{\partial \phi^2} + B(r,\phi) \frac{\partial}{\partial \phi} + \varepsilon \left[\frac{1}{2} a(r,\phi) \frac{\partial^2}{\partial r^2} + b(r,\phi) \frac{\partial}{\partial r}\right],$$

$0 < r < \infty$, $0 \le \phi < 2\pi$, and consider the Dirichlet problem

$$\begin{aligned} &(L_0 + \varepsilon L_1) u^\varepsilon(r,\phi) = 0, \quad (r,\phi) \in D, \\ &u^\varepsilon(r_1,\phi) = \psi_1, \quad u^\varepsilon(r_2,\phi) = \psi_2, \end{aligned} \tag{3.14}$$

where ψ_1 and ψ_2 are some constants. The coefficients of the operators

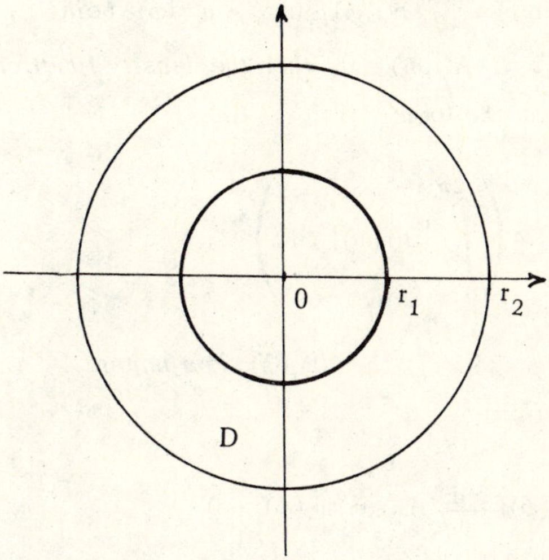

Fig. 7

L_0 and L_1 are assumed twice continuously differentiable. We shall suppose that these coefficients are extended onto the entire space $R^2 = \{-\infty < r < \infty\} \times \{-\infty < \phi < \infty\}$ in such a way that the boundedness, smoothness and periodicity in ϕ may be preserved as well as the non-negativity of $a(r,\phi)$ and $A(r,\phi)$. Let $(R_t^{\varepsilon;r,\phi}, \Phi_t^{\varepsilon;r,\phi}; P)$ be the Markov family in R^2 corresponding to the operator $\varepsilon^{-1}L_0 + L_1$. We shall also consider the process describing the fast movement when the slow variables are frozen (a counterpart of equation (3.2)):

$$d\Phi_t^{\phi}(r) = \tilde{\sigma}(r, \Phi_t^{\phi}(r))\, d\tilde{\tilde{W}}_t + B(r, \Phi_t^{\phi}(r))dt, \quad \Phi_0^{\phi} = \phi ,$$

where $\tilde{\sigma}(r,\phi) = \sqrt{A(r,\phi)}$.

LEMMA 3.1. *Suppose that* $A(r,\phi) + B^2(r,\phi) > 0$ *for* $\phi \in [0,2\pi]$, $r \in [r_1,r_2]$. *Then the coefficients of the operators* L_0 *and* L_1 *may be extended onto the entire space* R^2 *so that conditions (3.3) may be fulfilled: for any* $\delta > 0$, *uniformly in* $(r,\phi) \in R^2$, $t > 0$,

$$\lim_{T\to\infty} P\left\{\left|\frac{1}{T}\int_t^{t+T} b(r, \Phi_s^{\phi}(r))\, ds - \bar{b}(r)\right| > \delta\right\} = 0 ,$$

$$\lim_{T\to\infty} P\left\{\left|\frac{1}{T}\int_t^{t+T} a(r, \Phi_s^{\phi}(r))\, ds - \bar{a}(r)\right| > \delta\right\} = 0 , \tag{3.15}$$

where $\overline{a}(r) = \int_0^{2\pi} a(r,\phi)\,\mu_r(d\phi)$, $\overline{b}(r) = \int_0^{2\pi} b(r,\phi)\,\mu_r(d\phi)$, $\mu_r(d\phi)$ *being some normed measure on* $[0,2\pi]$. *If* $A(r,\phi) \equiv 0$, *then the density function* $m_r(\phi)$ *of the measure* $\mu_r(d\phi)$ *has the form*

$$m_r(\phi) = [B(r,\phi)]^{-1} \left(\int_0^{2\pi} [B(r,\phi)]^{-1} d\phi \right)^{-1} .$$

If $A(r,\phi) > 0$ *for some* r *and* $\phi \in [0,2\pi]$, *then* $m_r(\phi)$ *is a unique* 2π*-periodic solution of the equation*

$$\frac{1}{2}\frac{d^2}{d\phi^2}(A(r,\phi)m_r(\phi)) - \frac{d}{d\phi}(B(r,\phi)\,m_r(\phi)) = 0$$

$$\int_0^{2\pi} m_r(\phi)\,d\phi = 1 . \tag{3.16}$$

Proof. If $A(r,\phi) = 0$ for all ϕ, then $\Phi_s^\phi(r)$ is a function periodic in s with period $\hat{T} = \int_0^{2\pi} B^{-1}(r,\phi)\,d\phi$, and the claim of the lemma may be checked directly:

$$\frac{1}{T}\int_t^{t+T} b(r, \Phi_s^\phi(r)ds \to \hat{T}^{-1} \int_0^{\hat{T}} b(r, \Phi_s^\phi(r))\,ds = \frac{1}{\hat{T}} \int_0^{2\pi} \frac{b(r,\phi)}{B(r,\phi)}\,d\phi .$$

The second of equalities (3.15) may be verified in a similar way.

If $A(r_0,\phi) > 0$ for some $\phi_0 \in [0,2\pi]$, then one can find on the circle $r = r_0$ an open set $\mathcal{E}$ such that $A(r,\phi) \geq a > 0$ for $\phi \in \mathcal{E}$ and $P\{\Phi_{t_0}^\phi(r_0) \in \mathcal{E}\} \geq h > 0$ for some $t_0, h > 0$ and any $\phi \in [0,2\pi]$. From this one can deduce equalities (3.15) and the existence of the measure $\mu_{r_0}(\cdot)$ in the same way as was done when proving Lemma 3.7.2. The measure $\mu_{r_0}(\cdot)$ is the invariant normed measure of the process $\Phi_s^\phi(r_0)$. If $A(r,\phi) > 0$ for, $\phi \in [0,2\pi]$ then, as is known, this measure has a density function which is the solution of problem (3.16). □

REMARK. Under the conditions of Lemma 3.1, a more powerful assertion, than (3.15) holds. Namely, for any continuous function $f(\phi)$, periodic with period 2π,

$$P\left\{\lim_{T\to\infty} \frac{1}{T}\int_t^{T+t} f(\Phi_s^{\phi}(r))\,ds = \bar{f}_r = \int_0^{2\pi} f(\phi)\,\mu_r(d\phi)\right\} = 1 .$$

THEOREM 3.3. *Suppose that* $A(r,\phi) + B^2(r,\phi) > 0$ *for* $\phi \in [0,2\pi]$, $r \in [r_1, r_2]$, *and let* $\bar{a}(r)$ *and* $\bar{b}(r)$ *be the functions defined in Lemma 3.1. We shall assume that at least one of the following conditions is fulfilled:*

1. *For* $r \in \{r_1 \le r \le r_2 : \bar{a}(r) = 0\}$, *the function* $\bar{b}(r)$ *is distinct from zero and preserves its sign;*

2. *There exists an* $r_0 \in (r_1, r_2)$ *such that* $\bar{a}(r_0) > 0$; $\bar{b}(r) < 0$ *for* $r \in [r_1, r_0)$ *and* $\bar{b}(r) > 0$ *for* $r \in (r_0, r_2]$.

Moreover, suppose that at the endpoints of the interval $[r_1, r_2]$ *the following conditions hold:*

$$\bar{a}(r_1) > 0 \ \text{ or } \ \bar{b}(r_1) < 0; \quad \bar{a}(r_2) > 0 \ \text{ or } \ \bar{b}(r_2) > 0.$$

Then, for the function $u^{\varepsilon}(r,\phi)$ *which is the solution of problem (3.14),* $\lim_{\varepsilon\downarrow 0} u^{\varepsilon}(r,\phi) = \bar{u}(r)$ *exists, where* $\bar{u}(r)$ *is the solution of the boundary value problem*

$$\frac{1}{2}\,\bar{a}(r)\bar{u}''_{rr} + \bar{b}(r)\bar{u}'_r = 0, \quad r \in (r_1, r_2); \ u(r_1) = \psi_1, \ u(r_2) = \psi_2 .$$

Proof. Since, by the condition, $A(r,\phi) + B^2(r,\phi) > 0$, we conclude that by Lemma 3.1 equalities (3.15) hold for the family $(R_t^{\varepsilon;r,\phi}, \Phi_t^{\varepsilon;r,\phi}; P)$. Then, according to Theorem 3.1, for any $T > 0$, the measure in the space $C_{0,T}(R^1)$ corresponding to the processes $R_t^{\varepsilon;r,\phi}$ converge weakly as $\varepsilon \downarrow 0$ to the measure corresponding to the process $\bar{R}_t^r$:

$$d\bar{R}_t^r = \bar{\sigma}(\bar{R}_t^r)\,dW_t + \bar{b}(\bar{R}_t^r)\,dt, \quad \bar{R}_0^r = r, \ \bar{\sigma}(r) = \sqrt{\bar{a}(r)} .$$

We shall consider on the space $C_{0,\infty}(R^1)$ the following functionals

$$\tau(f) = \inf\{s : f(s) \notin (r_1, r_2)\} ;$$

$i(f) = 1$, if $f(\tau(f)) = r_1$, and $i(f) = 2$, if $f(\tau(f)) = r_2$; $\chi_{\tau(f)\le T}(f)$ is the indicator of the set $\{f \in C_{0,\infty}(R^1) : \tau(f) \le T\}$. The solution of problem (3.14) may be written in the form

$$u^{\varepsilon}(r,\phi) = E\psi_{i(R^{\varepsilon;r,\phi})} = E\psi_{i(R^{\varepsilon;r,\phi})}\cdot\chi_{\tau(R^{\varepsilon;r,\phi})\le T} + \tag{3.17}$$

$$+ E\psi_{i(R^{\varepsilon;r,\phi})}\cdot(1-\chi_{\tau(R^{\varepsilon;r,\phi})\le T}) .$$

The conditions of Theorem 3.3 imply that $P\{\tau(R^{\varepsilon;r,\phi}) > T\} \to 0$ uniformly in ε, r, and ϕ as $T \to \infty$. So, for any $\delta > 0$ there exists a T so large that the second summand on the right-hand side of equality (3.17) will be smaller than δ simultaneously for all $\varepsilon \in (0,1]$, $r \in [r_1, r_2]$, $\phi \in [0,2\pi]$. The first summand is the mathematical expectation of some bounded functional which may be considered on the space $C_{0,T}(R^1)$. Generally speaking, this functional is not continuous. Therefore, without additional considerations, one cannot conclude that the weak convergence of the processes $R^{\varepsilon;r,\phi}$ to $\bar{R}^r$ in $C_{0,T}(R^1)$ implies the convergence of the mathematical expectations. However, as it is known, the convergence of the expectations remains if the set of discontinuity points of the functional has probability zero for the limit process. It is not difficult to show that if the conditions of Theorem 3.3 hold, then, for T large enough, the set of functions belonging to $C_{0,T}(R^1)$, on which the functional $\psi_{i(f)}\chi_{\tau(\phi)\le T}$ is discontinuous, has probability zero for the process $\bar{R}_t^r$, $r \in (r_1, r_2)$.

This implies that

$$\lim_{\varepsilon\downarrow 0} E\psi_{i(R^{\varepsilon;r,\phi})}\chi_{\tau(R^{\varepsilon;r,\phi})\le T} = E\psi_{\tau(\bar{R}^r_{\cdot})}\chi_{\tau(\bar{R}^r_{\cdot})\le T} . \tag{3.18}$$

We remind that, at the expense of choosing T large enough, the second summand in (3.17) may be made smaller than any preassigned $\delta > 0$. From this, by (3.17) and (3.18) we conclude that $\lim_{\varepsilon \downarrow 0} u^{\varepsilon}(r,\phi) = E\psi_{i(\bar{R}^r)} = \bar{u}(r)$, which proves Theorem 3.3. □

We will make some remarks on this theorem.

1. If one adds to the operator L_1 (see (3.14)) terms involving the derivatives in the fast variables ϕ with sufficiently smooth coefficients, then the assertion of Theorem 3.3 will remain without any changes.

2. In Theorem 3.3, we have assumed that the boundary function does not depend on the fast variable. If the boundary conditions in problem (3.14) have the form $u^{\varepsilon}(r_i,\phi) = \psi_i(\phi)$, $i = 1,2$, then the boundary conditions for the limit function $\bar{u}(r)$ are obtained from $\psi_i(\phi)$ by means of averaging. For example, if in the operator L_1 the coefficient $a(r_i,\phi)$ is positive for $i = 1,2$, and $b(r,\phi)$ vanishes near the endpoints r_1, r_2, then the limit function $\bar{u}(r)$ is the solution of the problem $\frac{1}{2}\bar{a}(r)\cdot u'' + \bar{b}(r)u' = 0$ for $r \in (r_1, r_2)$, $u(r_i) = \bar{\psi}_i$, $i = 1,2$ where

$$\bar{\psi}_i = \int_0^{2\pi} \frac{\psi_i(\phi)\,\mu_{r_i}(d\phi)}{a(r_i,\phi)},$$

$\mu_{r_i}(\cdot)$ being the normed invariant measure for the Markov family $(\Phi_t^{\phi}(r_i), P)$ on the circle $r = r_i$ (Has'minskii [3]).

Under the conditions of Theorem 3.3, the process corresponding to the operator $L_0 + \varepsilon L_1$ leaves the domain in a time of order of ε^{-1}, $\varepsilon \downarrow 0$. Now we turn to the case when the velocity of the averaged motion along the slow variables vanishes. In this case, exit from the domain occurs at the expense of deviations from the averaged motion. Convergence to the averaged motion in the space of slow variables may be considered as a law of large numbers type result. Now we shall be interested in a central limit theorem type result. We will consider the simplest situation, where the effects we are interested in are displayed.

So, let $D = \tilde{D} \times S$ be in the domain represented in Fig. 3. Here $\tilde{D}$ is a bounded domain in the (x_1,x_2)-plane having smooth boundary; S is the circumference on which the angle coordinate ϕ varies. Consider the operator with 2π-periodic coefficients independent of x^1 and x^2:

$$\mathcal{L}^\varepsilon = L_0 + \varepsilon L_1 = \left(\frac{1}{2} A(\phi) \frac{\partial^2}{\partial \phi^2} + B(\phi) \frac{\partial}{\partial \phi}\right) + \varepsilon \sum_{i=1}^{2} b^i(\phi) \frac{\partial}{\partial x^i}, \quad A(\phi) > 0 .$$

Denote by $m(\phi)$ the density function of the invariant measure* of the process on the circumference S corresponding to the operator L_0. Suppose that

$$\int_0^{2\pi} b^i(\phi)\, m(\phi)\, d\phi = 0 ; \quad i = 1,2 . \tag{3.19}$$

From (3.19) it follows that 2π-periodic functions $u^i(\phi)$, $i = 1,2$, $\phi \in (-\infty, \infty)$, exist such that

$$L_0 u^i(\phi) = -b^i(\phi) ; \quad \phi \in (-\infty, \infty), \ i = 1,2 .$$

Denote by $n(x,\phi)$ the outward normal to ∂D at the point $(x,\phi) \in \partial D$, $\tilde{\Gamma}_1 = \{(x,\phi) \in \partial D; (b(\phi), n(x,\phi)) > 0\}$, Γ_1 being the closure of $\tilde{\Gamma}_1$. To avoid considering partly regular points, we will assume that all the points of the set Γ_1 are regular for the operator L^ε in the domain D (for all $\varepsilon > 0$). The set $\partial D \setminus \Gamma_1$ will be assumed inaccessible. Moreover, suppose that the set of all possible linear combinations of the vectors $b(\phi)$, $0 < \phi < 2\pi$, with non-negative coefficients fills out the whole plane (*Condition* V).

Consider the boundary value problem:

$$\mathcal{L}^\varepsilon u^\varepsilon(x,\phi) = 0, \ (x,\phi) \in D, \ u^\varepsilon(x,\phi)\Big|_{\Gamma_1} = \psi(x) . \tag{3.20}$$

It is not hard to check that, under the above assumptions, problem (3.20) for any continuous function $\psi(x)$, $x \in \partial D$, has a unique generalized solution.

THEOREM 3.4. *Suppose that the operator* $\mathfrak{L}^\varepsilon$ *and the domain* D *satisfy the above listed conditions. Then* $\lim_{\varepsilon \downarrow 0} u^\varepsilon(x,\phi) = \tilde{u}(x)$, *the function* $\tilde{u}(x)$ *being the solution of the Dirichlet problem*:

$$\tilde{L}\tilde{u}(x) = 0, \quad x \in \tilde{D}, \quad \tilde{u}(x)|_{\partial\tilde{D}} = \psi(x) , \tag{3.21}$$

where

$$\tilde{L} = \frac{1}{2} \sum_{i,j=1}^{2} \tilde{a}^{ij} \frac{\partial^2}{\partial x^i \partial x^j}, \quad \tilde{a}^{ij} = \int_0^{2\pi} A(\phi) \frac{\partial u^i}{\partial \phi}(\phi) \frac{\partial u^j}{\partial \phi}(\phi)\, m(\phi)\, d\phi .$$

For the *proof* of this theorem, consider the Markov family $(X_t^{\varepsilon;x,\phi}, \Phi_t^{\varepsilon;x,\phi}, P)$:

$$dX_s^{\varepsilon;x,y} = \varepsilon b(\Phi_{s/\varepsilon}^{\varepsilon;x,\phi})\, ds , \qquad X_0^{\varepsilon;x,\phi} = x ,$$

$$d\Phi_s^\phi = \tilde{\sigma}(\Phi_s^\phi)\, dW_s + B(\Phi_s^\phi)\, ds, \ \Phi_0^\phi = \phi \in S, \ \tilde{\sigma}(\phi) = A^{1/2}(\phi) .$$

Let us show that the processes $Y_t^\varepsilon = X_{t/\varepsilon^2}^{\varepsilon;x,\phi}$ converge weakly in the space $C_{0,T}(R^2)$, $T < \infty$, as $\varepsilon \downarrow 0$, to the Gaussian Markov process $(\tilde{X}_t, \tilde{P})$ in R^2 governed by the operator $\tilde{L}$.

We shall apply Ito's formula to $u^i(\Phi_t^\phi)$:

$$u^i(\Phi_t^\phi) = u^i(\phi) + \int_0^t \left(\frac{du^i}{d\phi}\right)(\Phi_s^\phi)\tilde{\sigma}(\Phi_s^\phi)\, dW_s - \int_0^t b^i(\Phi_s^\phi)\, ds .$$

With the help of this equality we obtain that

$$X^{\varepsilon;x,\phi}_{t/\varepsilon^2} - x = \varepsilon \int_0^{t/\varepsilon^2} b(\Phi_s^\phi)\,ds =$$

$$= \varepsilon \int_0^{t/\varepsilon^2} \left(\frac{du}{d\phi}\right)(\Phi_s^\phi)\,\tilde{\sigma}(\Phi_s^\phi)\,d\tilde{W}_s + 0(\varepsilon)\,, \quad \varepsilon \downarrow 0\,, \tag{3.22}$$

where $u(\phi) = (u^1(\phi), u^2(\phi))$.

Taking into account (3.22), one can easily show that the characteristic function $f^\varepsilon = f^\varepsilon_{t_1 \cdots t_n}(\lambda_1, \cdots, \lambda_n)$, $\lambda_k \in R^\varepsilon$, of the random vector $(Y^\varepsilon_{t_1}, \ldots, Y^\varepsilon_{t_n})$ can be written as follows ($\varepsilon \downarrow 0$):

$$f^\varepsilon = E \exp\left\{ i\varepsilon \sum_{m=1}^{n} \int_{\varepsilon^{-2} t_{m-1}}^{\varepsilon^{-2} t_m} \left(\lambda_k, \frac{du}{d\phi}\right) \tilde{\sigma}(\phi_s)\, d\tilde{W}_s + 0(\varepsilon) \right\} .$$

Whence, noting that the stochastic integral $\int_0^t g(s,\omega)\,dW_s$ has the same distribution as $W_{\int_0^t g^2 ds}$ has, it is not difficult to deduce that

$$\lim_{\varepsilon \downarrow 0} f^\varepsilon_{t_1 \cdots t_n}(\lambda_1, \cdots, \lambda_n) = \bar{f}_{t_1 \cdots t_n}(\lambda_1, \cdots, \lambda_n)\,,$$

where $\bar{f}_{t_1 \cdots t_n}(\lambda_1, \cdots, \lambda_n)$ is the characteristic function of the Gaussian vector $(\tilde{X}^x_{t_1}, \cdots, \tilde{X}^x_{t_n})$. The convergence of the characteristic functions implies the convergence of the corresponding distributions.

Next, from (3.22) one can deduce that for some $c > 0$

$$E|Y_t^\varepsilon|^4 < c < \infty, \quad \sup_{0 \le t \le T} E|Y^\varepsilon_{t+h} - Y^\varepsilon_t|^4 \le ch^2 + \alpha(\varepsilon)\,,$$

where $\alpha(\varepsilon) \downarrow 0$ as $\varepsilon \downarrow 0$, $h > 0$. This bound together with the convergence of the finite-dimensional distributions are sufficient for the weak convergence of Y_t^ε to $\tilde{X}_t^x$ in $C_{0,T}(R^2)$.

The rest of the proof of Theorem 3.4 is carried out in a standard way, if one uses the probabilistic representation of the functions $u^\varepsilon(x,\phi)$ and $\tilde{u}(x)$. One should take into account that by Condition V the matrix $(\tilde{a}^{ij})$ does not degenerate. □

We remark that, under the conditions of Theorem 3.4, the Markov process corresponding to the operator $L_0+\varepsilon L_1$ leaves the domain D in a time of order ε^{-2}. The problem considered in Theorem 3.4 is an example of "carrying over the stochasticness from some degrees of freedom to others": in problem (3.20) we have a non-degenerate diffusion only along the variable ϕ, while the limit function $\tilde{u}(x)$ satisfies the equation with a non-degenerate diffusion in the (x^1,x^2)-plane.

The general results on normal deviations in the case when the averaged movement has zero velocity, are available in the works of Stratonovich [1], Has'minskii [4], Borodin [1].

§4.4 *Leaving the domain at the expense of large deviations*

So, let the operator L^ε be the result of the perturbation of the operator L_0: $L^\varepsilon = L_0+\varepsilon L_1$. The previous section was concerned with the problems in which the operator L_0, in a sense, does not help, but also does not hinder the exit of the diffusion process $(X_t^\varepsilon, P_x^\varepsilon)$ corresponding to the operator L^ε from the domain. In such problems, the time needed for leaving the domain grew like a power: ε^{-1} or ε^{-2} as $\varepsilon \downarrow 0$.

Here we shall consider problems in which the operator L_0 hinders leaving the domain. We shall see that in these cases the time which is required for the process X^ε to reach the boundary of the domain, grows, as $\varepsilon \downarrow 0$, faster than any power. In the typical case, this time is of order $\exp\{\varepsilon^{-1}\cdot \mathrm{const}\}$. We have faced such a situation in Theorem 1.2, where L_0 was the first-order operator whose characteristics everywhere cross the boundary from the outside toward the interior (see Fig. 4). If the characteristics behave near the boundary in such a way, then no matter how they behave inside the domain, the exit of the process $(X_t^\varepsilon, P_x^\varepsilon)$ from the domain may occur only at the expense of large deviations of this process from the trajectories of the dynamical system corresponding to L_0. The case of a first-order operator L_0 is, in a sense, the basic one in regard to the questions considered in this section. Later on we shall see that the addition of terms with second-order derivatives to L_0, in essence, facilitates the exit from the domain. This section follows the works of Wentzell and Freidlin [1], Gärtner and Freidlin [1], and Gärtner [1].

Let $(X_t^{\varepsilon,x}, P)$ be a Markov family in R^r whose trajectories $X_t^{\varepsilon,x}$ are the solutions of the stochastic equation

$$dX_t^{\varepsilon,x} = \sigma(X_t^{\varepsilon,x})\, dW_t + B(X_t^{\varepsilon,x}) + \sqrt{\varepsilon}\, \tilde{\sigma}(X_t^{\varepsilon,x})\, d\tilde{W}_t \,, \tag{4.1}$$

$$X_0^{\varepsilon,x} = x \,.$$

Here W_t, $\tilde{W}_t$ are mutually independent r-dimensional Wiener processes, $\sigma(x)$, $\tilde{\sigma}(x)$ are square matrices of order r. Suppose that the matrix $\sigma(x)$ has the form:

$$\begin{pmatrix} \sigma^{(11)} & 0 \\ 0 & 0 \end{pmatrix},$$

where $\sigma^{(11)} = \sigma^{(11)}(x)$ is a square matrix of order $n < r$. Generally, we shall agree that, in this section, $A^{(11)}$ denotes the square submatrix of order n which is in the upper left-hand corner of the matrix A. Similar notations will be used for the other submatrices of the matrix A:

$$A = \begin{pmatrix} A^{(11)} & A^{(12)} \\ A^{(21)} & A^{(22)} \end{pmatrix}, \quad A^{(\cdot 1)} = \begin{pmatrix} A^{(11)} \\ A^{(21)} \end{pmatrix}, \quad A^{(1 \cdot)} = (A^{11} \;\; A^{12}) \,.$$

If $b = (b^1, \cdots, b^r)$, then $b^{(1)}$ denotes the n-dimensional vector $b^{(1)} = (b^1, \cdots, b^n)$; $b^{(2)} = (b^{n+1}, \cdots, b^r)$ denoting the (r-n)-dimensional vector.

We put $A(x) = (A^{ij}(x)) = \sigma(x)\sigma^*(x)$, $a(x) = (a^{ij}(x)) = \tilde{\sigma}(x)\tilde{\sigma}^*(x)$. We shall assume that, for some $m, M, K > 0$, the functions $A^{ij}(x)$, $B^i(x)$, and $a^{ij}(x)$ satisfy the following conditions:

$$\sum_{i,j=1}^{r} (|a^{ij}(x) - a^{ij}(y)|^2 + |A^{ij}(x) - A^{ij}(y)|^2) < K|x-y|^2, \quad x,y \in R^r ;$$

$$\sum_{i=1}^{r} |B^i(x) - B^i(y)|^2 < K|x-y| ; \quad |B(x)| = \left(\sum_{i=1}^{r} (B^i(x))^2\right)^{1/2} < M , \tag{4.2}$$

$$m\cdot(y,y) \le (a(x)y,y) \le M\cdot(y,y),\ y \in R^r;\ m(z,z) \le (A^{(11)}(x)z,z) \le M\cdot(z,z),\ z \in R^n .$$

The differential operator corresponding to the family $(X_t^{\varepsilon,x}, P)$ has the form

$$L^\varepsilon = L_0 + \varepsilon L_1 = \frac{1}{2}\sum_{i,j=1}^{n} A^{ij}(x)\frac{\partial^2}{\partial x^i \partial x^j} + \sum_{i=1}^{r} B^i(x)\frac{\partial}{\partial x^i} +$$

$$+ \frac{\varepsilon}{2}\sum_{i,j=1}^{r} a^{ij}(x)\frac{\partial^2}{\partial x^i \partial x^j} .$$

Therefore, the operator L_0 involves non-degenerate diffusion in the first $n < r$ coordinates. In particular, for $n = 0$ we get the classical case.

In the analysis of large deviations, a basic role is played by the bounds to be given in Theorems 4.1 and 4.2. To formulate them, we shall introduce the functionals $R_{0,T}(\phi)$, $S_{0,T}(\phi)$, $\phi \in C_{0,T}(R^r)$, $T > 0$:

$$R_{0T}(\phi) = \frac{1}{2}\int_0^T |[A^{(11)}(\phi_s)]^{-1/2}(\dot{\phi}^{(1)} - B^{(1)}(\phi_s))|^2 ds ,$$

if the component $\phi_s^{(1)}$ of the function $\phi_s = (\phi_s^{(1)}, \phi_s^{(2)}) \in C_{0,T}(R^r)$ is absolutely continuous; for the remaining $\phi \in C_{0,T}(R^r)$, we put $R_{0T}(\phi) = +\infty$;

$$S_{0T}(\phi) = \frac{1}{2} \int_0^T |[a_{(22)}(\phi_s)]^{1/2}(\dot{\phi}_s^{(2)} - B^{(2)}(\phi_s))|^2 \, ds \, ,$$

if the component $\phi_s^{(2)}$ of the function $\phi_s = (\phi_s^{(1)}, \phi_s^{(2)})$ is absolutely continuous; for the remaining $\phi \in C_{0T}(R^r)$, we set $S_{0T}(\phi) = +\infty$. Here $a_{(22)}(x)$ is the submatrix of the matrix

$$a^{-1}(x) = \begin{pmatrix} a_{(11)}(x) & a_{(12)}(x) \\ a_{(21)}(x) & a_{(22)}(x) \end{pmatrix} .$$

We shall, as usual, denote by $\rho_{0T}(\cdot,\cdot)$ the uniform metric in $C_{0,T}(R^r)$.

THEOREM 4.1. *Let conditions (4.2) hold. Then for any* $a > 0$, $0 < \delta \leq 1$, $0 < \varepsilon \leq 1$, $T > 0$, $x \in R^r$ *and for* $\phi \in C_{0,T}(R^r)$ *for which* $\phi_0 = x$,

$$P\{\rho_{0T}(X^{\varepsilon,x}, \phi) < \delta\} \geq C_1 \exp\left\{ -(1+a)\left[R_{0T}(\phi) + \frac{1}{\varepsilon} S_{0T}(\phi) \right] - \frac{C_2 T}{\delta^2} \right\}, \qquad (4.3)$$

where the positive constant C_1 *depends only on* a, *the constant* C_2 *depending only on* a, L, m, M *and on* r.

Proof. If $R_{0T}(\phi) = \infty$ or $S_{0T}(\phi) = \infty$, then inequality (4.3) is trivial; so, we shall assume that $R_{0T}(\phi)$, $S_{0T}(\phi) < \infty$. We put $Y_t^{\varepsilon,x} = X_t^{\varepsilon,x} - \phi_t$, $t \in [0,T]$. Clearly, $Y_t^{\varepsilon,x}$ satisfies the stochastic differential equation

$$dY_t^{\varepsilon,x} = \sigma(Y_t^{\varepsilon,x} + \phi_t)\,dW_t + [B(Y_t^{\varepsilon,x} + \phi_t) - \dot{\phi}_t]\,dt +$$

$$+ \sqrt{\varepsilon}\, \tilde{\sigma}(Y_t^{\varepsilon,x} + \phi_t)\, d\tilde{W}_t \, , \quad Y_0^{\varepsilon,x} = 0 \, .$$

Note that the addition of a bounded drift into a stochastic differential equation leads to an absolutely continuous change of measure in the space $C_{0,T}(R^r)$ (see §1.5). Hence, the distribution $\mu_{Y^{\varepsilon,x}}$ of the process $Y_t^{\varepsilon,x}$ in $C_{0,T}(R^r)$ is absolutely continuous with respect to the measure μ_{Z^ε}

in $C_{0T}(R^r)$ which corresponds to the solution Z_t^ε of the stochastic equation

$$dZ_t^\varepsilon = \sigma(Z_t^\varepsilon+\phi_t)\,dW_t + \sqrt{\varepsilon}\,\tilde{\sigma}(Z_t^\varepsilon+\phi_t)\,d\tilde{W}_t\,,\quad Z_0^\varepsilon = 0\,.$$

The density function $\dfrac{d\mu_{Y^{\varepsilon,x}}}{d\mu_{Z^\varepsilon}}$ is the following

$$\frac{d\mu_{Y^{\varepsilon,x}}}{d\mu_{Z^\varepsilon}}(Z^\varepsilon) = \exp\{-I_1 - I_2 - \varepsilon^{-1}(I_3+I_4)\}\,, \tag{4.4}$$

where

$$I_1 = \frac{1}{2}\int_0^T |[A^{(11)}(Z_t^\varepsilon+\phi_t)]^{-\frac{1}{2}}(\dot{\phi}_t^{(1)} - b^{(1)}(\phi_t))|^2\,dt\,,$$

$$I_2 = \int_0^T (\dot{\phi}_t^{(1)} - b^{(1)}(Z_t^\varepsilon+\phi_t),\,[\sigma^{(11)}(Z_t^\varepsilon+\phi_t)]^{-1}\,dW_t^{(1)})\,,$$

$$I_3 = \frac{1}{2}\int_0^T |a_{(22)}(Z_t^\varepsilon+\phi_t)\,[\dot{\phi}_t^{(2)} - b^{(2)}(Z_t^\varepsilon+\phi_t)]|^2\,dt\,,$$

$$I_4 = \int_0^T (\dot{\phi}_t^{(2)} - b^{(2)}(Z_t^\varepsilon+\phi_t),\,\tilde{\sigma}_{(2\cdot)}(Z_t^\varepsilon+\phi_t)\,d\tilde{W}_t)\,.$$

We set $\beta = \sqrt{1+\frac{\alpha}{2}} - 1$,

$$V_t^\varepsilon = (|Z_t^{\varepsilon,(1)}|^2 + \varepsilon^{-1}|Z_t^{\varepsilon,(2)}|^2)^{1/2}\,,$$

where, as always in this section, the indices (1), (2) denote that the vectors are taken which are made up out of the n first or the $(r-n)$ last components of the vector Z_t^ε respectively. By conditions (4.2), one can choose a positive constant $D \leq 1$, depending on α, K, m, M, and r, so small that the bound

$$(1+\beta)(I_1+\varepsilon^{-\frac{1}{2}}I_3) \leq (1+\alpha)[R_{0T}(\phi)+\varepsilon^{-1}S_{0T}(\phi)]+T$$

holds for $V_t^\varepsilon < D\delta$. On account of (4.4), this yields that, for any $\gamma > 0$

$$P\{\rho_{0T}(X^{\varepsilon,x},\phi)<\delta\} = E\chi_{\|z^\varepsilon\|_0^T<\delta} \exp\left\{-I_1-I_2-\frac{1}{\sqrt{\varepsilon}}(I_3+I_4)\right\} \geq$$

$$\geq E\chi_{\sup\limits_{0\leq t\leq T} V_t^\varepsilon<D\delta}\cdot\chi_{I_2+\frac{1}{\sqrt{\varepsilon}}I_4\leq\beta(I_1+\frac{1}{\varepsilon}I_3)+\gamma} \exp\left\{-\gamma-(1+\beta)(I_1+\frac{1}{\varepsilon}I_3)\right\} \geq \tag{4.5}$$

$$\geq \exp\{-(1+\alpha)[R_{0T}(\phi)+\varepsilon^{-1}S_{0T}(\phi)]-T-\gamma\}\times$$

$$\times P\{\sup_{0\leq t\leq T} V_t^\varepsilon<D\delta, I_2+\varepsilon^{-\frac{1}{2}}I_4<\beta(I_1+\varepsilon^{-1}I_3)+\gamma\},$$

where $\|\cdot\|_0^T$ is the uniform norm in the space $C_{0T}(R^r)$, χ_B denoting the indicator of the set B.

The probability on the right-hand side of (4.5) may be bounded from below with the expression

$$P\{\sup_{0\leq t\leq T} V_t^\varepsilon<D\delta\} - P\{I_2+\varepsilon^{-\frac{1}{2}}I_4>\beta(I_1+\varepsilon^{-1}I_3)+\gamma\}.$$

Next, taking into account that $E\exp\{-\beta^2 I_1+\beta I_2-\beta^2\varepsilon^{-1}I_3+\beta\varepsilon^{-\frac{1}{2}}I_4\}=1$ for any β and using the exponential Chebyshev inequality, we arrive at the bound

$$P\{I_2+\varepsilon^{-\frac{1}{2}}I_4>\beta(I_1+\varepsilon^{-1}I_3)+\gamma\} \leq$$

$$\leq c^{-\beta\gamma} E\exp\{-\beta^2 I_1+\beta I_2-\beta^2\varepsilon^{-1}I_3+\beta\varepsilon^{-\frac{1}{2}}I_4\} = e^{-\beta\gamma}.$$

Let us pick up γ from the condition $2e^{-\beta\gamma} = P\{\sup\limits_{0\leq t\leq T} V_t^\varepsilon < D\delta\}$. Then (4.5) and the bounds following (4.5) imply that

$$P\{\rho_{0T}(X^{\varepsilon,x},\phi)<\delta\}\geq \tag{4.6}$$

$$\geq \exp\{-(1+\alpha)[R_{0T}(\phi)+\varepsilon^{-1}S_{0T}(\phi)]-T\}\left[\frac{1}{2}P\{\sup_{0\leq t\leq T} V_t^\varepsilon<D\delta\}\right]^{\frac{1+\beta}{\beta}}.$$

To deduce inequality (4.2) from (4.6), one should also show that

$$P\{\sup_{0\leq t\leq T} V_t^\varepsilon<D\delta\}>\exp\left\{\frac{-\kappa T}{\delta^2}\right\} \tag{4.7}$$

for some κ depending only on α, K, m, M, and r.

From the definition of V_t^ε and Z_t^ε, with the help of Ito's formula, we derive the following equation for V_t^ε, $t\in[0,T]$:

$$dV_t^\varepsilon=\frac{1}{2V_t^\varepsilon}[Q_t^\varepsilon-R_t^\varepsilon]dt+\sqrt{R_t^\varepsilon}\,d\tilde{\tilde{W}}_t\,, \tag{4.8}$$

where $\tilde{\tilde{W}}_t$ is a one-dimensional Wiener process,

$$Q_t^\varepsilon=\operatorname{Tr}A^{(11)}(Z_t^\varepsilon+\phi_t)+\varepsilon\operatorname{Tr}a^{(11)}(Z_t^\varepsilon+\phi_t)+\operatorname{Tr}a^{(22)}(Z_t^\varepsilon+\phi_t)\,,$$

$$R_t^\varepsilon=(V_t^\varepsilon)^{-2}[(Z_t^{\varepsilon,(1)},A^{(11)}(Z_t^\varepsilon+\phi_t)Z_t^{\varepsilon,(1)})+\varepsilon(Z_t^{\varepsilon,(1)},a^{(11)}(Z_t^\varepsilon+\phi_t)Z_t^{\varepsilon,(1)})+$$

$$+2(Z_t^{\varepsilon,(1)},a^{(12)}(Z_t^\varepsilon+\phi_t)Z_t^{\varepsilon,(2)})+\varepsilon^{-1}(Z_t^{\varepsilon,(2)},a^{(22)}(Z_t^\varepsilon+\phi_t)Z^{\varepsilon,(2)})]\,.$$

Conditions (4.2) imply that one can find $\kappa_1,\kappa_2>0$ depending merely on m, M, and r, such that

$$Q_t^\varepsilon-R_t^\varepsilon\leq\kappa_1R_t^\varepsilon,\quad R_t^\varepsilon\leq\kappa_2\,. \tag{4.9}$$

Without loss of generality it is possible to assume that $\frac{1}{2}\kappa_1+1=m$ is an integer. Together with equation (4.8), we shall consider the equation

$$dU_t^\varepsilon=\frac{(m-1)R_t^\varepsilon}{U_t^\varepsilon}+\sqrt{R_t^\varepsilon}\,d\tilde{\tilde{W}}_t^\varepsilon\,. \tag{4.10}$$

From equations (4.8), (4.10) and inequalities (4.9), one can easily conclude that $U_t^\varepsilon \geq V_t^\varepsilon$, $t \geq 0$, whenever $U_0^\varepsilon \geq V_0^\varepsilon \geq 0$.

We shall define the Markov times τ_t, $t \geq 0$, with the aid of the equation

$$\int_0^{\tau_t} R_s^\varepsilon \, ds = D^2 \delta^2 t .$$

The random process $\widetilde{U}_t^\varepsilon = \frac{1}{D\delta} U_{\tau_t}^\varepsilon$ obeys the equation

$$d\widetilde{U}_t^\varepsilon = \frac{m-1}{\widetilde{U}_t^\varepsilon} + d\hat{W}_t^\varepsilon , \tag{4.11}$$

where $\hat{W}_t^\varepsilon$ is a one-dimensional Wiener process. Taking into account that $\tau_t \geq \frac{D^2\delta^2}{\kappa_2} t$, we obtain the bound:

$$P\{ \sup_{0 \leq t \leq T} V_t^\varepsilon < D\delta \} > P\{ \sup_{0 \leq t \leq \frac{\kappa_2 T}{D^2 \delta^2}} \widetilde{U}_t^\varepsilon < 1 \} . \tag{4.12}$$

We observe now that (4.11) is the equation of the radial component of the Wiener process $\overline{W}_t$ in R^m. Denote by λ the first eigenvalue of the operator $-\frac{1}{2}\Delta$ in a unit ball in R^m with zero boundary conditions. Then for the Wiener process in R^m, the bound holds:

$$P\{ \sup_{0 \leq s \leq t} |\overline{W}_t| < 1 \} \geq \exp\{-\lambda t\} .$$

Relying on this bound, we deduce (4.7) from (4.12):

$$P\{ \sup_{0 \leq t \leq T} V_t^\varepsilon < D\delta \} > P\{ \sup_{0 \leq t \leq \frac{\kappa_2 T}{D^2 \delta^2}} |\overline{W}_t| < 1 \} \geq$$

$$\geq \exp\left\{ -\frac{\kappa T}{\delta^2} \right\}, \quad \kappa = \frac{\lambda \kappa_2}{D^2} .$$

Thereby, Theorem 4.1 is proved. □

For any ε, $T > 0$, $s \geq 0$, we put

$$\Phi^{\varepsilon}_{0T}(s) = \{\phi \in C_{0,T}(R^r) : S_{0T}(\phi) + \varepsilon R_{0T}(\phi) \leq s\}.$$

THEOREM 4.2. *Suppose that conditions (4.2) hold true. Then, for any* $\delta > 0$, $s \geq 0$, $T > 0$, $\theta > 0$, *and* $x \in R^r$, *one can find* $\varepsilon_0 > 0$ *depending merely on* m, M, θ *and* λ *depending only on* δ, K, m, M, r, *and* θ *such that the bound*

$$P\{\rho_{0T}(X^{\varepsilon,x}, \Phi^{\varepsilon}_{0T}(s)) > \delta\} \leq 2\exp\left\{\frac{-(1-\theta)s + \theta T}{\varepsilon} + \lambda T\right\} \tag{4.13}$$

holds for $\varepsilon \in (0, \varepsilon_0]$.

Proof. For every $\alpha \in (0,1)$, one can find a natural number N and unit vectors $e_1, \cdots, e_N$ from R^r such that

$$\bigcap_{i=1}^{N} \{x \in R^r : (x, e_i) \leq \alpha\} \subset \{x \in R^r : |x| \leq 1\}.$$

We set

$$Z_i(s,t) = \int_s^t (e_i^{(1)}; [\sigma^{(11)}(X_u^{\varepsilon,x})]^{-1} [\sigma^{(11)}(X_u^{\varepsilon,x})\, dW_u^{(1)} + \varepsilon\, \tilde{\sigma}^{(1\cdot)}(X_u^{\varepsilon,x})\, d\tilde{W}_u]) +$$

$$+ \int_s^t (e_i^{(2)}, [a_{(22)}(X_s^{\varepsilon,x})]^{1/2} \tilde{\sigma}^{(2\cdot)}(X_s^{\varepsilon,x})\, d\tilde{W}_u ;$$

$$\eta_i(u) = \min\{t \geq 0 : Z_i[u, u+t] \geq \alpha\kappa\}$$

Here $0 \leq s \leq t < \infty$; $i = 1, 2, \cdots, n$; $\kappa > 0$, $u \geq 0$.

For an arbitrary $t_0 > 0$, we shall define an increasing sequence of Markov times $\{\tau_n\}$ with the help of the following equalities

$$\tau_0 = 0,\ \tau_{k+1} = \tau_k + \min(t_0, \eta_1(\tau_k), \cdots, \eta_N(\tau_k))$$

and we put $\nu = \max\{n \geq 1 : \tau_n \leq T\}$.

We will define the continuous random process ℓ_t^ε, $t \in [0,T]$, by the following relations:

$$\ell_{\tau_\nu}^\varepsilon = X_{\tau_\nu}^{\varepsilon,x},\ \ell_t^\varepsilon = \ell_{\tau_\nu}^\varepsilon + \int_{\tau_\nu}^t B(\ell_s^\varepsilon)\,ds \quad \text{for} \quad s \in [\tau_\nu, T],$$

$$\ell_t^\varepsilon = \ell_{\tau_{n+1}}^\varepsilon - \int_t^{\tau_{n+1}} B(\ell_s^\varepsilon)\,ds -$$

$$-\frac{\tau_{n+1}-t}{\tau_{n+1}-\tau_n}\left[\ell_{\tau_{n+1}}^\varepsilon - X_{\tau_{n+1}}^{\varepsilon,x} + \int_{\tau_n}^{\tau_{n+1}} \sigma(X_u^{\varepsilon,x})\,dW_u + \sqrt{\varepsilon}\int_{\tau_n}^{\tau_{n+1}} \widetilde{\sigma}(X_u^{\varepsilon,x})\,d\widetilde{W}_u\right]$$

for $t \in [\tau_n, \tau_{n+1}]$ and $n = \nu-1,\ \nu-2, \cdots, 2,1,0$. It is easy to see that

$$|X_{\tau_n}^{\varepsilon,x} - \ell_{\tau_n}^\varepsilon| \leq 2M\cdot(\tau_{n+1}-\tau_n) \leq 2M\cdot t_0 \quad \text{for} \quad n = 0,1,\cdots,\nu. \tag{4.14}$$

Next, from the definition of the Markov times τ_n it follows that

$$\left|[\sigma^{(11)}(X_{\tau_n}^{\varepsilon,x})]^{-1}\left[\int_{\tau_n}^t \sigma^{(11)}(X_u^{\varepsilon,x})\,dW_u^{(1)} + \sqrt{\varepsilon}\int_{\tau_n}^t \widetilde{\sigma}^{(1\cdot)}(X_u^{\varepsilon,x})\,d\widetilde{W}_u\right]\right|^2 + \tag{4.15}$$
$$+\left|[a_{(22)}(X_{\tau_n}^{\varepsilon,x})]^{1/2}\int_{\tau_n}^t \sigma^{(2\cdot)}(X_u^{\varepsilon,x})\,d\widetilde{W}_u\right|^2 < \kappa^2$$

for $t \in [\tau_n, \tau_{n+1}]$. Consequently, for $\in [0,1]$,

$$\left| \int_{\tau_n}^{t} \sigma(X_u^{\varepsilon,x})\, dW_u + \sqrt{\varepsilon} \int_{\tau_n}^{t} \tilde{\sigma}(X_u^{\varepsilon,x})\, d\tilde{W}_u \right| \le \sqrt{M}\,\kappa, \quad t \in [\tau_n, \tau_{n+1}] .$$

Relying on this bound one can easily check that

$$|X_t^{\varepsilon,x} - X_{\tau_n}^{\varepsilon,x}| \le Mt_0 + \kappa\sqrt{M} ,$$

$$|\ell_t^\varepsilon - \ell_{\tau_n}^\varepsilon| \le 3Mt_0 + \kappa\sqrt{M}, \quad t \in [\tau_n, \tau_{n+1}], \quad n = 0,1,\cdots,\nu-1 . \tag{4.16}$$

This, along with (4.14), implies the bound

$$\rho_{0T}(X^{\varepsilon,x}, \ell^\varepsilon) \le \delta', \quad \delta' = 6Mt_0 + 2\kappa\sqrt{M} . \tag{4.17}$$

Noting that $B(x)$ is Lipschitz continuous, we conclude from (4.17) that

$$|X_{\tau_n}^{\varepsilon,x} - \ell_{\tau_n}^\varepsilon| \le K\delta'(\tau_{n+1} - \tau_n) \tag{4.18}$$

for $n = 0,1,2,\cdots,\nu-1$.

Let us choose $t_0 = t_0(M,\delta)$ and $\kappa = \kappa(M,\delta)$ so small that $\delta' < \delta$. Then using (4.17), by the Chebyshev inequality we derive:

$$P\{\rho_{0T}(X^{\varepsilon,x}, \Phi_{0T}^\varepsilon(s)) > \delta\} \le P\{\ell^\varepsilon \notin \Phi_{0T}^\varepsilon(s)\} =$$

$$= P\{S_{0T}(\ell^\varepsilon) + \varepsilon R_{0T}(\ell^\varepsilon) > s\} \le \tag{4.19}$$

$$\le \exp\left\{-\frac{(1-\theta)s}{\varepsilon}\right\} E \exp\{(1-\theta)\varepsilon^{-1} S_{0T}(\ell^\varepsilon) + (1-\theta) R_{0T}(\ell^\varepsilon)\} .$$

From (4.15), (4.17), and (4.18) it results that for $n = 0,1,2,\cdots, \nu-1$ and for an arbitrary $\gamma > 0$,

$$R_{\tau_n,\tau_{n+1}}(\ell^\varepsilon) + \varepsilon^{-1} S_{\tau_n,\tau_{n+1}}(\ell^\varepsilon) \leq$$
$$\leq \frac{1+\gamma}{2}\left(1 + \frac{KM}{m^2}\delta'\right)\frac{\kappa^2}{\tau_{n+1}-\tau_n} + \frac{K^2}{2\varepsilon m}\left(1+\frac{1}{\gamma}\right)\delta'^2 \cdot (\tau_{n+1}-\tau_n)\,. \tag{4.20}$$

Depending on δ, K, m, M and θ, one can choose constants γ, κ and t_0 (and thereby $\delta' < \delta$ too) so that

$$(1-\theta)(1+\gamma)\left(1 + \frac{KM}{m^2}\delta'\right) < 1 - \frac{1}{2}\theta\,, \quad \left(1+\frac{1}{\gamma}\right)\frac{K^2\delta'^2}{m} < 2\theta\,.$$

Then (4.19) and (4.20) imply the bound

$$P\{\rho_{0T}(X^{\varepsilon,x}, \Phi^\varepsilon_{0T}(s)) > \delta\} \leq \exp\left\{\frac{-(1-\theta)s + \theta T}{\varepsilon}\right\} \times$$
$$\times E \exp\left\{\frac{1}{2}\left(1-\frac{\theta}{2}\right)\sum_{k=1}^{\nu}\frac{\kappa^2}{\tau_k - \tau_{k-1}}\right\}. \tag{4.21}$$

To bound from above the expectation on the right-hand side of (4.21), first we shall show that one can find an $\varepsilon_0 = \varepsilon_0(m,M,\theta)$ such that, for $\varepsilon \in [0, \varepsilon_0]$,

$$E \exp\left\{\frac{1}{2}\left(1-\frac{\theta}{3}\right)\frac{\kappa}{\tau_1}\right\} < C < \infty\,, \tag{4.22}$$

where the constant C depends only on δ, K, m, M, r, and θ. To this end, we shall define the Markov times $\mu_i(t)$, $i = 1, \cdots, N$, $t \geq 0$, with the help of the following equalities:

$$t = \int_0^{\mu_i(t)} |\sigma^{(11)}(X_u^{\varepsilon,x})\,[\sigma^{(11)}(x)]^{-1} e_i^{(1)}|\,du +$$
$$+ \int_0^{\mu_i(t)} |\sqrt{\varepsilon}\,\tilde{\sigma}^{(\cdot 1)}(X_u^{\varepsilon,x})\,[\sigma^{(11)}(x)]^{-1} e_i^{(1)} + \tilde{\sigma}^{(\cdot 2)}(X_u^{\varepsilon,x})[a_{(22)}(x)]^{1/2} e_i^{(2)}|^2\,du\,. \tag{4.23}$$

Then the processes $\widehat{W}_i(t) = Z_i[0, \mu_i(t)]$ are one-dimensional Wiener processes for $i = 1, 2, \cdots, N$. Put $\widetilde{\eta}^i = \min\{t \geq 0 : \widehat{W}_i(t) \geq \alpha\kappa\}$, $i=1,2,\cdots,N$. It is easy to see that $\eta_i = \eta_i(0) = \mu_i(\widetilde{\eta}_i)$. On account of (4.16), we derive from (4.23), for $t \in [0, t_0]$ and $i = 1, 2, \cdots, N$, that $\mu_i(t) \geq \beta t$, where $\beta = \left(1 + \frac{\kappa}{m}\delta' + 2\sqrt{\varepsilon}\,\frac{M}{m}\right)^{-1}$. Therefore $(t_0 \wedge \eta_i) \geq \beta(t_0 \wedge \widetilde{\eta}_i)$ and, for for $\varepsilon \in [0, \varepsilon_0]$, we arrive at the bound:

$$E \exp\left\{\frac{1}{2}\left(1 - \frac{\theta}{3}\right)\frac{\kappa^2}{\tau_1}\right\} \leq \sum_{i=1}^{N} E \exp\left\{\frac{1}{2}\left(1 - \frac{\theta}{3}\right)\frac{\kappa^2}{t_0 \wedge \eta_i}\right\} \leq$$

$$\leq \sum_{i=1}^{N} E \exp\left\{\frac{1 - \frac{1}{3}\theta}{\alpha^2\beta} \cdot \frac{\alpha^2 \frac{\kappa^2}{2}}{t_0 \wedge \widetilde{\eta}_i}\right\} = NE \exp\left\{\frac{1 - \frac{\theta}{3}}{\alpha^2\beta} \cdot \frac{\alpha^2 \frac{\kappa}{2}}{t_0 \wedge \widetilde{\eta}_1}\right\} < C < \infty$$

whenever $\alpha^2\beta > 1 - \frac{\theta}{3}$. The last relation may be ensured at the expense of the choice of N, t_0, κ, and ε_0; the choice of N, t_0, and κ depends only on δ, K, m, r, M, θ, the choice of ε_0 depending only on m, M and θ. Thereby, inequality (4.22) is proved.

By (4.22), using the inequality

$$\frac{1}{\tau} - \beta\tau \leq \frac{1 + \alpha}{\tau} - 2\sqrt{\alpha\beta}$$

which is valid for any α, β, $\tau > 0$, we conclude that there exists a $\lambda = \lambda(\delta, K, m, M, r, \theta)$ such that

$$E \exp\left\{\frac{1}{2}\left(1 - \frac{\theta}{2}\right)\frac{\kappa^2}{\tau_1} - \lambda\tau_1\right\} \leq \frac{1}{2}.$$

Whence, relying on the strong Markov property, we obtain the bound

$$E \exp\left\{\frac{1}{2}\left(1 - \frac{\theta}{2}\right)\sum_{k=1}^{\nu}\frac{\kappa^2}{\tau_k - \tau_{k-1}}\right\} \leq 2e^{\lambda T},$$

which, together with (4.21), implies (4.13) for $\varepsilon \in [0, \varepsilon_0]$. □

In particular, Theorems 4.1 and 4.2 hold true when there are no second-order terms in the operator L_0. In this case, the proof is simpler: the matrices $\sigma^{(11)}$, $(\sigma^{(11)})^{-1}$ and $R_{0T}(\phi)$ should be assumed to be identically zero. In doing so, the integrals I_1, I_2 introduced in the proof of Theorem 4.1, will vanish, and $V_t^\varepsilon = |Z_t^\varepsilon|$. The set $\Phi_{0T}^\varepsilon(s) = \Phi_{0T}(s) = \{\phi \in C_{0T}(R^r): S_{0T}(\phi) \le s\}$ will no longer depend on ε. Also the determination of $Z_i(s,t)$, ℓ_t^ε will become simpler.

Let us formulate separately a somewhat rougher version of bounds (4.3) and (4.13) intended especially for the case when L_0 is a first-order operator, that is when the matrix $\sigma(x)$ in equation (4.1) is equal to zero identically.

THEOREM 4.3. *Suppose that the matrix* $\sigma(x)$ *in (4.1) is equal to zero identically, and let conditions (4.2) be fulfilled (except for the condition on the square form* $(A^{(11)}(z,z))$ *). Then the following bounds hold:*

1. *for any* δ, h, $H > 0$, *one can find* $\varepsilon_0 > 0$ *such that, for* $\varepsilon \in (0, \varepsilon_0]$

$$P\{\rho_{0T}(X^{\varepsilon,x}, \phi) < \delta\} \ge \exp\{-\varepsilon^{-1}[S_{0T}(\phi)+h]\}, \tag{4.24}$$

where $T > 0$ *and* $\phi \in C_{0T}(R^r)$ *are such that* $\phi_0 = x$, $T + S_{0T}(\phi) \le H$;

2. *for any* $\delta > 0$, $h > 0$, $s_0 > 0$, *one can find* $\varepsilon_0 > 0$ *such that*

$$P\{\rho_{0T}(X^{\varepsilon,x}, \Phi_{0T}(s)) > \delta\} \le \exp\{-\varepsilon^{-1}(s-h)\} \tag{4.25}$$

for $\varepsilon \in (0, \varepsilon_0]$ *and* $s \le s_0$.

The functional $S_{0T}(\phi)$ *is lower semi-continuous in the sense of uniform convergence; i.e. if* $\lim_{n\to\infty} \rho_{0T}(\phi^{(n)}, \phi) = 0$, $\phi^{(n)} \in C_{0T}(R^r)$, *then* $S_{0T}(\phi) \le \underline{\lim}_{n\to\infty} S_{0T}(\phi^{(n)})$.

For any compact $F \subset R^r$, *the set* $\Phi_{0T}(F,s) = \{\phi \in C_{0T}(R^r): \phi_0 \in F, S_{0T}(\phi) \le s_0 < \infty\}$ *is compact in* $C_{0T}(R^r)$.

Proof. Inequalities (4.24) and (4.25) come from Theorems 4.1 and 4.2 respectively. The proof of the lower semi-continuity of the functional

$S_{0T}(\phi)$ and the compactness of the set $\Phi_{0T}(F,s)$ may be found in Wentzell and Freidlin [1]. □

If the conditions of Theorem 4.3 hold, then the functional $S_{0T}(\phi)$ is called the normed action functional for the family of processes (X_t^ε, P_x) corresponding to the operators

$$L^\varepsilon = \frac{\varepsilon}{2} \sum_{i,j=1}^{r} a^{ij}(x) \frac{\partial^2}{\partial x^i \partial x^j} + \sum_{i=1}^{r} B^i(x) \frac{\partial}{\partial x^i} = \varepsilon L_1 + L_0 . \qquad (4.26)$$

In the classical case (i.e. when L_0 is a first-order operator), one can easily deduce from Theorem 4.3 some results concerning the asymptotics of the solutions of Cauchy's problem and mixed problems for the parabolic equation $\frac{\partial u^\varepsilon(t,x)}{\partial t} = L^\varepsilon u^\varepsilon(t,x)$ as $\varepsilon \downarrow 0$ (see Wentzell and Freidlin [1], [2]). Similar results in the general case may be obtained with the help of Theorems 4.1 and 4.2. We shall not dwell on these questions here and turn to the Dirichlet problem for the equation $L^\varepsilon u(x) = 0$. First, we consider the classical case and outline the proof of Theorem 1.2. For a more detailed investigation, see Wentzell and Freidlin [1]. The construction used in the proof of this theorem, together with bounds (4.3) and (4.13), enables one to consider a more general case too.

Given a bounded domain $D \subset R^r$ with the smooth boundary ∂D, denote by $n(x)$ the unit vector of the outward normal to ∂D at the point $x \in \partial D$. Along with the dynamical system

$$\dot{X}_t^x = B(X_t^x), \quad x \in R^r, B(x) = (B^1(x), \cdots, B^r(x)) \qquad (4.27)$$

corresponding to the operator L_0, let us consider the Markov process $(X_t^\varepsilon, P_x^\varepsilon)$ governed by the operator L^ε.

Suppose that in the domain D there is a unique asymptotic stable equilibrium point 0 of system (4.27), and that all the trajectories X_t^x, $x \in D$, tend to 0 without leaving the domain D as $t \to \infty$. Moreover, let $(B(x), n(x)) < 0$ for $x \in D$ (Fig. 8).

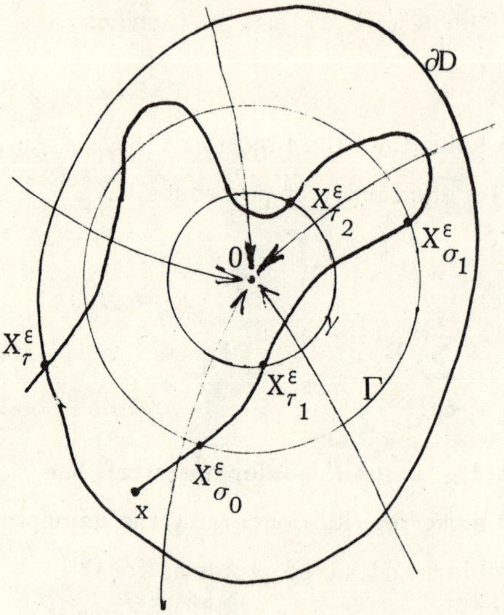

Fig. 8

How will a trajectory, starting from a point $x \in D$, behave for small ε? First of all, with probability tending to 1 as $\varepsilon \downarrow 0$, this trajectory will get into a small neighborhood of the equilibrium point 0. In particular, this follows from the considerations of §4.1, if one takes into account that the trajectories X_t^x, $x \in D$, of system (4.27) are attracted to the point 0 (see also the first claim of Lemma 4.2). Then the trajectories X_t^{ε} will make "excursions" to the periphery of the domain D and return again and again into the neighborhood of the point 0. (We assume the process $(X_t^{\varepsilon}, P_x^{\varepsilon})$ to be recurrent. Otherwise, one can change the field B(x) outside the domain D so that this process can be recurrent for any $\varepsilon \in (0,1]$.) At last during one of its "excursions", the diffusing particle will touch ∂D.

To describe these excursions more precisely, we shall choose a small number $\mu > 0$ (later on the choice of μ will be made precise) and put $\gamma = \left\{x \in R^r : |x| = \frac{1}{2}\mu\right\}$, $\Gamma = \{x \in R^r : |x| = \mu\}$ (Fig. 8). Introduce the increasing sequence of Markov times $\tau_0, \sigma_0, \tau_1, \sigma_1, \tau_2, \cdots$ as follows: $\tau_0 = 0$, $\sigma_n = \inf\{t > \tau_n : X_t^{\varepsilon} \in \Gamma\}$, $\tau_{n+1} = \inf\{t > \sigma_n : X_t^{\varepsilon} \in \gamma \cup \partial D\}$. Since we assume the process $(X_t^{\varepsilon}, P_x^{\varepsilon})$ to be recurrent, all these Markov times are finite with probability 1 starting from any $x \in R^r$.

Consider the Markov chain $Z_n = X^{\varepsilon}_{\tau_n}$, $n = 1, 2, \cdots$, on the state space $\gamma \cup \partial D$. With probability approaching 1 for small ε, the chain Z_n jumps from any point $x \in \gamma \cup \partial D$ onto γ in one step. However, for $\varepsilon > 0$, sooner or later, the chain will, without fail, hit ∂D. It turns out that if at some step the trajectory makes a transition onto ∂D, then, with probability approaching 1 for small ε, this will be a transition into a neighborhood of the point x_0 at which the quasi-potential $V(x)$ attains its minimum on ∂D. We recall that the quasi-potential $V(x)$ of the vector field $B(x)$ has been defined by the formula

$$V(x) = \inf \{S_{T_1 T_2}(\phi) : \phi \in C_{T_1, T_2}(R^r), \phi_{T_1} = 0, \phi_{T_2} = x, T_1 < T_2\} .$$

By the condition of Theorem 1.2, the point of the minimum of the quasi-potential $V(x)$ on ∂D is unique. More precisely, for every $\delta > 0$

$$\lim_{\varepsilon \downarrow 0} P_z^\varepsilon \{|Z_1 - x_0| < \delta | Z_1 \in \partial D\} = 1 \tag{4.28}$$

uniformly in $z \in \gamma$. From (4.28) one can easily deduce the claim of Theorem 1.2: $\lim_{\varepsilon \downarrow 0} u^\varepsilon(x) = \psi(x_0)$, where $u^\varepsilon(x)$ is the solution of the Dirichlet problem

$$L^\varepsilon u^\varepsilon(x) = 0, \quad x \in D, \quad u^\varepsilon(x)\Big|_{\partial D} = \psi(x) . \tag{4.29}$$

Really, the function $u^\varepsilon(x)$ can be represented in the form $u^\varepsilon(x) = E_x \psi(X^\varepsilon_{\tau^\varepsilon})$, where $\tau^\varepsilon = \inf\{t : X_t^\varepsilon \notin D\}$. Further, denoting, as usual, by χ_B the indicator of the set B and using the continuity of $\psi(x)$ and (4.28), we have that for any $\gamma > 0$, and for sufficiently small ε:

$$|u^\varepsilon(x) - \psi(x_0)| = |E_x \psi(X^\varepsilon_{\tau^\varepsilon}) - \psi(x_0)| =$$

$$= \left| \sum_{k=0}^\infty E_x \chi_{\tau_k < \tau^\varepsilon} E_{X^\varepsilon_{\tau_k}} [\psi(Z_1) - \psi(x_0)] \chi_{Z_1 \in \partial D} \right| =$$

$$= \sum_{k=0}^\infty E_x \chi_{\tau_k < \tau^\varepsilon} E_{X^\varepsilon_{\tau_k}} ((\psi(Z_1) - \psi(x_0)) | Z_1 \in \partial D) E_{X^\varepsilon_{\tau_k}} \chi_{Z_1 \in \partial D} \le \gamma \sum_{k=0}^\infty E_x \chi_{\tau_k < \tau^\varepsilon \le \tau_{k+1}} =$$

Therefore, to complete the proof of Theorem 1.2, one should establish (4.28). For this, we shall need the two following lemmas.

LEMMA 4.1. *Suppose that the conditions of Theorem 4.3 are fulfilled. Then:*

1. *There exists an* $\ell > 0$ *such that, for any* $x, y \in R^r$, *one can find a smooth function* ϕ_t, $0 \le t \le T$, $T = |x-y|$, $\phi_0 = x$, $\phi_T = y$, *for which* $S_{0T}(\phi) \le \ell \cdot |x-y|$.

2. *In addition, let the conditions of Theorem 1.2 hold as well. Then, for any* $\alpha > 0$, *one can find* $a, T_0 > 0$ *such that the inequality*

$$S_{0T}(\phi) > a(T - T_0)$$

is valid for any T *and every function* ϕ_s, $0 \le s \le T$, *taking values in* $D \cup \partial D \setminus U_\alpha(0)$, *where* $U_\alpha(0) = \{x \in R^r, |x| < \alpha\}$.

Proof. 1. As such a function ϕ, it is possible to take $\phi_t = x + \dfrac{t(y-x)}{|x-y|}$.

2. Note that the solutions of system (4.27) depend continuously on the initial conditions, which implies that $T_0 = \max_{x \in D \cup \partial D} \inf\{t : |X_t^x| < \alpha\} < \infty$. Put

$$\mathcal{F} = \{\phi \in C_{0T}(R^r) : \phi_s \in D \cup \partial D, |\phi_s| \ge \alpha \text{ for } 0 \le s \le T_0\} .$$

We observe that the trajectories X_s^x, $0 \le s \le T_0$, of the dynamical system (4.27) do not belong to $\mathcal{F}$. Hence, taking into account that the functional $S_{0T_0}(\phi)$ is semi-continuous and vanishes only on the trajectories of system (4.27), we obtain that $\inf\{S_{0T_0}(\phi) : \phi \in \mathcal{F}\} = A > 0$. From this, by the additivity of the functional $S_{0T}(\phi)$, we deduce the second claim of Lemma 4.1:

$$S_{0T}(\phi) > A\left(\frac{T}{T_0} - 1\right) = a(T - T_0) . \quad \square$$

LEMMA 4.2. *Suppose that the conditions of Theorem 4.3 are fulfilled. Then:*

1. *For any* $T, \delta > 0$ *and any compact set* $F \subset R^r$, *one can find positive numbers* ε_0, β *such that*

$$\sup_{x \in F} P_x^\varepsilon \{\rho_{0T}(X_\cdot^\varepsilon, X_\cdot^x) \geq \delta\} \leq \exp\{-\beta\varepsilon^{-1}\}, \quad \varepsilon \leq \varepsilon_0$$

2. *In addition, let the conditions of Theorem 1.2 hold as well. Then, for any* $a > 0$, *one can find* c, $T_0 > 0$ *such that, for sufficiently small* ε *and for any* $x \in (D \cup \partial D) \setminus U_a(0)$

$$P_x^\varepsilon \{\zeta_a > T\} \leq \exp\{-\varepsilon^{-1} c \cdot (T - T_0)\}, \tag{4.30}$$

where $\zeta_a = \inf\{t : X_t^\varepsilon \notin D \setminus U_a(0)\}$.

Proof. 1. Put

$$\widetilde{\mathcal{F}} = \{\phi \in C_{0,T}(R^r) : \phi_0 \in F, \rho_{0T}(\phi, X^x) \geq \delta\}.$$

By virtue of the lower semi-continuity of the functional $S_{0T}(\phi)$, we have that $d = \inf\{S_{0T}(\phi) : \phi \in \widetilde{\mathcal{F}}\} > 0$, because $S_{0T}(\phi)$ vanishes only if ϕ is a trajectory of system (4.27). Let $d' < d$, $\delta' = \rho_{0T}(\widetilde{\mathcal{F}}, \widetilde{\Phi})$, where $\widetilde{\Phi} = \{\phi \in C_{0,T}(R^r), \phi_0 \in F, S_{0T}(\phi) < d'\}$. Since $\widetilde{\mathcal{F}}$ is closed, $\widetilde{\Phi}$ is compact, and $\widetilde{\Phi} \cap \widetilde{\mathcal{F}} = \emptyset$, we get $\delta' > 0$. By bound (4.25), for any $h > 0$ and $\varepsilon > 0$ small enough, we derive that

$$P_x^\varepsilon \{\rho_{0T}(X_\cdot^\varepsilon, X_\cdot^x) \geq \delta\} \leq P_x^\varepsilon \{\rho_{0T}(X_\cdot^\varepsilon, \widetilde{\Phi}) > \delta\} \leq$$

$$\leq \exp\{-\varepsilon^{-1}(d' - h)\} = \exp\{-\varepsilon^{-1}\beta\}.$$

2. We notice that the domain D is attracted to 0 and $(b(x), n(x)) < 0$ for $x \in \partial D$. Hence it follows that the δ-neighborhood of the domain D possesses the same properties, provided δ is small enough. Pick $\delta \in \left(0, \frac{1}{2}\mu\right)$. By the second of the claims of Lemma 4.1 one can choose positive numbers T_0 and A such that $S_{0T_0}(\phi) > A$ for all the functions ϕ which, on the interval $[0, T_0]$, never move away from D by more than δ and never visit $U_{a/2}(0)$. Whence, with the help of (4.25) we obtain,

that for any $h > 0$ and for sufficiently small $\varepsilon > 0$,

$$P_x^\varepsilon\{\zeta_\alpha > T\} \le P_x^\varepsilon\{\rho_{0T}(X_\cdot^\varepsilon, \Phi_{0T}(A)) \ge \delta\} \le \le \exp\{-\varepsilon^{-1}(A-h)\}\,. \tag{4.31}$$

Using the Markov property, (4.31) leads to (4.30):

$$P_x^\varepsilon\{\zeta_\alpha > T\} \le P_x^\varepsilon\left\{\zeta_\alpha > \left[\frac{T}{T_0}\right] T_0\right\} \le [\sup_{y\in D} P_y^\varepsilon\{\zeta_\alpha > T_0\}]^{\left[\frac{T}{T_0}\right]} \le$$

$$\le \exp\left\{-\varepsilon^{-1}\left(\frac{T}{T_0}-1\right)(A-h)\right\} = \exp\{-\varepsilon^{-1} c\cdot(T-T_0)\}\,. \quad \square$$

Now we can turn directly to proving (4.28). Choose a small $\delta > 0$ and put: $V_0 = \min_{x\in\partial D} V(x) = V(x_0)$, $V_1 = \min_{x\in\partial D\setminus U_{\delta/2}(x_0)} V(x)$, $h = \frac{1}{10}(V_1 - V_0)$. The definition of V_0 implies that one can find $T_1 > 0$ and $\tilde{\phi}\in C_{0,T_1}(R^r)$ such that $\tilde{\phi}_0 = 0$, $\tilde{\phi}_{T_1} = x_0$, $S_{0T_1}(\tilde{\phi}) < V_0 + h$. Now let us choose a $\mu < \delta$ so small that, for any point $x \in \gamma = \left\{x\in R^r : |x| = \frac{1}{2}\mu\right\}$, one can find $T_2 > 0$ and $\bar{\phi}\in C_{0,T_2}(R^r)$ such that $\bar{\phi}_0 = x$, $\rho(\bar{\phi}_{T_2}, D) > \mu$, $S_{0T_2}(\bar{\phi}) \le V_0 + 2h$. Besides, suppose that the curve $\bar{\phi}_s$, $0 \le s \le T_2$, does not return to $U_\mu(0)$ after crossing $\Gamma = \{x\in R^r : |x| = \mu\}$ and let the pipe of radius $\frac{1}{4}\mu$ about the curve $\bar{\phi}$ never intersect $\partial D\setminus U_{\delta/2}(x_0)$: $\left\{x\in R^r : \inf_{0\le s\le T_2} \rho(x,\bar{\phi}_s) \le \frac{1}{4}\mu\right\} \cap (\partial D\setminus U_{\delta/2}(x_0)) = \emptyset$. Such a function $\bar{\phi}$ may be constructed from $\tilde{\phi}$ with the help of the first assertion of Lemma 4.1.

It is easy to see that all the trajectories X_t^ε lying in the $\frac{\mu}{4}$-neighborhood of the function $\bar{\phi}$ cross ∂D before time T_2 at a point $X_{\tau^\varepsilon}^\varepsilon$ which lies in the $\frac{\delta}{2}$-neighborhood of the point x_0. Moreover, these trajectories reach ∂D, without returning to γ, after crossing Γ. From this, applying (4.24) we obtain: for any $x\in\gamma$,

$$P_x^\varepsilon\{|Z_1 - x_0| < \delta\} > P_x^\varepsilon\left\{\rho_{0T_2}(X^\varepsilon, \bar{\phi}) < \frac{\mu}{4}\right\} \geq$$
$$\geq \exp\{-\varepsilon^{-1}(S_{0T_2}(\bar{\phi}) + h)\} \geq \exp\{-\varepsilon^{-1}(V_0 + 3h)\}\,, \tag{4.32}$$

if ε is small enough.

We proceed to derive an upper bound for $P_z^\varepsilon\{Z_1 \in \partial D \setminus U_\delta(x_0)\}$, $z \in \gamma$. In accordance with (4.30), one can choose T so large that

$$P_x^\varepsilon\{\zeta_{\mu/2} > T\} \leq \exp\{-\varepsilon^{-1}(V_0 + 11h)\} \tag{4.33}$$

for $x \in D \setminus U_{\mu/2}$. By the first assertion of Lemma 4.1, it is easy to see that

$$\inf\{S_{0T}(\phi) : \phi \in C_{0T}(R^2), \phi_0 = x \in \gamma, \phi_T \in \partial D \setminus U_{\delta/2}(x_0)\} =$$
$$= V_2 > V_0 + 9h \tag{4.34}$$

whenever μ is small enough. Otherwise, V_1 would be smaller than $V_0 + 10h$, which is contrary to the definition of h. Taking into account bounds (4.33), (4.34), and (4.25), for $z \in \gamma$ and δ' fairly small, we have:

$$P_z^\varepsilon\{Z_1 \in \partial D \setminus U_\delta(x_0)\} < P_z^\varepsilon\{\zeta_{\mu/2} > T\} +$$
$$+ P_z^\varepsilon\{\rho_{0T}(X^\varepsilon_\cdot, \Phi_{0T}(V_0 + 8h)) > \delta'\} \leq$$
$$\leq \exp\{-\varepsilon^{-1}(V_0 + 11h)\} + \exp\{-\varepsilon^{-1}(V_0 + 7h)\} \leq$$
$$\leq 2\exp\{\varepsilon^{-1}(V_0 + 7h)\}\,,$$

provided ε is small enough. This, along with (4.32), implies (4.28):

$$1 \geq P_z^\varepsilon\{|Z_1 - x_0| < \delta | Z_1 \in \partial D\} = \frac{P_z^\varepsilon\{|Z_1 - x_0| < \delta\}}{P_z^\varepsilon\{Z_1 \in \partial D\}} >$$
$$> \frac{P_z^\varepsilon\{|Z_1 - x_0| < \delta\}}{P_z^\varepsilon\{|Z_1 - x_0| < \delta\}(1 + o_\varepsilon(1))} \to 1, \quad \varepsilon \downarrow 0\,.$$

As was pointed out previously, relation (4.28) implies the claim of Theorem 1.2. □

In the conclusion of this section, we shall dwell briefly on the case when the trajectories of the dynamical system (4.27) corresponding to the first-order operator L_0 enter the domain D and have in D several asymptotically stable equilibrium points or some ω-limit sets of a more general form. For example, the dynamical system in Fig. 9 has two stable

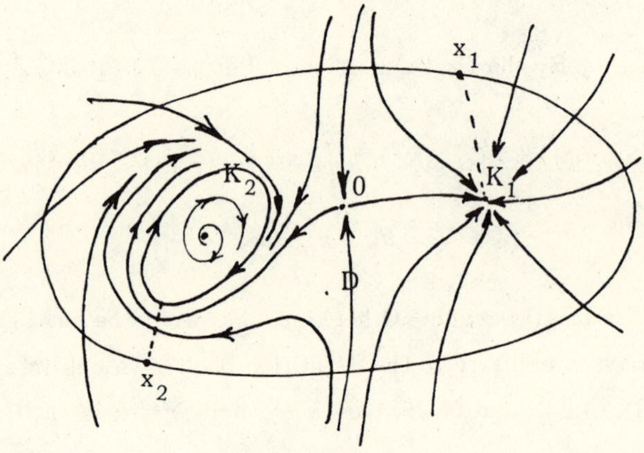

Fig. 9

ω-limit sets: the equilibrium point K_1 and the limit cycle K_2. The trajectories entering the saddle point 0 separate this domain into two parts. One of them (call it D_1) is attracted to K_1, the other– D_2 –is attracted to the cycle K_2. How will the trajectory of the process $(X_t^\varepsilon, P_x^\varepsilon)$ corresponding to operator (4.26) behave for small $\varepsilon > 0$? If the initial point x belongs to D_i, then, with overwhelming probability, the trajectory will first of all get into a small neighborhood of K_i, $i = 1,2$. Then it will make excursions to other points of D_i and again return to a small neighborhood of K_i. When there was only one equilibrium point, some of these excursions finally reached the boundary at a point where the quasi-potential was minimal. (We consider the case of general position when such a point is unique.) In the case of several ω-limit sets, a somewhat different behavior is possible: before hitting the boundary, the trajectory goes over to a neighborhood of the other set K_j. To describe the behavior of the trajectory more precisely, we shall introduce the functions $V(K_i,x) = V_D(K_i,x) = \inf\{S_{T_1,T_2}(\phi) : \phi_{T_1} \in K_i, \phi_{T_2} = x, T_1 < T_2, \phi_s \in D$ for $s \in (T_1,T_2)\}$, $x \in D \cup \partial D$, $i = 1,2$. Suppose that for every $i = 1,2$, a unique point $x_i \in \partial D$ exists such that $V_D(K_i, x_i) = \min_{x \in \partial D} V(K_i, x)$. Note that $V(K_i, x)$ takes one and the same value for $x \in K_j$. We shall denote this value by $V(K_i, K_j)$. It is possible to prove that if $V(K_i, x_i) < V(K_i, K_j)$, $i \neq j$, then with overwhelming probability for small ε, starting from $x \in D_i$ trajectories leave D near the point x_i. This implies that in this case $\lim_{\varepsilon \downarrow 0} u^\varepsilon(x) = \psi(x_i)$ for $x \in D_i$. If $V(K_i, x_i) > V(K_i, K_j)$, then the trajectory will go over into a neighborhood of the set K_j, before it hits ∂D. Having gone over into a neighborhood of the other set K_j, the trajectory starts making excursions over D_j. If $V(K_j, x_j) < V(K_j, K_i)$, then with overwhelming probability for small ε, the trajectory will hit ∂D near x_j before it returns to D_i. But if $V(K_j, x_j) > V(K_j, K_i)$, then the trajectory will again return into K_i and hitting ∂D will occur after many transitions from D_i to D_j and back. Therefore, up to when it hits the boundary, the process (X_t^ε, P_x) may be approximated, for small ε, by some Markov chain. The states of this chain are ω-limit sets (in our example, these are K_1 and K_2) and another extra state " ∂ " corresponding to the boundary of the domain. In the case of the general position, such an accuracy of approximation is sufficient for calculating $\lim_{\varepsilon \downarrow 0} u^\varepsilon(x)$, where $u^\varepsilon(x)$ is the solution of problem (4.29).

If, finally, at some step, the chain for the first time gets into " ∂ ", after leaving K_i then, for small ε with probability approaching 1, the trajectory X_t^ε will, for the first time, hit ∂D near the point x_i. With the aid of Theorem 4.3, one can show that the transition probabilities in one step for this chain are of logarithmic order:

$$P(K_i, K_j) \sim \exp\{-\varepsilon^{-1} V_D(K_i, K_j)\}, \quad i \neq j,$$

$$P(K_i, \text{``}\partial\text{''}) \sim \exp\{-\varepsilon^{-1} V_D(K_i, x_i)\}.$$

On having such asymptotic expression, one already can easily evaluate, how such a chain will, for the first time, get into the state " ∂." Thereby, one can find out near what point $x_i \in \partial D$, with probability overwhelming for small ε, the first exit of the trajectory X_t^ε from the domain D will happen. From this, noting that $u^\varepsilon(x) = E_x \psi(X^\varepsilon_{\tau^\varepsilon})$, one can readily calculate $\lim_{\varepsilon \downarrow 0} u^\varepsilon(x)$. Note that this limit is constant in the domain of attraction of each of the sets K_i; for different domains of attraction these limits are, generally speaking, different. For instance, in our example, in the case when $V_D(K_i, K_j) > V_D(K_i, x_i)$ and $V_D(K_j, K_i) > V_D(K_j, x_j)$, $\lim_{\varepsilon \downarrow 0} u^\varepsilon(x)$ is equal to $\psi(x_1)$ for $x \in D_1$ and it is equal to $\psi(x_2)$ for $x \in D_2$.

In a rather general situation, the case when there are several ω-limit sets in D is examined in detail by Wentzell and Freidlin in [1], [2].

§4.5 *Large deviations. Continuation*

Now we turn to the process $(X_t^\varepsilon, P_x^\varepsilon)$ of a more general form which is described by equations (4.1). The corresponding generator is the following

$$L^\varepsilon = L_0 + \varepsilon L_1 = \frac{1}{2} \sum_{i,j=1}^{n} A^{ij}(x) \frac{\partial^2}{\partial x^i \partial x^j} + \sum_{i=1}^{n} B^i(x) \frac{\partial}{\partial x^i} + \frac{\varepsilon}{2} \sum_{i,j=1}^{r} a^{ij}(x) \frac{\partial^2}{\partial x^i \partial x^j}, \quad n < r$$

In order to describe the large deviations of the process $(X_t^\varepsilon, P_x^\varepsilon)$ from the process (X_t^0, P_x^0) corresponding to the operator L_0, we shall introduce the functional $\widetilde{S}_{0T}(\psi)$, $\psi \in C_{0,T}(R^{r-n})$,

$$\widetilde{S}_{0T}(\psi) = \inf\{S_{0T}(\phi, \psi) : \phi \in C_{0,T}(R^n)\}.$$

The functional $\widetilde{S}_{0T}(\psi)$, defined in such a way, is not, generally speaking, lower semi-continuous in $C_{0T}(R^{r-n})$. Consider its semi-continuous regularization

$$S^*_{0T}(\psi) = \varliminf_{\widetilde{\psi}\in C_{0T}(R^{r-n}),\rho_{0T}(\widetilde{\psi},\psi)\to 0} \widetilde{S}_{0T}(\widetilde{\psi}) .$$

It is easy to see that the functional $S^*_{0T}(\psi)$ is lower semi-continuous. We set

$$\Phi^*_{0T}(s) = \{\psi \in C_{0T}(R^{r-n}) : S^*_{0T}(\psi) \le s\} .$$

THEOREM 5.1. *Suppose that conditions (4.2) are fulfilled. Then the following assertions hold:*

1. *For any* δ, h, $T > 0$, $x = (x^{(1)}, x^{(2)}) \in R$, *and for all* $\psi \in C_{0,T}(R^{r-n})$ *for which* $\psi_0 = x^{(2)}$, *one can find* $\varepsilon_0 > 0$ *such that the bound*

$$P^\varepsilon_x\{\rho_{0T}(X^{\varepsilon,(2)},\psi) < \delta\} \ge \exp\left\{-\frac{1}{\varepsilon}(S^*_{0T}(\psi)+h)\right\} . \tag{5.1}$$

will be valid for $\varepsilon \in [0, \varepsilon_0]$;

2. *For any* δ, T, $\theta > 0$, $s \ge 0$, $x \in R^r$,

$$P^\varepsilon_x\{\rho_{0T}(X^{\varepsilon,(2)},\Phi^*_{0T}(s)) > \delta\} \le 2\exp\left\{-\frac{(1-\theta)s-\theta T}{\varepsilon} + \lambda T\right\} , \tag{5.2}$$

whenever $0 \le \varepsilon \le \varepsilon_0$, *where* $\varepsilon_0 > 0$ *depends only on* m, M, θ, *and the constant* λ *depends only on* δ, K, m, M, r, θ.

The proof of inequalities (5.1) and (5.2) comes from bounds (4.3) and (4.13) respectively. □

Thus, the functional $S^*_{0T}(\psi)$ is defined as the lower semi-continuous regularization of the lower bound of the functional $S_{0T}(\phi,\psi)$ in $\phi \in C_{0,T}(R^n)$. It is possible to suggest a somewhat more convenient representation of this functional. For $y, z \in R^{r-n}$, we put

$$L(y,z) = \inf\{(a_{(22)}(x,y)(z-B^{(2)}(x,y)), (z-B^{(2)}(x,y)) : x \in R^n\} ,$$

and let, for every fixed $y \in R^{r-n}$, $L^c(y,\cdot)$ be the closed convex hull of the function $L(y,\cdot)$ on R^{r-n}, which means that $L^c(y,\cdot)$ is the maximal, convex, lower semi-continuous function not exceeding $L(y,\cdot)$. Denote

$$S^c_{0T}(\psi) = \begin{cases} \frac{1}{2}\int_0^T L^c(\psi_s, \dot{\psi}_s)\,ds, & \text{if } \psi \in C_{0,T}(R^{r-n}) \text{ is absolutely continuous,} \\ +\infty & \text{for the remaining } \psi \in C_{0,T}(R^{r-n}) . \end{cases} \tag{5.3}$$

LEMMA 5.1. *Suppose that conditions (4.2) are fulfilled. Then the functional* $S^c_{0T}(\psi)$ *is semi-continuous in* $C_{0,T}(R^{r-n})$. *For any* $s \in [0,\infty)$ *and any compact* $F \in R^{r-n}$, *the set* $\{\psi \in C_{0,T}(R^{r-n}) : \psi_0 \in F, S^c_{0T}(\psi) \le s\}$ *is compact in* $C_{0,T}(R^{r-n})$ *and* $S^*_{0T}(\psi) = S^c_{0T}(\psi)$.

The proof of this lemma needs some special information from convex analysis and we will not cite it here. This proof may be arranged on the basis of Theorems 3 and 5 in §5.1 of the monograph by Ioffe and Tihomirov [1]. It is also available in Gärtner [1].

Now we turn to some applications of Theorem 5.1 to differential equations with small parameter. Let us consider the random process on the plane R^2 corresponding to the differential operator L^ε which has the form

$$L^\varepsilon = L_0 + L_1 = \frac{1}{2} A(r,\phi) \frac{\partial^2}{\partial\phi^2} + B^1(r,\phi)\frac{\partial}{\partial\phi} + B^2(r,\phi)\frac{\partial}{\partial r} + \frac{\varepsilon}{2}\left(a^{11}(r,\phi)\frac{\partial^2}{\partial\phi^2} + 2a^{12}(r,\phi)\frac{\partial^2}{\partial r\,\partial\phi} + a^{22}(r,\phi)\frac{\partial^2}{\partial r^2}\right) \tag{5.4}$$

in the polar coordinates (r,ϕ). It is assumed that the coefficients of the operators L_0 and L_1 are bounded and sufficiently smooth, $A(r,\phi) > 0$, $\sum_{i,j=1}^{2} a^{ij}(r,\phi)\lambda_i\lambda_j > m(\lambda_1^2 + \lambda_2^2)$. The Markov process in R^2 corresponding to the operator L^ε will be denoted by $X^\varepsilon = (r_t^\varepsilon, \phi_t^\varepsilon; P^\varepsilon_{r,\phi})$; $X^0 = (r_t^0, \phi_t^0; P^0_{r,\phi})$ denoting the process corresponding to the operator L_0. With the process

X^ε, one can associate the functional $S^c_{0T}(\psi)$, $T > 0$, $\psi \in C_{0T}(R^1)$ defined by formula (5.3).

In this section we shall deal only with homogeneous equations of the form $L^\varepsilon u(r,\phi) = 0$. Hence, without loss of generality, one can assume that $a^{22}(r,\phi) \equiv 1$. Otherwise, one can reduce by $a^{22}(r,\phi)$.

If $a^{22}(r,\phi) \equiv 1$, then the kernel $L^c(r,z)$ of the functional $S^c_{0T}(\psi)$ may be written in the form

$$L^c(r,z) = \min_{0 \le \phi \le 2\pi} (z - B^2(r,\phi))^2 . \tag{5.5}$$

On the plane R^2, we shall consider a bounded domain D which is homeomorphic to a ring and has a smooth boundary $\partial D = \Gamma_1 \cup \Gamma_2$ (Fig. 10).

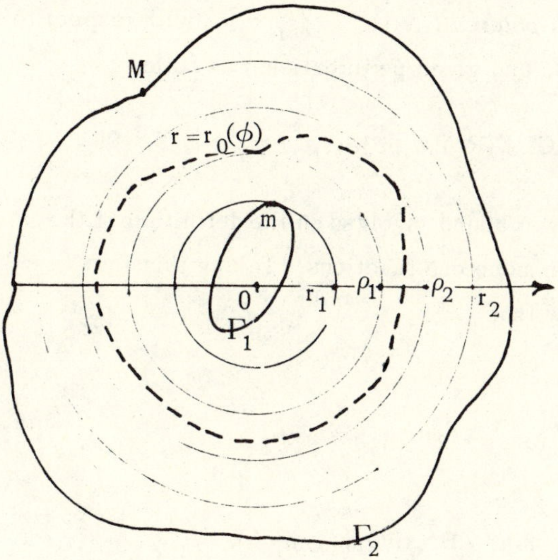

Fig. 10

We shall assume that the origin 0 lies inside the smaller contour Γ_1 and let the following conditions hold:

1. On the contour Γ_1 there is only one point M such that $\rho(0,M) = \min\{\rho(0,x) : x \in \Gamma_2\} = r_2$;
2. Suppose that, for every $\phi \in [0,2\pi]$, there is a number $r_0(\phi) \in (r_1, r_2)$ such that $B^2(r,\phi) > 0$ for $r < r_0(\phi)$ and $B^2(r,\phi) < 0$ for $r > r_0(\phi)$.

Put $\rho_1 = \min_{0 \le \phi \le 2\pi} r_0(\phi)$, $\rho_2 = \max_{0 \le \phi \le 2\pi} r_0(\phi)$; $0 < r_1 < \rho_1 \le \rho_2 < r_2$. In Fig. 10, the curve $r = r_0(\phi)$, $0 \le \phi \le 2\pi$, is represented by the dashed line.

For every fixed r, the function $L^c(r,z)$, defined by (5.5), has the form represented in Fig. 11. It is zero for $z \in [\underline{b}(r), \overline{b}(r)$; $L^c = (z - \overline{b}(r))^2$ for

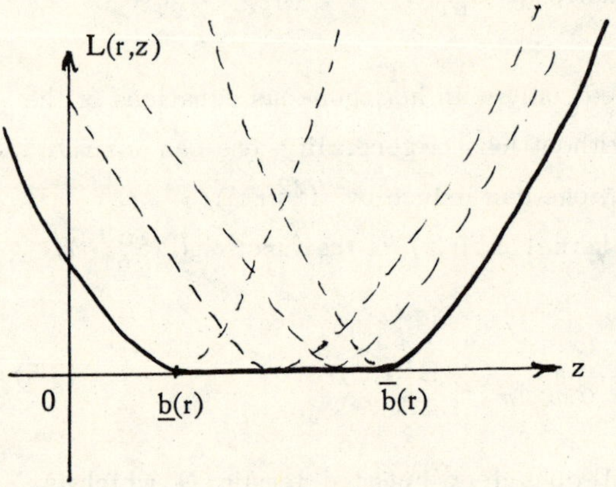

Fig. 11

$z > \overline{b}(r)$ and $L^c = (z - \underline{b}(r))^2$ for $z < \underline{b}(r)$. Here $\underline{b}(r) = \min_{0 \le \phi \le 2\pi} B^2(r,\phi)$, $\overline{b}(r) = \max_{0 \le \phi \le 2\pi} B^2(r,\phi)$. If condition 2 is fulfilled, then, for $r > \rho_2$, the interval $[\underline{b}(r), \overline{b}(r)]$ is to the left of zero; for $r < \rho_1$, this interval lies to the right of zero. And $\underline{b}(r) < 0 < \overline{b}(r)$ for $r \in (\rho_1, \rho_2)$.

We shall define the *quasi-potential* $V(r)$, $r \in [r_1, r_2]$, (with respect to the curve $r_0(\phi)$, $0 \le \phi \le 2\pi$, and given perturbations) as follows:

$$V(r) = \inf \{S^c_{0T}(\psi) : \psi \in C_{0,T}(R^1), \psi_0 \in [\rho_1, \rho_2], \psi_T = r, T \ge 0\}.$$

It is readily seen that the lower bound involved in the definition of the quasi-potential is attained on monotone functions. Taking this into account, one can easily show that

$$V(r) = \begin{cases} 0, & \rho_1 \le r \le \rho_2, \\ \displaystyle\int_{\rho_2}^{r} \min_{0 \le \phi \le 2\pi} |B^2(x,\phi)| dx, & \rho_2 \le r < r_2, \\ \displaystyle\int_{r}^{\rho_1} \min_{0 \le \phi \le 2\pi} B^2(x,\phi) dx, & r_1 \le r \le \rho_1. \end{cases}$$

THEOREM 5.2. *Consider the Dirichlet problem*

$$L^\varepsilon u^\varepsilon(r,\phi) = 0,\ (r,\phi) \in D,\ u^\varepsilon(r,\phi)|_{\partial D} = \psi(x)\ , \tag{5.6}$$

where $\psi(x)$ *is a continuous function on* $\partial D = \Gamma_1 \cup \Gamma_2$ *(see Fig. 10). Suppose that Conditions 1 and 2 are fulfilled,* $a_{22}(r,\phi) \equiv 1$ *and let* $r_1 < r < r_2$. *Then* $\lim_{\varepsilon \downarrow 0} u^\varepsilon(r,\phi) = \psi(M)$, *provided* $V(r_1) > V(r_2)$. *If* $V(r_1) < V(r_2)$, *then* $\lim_{\varepsilon \downarrow 0} u^\varepsilon(r,\phi) = \psi(m)$.

The proof of this theorem is carried out according to the same scheme as the proof of Theorem 1.2 was (it is given in the previous section). Denote by Π_a the circumference of radius a with center at the origin. We put $\Gamma = \Pi_{\rho_2+\mu} \cup \Pi_{\rho_1-\mu}$, $\gamma = \Pi_{\rho_2+\frac{1}{2}\mu} \cup \Pi_{\rho_1-\frac{1}{2}\mu}$, μ being a small positive number. Just as in §4.4, we consider the Markov times $\tau_0 = 0$, $\sigma_n = \inf\{t > \tau_n : (r_t^\varepsilon, \phi_t^\varepsilon) \in \Gamma\}$, $\tau_{n+1} = \inf\{t > \sigma_n : (r_t^\varepsilon, \phi_t^\varepsilon) \in \gamma \cup \Pi_{r_2+\kappa} \cup \Pi_{r_1}\}$, where $\kappa > 0$ will be chosen later. Let us consider the Markov chain $Z_n = (r_{\tau_n}^\varepsilon, \phi_{\tau_n}^\varepsilon)$ on the state space $\mathcal{E}_\kappa = \gamma \cup \Pi_{r_1} \cup \Pi_{r_2+\kappa}$. Condition 2 implies that, starting from any $z \in \mathcal{E}_\kappa$, for small ε, the chain Z_n will, in one step, get into γ with overwhelming probability. By Theorem 5.1, analogously to the way it was done when proving equality (4.28), one can establish that if $V(r_1) > V(r_2)$, then

$$\lim_{\varepsilon \downarrow 0} P_z^\varepsilon\{Z_1 \in \Pi_{r_2+\kappa} | Z_1 \in \Pi_{r_1} \cup \Pi_{r_2+\kappa}\} = 1 \tag{5.7}$$

uniformly in $z \in \gamma$, for κ small enough. Whence, noting that the diffusion in ϕ does not degenerate, it is not difficult to deduce that, for any $\delta > 0$

$$\lim_{\varepsilon \downarrow 0} P_z^\varepsilon\{(r_{\tau^\varepsilon}^\varepsilon, \phi_{\tau^\varepsilon}^\varepsilon) \in U_\delta(M)\} = 1\ , \tag{5.8}$$

where $\tau^\varepsilon = \inf\{t : (r_t^\varepsilon, \phi_t^\varepsilon) \notin D$. From (5.8), using the equality $u^\varepsilon(r,\phi) = E_{r,\phi}^\varepsilon \psi(r_{\tau^\varepsilon}^\varepsilon, \phi_{\tau^\varepsilon}^\varepsilon)$, we arrive at the assertion of Theorem 5.2. □

REMARK. Theorem 5.2 deals solely with the limit behavior of $u^{\varepsilon}(r,\phi)$ for $r \in (r_1, r_2)$. In the remaining part of the domain D, the generalized Levinson conditions are, in essence, fulfilled, and $\lim_{\varepsilon \downarrow 0} u^{\varepsilon}$ is defined by the reasoning cited in §4.2.

Theorem 5.2 describes the limit behavior of $u^{\varepsilon}(x)$, as $\varepsilon \downarrow 0$, in the case of the general position: when there is only one point M on Γ_2 for which $\rho(0,M) = \min(\rho(0,x) : x \in \Gamma_1)$; we exclude the case $V(r_1) = V(r_2)$. We shall cite without proof a result for the case when $\Gamma_2 = \Pi_{r_2}$.

THEOREM 5.3. *Suppose that the conditions of Theorem 5.1 are fulfilled, except for Condition 1. Assume that* $\Gamma_2 = \Pi_{r_2}$, $V(r_2) < V(r_1)$ *and let the function* $B^2(r,\phi)$ *attain on* Γ_2 *its maximum only when* $\phi = \phi^* : -B^2(r_2,\phi^*) < -B^2(r_2,\phi)$ *for* $\phi = \phi^*$. *Then* $\lim_{\varepsilon \downarrow 0} u^{\varepsilon}(r,\phi) = \psi(r_2,\phi^*)$ *for points* (r,ϕ) *such that* $r_1 < r < r_2$.

The proof of this theorem is available in the note of Gärtner and Freidlin [1]. □

Now we turn to a somewhat different situation, where the exit from the domain occurs at the expense of large deviations too. This example will illustrate the peculiarities arising in the case when the operator L_1 degenerates.

In the plane R^2, consider the differential operator

$$L^{\varepsilon} = L_0 + \varepsilon L_1 = b(x)\frac{\partial}{\partial x} + (x-y)\frac{\partial}{\partial y} + \frac{\varepsilon}{2}\frac{\partial^2}{\partial x^2}. \tag{5.9}$$

Let $(X_t^{\varepsilon;x,y}, Y_t^{\varepsilon;x,y}, P)$ be the corresponding Markov family.

Suppose that $b(x)$ is a bounded smooth function, positive for $x < 0$ and negative for $x > 0$. We put $G = \{(x,y) \in R^2 : |y| < 1\}$ (Fig. 12).

Fig. 12 represents the trajectories of the dynamical system

$$\begin{cases} \dot{X} = b(X) \\ \dot{Y} = X - Y \end{cases} \tag{5.10}$$

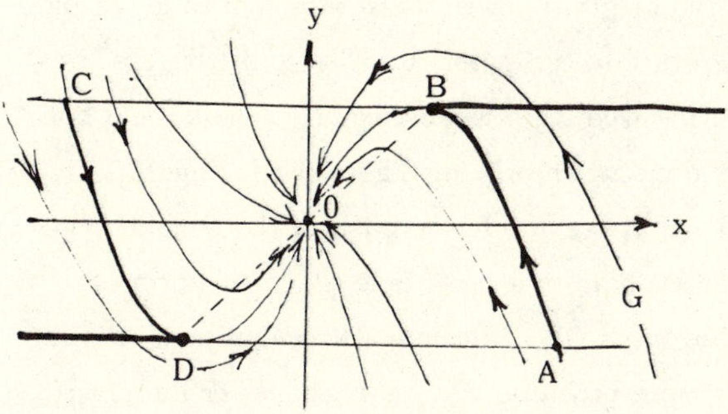

Fig. 12

corresponding to the operator L_0. It is clear that $B = (1,1)$ and $D = (-1,-1)$ are the points at which the trajectories of this system are tangent to the straight lines $y = +1$ and $y = -1$ respectively. Let C and A be the points where the trajectories, passing respectively through D and B, enter the domain G. The regular part Γ_1 of the boundary ∂G of the domain G for the operator $L^\varepsilon, \varepsilon > 0$, consists of the ray $(1, \infty)$ of the half-line $y = 1$ and of the ray $(-\infty, -1)$ of the half-line $y = -1$. The set $\partial G \setminus \Gamma_1$ is inaccessible for $\varepsilon > 0$.

In accordance with the results of Chapter III, the first boundary value problem for the operator L^ε in the domain G should be set as follows:

$$L^\varepsilon u^\varepsilon(x,y) = 0, \ (x,y) \in G, \ u^\varepsilon(x,y)\Big|_{\Gamma_1} = \psi(x,y) , \tag{5.11}$$

where $\psi(x,y)$ is a bounded continuous function on Γ_1. For $\varepsilon > 0$, problem (5.11) has a unique generalized solution.

It is not difficult to understand that, with probability approaching 1 for small ε, the trajectories, starting from the points of the quadrangle ABCD, leave the domain G for the first time either near the point B or near the point D. This comes from the fact that the exit of the trajectories of the Markov family may happen only through those points of the boundary through which the trajectories of system (5.10) leave the domain. It is these points B and D which are accessible most easily at the expense of the diffusion along the variable x.

To understand which of these two possible ways of leaving will take place, one should make use of the action functional. Just as in the problems considered before in this section and in §4.4, a trajectory of the

Markov family starting from the origin 0, will make attempts to leave the domain and again and again return to the point 0. The exit from the domain in one attempt (to define what this means exactly, one should consider a Markov chain as was done when proving Theorems 1.2 and 5.2) is extremely unlikely for small ε. However, in the case of the general position, under the condition that such an exit does take place, it turns out that one of these two possible ways (the exit near B or that near D) is far more probable. It is this more probable way that, sooner or later, will be realized with probability approaching 1 for small ε. Estimating the probabilities of these two ways is accomplished with Theorem 4.3. It is easily checked that

$$Y_t^{\varepsilon;0,0} = e^{-t} \int_0^t e^s X_s^{\varepsilon;0,0} ds . \tag{5.12}$$

By Theorem 4.3, this implies that, for small ε, the probability of hitting in one attempt the point B with the coordinates $(1,1)$ or the point D with the coordinates $(-1,-1)$ is estimated by the quantities $\exp\{-\varepsilon^{-1}(V_+ \pm h)\}$ and $\exp\{-\varepsilon^{-1}(V_- \pm h)\}$ respectively, where h is a small positive number and

$$V_{\pm} = \inf\left\{ \int_0^T |\dot\phi_s - b(\phi_s)|^2 ds : \phi \in C_{0,T}(R^r), \phi_0 = 0, \phi_T = \pm 1, \right.$$
$$\left. e^{-T} \int_0^T e^s \phi_s ds = \pm 1, T > 0 \right\} .$$

This reasoning leads us to the following result.

THEOREM 5.4. *Suppose that* L^ε *is the operator defined by formula (5.9),* $u^\varepsilon(x,y)$ *is the solution of problem (5.11), and let* (x,y) *be a point belonging to the curvilinear quadrangle* ABCD. *Then* $\lim_{\varepsilon \downarrow 0} u^\varepsilon(x,y)$ *is equal to* $\psi(1,1)$, *if* $V_+ < V_-$, *and is equal to* $\psi(-1,-1)$ *if* $V_- < V_+$. *If the point*

(x,y) *does not belong to* ABCD, *then the trajectory of system (5.10), starting from* (x,y), *leaves the domain* G *through some point* $F = F(x,y) \in \Gamma_1$ *and* $\lim_{\varepsilon \downarrow 0} u^\varepsilon(x,y) = \psi(F(x,y))$.

The proof of this theorem is arranged according to the above scheme, and we leave it to the reader. □

In the problems dealt with in this and in the preceding sections, the time required for a trajectory to leave the domain, grew like $\exp\{\varepsilon^{-1} \cdot \text{const.}\}$ as $\varepsilon \downarrow 0$. This follows, for example, from the fact that the probability of reaching the boundary "in one attempt" is of the order (logarithmic) of $\exp\{-\varepsilon^{-1} \cdot \text{const.}\}$.

Concluding this section we will discuss the situation when the exit from the domain in fact also occurs at the expense of large deviations, but the time required for leaving grows somewhat more slowly. Let $D = (-1,1) \in R^1$, $L_0 = x^2(1-x)^2 \frac{d^2}{dx^2} + b(x) \frac{d}{dx}$, where $b(x) > 0$ for $x < 0$ and $b(x) < 0$ for $x > 0$. Consider the problem

$$L^\varepsilon u^\varepsilon(x) = \frac{\varepsilon}{2} \frac{d^2 u^\varepsilon}{dx^2} + L_0 u^\varepsilon = 0, x \in (-1,1), u(\pm 1) = \psi_\pm .$$

This problem may be solved explicitly. It appears that, for the corresponding process with probability approaching 1 for small ε, the first exit from $(-1,1)$ occurs through the point 1, if $|b(1)| < b(-1)$, and it occurs through -1, if $b(-1) < |b(1)|$. Consequently, $\lim_{\varepsilon \downarrow 0} u^\varepsilon(x)$ is equal to ψ_+ or ψ_-. The time required for leaving the interval grows like $\exp\{\varepsilon^{1/2} \cdot \text{const.}\}$. Similar multi-dimensional problems were considered by Sarafian [3].

In the next section we shall also discuss some problems connected with large deviations.

§4.6. *Small parameter in problems with mixed boundary conditions*

This section will be concerned with some problems connected with the averaging principle in which the exit from the domain occurs at the expense of large deviations from an averaged motion. Similar effects also arise in a number of cases when considering Dirichlet's problem in a ring (see Theorem 3.3), but here, just for a change, we will deal with the problem with mixed boundary value conditions. For brevity, a fairly special example will be discussed. Here we follow the works of Sarafian, Safarian and Freidlin [1], Safarian [1].

So, consider the differential operator

$$L^{\varepsilon} = \frac{1}{2}\frac{\partial^2}{\partial x^2} + \varepsilon b(x,y)\frac{\partial}{\partial y}, \quad (x,y) \in R^2 .$$

We assume $b(x,y)$ to be a function, bounded along with its second-order derivatives. We put $D = \{(x,y) \in R^2 : |x| < 1,\ |y| < 1\}$. Denote $\Gamma_+ = \{x \in [-1,1] : b(x,1) > 0\}$, $\Gamma_- = \{x \in [-1,1] : b(x,-1) < 0\}$, and consider the boundary value problem

$$L^{\varepsilon}u^{\varepsilon}(x,y) = 0, \quad (x,y) \in D, \quad \frac{\partial u^{\varepsilon}}{\partial x}(x,y)\Big|_{|x|=1} = 0,$$
$$u^{\varepsilon}(x,1)\big|_{x\in\Gamma_+} = \psi(x,1), \quad u^{\varepsilon}(x,-1)\big|_{x\in\Gamma_-} = \psi(x,-1), \tag{6.1}$$

where $\psi(x,1)$ and $\psi(x,-1)$ are continuous functions on $[-1,1]$.

A solution of problem (6.1), at least a generalized one, exists. To construct and study this generalized solution, it will be convenient for us to introduce into consideration the Markov process $(X_t^{\varepsilon}, Y_t^{\varepsilon}; P_{x,y}^{\varepsilon})$ in the strip $\{(x,y) : |x| \le 1, -\infty < y < \infty\}$ which is governed by the operator $\varepsilon^{-1}L^{\varepsilon}$ inside this strip and is subject to the reflection along the normal on the boundary. The Markov family corresponding to this process will be denoted by $(X_t^{\varepsilon;x,y}, Y^{\varepsilon;x,y}; P)$. Such a Markov family and process do exist. In order to construct, say, the family $(X_t^{\varepsilon;x,y}, Y_t^{\varepsilon;x,y}; P)$, we shall designate by (ξ_t^x, P) the Markov family in $[-1,1]$ governed inside $(-1,1)$ by the operator $\frac{1}{2}\frac{d^2}{dx^2}$ with reflection at the endpoints of the interval $[-1,1]$ (see §1.6). We set

$$X_t^{\varepsilon;x,y} = \xi_{t/\varepsilon}^x ,$$
$$\dot{Y}_t^{\varepsilon;x,y} = b(\xi_{t/\varepsilon}^x, Y_t^{\varepsilon;x,y}),\ Y_0^{\varepsilon;x,y} = y . \tag{6.2}$$

Then, it is $X_t^{\varepsilon;x,y}$ and the solution $Y_t^{\varepsilon;x,y}$ of equation (6.2) which, together with the measure P, form the needed Markov family.

Denote by A^ε the infinitesimal operator of the family $(X_t^{\varepsilon;x,y}, Y_t^{\varepsilon;x,y}; P)$.

By a generalized solution of problem (6.1), we mean a function $u^\varepsilon(x,y)$ which satisfies the conditions

$$A^\varepsilon u^\varepsilon = 0,\ \lim_{y\to 1} u^\varepsilon(x,y)\big|_{x\epsilon\Gamma_+} = \psi(x,1),\ \lim_{y\to -1} u^\varepsilon(x,y)\big|_{x\epsilon\Gamma_-} = \psi(x,-1) .$$

Analogous to the way it was done in Chapter III, one can show that such a notion of the generalized solution is correct.

We say that Condition 1 is fulfilled if, for every $y_0 \epsilon [-1,1]$, one can find $x_0 = x_0(y_0) \epsilon [-1,1]$ such that either $b(x_0,y) > 0$ for $y \geq y_0$, or $b(x_0,y) < 0$ for $y \leq y_0$.

Under Condition 1, with probability 1 the trajectories of the process $(X_t^\varepsilon, Y_t^\varepsilon; P_{x,y}^\varepsilon)$, $\varepsilon > 0$, starting from any point $(x,y) \epsilon D$, leave D. As was done when proving Theorems 3.5.1 and 3.5.2, one can prove the following result.

THEOREM 6.1. *The generalized solution of problem (6.1) exists. If Condition 1 holds true, then the generalized solution is unique in the class of bounded functions and may be written down in the form* $u^\varepsilon(x,y) = E_{x,y}\psi(X_{\tau^\varepsilon}^\varepsilon, Y_{\tau^\varepsilon}^\varepsilon)$, *where* $\tau^\varepsilon = \inf\{t : Y_t^\varepsilon \notin (-1,1)\}$.

In this section we shall study the behavior of the generalized solution of problem (6.1) as $\varepsilon \downarrow 0$.

We observe that the uniform distribution on $[-1,1]$ is invariant for the Markov family (ξ_t^x, P). The transition function $P(t,x,\Gamma) = P\{\xi_t^x \in \Gamma\}$ converges to $\frac{1}{2}\int_\Gamma dx$, $\Gamma \subseteq [-1,1]$ as $t \to \infty$. For any bounded measurable function $f(x)$

$$P\left\{\lim_{t\to\infty} \frac{1}{t} \int_0^t f(\xi_s^x)\,ds = \frac{1}{2} \int_{-1}^1 f(x)\,dx\right\} = 1, \quad x \in [-1,1]. \tag{6.3}$$

Put

$$\overline{b}(y) = \frac{1}{2} \int_{-1}^1 b(x,y)\,dx, \quad -\infty < y < \infty,$$

and consider the differential equation

$$\dot{\overline{Y}}_t^y = \overline{b}(\overline{Y}_t^y), \quad \overline{Y}_0^y = y. \tag{6.4}$$

THEOREM 6.2. *For arbitrary* $x \in [-1,1]$ *and* $y \in (-\infty,\infty)$, *for any* $\delta, T > 0$,

$$\lim_{\varepsilon \downarrow 0} P\{\rho_{0T}(Y_\cdot^{\varepsilon;x,y}, \overline{Y}^y) > \delta\} = 0.$$

Proof. The definition of $Y_t^{\varepsilon;x,y}$ and Y_t^y implies that

$$\begin{aligned} Y_t^{\varepsilon;x,y} - \overline{Y}_t^y &= \int_0^t [b(\xi_{s/\varepsilon}^x, Y_s^{\varepsilon;x,y}) - b(\xi_{s/\varepsilon}^x, \overline{Y}_s^y)]\,ds + \\ &+ \int_0^t [b(\xi_{s/\varepsilon}^x, \overline{Y}_s^y) - \overline{b}(\overline{Y}_s^y)]\,ds. \end{aligned} \tag{6.5}$$

We put $m(t) = \sup_{0\le s\le t} |Y_s^{\varepsilon;x,y} - \overline{Y}_s^y|$. From (6.5) it follows:

$$m(t) \le K \int_0^t m(s)\,ds + \sup_{0\le s\le t} \left| \int_0^s [b(\xi_{s_1}^x, \overline{Y}_{s_1}^y) - \overline{b}(\overline{Y}_{s_1}^y)\,ds_1 \right|,$$

where K is the Lipschitz constant of the function $b(x,y)$. From this it results that

$$m(T) \le \exp\{KT\} \cdot \sup_{0 \le t \le T} \left| \int_0^t [b(\xi^x_{s/\varepsilon}, \bar{Y}^y_s) - \bar{b}(\bar{Y}^y_s)]ds \right| \qquad (6.6)$$

Denote by f^ε_s the expression under the integral sign in (6.6). For any integer n we have

$$\int_0^t f^\varepsilon_s\, ds = \sum_{k=0}^{n-1} \int_{\frac{kt}{n}}^{\frac{t(k+1)}{n}} f^\varepsilon_s ds = \sum_{k=0}^{n-1} \int_{\frac{kt}{n}}^{\frac{t(k+1)}{n}} [b(\xi^x_{s/\varepsilon}, \bar{Y}^y_{\frac{kt}{n}}) - \bar{b}(\bar{Y}^y_{\frac{kt}{n}})]ds +$$

$$+ \sum_{k=0}^{n-1} \int_{\frac{kt}{n}}^{\frac{(k+1)}{n}t} [b(\xi^x_{s/\varepsilon}, \bar{Y}^y_s) - b(\xi^x_{s/\varepsilon}, \bar{Y}^y_{\frac{kt}{n}})]ds + \sum_{k=0}^{n-1} \int_{\frac{kt}{n}}^{\frac{(k+1)}{n}t} [\bar{b}(\bar{Y}^y_{\frac{kt}{n}}) - \bar{b}(\bar{Y}^y_s)]ds =$$

$$= \sum_{k=0}^{n-1} \int_{\frac{kt}{n}}^{\frac{(k+1)}{n}t} [b(\xi^x_{s/\varepsilon}, \bar{Y}^y_{\frac{kt}{n}}) - \bar{b}(\bar{Y}^y_{\frac{kt}{n}})]ds + \rho_{n,t}, \qquad (6.7)$$

where $|\rho_{n,t}| < \frac{ct^2}{n}$, c is a constant depending on the Lipschitz constant and on the maximum modulus of the function $b(x,y)$. By (6.3), for any $a \in (-\infty, \infty)$

$$\int_{\frac{kt}{n}}^{(k+1)\frac{t}{n}} [b(\xi^x_{s/\varepsilon}, a) - \bar{b}(a)]ds = \varepsilon \int_0^{\frac{t}{n\varepsilon}} [b(\xi^x_{\frac{kt}{n}+s}, a) - \bar{b}(a)]ds \to 0$$

as $\varepsilon \downarrow 0$ P-a.s. Hence, for a fixed n, as $\varepsilon \downarrow 0$, the sum on the right-hand side of equality (6.7) tends to zero. From this, noting that the functions $b(x,y)$ and $b(y)$ are bounded, we conclude that the right-hand side of (6.6) tends to zero, which implies the claim of the theorem. □

Put $T(y) = \inf\{t : \bar{Y}_t^y \notin (-1,1)\}$, $y \in [-1,1]$.

THEOREM 6.3. *Suppose that a point* $y \in (-1,1)$ *is such that* $T(y) < \infty$ *and let, for the sake of definiteness,* $\bar{Y}^y_{T(y)} = 1$. *Moreover assume that* $b(x,1) > 0$ *for* $x \in [-1,1]$ *and let the function* $\psi(x,1)$ *be twice continuously differentiable. Then*

$$\lim_{\varepsilon \downarrow 0} u^{\varepsilon}(x,y) = \int_{-1}^{1} \psi(x,1)\, b(x,1)\, dx \left(\int_{-1}^{1} b(x,1)\, dx \right)^{-1} = \bar{\psi}\,.$$

Proof. Suppose that $h_0 > 0$ is chosen so that $b(x,y) > 0$ for $y \in [1-h_0, 1]$, $-1 \le x \le 1$. One can easily see that it is sufficient to prove the claim of the theorem for $y \in [1-h, 1]$, $h \in (0, h_0)$. We shall introduce the new variable $t = \frac{1-y}{\varepsilon}$. The function $v^{\varepsilon}(t,x) = u^{\varepsilon}(x, 1-\varepsilon t)$ is the solution of the following problem

$$\frac{\partial v^{\varepsilon}}{\partial t} = \frac{1}{2}\, a^{\varepsilon}(t,x) \frac{\partial^2 v^{\varepsilon}}{\partial x^2}, \quad \frac{\partial v^{\varepsilon}}{\partial x}\bigg|_{|x|=1} = 0, \quad v^{\varepsilon}(0,x) = \psi(x,1)\,, \tag{6.8}$$

where $a^{\varepsilon}(t,x) = b^{-1}(x, 1-\varepsilon t)$.

Let δ be a small positive number. Choose $h \in (0, h_0)$ so that

$$\sup_{-1 \le x \le 1,\, t \in [0, h\varepsilon^{-1}]} |a^{\varepsilon}(t,x) - a^0(x)| < \delta\,,$$

where $a^0(x) = b^{-1}(x,1)$. Along with problem (6.8), we consider the boundary value problem

$$\frac{\partial v^0(t,x)}{\partial t} = \frac{1}{2}\, a^0(x) \frac{\partial^2 v^0}{\partial x^2}, \quad \frac{\partial v^0}{\partial x}\bigg|_{|x|=1} = 0\,, \quad v^0(0,x) = \psi(x,1)\,. \tag{6.9}$$

LEMMA 6.1. *A constant* c *exists such that, for all* $\varepsilon \in (0,1)$,

$$\sup_{|x| \le 1,\, 0 \le t \le \frac{h}{\varepsilon}} |v^{\varepsilon}(t,x) - v^0(t,x)| < c\,\delta\,.$$

Proof. Let us differentiate equation (6.9) in x. For $v'(t,x) = \frac{\partial v^0(t,x)}{\partial x}$, we arrive at the problem

$$\frac{\partial v'}{\partial t} = \frac{\partial}{\partial x}\left(a^0(x)\frac{\partial v'}{\partial x}\right), \quad v'(t,x)\Big|_{|x|=1} = 0, \quad v'(0,x) = \psi' = \frac{\partial \psi(x,1)}{\partial x}. \tag{6.10}$$

The Sturm-Liouville operator $L = \frac{\partial}{\partial x}\left(a^0(x)\frac{\partial}{\partial x}\right)$ in $[-1,1]$ with zero boundary conditions has a complete set of orthonormalized eigenfunctions $e_1(x), e_2(x), \cdots; \lambda_1, \lambda_2, \cdots$ being the corresponding eigenvalues. The solution of problem (6.10) may be written in the form

$$v'(t,x) = \sum_{k=1}^{\infty} c_k e_k(x) e^{-\lambda_k t}, \quad c_k = \int_{-1}^{1} \psi'(x,1) e_k(x)\, dx .$$

The fact that $\psi'(x,1)$ is continuously differentiable, implies that the series for $v'(t,x)$ may be differentiated termwise in x

$$v''(t,x) = \sum_{k=1}^{\infty} c_k e^{-\lambda_k t} e_k'(x) .$$

This results in

$$\max_{|x|\le 1} \left| \frac{\partial^2 v^0(t,x)}{\partial x^2} \right| \le e^{-\lambda_1 t} \cdot \text{const.} \tag{6.11}$$

Next, for the difference $w^\varepsilon(t,x) = v^\varepsilon(t,x) - v^0(t,x)$, (6.8) and (6.9) yield the equation

$$\frac{\partial w^\varepsilon}{\partial t} = \frac{1}{2} a^\varepsilon(t,x) \frac{\partial^2 w^\varepsilon}{\partial x^2} + \frac{1}{2}[a^\varepsilon(t,x) - a^0(t,x)] \frac{\partial^2 v^0}{\partial x^2},$$

$$w^\varepsilon(0,x) = 0, \quad \frac{\partial w^\varepsilon}{\partial x}\Big|_{|x|=1} = 0 . \tag{6.12}$$

On account of (6.11) and of the fact that $|a^\varepsilon(t,x) - a^0(t,x)| < \delta$, we obtain from (6.12):

$$\sup_{|x|\le 1,\, t\in[0,\varepsilon^{-1}h]} |w^{\varepsilon}(t,x)| \le \text{const.}\ \delta \cdot \int_0^{\infty} e^{-\lambda_1 s}\, ds \le c\,\delta\ .$$

Now, to complete the proof of the theorem, we observe that $\lim_{t\to\infty} v^0(t,x) = \overline{\psi}$. This comes, for instance, from the fact that $(a^0(x))^{-1}\cdot$ $(\int_{-1}^{1} (a^0(x))^{-1}\,dx)^{-1}$ is the density function of the limit distribution for the process with reflection on $[-1,1]$ corresponding to the operator $a^0(x)\frac{\partial^2}{\partial x^2}$. This observation together with Lemma 6.1 leads to the assertion of Theorem 6.3: $\lim_{\varepsilon\downarrow 0} u^{\varepsilon}(x, 1-h) = \lim_{\varepsilon\downarrow 0} v^{\varepsilon}(h\varepsilon^{-1},x) = \overline{\psi}$. □

Therefore, if an averaged trajectory leaves $(-1,1)$ in a finite time, then we have a counterpart of Theorem 3.3. The field $\overline{b}(y)$ might vanish identically. Then, under certain extra assumptions, the exit from the domain happens at the expense of normal deviations. Here we have the situation similar to that we have faced in Theorem 3.4. Finally, if the field $\overline{b}(y)$ has a stable equilibrium point in $(-1,1)$ (one or a few), then the limit behavior of a solution of problem (6.1) as $\varepsilon\downarrow 0$ is defined by large deviations.

For these large deviations to be described, we shall consider the eigenvalue problem:

$$\begin{gathered} \frac{1}{2}\frac{d^2u(x)}{dx^2} + \beta\, b(x,y)\, u(x) = \mu\cdot u(x),\quad x\in(-1,1)\ , \\ \frac{du}{dx}(-1) = \frac{dv}{dx}(1) = 0\ . \end{gathered} \tag{6.13}$$

Here β is a number parameter. Let $\mu = \mu(y,\beta)$ be the first eigenvalue of problem (6.13). One can show that such an eigenvalue has multiplicity one and moreover is a differentiable function, convex in the parameter β (see, e.g. Wentzell and Freidlin [2]). We define the function $L(y,\beta)$ as the Legendre transform of the function $\mu(y,\beta)$ with respect to the parameter β:

$$L(y,\lambda) = \sup_{\beta} [\lambda\beta - \mu(y,\beta)] .$$

The function $L(y,\lambda)$ takes non-negative values and $+\infty$. It is convex in λ. For absolutely continuous $\phi \in C_{0,T}(R^1)$, we put

$$S_{0T}(\phi) = \int_0^T L(\phi_s, \dot{\phi}_s)\,ds \,;$$

for the remaining $\phi \in C_{0,T}(R^1)$, we put $S_{0T}(\phi) = +\infty$.

THEOREM 6.4. *The functional* $\varepsilon^{-1}S_{0T}(\phi)$ *is the action functional for the family of processes* $Y_t^{\varepsilon;x,y}$ *in the space* $C_{0,T}(R^1)$, $T > 0$, *as* $\varepsilon \downarrow 0$, *i.e. for any* $s, \delta, h > 0$, *for arbitrary* $\phi \in C_{0,T}(R^1)$, $\phi_0 = y$, *and for arbitrary* $x \in [-1,1]$, *one can find* $\varepsilon_0 > 0$ *such that for* $\varepsilon \in (0, \varepsilon_0]$

$$P\{\rho_{0T}(Y^{\varepsilon;x,y}, \phi) < \delta\} \geq \exp\{-\varepsilon^{-1}(S_{0T}(\phi) + h)\} ,$$

$$P\{\rho_{0T}(Y^{\varepsilon;x,y}, \Phi_{0T}(s)) > \delta\} \leq \exp\{-\varepsilon^{-1}(s-h)\} ,$$

where $\Phi_{0T}(s) = \{\phi \in C_{0,T}(R^1) : \phi_0 = y,\ S_{0T}(\phi) \leq s\}$.

The functional $S_{0T}(\phi)$ *is lower semi-continuous, the set* $\Phi_{0T}(s)$ *being compact in the space* $C_{0T}(R^1)$.

The *proof* of this theorem may be found in Freidlin [14], Wentzell and Freidlin [2]. □

Let the function $\overline{b}(y)$ on $[-1,1]$ have its only zero at the point $q \in (-1,1)$, $\overline{b}(y)$ being positive for $y < q$ and negative for $y > q$. This means that an averaged trajectory $\overline{Y}_t^y$, $y \in [-1,1]$, does not leave the interval $[-1,1]$ and is attracted to q as $t \to \infty$. On every finite time interval, for small ε, the trajectories of the process $Y_t^{\varepsilon;x,y}$ will be close to $\bar{Y}_t^y$. However, if Condition 1 is fulfilled, then for any $\varepsilon > 0$, the trajectories of $Y_t^{\varepsilon;x,y}$ will, sooner or later, leave $[-1,1]$. Just as in the case of Theorem 1.2, this leaving happens at the expense of large deviations from the limit dynamical system.

Put

$$V(y) = \inf\{S_{0T}(\phi) : \phi \in C_{0,T}(R^1), \phi_0 = q, \phi_T = y, T \geq 0\}.$$

THEOREM 6.5. *Suppose that Condition 1 holds and there is a* $q \in (-1,1)$ *such that* $\bar{b}(q) = 0$, $\bar{b}(y) > 0$ *for* $y < q$, *and* $\bar{b}(y) < 0$ *for* $y > q$. *Let* $V(1) < V(-1)$ *and* $\psi(x,1) = \psi_1 = \text{const}$. *Then the solution* $u^\varepsilon(x,y)$ *of problem (6.1) tends to* ψ_1 *as* $\varepsilon \downarrow 0$.

Proof of this theorem is arranged according to the same scheme as that of Theorem 1.2 in §4.4. Instead of bounds (4.24) and (4.25), one should make use of the bounds given by Theorem 6.4. □

At last, we will formulate a result concerning the case when there are many stable equilibrium points on $(-1,1)$.

Let $y_1, y_2, \cdots, y_\ell \in (-1,1)$ be the stable zeroes of the function $\bar{b}(y)$ ranged according to magnitude. By a stable zero we mean a point where $b(y)$ changes its sign from plus to minus when y grows (Fig. 13).

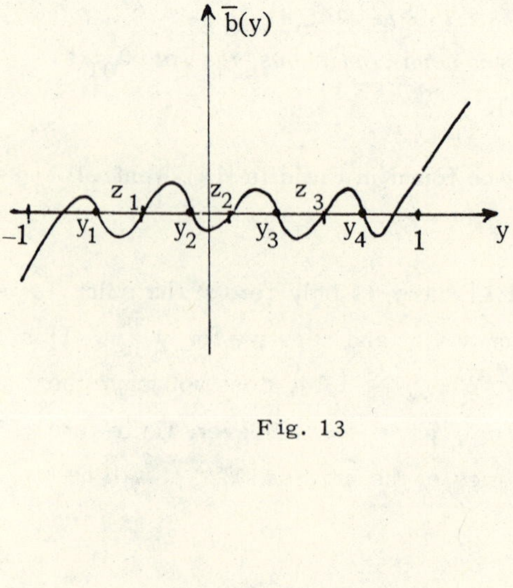

Fig. 13

Between any neighboring y_k and y_{k+1} there is an unstable zero z_k. At the other points, $\bar{b}(y)$ does not vanish. It will be suitable for us to put $y_0 = -1$, $y_{\ell+1} = 1$. We put $V_{ij} = \inf\{S_{0T}(\phi) : \phi \in C_{0,T}(R^1)$, $\phi_0 = y_i$, $\phi_T = y_j$, $T \geq 0\}$ for $i \in \{1,2,\cdots,\ell\}$, $j \in \{0,1,\cdots,\ell+1\}$.

THEOREM 6.6. *Suppose that the function* $b(x,y)$ *satisfies Conditions 1 and* $\overline{b}(y)$ *satisfies the above conditions (see Fig. 14). Let* $\psi(x,1) = \psi_1 = \text{const.}$, $\psi(x,-1) = \psi_{-1} = \text{const.}$ *Suppose that*

$$\min_{k \in \{0,1,\cdots,\ell\}} \left(\sum_{m=k+1}^{\ell} V_{m,m+1} + \sum_{m=1}^{k} V_{m,m-1} \right)$$

is attained for the only value $k = k^*$. *Then, if* z_{k^*} *is the unstable equilibrium point separating the point* y_{k^*} *and* y_{k^*+1}, *we have*

$$\lim_{\varepsilon \downarrow 0} u^{\varepsilon}(x,y) = \begin{cases} \psi_{-1}, & y < z_{k^*}, \\ \psi_1, & y > z_{k^*}, \end{cases}$$

where $u^{\varepsilon}(x,y)$ *is the solution of problem (6.1).*

As was explained in the remainder of §4.4, in the case of several stable equilibrium points the trajectory of the process governed by the operator L^{ε} is approximated by a certain Markov chain. In our problem, the points $\{-1, y_1, y_2, \cdots, y_\ell, 1\} \in [-1,1]$ are the states of this chain. The state " ∂ " will be divided into two states -1 and 1. The peculiarity of the one-dimensional case is that transitions in one step in this chain are possible solely between neighboring states. Hence, only the numbers V_{ij}, $|i-j| = 1$, are involved in the formulation of the result. For the approximating chain, the transition probability from y_i to y_j, $|i-j| = 1$, in one step is of order $\exp\{-\varepsilon^{-1} V_{ij}\}$, as $\varepsilon \downarrow 0$. From this one can deduce that, with probability tending to 1 as $\varepsilon \downarrow 0$, starting from the points y_k, $k < k^*$, the trajectory of the chain will reach the point -1 before 1; and conversely, starting from y_k, $k > k^*$, it will reach the point 1 before -1. This implies the claim of Theorem 6.6. The detailed proof of this theorem is available in the article of Safarian [1].

Chapter V

QUASI-LINEAR PARABOLIC EQUATIONS WITH NON-NEGATIVE CHARACTERISTIC FORM

§5.1 *Generalized solution of Cauchy's problem. Local solvability*

Let us consider the Cauchy problem

$$\frac{\partial u(t,x)}{\partial t} = \frac{1}{2}\sum_{i,j=1}^{r} a^{ij}(x,u)\frac{\partial^2 u}{\partial x^i \partial x^j} + \sum_{i=1}^{r} b^i(x,u) + F(x,u) = \tag{1.1}$$

$$= L(u) + F,\ t > 0,\ x \in R^r;\quad u(0,x) = g(x)\ .$$

The form $\sum_{i,j=1}^{r} a^{ij}(x,u)\lambda_i\lambda_j$ is assumed to be non-negative definite. The following assumptions will be made on the coefficients and initial function.

1. The matrix $(a^{ij}(x,u)) = a(x,u)$ may be represented in the form $a(x,u) = \sigma(x,u)\,\sigma^*(x,u)$, the elements $\sigma^i_j(x,u)$ of the matrix $\sigma(x,u)$ being bounded and Lipschitz continuous in x and u. As was shown in §3.2, such a matrix exists, at least, if the functions $a^{ij}(x,u)$ are bounded and possess bounded second-order derivatives.

2. The functions $b^i(x,u)$, $g(x)$ are bounded and have bounded first-order derivatives in x and u.

3. We put $F_0(x) = F(x,0)$, $c(x,u) = u^{-1}[F(x,u) - F_0(x)]$. The functions $F_0(x)$ and $c(x,u)$ are assumed bounded and having bounded derivatives. We will denote by K_1, K_2 constants such that

$$\sum_{i,j=1}^{r} |\sigma_j^i(x,u) - \sigma_j^i(y,v)|^2 + \sum_{i=1}^{r} |b^i(x,u) - b^i(y,v)|^2 +$$

$$+ |F_0(x) - F_0(y)|^2 + |c(x,u) - c(y,v)|^2 \le$$

$$\le K_1 |x-y|^2 + K_2 |u-v|^2 ,$$

and let K_g be the Lipschitz constant of the initial function; $\|f(x,u)\| = \sup\{|f(x,u)| : x \in R^r, -\infty < u < \infty\}$.

In particular, if $a^{ij}(x,u) \equiv 0$, $F(x,u) \equiv 0$, then problem (1.1) becomes the Cauchy problem for the first-order equation

$$\frac{\partial u}{\partial t} = \sum_{i=1}^{r} b^i(x,u) \frac{\partial u}{\partial x^i} \; ; \quad u(0,x) = g(x) . \tag{1.2}$$

As is known (see Gelfand [1]), problem (1.2) may have no smooth or even continuous solution. This is not due to insufficient smoothness of the coefficients or the initial function. It is not difficult to provide examples of equations with infinitely differentiable coefficients and initial function whose solutions have discontinuities outside some neighborhood of the initial plane $t = 0$. Hence, without making supplementary assumptions, one cannot expect the existence of a continuous (even generalized) solution of problem (1.1) for all $t > 0$.

In this section, we shall introduce the notion of the generalized solution of problem (1.1) and provide a local existence theorem. In later sections we shall cite conditions ensuring the solvability "in the large" of problem (1.1), that is, for all $t > 0$, in the class of continuous functions. We shall consider two types of such conditions. First, one can attain the solvability in the large in the class of continuous functions at the expense of sufficiently rapid killing, i.e. at the expense of having $c(x,u)$ be large in magnitude and negative. Secondly, one can attain this at the expense of the non-linear terms being subordinate to the linear ones.

To introduce the notion of a generalized solution of problem (4.1), we assume at first that the classical solution $u(t,x)$ of problem (1.1) exists. If this function is substituted into the coefficients, then to the operator

$$\mathfrak{L} = -\frac{\partial}{\partial t} + L(u) = -\frac{\partial}{\partial t} + \frac{1}{2} \sum_{i,j=1}^{r} a^{ij}(x,u) \frac{\partial^2}{\partial x^i \partial x^j} + \sum_{i=1}^{r} b^i(x,u) \frac{\partial}{\partial x^i},$$

there corresponds a Markov family and corresponding process (homogeneous in time) in the state space $(-\infty, T] \times R^r$, $T > 0$. These are defined by the stochastic differential equations

$$X_t^{t',x} - x = \int_0^t \sigma(X_s^{t',x}, u(t'-s, X_s^{t',x}))\, dW_s + \int_0^t b(X_s^{t',x}, u(t'-s, X_s^{t',s}))\, ds, \tag{1.3}$$

$$t_t^{t'} = t' - t.$$

For negative t, we define $u(t,x)$ by putting $u(t,x) = g(x)$ for $t \leq 0$. As follows from §2.1, for its part, the solution of problem (1.1) may be written down in the form

$$u(t,x) = E_{t,x}\, g(X_t) \exp\left\{ \int_0^t c(X_s, u(t_s, X_s))\, ds \right\} + \tag{1.4}$$
$$+ E_{t,x} \int_0^t F_0(X_s) \exp\left\{ \int_0^s c(X_{s'}, u(t_{s'}, X_{s'}))\, ds' \right\} ds.$$

Thus, the classical solution of problem (1.1), if it exists, satisfies, together with the functions $X_t^{t',x}$, the system of equations (1.3) - (1.4). This enables us to introduce the generalized solution of the Cauchy problem (1.1) as follows.

The function $u(t,x)$, $t \in [0, \infty)$, $x \in R^r$, is called the *generalized solution* of problem (1.1), provided it, together with some $X_t^{t',x}(\omega)$, satisfies the system of equations (1.3) - (1.4).

Note that the generalized solution does not depend on the choice of the factorization $a(x,u) = \sigma(x,u)\sigma^*(x,u)$. Indeed, if $\sigma_1(x,u)\sigma_1^*(x,u) = \sigma_2(x,u)\sigma_2^*(x,u) = a(x,u)$, then one can find an orthogonal matrix $\theta = \theta(x,u)$ such that $\sigma_2 = \sigma_1\theta$. This implies immediately that the generalized solutions defined via two systems of (1.3)-(1.4) type, using different factorizations, coincide.

We proceed to list the basic properties of the generalized solution. These properties, in particular, imply the correctness of the definition of the generalized solution of problem (1.1).

a) A classical solution is a generalized one.

b) If the generalized solution has bounded first- and second-order derivatives in x and first-order derivatives in t, and if these derivatives are uniformly continuous, then the generalized solution is a classical one.

c) The generalized solution satisfies the initial conditions at the continuity points of the initial function.

d) If $F_0(x) \equiv 0$ and $c(x,u) < 0$, then the generalized solution obeys the maximum principle.

Property a) has been clarified previously. To verify b), we shall introduce the infinitesimal operator $\mathfrak{A}$ of the Markov process $(t_s, X_s; P_{t,x})$ generated by equations (1.3). If the function $u(t,x)$ has bounded, uniformly continuous first-order derivatives in t and second-order ones in x, then, by Ito's formula (see §1.4), $u \in D_{\mathfrak{A}}$ and

$$\mathfrak{A}u = \mathfrak{L}u = -\frac{\partial u}{\partial t} + L(u) .$$

On the other hand, one can calculate $\mathfrak{A}u$ relying on the definition of the operator $\mathfrak{A}$ and on equality (1.4) with the aid of the Markov property:

$$\mathfrak{A}u(t,x) = \lim_{h \downarrow 0} \frac{E_{t,x}u(t_h, X_h) - u(t,x)}{h} =$$

$$= -c(x,u(t,x))\,u(t,x) - F_0(x) .$$

Comparing these two expressions for $\mathcal{A}u$, we obtain that the function $u(t,x)$ satisfies equation (1.1).

To complete proving property b), we should also check property c): the generalized solution obeys the initial conditions at the continuity points of the initial function. This is straightforward from (1.4) on account of the fact that the coefficients of equation (1.3) are bounded.

Property d) (maximum principle) follows from formula (1.4).

Later on, we shall demonstrate that, under some additional assumptions, the generalized solution under consideration is also the generalized solution in the small parameter sense.

Now we will cite a result concerning the local solvability of problem (1.1) (Blagoveščenskii [1], Tanaka [1]).

THEOREM 1.1. *Let conditions 1-3 be fulfilled. Then there is a* $t_0 > 0$ *such that problem (1.1) has a unique generalized solution for* $0 < t < t_0$. *This generalized solution satisfies Lipschitz condition in* x.

For any integer n, one can prove that the generalized solution is n times differentiable in a sufficiently small neighborhood of the plane $t = 0$, whenever the coefficients of the operator $L(u)$, of the function $F(x,u)$ and the initial function are smooth enough.

We shall not cite here the proof of Theorem 1.1 nor that of the last statement. The reader can easily carry out these proofs via the method of successive approximations using bounds similar to those to be obtained in the next section.

§5.2 *Solvability in the large at the expense of absorption. The existence conditions for derivatives*

The solvability of equations (1.3)-(1.4) may be established in various ways. Under certain assumptions one can introduce new variables $(\tilde{X}_t, u)$ so that this system of equations will reduce to "triangular" form, i.e. the equation for $\tilde{X}_t$ will not depend on $u(t,x)$. We shall utilize such a method in the next section. Here we shall construct the solution of equations (1.3) - (1.4) via successive approximations. We set $u^{(1)}(t,x) = g(x)$,

$$X_t^{(n)} = X_t^{(n),t',x} = x + \int_0^t \sigma(X_s^{(n)}, u^{(n)}(t'-s, X_s^{(n)}))\, dW_s + $$

$$+ \int_0^t b(X_s^{(n)}, u^{(n)}(t'-s, X_s^{(n)}))\, ds\ ,$$

$$u^{(n+1)}(t,x) = E_{t,x} g(X_t^{(n)}) \exp\left\{\int_0^t c(X_s^{(n)}, u^{(n)}(t-s, X_s^{(n)}))\, ds\right\} + \tag{2.1}$$

$$+ E_{t,x} \int_0^t F_0(X_s^{(n)}) \exp\left\{\int_0^s c(X_s^{(n)}, u^{(n)}(t-s', X_{s'}^{(n)}))\, ds\right\} ds\ .$$

Note that for $X_t^{(n)}$ to be defined, one has to solve the stochastic equation. For the solution of this equation to exist and be unique, the coefficients of this equation must possess sufficiently good properties. For example, it is sufficient that they be continuous in (t,x) and be Lipschitz continuous in x.

Under some special assumptions, these properties are guaranteed by the following lemma.

We shall introduce the notations: $\kappa(z) = (K_1 + K_2 z)^2 + 2(K_1 + K_2 z)$, $\mu = K_g + K_{F_0} + 4(\|g\| + \|F_0\|)$.

LEMMA 2.1. *Suppose that* $\sup\{c(x,u) : x \in R^r, -\infty < u < \infty\} = -c < -\left(\frac{\kappa(\mu)}{2} + 1\right)$. *Then, for all* $t,h > 0$, $x \in R^r$, *and any integer* n *the following bounds hold*:

$$|u^{(n)}(t,x) - u^{(n)}(t,y)| < \mu|x-y|\ ,$$

$$|u^{(n)}(t+h,x) - u^{(n)}(t,x)| < \lambda h^{1/2}\ ,$$

where the constant λ *is determined by* μ *and by the maximum moduli of the coefficients of the operator* $\mathfrak{L}$.

Proof. First, we shall derive the bound of the Lipschitz constant in x. Let us denote by $K^{(n)}$ the Lipschitz constant of the function $u^{(n)}(t,x)$: $K^{(n)} = \sup_{t\geq 0;\, x,y\in R^r} |u^{(n)}(t,x) - u^{(n)}(t,y)|\,(|x-y|)^{-1}$. Let $\kappa_n = \kappa(K^{(n)})$. We will consider the Markov family $(X_s^{(n),t',x}, t_s^{t'}, P)$ defined by equations (2.1). We put $X_s^{(n)} = X_s^{(n),t,x}$, $Y_s^{(n)} = X_s^{(n),t,y}$. From the definition of the generalized solution it follows that

$$
\begin{aligned}
|u^{(n+1)}(t,x) - u^{(n+1)}(t,y)| &\leq E\Bigg[f(X_t^{(n)}) \exp\Bigg\{ \int_0^t c(X_s^{(n)}, u^{(n)}(t-s, X_s^{(n)}))\, ds \Bigg\} - \\
&\quad - f(Y_t^{(n)}) \exp\Bigg\{ \int_0^t c(Y_s^{(n)}, u^{(n)}(t-s, Y_s^{(n)}))\, ds \Bigg\} \Bigg] + \\
&\quad + E \int_0^t \Bigg[F_0(X_s^{(n)}) \exp\Bigg\{ \int_0^s c(X_{s'}^{(n)}, u^{(n)}(t-s', X_{s'}^{(n)}))\, ds' \Bigg\} - \\
&\quad - F_0(Y_s^{(n)}) \exp\Bigg\{ \int_0^s c(Y_{s'}^{(n)}, u^{(n)}(t-s', Y_{s'}^{(n)}))\, ds' \Bigg\} \Bigg] ds \leq \\
&\leq K_g\, E|X_t^{(n)} - Y_t^{(n)}| e^{-ct} + \|g\|\, E\Bigg|\exp\Bigg\{ \int_0^t c(X_s^{(n)}, u^{(n)}(t-s, X_s^{(n)}))\, ds \Bigg\} - \\
&\quad - \exp\Bigg\{ \int_0^t c(Y_s^{(n)}, u^{(n)}(t-s, Y_s^{(n)}))\, ds \Bigg| + K_{F_0} \int_0^t e^{-cs}\, E|X_s^{(n)} - Y_s^{(n)}|\, ds \Bigg\} + \\
&\quad + \|F_0\| \int_0^t E\Bigg| \exp\Bigg\{ \int_0^s c(X_{s'}^{(n)}, u^{(n)}(t-s', X_{s'}^{(n)}))\, ds' \Bigg\} -
\end{aligned}
\tag{2.2}
$$

$$- \exp\left\{\int_0^s c(Y_{s'}^{(n)}, u^{(n)}(t-s', Y_{s'}^{(n)}))\, ds'\right\}\Big|\, ds \le$$

$$\le K_g\, e^{-ct} E|X_t^{(n)} - Y_t^{(n)}| + 4\|g\|(K_1 + K_2 K^{(n)}) \int_0^t E|X_s^{(n)} - Y_s^{(n)}|\, ds\, e^{-ct} +$$

$$+ K_{F_0} \int_0^t e^{-cs} E|X_s^{(n)} - Y_s^{(n)}|\, ds +$$

$$+ 4\|F_0\|(K_1 + K_2 K^{(n)}) \int_0^t e^{-cs} E\left(\int_0^s |X_{s'}^{(n)} - Y_{s'}^{(n)}|\, ds'\right) ds\,.$$

Now we shall make use of the bound

$$E|X_t^{(n)} - Y_t^{(n)})|^2 \le |x-y|^2 \exp\{\kappa_n t\}$$

which follows from Lemma 3.6.1. With the help of this bound and (2.2) we obtain:

$$|u^{(n+1)}(t,x) - u^{(n+1)}(t,y)|(|x-y|)^{-1} \le$$

$$\le K_g \exp\left\{\left(\frac{\kappa_n}{2} - c\right) t\right\} + 4\|g\| \frac{2(K_1 + K_2 K^{(n)})}{\kappa_n} e^{-ct} \int_0^t e^{\frac{\kappa_n s}{2}}\, d\,\frac{\kappa_n s}{2} +$$

$$+ K_{F_0} \int_0^t \exp\left\{\left(\frac{\kappa_n}{2} - c\right) s\right\} ds + \tag{2.3}$$

$$+ 4\|F_0\| \frac{2(K_1 + K_2 K^{(n)})}{\kappa_n} \cdot \int_0^t e^{-cs}\left(\int_0^s e^{\kappa_n \frac{s'}{2}}\, ds'\right) ds \le$$

$$\leq K_g \exp\left\{\left(\frac{\kappa_n}{2}-c\right)t\right\} + 4\|g\| \exp\left\{\left(\frac{\kappa_n}{2}-c\right)t\right\} +$$

$$+ \frac{K_{F_0}}{\left|\frac{\kappa_n}{2}-c\right|} \cdot \left|e^{\left(\frac{\kappa_n}{2}-c\right)t}-1\right| + \frac{4\|F\|}{\left|\frac{\kappa_n}{2}-c\right|} \cdot \left|e^{\left(\frac{\kappa_n}{2}-c\right)t}-1\right| .$$

When deducing this inequality, we have used the fact that $\dfrac{K_1+K_2K^{(n)}}{\kappa_n} < \dfrac{1}{2}$.

From (2.3) it follows that if $\frac{\kappa_n}{2}+1 < c$, then

$$K^{(n+1)} \leq K_g + K_{F_0} + 4(\|g\|+\|F_0\|) = \mu \tag{2.4}$$

and, thereby, $\frac{\kappa_{n+1}}{2}+1 < c$, because $K^{(n+1)} < \mu$ and $\kappa_{n+1} = \kappa(K^{(n+1)}) < \kappa(\mu) < 2(c-1)$. The last inequality is involved in the condition of the lemma.

Therefore, we have demonstrated that if the conditions of the lemma are fulfilled and $K^{(n)} < \mu$, then $K^{(n+1)} < \mu$. Since for $u^{(1)}(t,x) = g(x)$ this assumption holds, we conclude that $K^{(n)} < \mu$ for all n.

We proceed to deduce the bound for the modulus of continuity in t of the functions $u^{(n)}(t,x)$. From (2.1) it follows that

$$u^{(n)}(t+h,x) = Eu^{(n)}(t,X_h^{(n-1),t,x}) \exp\left\{\int_0^h c(X_s^{(n-1),t,x},u^{(n)}(t+h-s,X_s^{(n-1),t,x}))\,ds\right\}$$

Whence we conclude that

$$\begin{aligned} &|u^{(n)}(t+h,x)-u^{(n)}(t,x)| \leq \\ &\leq K^{(n)}E|X_h^{(n-1),t,x}-x| + \|u^{(n)}\|\cdot\|c\|\cdot h \leq \\ &\leq K^{(n)}(E|X_h^{(n-1),t,x}-x|^2)^{1/2} + \lambda_1 h \leq \lambda h^{1/2} . \end{aligned} \tag{2.5}$$

The constant λ is determined by the maximum of the moduli of the coefficients, by $\|F_0\|$, by the dimension of the space, and by the constant $K^{(n)}$.

If the conditions of the lemma are valid, then $K^{(n)} < \mu$ for all n, which together with (2.5) implies the claim of the lemma. □

REMARK. If $F_0(x) \equiv 0$ and $-c(x,u) \equiv c = \text{const.}$ is large enough, then the Lipschitz constant for the function $u(t,x)$ does not grow with t. In this case it is possible to formulate the following statement which may be established just as Lemma 2.1.

LEMMA 2.1′. *Suppose that* $F_0(x) \equiv 0$, $c(x,u) = -c < \frac{\kappa(K_g)}{2} < 0$. *Then, for all integers* n, *the bounds hold*:

$$|u^{(n)}(t,x) - u^{(n)}(t,y)| \leq K_g\, |x-y| \,,$$

$$|u^{(n)}(t+h,x) - u^{(n)}(t,x)| \leq \lambda\, h^{1/2} \,.$$

THEOREM 2.1 (Freidlin [12]). *Suppose that*

$$\sup\{c(x,u) : x \in R^r, -\infty < u < \infty\} = -c < -\frac{\kappa(\mu)}{2} - 1 \,,$$

where $\mu = K_g + K_{F_0} + 4(\|g\| + \|F_0\|)$. *Then a continuous generalized solution of problem (1.1) exists which is defined for all* $t > 0$ *and is Lipschitz continuous in* x *with a constant independent of* t. *This generalized solution is unique in the class of bounded measurable functions and may be constructed with the successive approximations indicated above.*

Proof. Let us introduce the notations

$$\delta_n(t) = \sup_{x \in R^r} |u^{(n)}(t,x) - u^{(n-1)}(t,x)|, \quad m_n(t) = \sup_{\substack{x \in R^r \\ s \leq t,\, t' > 0}} (E|X_s^{(n)} - X_s^{(n-1)}|^2)^{1/2} \,.$$

Relations (2.1) and Lemma 2.1 imply that if the conditions of the theorem are valid, one can find constants A_k (independent of n) such that the following relations hold

$$\delta_{n+1}(t) \le A_1 m_n(t) + A_2 \int_0^t m_n(s)\,ds + A_3 \int_0^t \delta_n(t-s)\,ds + \tag{2.6}$$
$$+ A_4 \int_0^t \int_0^s m_n(s')\,ds'\,ds + A_5 \int_0^t \int_0^s \delta(t-s')\,ds'\,ds ,$$

$$m_n^2(t) \le A_6 \int_0^t m_n^2(s)\,ds + A_7 \int_0^t \delta_n^2(t-s)\,ds . \tag{2.7}$$

Squaring (2.6) and using the inequality $(\sum_{i=1}^k x_i)^2 \le k \sum_1^k x_i^2$ and the Schwarz inequality we obtain

$$\delta_{n+1}^2(t) \le A_8 m_n^2(t) + (A_9 + A_{10} t^3) \int_0^t m_n^2(s)\,ds + \tag{2.8}$$
$$+ (A_{11} + A_{12} t^3) \int_0^t \delta_n^2(s)\,ds .$$

From (2.7) follows the bound

$$m_n^2(t) \le A_7 \int_0^t \delta_n(s)\,ds \cdot \exp\{A_6 t\} . \tag{2.9}$$

Given some $T > 0$, from (2.8) and (2.9) it follows that one can find $A_1' = A_1(T)$ such that

$$\delta_{n+1}^2(t) \le A_1' \int_0^t \delta_n^2(s)\,ds , \quad 0 \le t \le T .$$

$$\sup_{0\le t\le T} \delta^2_{n+1}(t) \le \frac{(A_1'T)^{n+1}}{(n+1)!} \cdot A_2' . \qquad (2.10)$$

Here and henceforth, A_i' are numbers independent of n, but depending on T.

Notice that

$$u^{(n+1)}(t,x) = u^{(1)} + (u^{(2)} - u^{(1)}) + \cdots + (u^{(n+1)} - u^{(n)}) .$$

Next, by virtue of (2.10), $\sup_{x,t\le T} |u^{(n)}(t,x) - u^{(n-1)}(t,x)| < A_2' \frac{(A_1'T)^n}{n!}$. From this we obtain that $\lim_{n\to\infty} u^{(n)}(t,x) = u(t,x)$ exists, and this convergence is uniform on the set $\{x \in R^r, 0 \le t \le T\}$.

Next, (2.9) and (2.10) imply that

$$m^2_{n+1}(T) \le A_3' \frac{(A_1'T)^{n+1}}{(n+1)!} . \qquad (2.11)$$

From (2.1), with the aid of the Kolmogorov inequality for stochastic integrals and Chebyshev's inequality we conclude that

$$P\left\{ \sup_{0\le t\le T} |X_s^{(n+1),t',x} - X^{(n),t',x}| > \frac{1}{2^n} \right\} \le 2^n A_4' m_{n+1}(T) . \qquad (2.12)$$

From (2.11) and (2.12), by the Borel-Cantelli lemma we deduce that the series $\sum_{n=1}^{\infty} (X_s^{(n+1),t',x} - X_s^{(n),t',x})$ converges with probability 1 uniformly on $[0,T]$. Hence it appears clear that with probability 1, $X_s^{(n),t',x}$ uniformly converge to some random functions $X_s^{t',x}$ as $n\to\infty$. Passing to the limit in (2.1) as $n\to\infty$ we infer that the pair $(u(t,x), X_t^{t',x})$ satisfies system (1.3)-(1.4). Hence, the function $u(t,x)$ is the generalized solution of problem (1.1). The construction of $u(t,x)$ and lemma 2.1 imply that this generalized solution is Lipschitz continuous in x and Hölder continuous in t with exponent 0,5.

Now we shall prove the uniqueness of the generalized solution. Suppose that, besides the just constructed solution $(X_t^{t',x}, u(t,x))$, system (1.3)-(1.4) has another solution $(\bar{X}_t^{t',x}, v(t,x))$. For the difference $X_t^{t',x} - \bar{X}_t^{t',x}$ we derive the relation

$$X_t^{t',x} - \bar{X}_t^{t',x} = \int_0^t [\sigma(X_s^{t',x}, u(t'-s, X_s^{t',x})) - \sigma(\bar{X}_s^{t',x}, v(t'-s, \bar{X}_s^{t',x}))]dW_s + \\ + \int_0^t [b(X_s^{t',x}, u(t'-s, X_s^{t',x})) - b(\bar{X}_s^{t',x}, v(t'-s, \bar{X}_s^{t',x}))]ds .$$

From this, for $n(t) = \sup_x E|X_t^{t',x} - \bar{X}_t^{t',x}|^2)^{1/2}$, follows the inequality

$$n^2(t) \le A_5' \int_0^t n^2(s)\,ds + A_6' \int_0^t \max_x |u(s,x) - v(s,x)|^2\,ds . \tag{2.13}$$

Further, (1.4) yields for $\ell(t) = \max_{0 \le s \le t, x \in R^r} |u(s,x) - v(s,x)|$ that

$$\ell(t) \le A_7' \int_0^t n(s)\,ds + A_8' \int_0^t \ell(s)\,ds . \tag{2.14}$$

From (2.13) and (2.14) it is easily deduced that $\ell(t) \le A_9' \cdot t \cdot \ell(t)$ for $0 \le t \le T$. The last inequality implies that $\ell(t)$ vanishes at least for $t < t_0 = (A_9')^{-1}$. Since A_9' does not depend on t (A_9' depends on T which has been chosen beforehand), one can conclude that $u(t,x) = v(t,x)$ for $0 \le t \le T$, $x \in R^r$. As T is an arbitrary positive number, we deduce that $u(t,x) \equiv v(t,x)$ for all $t > 0$ and $x \in R^r$. □

COROLLARY. *Suppose that* $F_0(x) \equiv 0$ *and the coefficients of the operator* $L(u)$ *do not depend explicitly on* x. *Let* $c(u) < -c < 0$. *Then a unique continuous generalized solution exists for any initial function possessing sufficiently small norm in* $C^1_{R^r}(R^1)$. *This follows from the fact that, in the indicated case,* $\mu = K_g + 4\|g\|$ *and* $\kappa(\mu) = (K_2\mu)^2 + 2K_2\mu$ *may be made arbitrarily small at the expense of the smallness of the function* $g(x)$ *and its derivatives.*

One can show that if the conditions of Theorem 2.1 hold, then the above constructed generalized solution of problem (1.1) is a solution in the small parameter sense as well. To make this assertion more precise, we shall consider the strictly elliptic operator

$$\tilde{L}^\varepsilon(u) = \frac{\varepsilon}{2} \sum \tilde{a}^{ij}(x,u) \frac{\partial^2 u}{\partial x^i \partial x^j} + \sum \tilde{b}^i(x,u) \frac{\partial u}{\partial x^i},$$

whose coefficients are bounded together with their first-order derivatives. Let $u^\varepsilon(t,x)$ *be the solution of the Cauchy problem*

$$\frac{\partial u^\varepsilon}{\partial t} = L(u^\varepsilon) + \tilde{L}^\varepsilon(u^\varepsilon) + F(x,u^\varepsilon), \quad u^\varepsilon(0,x) = g(x). \tag{2.15}$$

The solution of problem (2.15) exists and is unique (see, e.g. Ladyženskaja, Solonnikov and Ural'ceva [1]).

THEOREM 2.2. *Suppose the conditions of Theorem 2.1 are fulfilled. Then* $u(t,x) = \lim_{\varepsilon \downarrow 0} u^\varepsilon(t,x)$ *uniformly on the set* $\{(t,x) : 0 \le t \le T,\ x \in R^r\}$ *for any* $T < \infty$.

The proof of this theorem is dropped. This proof is not difficult to arrange using the bounds given in Lemma 2.1. □

In the conclusion of this section we shall state without proof a result concerning the existence of derivatives of the generalized solution. For brevity, we shall assume that $F(x,u) = -cu$, $c = \text{const}$.

THEOREM 2.3. *An increasing sequence* $\theta_1, \cdots, \theta_r, \cdots$ exists, *defined by* K_1, K_2, K_g, *such that if* $g(x)$, $\sigma^i_j(x,u)$, $b^i(x,u)$ *have* $m+1$ *bounded derivatives in* x,u *and if* $c > \theta_m$, *then the generalized solution has* m *continuous derivatives in* x *and* $\left[\frac{m}{2}\right]$ *derivatives in* t.

The proof of this theorem is based on the bounds given by Lemma 2.1 and Lemma 3.6.2. This proof may be found in the work of Freidlin [12].

§5.3 *On equations with subordinate non-linear terms*

Let us consider two differential operators

$$L_1 = \frac{1}{2} \sum_{i,j=1}^{r} a^{ij}(x) \frac{\partial^2}{\partial x^i \partial x^j} + \sum_{i=1}^{r} b^i(x) \frac{\partial}{\partial x^i}, \quad L_2^v = \sum_{i=1}^{r} B^i(x,v) \frac{\partial}{\partial x^i}, \quad x \in R^r,\ v \in R^1 .$$

The coefficients of these operators are assumed to be bounded functions, Lipschitz continuous in x and v. Moreover, as always, we suppose that there is a Lipschitz factorization $a(x) = (a^{ij}(x)) = \sigma(x)\,\sigma^*(x)$ (see §3.2).

We say that the operator L_2^v in a domain $G = D \times [a,b]$, $D \subset R^r$, $-\infty < a < b < \infty$, is *subordinate* to the operator $L_1 (L_2^v \prec L_1)$, if there exists a solution $\phi = (\phi^1(x,v), \phi^2(x,v), \cdots, \phi^r(x,v))$ of the system of linear algebraic equations $\sum_{j=1}^{r} \sigma_j^i(x)\,\phi^j(x,v) = B^i(x,v);\ i = 1, 2, \cdots, r$, which is bounded in G. In particular, if the operator L_1 is uniformly elliptic, then any first-order operator with bounded coefficients is subordinate to it.

The representation of the matrix $a(x)$ in the form $a(x) = \sigma(x)\,\sigma^*(x)$ is not unique; however, the subordination of operators does not depend on the choice of the matrix $\sigma(x)$ in this representation. In fact, if $\sigma_1(x)\sigma_1^*(x) = \sigma_2(x)\sigma_2^*(x) = a(x)$, then $\sigma_2(x) = \sigma_1(x)\,\theta(x)$, where $\theta(x)$ is some orthogonal matrix. Therefore, if $\sigma_1(x)\,\phi(x,v) = B(x,v)$, then the vector $\tilde{\phi}(x,v) = \theta^{-1}\phi(x,v)$ satisfies the relation $\sigma_2\phi = \sigma_1\theta\theta^{-1}\phi = \sigma_1\phi = B(x,v)$, and obviously has the same length as does the vector $\phi(x,v)$, because $\theta(x)$ is an orthogonal matrix.

Consequently, the subordination is actually defined by the matrix $a(x)$ and the vector $B(x,v)$, rather than by what factorization has been chosen.

Let us consider the Cauchy problem

$$\frac{\partial u(t,x)}{\partial t} = L_1 u + L^u u; \quad u(0,x) = g(x) . \tag{3.1}$$

System, (2.1) for such a problem is written down as follows:

$$\tilde{X}_t^{t',x} = x + \int_0^t \sigma(\tilde{X}_s^{t',x})\, dW_s + \int_0^t [b(\tilde{X}_s^{t',x}) + B(\tilde{X}_s^{t',x}, u(t'-s, \tilde{X}_s^{t',x}))]ds$$

$$u(t,x) = E\, g(\tilde{X}_t^{t,x}) \,. \tag{3.2}$$

We put $\Lambda = [\inf_{x\epsilon R^r} g(x),\ \sup_{x\epsilon R^r} g(x)]$. If $L_2^v \prec L_1$ in the domain $G = R^r \times \Lambda$, then system (3.2) may be reduced to "triangular" form. To this end, we introduce the Markov family (X_t^x, P), defined by the stochastic equation

$$X_t^x - x = \int_0^t \sigma(X_s^x)\, dW_s + \int_0^t b(X_s^x)\, ds \,. \tag{3.3}$$

The assumption of operator subordination implies (see §1.5) that the measures μ and $\tilde{\mu}$, induced in the space $C_{0t}(R^r)$ by the processes X_t^x and $\tilde{X}_t^{t',x}$ respectively, are absolutely continuous with respect to each other. By the Cameron-Martin-Girsanov formula the density function of one measure with respect to other has the form:

$$\frac{d\tilde{\mu}}{d\mu} = \exp\left\{ \int_0^t (\phi(X_s^x, u(t'-s, X_s^x)), dW_s) - \frac{1}{2} \int_0^t |\phi(X_s^x, u(t'-s, X_s))|^2 ds \right\} .$$

This enables us to rewrite the second of equalities (3.2) as follows

$$u(t,x) = E\, g(X_t^x) \exp\left\{ \int_0^t (\phi(X_s^x, u(t-s, X_s^x)), dW_s) - \frac{1}{2} \int_0^t |\phi(X_s^x, u(t-s, X_s^x))|^2 ds \right\} . \tag{3.4}$$

Therefore, if the function $u(t,x)$, along with $\tilde{X}_t^{t',x}$, satisfies system (3.2), then $u(t,x)$ together with X_t^x is a solution of equations (3.3) - (3.4). The first of these equations does not depend on the unknown function $u(t,x)$ and may be solved separately. To find the function $u(t,x)$, equation (3.4) remains in which the process X_t^x may already be looked upon as a known one. The converse is also true: namely, if $(X_t^x, u(t,x))$ is a solution of system (3.3) - (3.4), then one can find a process $\tilde{X}_t^{t',x}$ such that the pair $(X_t^{t',x}, u(t,x))$ obeys equations (3.2). Indeed, let us define the new measure $\tilde{P}$:

$$\tilde{P}(d\omega) = P(d\omega) \exp\left\{ \int_0^t (\phi(X_s^x, u(t'-s, X_s^x)), dW_s) - \frac{1}{2} \int_0^t |\phi(X_s^x, u(t'-s, X_s^x))|^2 \, ds \right\}$$

in the space Ω on which the underlying Wiener process is defined. As follows from Lévy's theorem (see §1.3), the process

$$\tilde{W}_t = W_t - \int_0^t \phi(X_s^x, u(t'-s, X_s^x)) \, ds\,, \quad t \geq 0\,,$$

is a Wiener process with respect to the new measure $\tilde{P}$. Now it remains to note that since $\sigma\phi = B$, the process X_t^x satisfies the equality

$$X_t^x - x = \int_0^t \sigma(X_s^x) \, d\tilde{W}_s + \int_0^t [b(X_s^x) + B(X_s^x, u(t'-s, X_s^x))] \, ds\,.$$

The equivalence of systems (3.2) and (3.3) - (3.4) implies, in particular, that the totality of solutions of equation (3.4) does not depend on the choice of the vector ϕ out of the solutions of the linear system $\sigma\phi = B$.

So, to construct the generalized solution of problem (3.1), we must solve equation (3.4) for the known process X_t^x. The properties of this equation are similar to those of the Volterra-type integral equation. We shall put

$$\zeta_0^t(\phi) = \int_0^t (\phi(X_s^x, u(T-s,X_s^x)), dW_s) -$$

$$-\frac{1}{2}\int_0^t |\phi(X_s^x, u(T-s,X_s^x))|^2 ds, \quad 0 \le t \le T < \infty .$$

LEMMA 3.1 (Girsanov [1]). *If* $c_1 < |\phi(x,u)| < c_2 < \infty$ *for* $(x,u) \in G$, *then for* $a \ge 1$, *the bounds hold*

$$\exp\left\{\frac{a^2-a}{2} c_1^2 t\right\} \le E \exp\{a\zeta_0^t(\phi)\} \le$$
$$\le \exp\left\{\frac{a^2-a}{2} c_2^2 t\right\} . \tag{3.5}$$

Proof. Let us put $\tau_N = \inf\{s : \zeta_0^s(\phi) = N\}$. Applying Ito's formula to the function $\exp\{a\zeta_0^t(\phi)\}$ we obtain:

$$\exp\{a\zeta_0^{t\wedge\tau_N}(\phi)\} = 1 + a\int_0^{t\wedge\tau_N} \exp\{a\zeta_0^s(\phi)\}(\phi(X_s^x, u(T-s,X_s)), dW_s) +$$
$$+ \frac{a^2-a}{2}\int_0^{t\wedge\tau_N} |\phi(X_s^x, u(T-s,X_s^x))|^2 \exp\{a\zeta_0^s(\phi)\} ds . \tag{3.6}$$

The integrand in the stochastic integral is bounded (just to this end, $\zeta_0^{t\wedge\tau_N}$ is considered rather than ζ_0^t). Therefore, the expectation of this integral is zero. Let us find the expectation of both sides of equality (3.6):

$$m_N(t) = E \exp\{\alpha \zeta_0^{t\wedge\tau_N}(\phi)\} \le 1 + \frac{\alpha^2-\alpha}{2} E \int_0^{t\wedge\tau_N} |\phi|^2 \exp\{\zeta_0(\phi)\}\, ds \le$$

$$\le 1 + \frac{\alpha^2-\alpha}{2} c_2^2 \int_0^t m_N(s)\, ds \, .$$

This permits us to conclude that

$$m_N(t) \le \exp\left\{\frac{t\, c_2^2(\alpha^2-\alpha)}{2}\right\} . \tag{3.7}$$

Taking into account that $\zeta_0^{t\wedge\tau_N} \to \zeta_0^t$ P-a.s. as $N \to \infty$, by Fatou's lemma, we deduce from (3.7) the right-hand inequality of (3.5). Next, by Ito's formula, on the basis of the above derived upper bound for $E \exp\{\alpha \zeta_0^t(\phi)\}$, we see that

$$m(t) = E \exp\{\alpha \zeta_0^t(\phi)\} = 1 + \frac{\alpha^2-\alpha}{2} \int_0^t E|\phi(X_s^x, u(t-s, X_s^x))|^2 \exp\{\alpha \zeta_0^s(\phi)\} ds \ge$$

$$\ge 1 + \frac{\alpha^2-\alpha}{2} c_1^2 \int_0^t m(s)\, ds \, .$$

This implies the inequality

$$\frac{A\, m(t)}{1 + A \int_0^t m(s)\, ds} = \frac{d}{dt} \ln\left(1 + A \int_0^t m(s)\, ds\right) \ge A \, , \tag{3.8}$$

where $A = \dfrac{c_1^2(\alpha^2-\alpha)}{2}$. From (3.7) it follows that

$$m(t) \geq 1 + A \int_0^t m(s)\,ds \geq \exp\{At\}$$

which is none other than the left-hand side of inequality (3.5). □

THEOREM 3.1 (Freidlin [8]). *Suppose that* $L_2^v \prec L_1$ *in a domain* $G = R^r \times \Lambda$, *where* $\Lambda = [\inf_{x \in R^r} g(x), \sup_{x \in R^r} g(x)]$. *Moreover, let the functions* $\phi^i(x,u)$ *be Lipschitz continuous in* x *and* u. *Then for any bounded measurable initial function* $g(x)$, *the generalized solution of problem (3.1) exists for all* $t > 0$. *This solution is unique in the class of bounded measurable functions. If the initial function is continuous, then the generalized solution is continuous for all* $t \geq 0$.

Proof. Let T be an arbitrary positive number, B_T being the Banach space of bounded measurable functions on $[0,T] \times R^r$ with the norm $\sup_{t \in [0,T], x \in R^r} |v(t,x)| = \|v\|$. We shall consider in B_T the following operator

$$F[v] = F[v](t,x) = E\, g(X_t^x) \exp\{\mathcal{T}_0^t(\phi(v))\},$$

where

$$\mathcal{T}_0^t(\phi(v)) = \int_0^t (\phi(X_s^x, v(t-s, X_s^x)), dW_s) - \frac{1}{2} \int_0^t |\phi(X_s^x, v(t-s, X_x^x))|^2\, ds.$$

Since the functions $g(x)$ and $|\phi(x)|$ are bounded, the operator F transforms the space B_T into itself. This results from Lemma 3 1.

Now we will show that one can find $t_0 > 0$ depending only on $\sup_{(x,v) \in G} |\phi(x,v)|$, the Lipschitz constant K_ϕ of the function $\phi(x,v)$ in v, and $\sup_{x \in R^r} |g(x)|$, such that the operator F in the space B_{t_0} will be a contraction. For this, let us consider the difference

$$|F(u)-F(v)| = |E\,g(X_t^x)\,[\exp\{\mathcal{T}_0^t(\phi(u))\}-\exp\{\mathcal{T}_0^t(\phi(v))\}]| \leq$$
$$\leq \|g\|\,(E\,\exp\{2\,\mathcal{T}_0^t(\phi(v))\})^{1/2}(E\,(\exp\{\mathcal{T}_0^t(\phi(u))-\mathcal{T}_0^t(\phi(v))\}-1)^2)^{1/2}\,. \tag{3.9}$$

Let $\chi_1 = \chi_1(\omega)$ be the indicator of the set $\{\omega \subset \Omega : |\mathcal{T}_0^t(\phi(u))-\mathcal{T}_0^t(\phi(v))| \leq 1\}$. We shall make use of the fact that $|e^x-1| < 2\,|x|$ for $|x| < 1$:

$$E\,(\exp\{\mathcal{T}_0^t(\phi(u))-\mathcal{T}_0^t(\phi(v))\}-1)^2 = E\,[\exp\{\mathcal{T}_0^t(\phi(u)-\mathcal{T}_0^t(\phi(v))\}-1\,]^2\chi_1 +$$
$$+\,E\,[\exp\{\mathcal{T}_0^t(\phi(u))-\mathcal{T}_0^t(\phi(v))\}-1\,]^2(1-\chi_1) \leq \tag{3.10}$$
$$\leq 4E\,(\mathcal{T}_0^t(\phi(u))-\mathcal{T}_0^t(\phi(v)))^2 + [E_x(\exp\{\mathcal{T}_0^t(\phi(u))-\mathcal{T}_0^t(\phi(v))\}-1)^4]^{1/2}[E(1-\chi_1)]^{1/2}\,.$$

Lemma 3.1 implies the following bounds

$$E\,(\exp\{\mathcal{T}_0^t(\phi(u))-\mathcal{T}_0^t(\phi(v))\}-1)^4 \leq \exp\{A_1\|\phi\|\} + A_2\,,$$
$$E\,\exp\{2\,\mathcal{T}_0^t(\phi(v))\} \leq \exp\{A_3 t\,\|\phi\|\}\,. \tag{3.11}$$

Here and henceforth A_i are some constants. Denoting by K_ϕ the Lipschitz constant of the function $\phi(x,u)$ in u, we have

$$E(1-\chi_1) = P\{|\mathcal{T}_0^t(\phi(u))-\mathcal{T}_0^t(\phi(v))| > 1\} \leq$$
$$\leq 4E\left(\int_0^t ([\phi(u)-\phi(v)],\, dW_s)\right)^4 + 4E\left(\int_0^t (|\phi(u)|^2-|\phi(v)|^2)\,ds\right)^4 \leq$$
$$\leq A_4(t^2+t^4)K_\phi^4(1+\|\phi\|)\,\|u-v\|^4\,; \tag{3.12}$$

$$E\,(\mathcal{T}_0^t(\phi(u))-\mathcal{T}_0^t(\phi(v)))^2 \leq 2\int_0^t E\,|\phi(u)-\phi(v)|^2\,ds +$$
$$+\,2E\left(\int_0^t (|\phi(u)|^2-|\phi(v)|^2)\,ds\right)^2 \leq A_5(t+t^2)K_\phi^2(1+\|\phi\|)\,\|u-v\|^2\,.$$

In the first of these inequalities, we have used Lemma 3.6.3 for bounding the fourth moment of the stochastic integral.

On gathering bounds (3.10)-(3.12), inequality (3.9) yields:

$$\|F[u]-F[v]\| \le A_6 \exp\{A_7 t\|\phi\|\} \|g\| (1+\|\phi\|)(t+t^4)^{1/2}\|u-v\| .$$

This inequality implies that, for $t_0 = 1 \wedge [(A_6\|g\|(1+\|\phi\|)^{-1} \times \exp\{-A_7\|\phi\|\}]$, the mapping F in the space B_{t_0} is a contraction. Whence we conclude that, in B_{t_0}, the operator F has only one fixed point which obviously is a solution of equation (3.4). This solution is unique.

In order to get the existence and uniqueness "in the large" for all $T > 0$, we note that $|u(t,x)| < \|g\|$, and hence, starting at time t_0 from the function $u(t_0, x)$ one can again solve the equation for $t - t_0 < t_0$. Using the steps of the magnitude t_0 it is possible to reach any $t > 0$.

To prove the last statement of the theorem, it suffices to check that the operator F transforms the space of bounded continuous functions into itself. This fact may be readily established provided one takes into account the following:

1) the continuity of the functions $\phi(x,u)$ and $g(x)$;

2) the continuity a.s. of the trajectories X_t^x;

3) the uniform convergence on every finite time interval of the trajectories X_t^x to $X_t^{x_0}$ as $x_0 \to x$;

4) the bounds given by Lemma 3.1.

Detailed calculations are dropped. □

Now we shall consider the existence of derivatives of the generalized solution.

LEMMA 3.2. *Suppose that the functions* $\sigma_j^i(x)$, $b^i(x)$, $\phi(x,v)$, $g(x)$ *are continuous and have* $k > 1$ *derivatives in the variables* x *and* v. *Then, for any integers* $\ell, k_1, k_2, \cdots, k_r$ *such that* $k_1 + k_2 \cdots + k_r = \ell < k$ *and for any* $T > 0$, *the derivatives* $\dfrac{\partial^\ell u(t,x)}{(\partial x^1)^{k_1} \cdots (\partial x^r)^{k_r}}$ *(if they exist) are bounded in modulus for* $t \in [0,T]$, $x \in R^r$, *by a constant which depends on* k, *on the maximum of the moduli of* σ_i^j, b^i, ϕ^i, g *and their derivatives in* x *and* u *up to the order* k *inclusively.*

Proof is carried out by induction on k using the bound

$$E\left|\frac{\partial^k X_t^x}{(\partial x^1)^{\ell_1}\cdots(\partial x^r)^{\ell_r}}\right|^{2m} < c_{k,m}\exp\{\beta_{k,m}t\}. \tag{3.13}$$

Here the numbers $c_{k,m}$ are determined by the norm of $\sigma_j^i(x)$ and $b^i(x)$ in $C_{R^r}^{(k)}(R^r)$; $\beta_{k,m}$ are determined only by k, m, r, and by the Lipschitz constant of the coefficients of the stochastic equation. These bounds follow from Lemma 3.6.2. For $k=1$, the bound given by Lemma 3.2, may be obtained by differentiating (3.4) in x^i. In fact, for every $i = 1, \cdots, r$,

$$\frac{\partial u(t,x)}{\partial x^i} = E\sum_{j=1}^r \frac{\partial g}{\partial x^j}(X_t^x)\frac{\partial X_t^{x,j}}{\partial x^i}\exp\{\mathcal{T}_0^t(\phi(u))\} + \tag{3.14}$$

$$+ E\, g(X_t^x)\exp\{\mathcal{T}_0^t(\phi(u))\}\frac{\partial}{\partial x^i}\mathcal{T}_0^t(\phi(u)).$$

We shall let $v_1(t) = \sup_{0\le s\le t, x\in R^r}\sum_{i=1}^r\left|\frac{\partial u(s,x)}{\partial x^i}\right|$. From (3.13), (3.14) and Lemma 3.1 we derive

$$v_1^2(t) \le c_1(t) + d_1(t)\int_0^t v_1^2(s)\,ds, \tag{3.15}$$

where $c_1(t)$ and $d_1(t)$ are some continuous functions growing with t and defined by the maximum of the moduli of σ_j^i, b^i, ϕ_i, g, and their first-order derivatives in x and u. From (3.15) follows the bound for $\sup_{0\le t\le T,\, x\in R^r}|\nabla_x u(t,x)|$ pointed out in Lemma 3.2.

To obtain the bounds of (k+1)-order derivatives in x, provided the first k lower derivatives have already been estimated, one should differentiate equality (3.4) (k+1) times. In doing so, noting bound (3.13) for

$$v_{k+1}(t) = \sup_{0\le s\le t,\ x\in R^r} \sum_{\substack{\ell_1+\cdots+\ell_r=k+1\\ \ell_i\ge 0}} \left| \frac{\partial^{k+1}u(s,x)}{(\partial x^1)^{\ell_1}\cdots(\partial x^r)^{\ell_r}} \right|$$

we arrive at the inequality

$$v_{k+1}^2(t) \le c_{k+1}(t) + d_{k+1}(t) \int_0^t v_{k+1}^2(s)\,ds ,$$

where $c_{k+1}(t)$ and $d_{k+1}(t)$ are continuous functions determined by the maximum of the moduli of σ_j^i, b^i, ϕ^i, g, and their first (k+1) derivatives. This inequality implies the desired bound on every finite time interval. □

Note that if the coefficients of differential equation (3.1) are sufficiently smooth functions and there are already bounds for the derivatives in the space variables up to the k-th order inclusively, then, by differentiating the equation in t, one can obtain the bound of mixed derivatives and derivatives in t up to the order $\frac{k}{2}$. If $k = 1$, then, just as was done when proving Lemma 2.1, one can derive the bound of the constant in the Hölder condition with exponent $1/2$.

Using Lemma 3.2 and the last remark it is not difficult to prove the following

THEOREM 3.2. *Suppose that the hypotheses of Theorem 3.1 are fulfilled and, moreover, let the functions* σ_j^i, b^i, ϕ^i, g *have* k *bounded, uniformly continuous derivatives in* x *and in* u. *Then the generalized solution has* k *bounded derivatives in* x *and* $\left[\frac{k}{2}\right]$ *derivatives in* t.

Consider now the following problem

$$\frac{\partial u(t,x)}{\partial t} = \frac{a^2}{2} \Delta u(t,x) + \sum_{i=1}^{r} B^i[u](t,x) \frac{\partial u(t,x)}{\partial x^i}, \quad t > 0, \; x \in R^r,$$

$$u(0,x) = g(x). \tag{3.16}$$

Here $B^i[u](t,x)$ are operators acting on the functions $u(s,y)$, $0 \leq s \leq T$, $y \in R^r$, $B^i[u](t,x)$ being defined only by the values of the function $u(s,x)$ for $0 \leq s \leq t$, $y \in R^r$.

A number of problems can be reduced to equations of this kind. In particular, the Navier-Stokes equations for two-dimensional incompressible fluid can be written in such a form. In this case $r = 2$,

$$B^1[u](t,x) = \int_{R^2} \frac{\partial h}{\partial x^2}(|x-y|)\, u(t,y)\, dy, \quad B^2[u](t,x) = \int_{R^2} \frac{\partial h}{\partial x^1}(|x-y|)\, u(t,y)\, dy,$$

$$h(s) = -\frac{1}{2\pi} \ln s, \; s > 0, \; x,y \in R^2. \tag{3.17}$$

By a *generalized solution of problem* (3.16), we mean a function $u(t,x)$, $t \geq 0$, $x \in R^r$, satisfying the equation

$$u(t,x) = Eg(X_t^x) \exp\left\{\frac{1}{a} \int_0^t (B[u](t-s, X_s^x), dW_s) - \frac{1}{2a^2} \int_0^t |B[u](t-s, X_s^x)|^2 ds\right\},$$

$$X_s^x = x + aW_s.$$

It is not hard to check the correctness of such a definition.

If the operators $B^i[u]$ obey the Lipschitz condition

$$\sup_{\substack{0 \leq s \leq t \\ x \in R^r}} |B^i[u](s,x) - B^i[v](s,x)| < K \sup_{\substack{0 \leq s \leq t \\ x \in R^r}} |u(s,x) - v(s,x)|,$$

then an existence and uniqueness theorem for the generalized solution can be established just as Theorem 3.1 was proved. In the case of the Navier-Stokes equations, the operators $B^i[u]$ do not satisfy the Lipschitz condition.

If the function $h(r)$ in (3.17) is replaced by the function $h_\varepsilon(r)$ equal to $h(r)$ for $r > \varepsilon$ and such that $|h'_\varepsilon(r)| \le |h'(r)|$, $|h''_\varepsilon(r)| \le |h''(r)|$, then the new operators $B^i_\varepsilon[u]$ constructed with the help of the kernel h_ε already obey the Lipschitz condition. Equation (3.16) with such coefficients $B^i_\varepsilon[u]$ already has a unique generalized solution $u^\varepsilon(t,x)$. Now passing to the limit as $\varepsilon \downarrow 0$, one can construct the generalized solution of the Navier-Stokes equation. Certainly, for this, we shall need some bounds uniform in ε for the operators $B^i_\varepsilon[u]$. The related construction of the generalized solution of the Navier-Stokes equations is carried out in the work of Marchioro and Pulvirenti [1]. We will also mention the pioneer works of McKean (McKean [2]), where a similar approach is applied to some other non-linear problems.

Now we proceed to examine mixed problems for equations with subordinate non-linear terms (Freidlin [9]).

Let D be a domain in R^r, ∂D being its boundary. We shall consider the first mixed problem for the equation $\frac{\partial u(t,x)}{\partial t} = L_1 u + L_2^u u$ in the cylinder $\pi = [0,\infty) \times D$. Just as in the linear case, because of the operator L_1 may be degenerate, boundary conditions should be assigned, generally speaking, not on the entire boundary. In the case of quasi-linear equations, the question of what boundary points should be considered regular, i.e. where boundary conditions should be assigned, depends, generally speaking, on the boundary function. However, in our case $(L_2^v \prec L_1)$, it turns out that a boundary point should be considered regular if and only if it is regular for the linear operator $-\frac{\partial}{\partial t} + L_1$.

So, let $\partial\tilde{\pi}$ be the set of points of the lateral boundary $\partial\pi$ of the cylinder π, which are regular for the operator $L_1 - \frac{\partial}{\partial t}$ in π. It is readily checked that $\partial\tilde{\pi}$ consists of whole rays of the cylinder π, and the projection $\partial\tilde{\pi}$ on ∂D coincides with the set of points which are t-regular for the operator L_1 in the domain D (see §3.4). The set $\partial\pi \setminus \partial\tilde{\pi}$ will be assumed inaccessible for the operator $L_1 - \frac{\partial}{\partial t}$ in the domain $(0,\infty)\times D$.

Let $g(x)$ and $\psi(t,x)$ be bounded, measurable functions defined on D and on $\partial\pi$ respectively. We shall consider the boundary value problem

$$\frac{\partial u(t,x)}{\partial t} = L_1 u + L_2^u u\,, \quad (t,x)\in(0,\infty)\times D\,;$$
$$\lim_{t\downarrow 0, x\in D} u(t,x) = g(x), \qquad \lim_{(t,x)\to(t_0,x_0)\in\partial\tilde{\pi}} u(t,x) = \psi(t_0,x_0)\,. \tag{3.18}$$

We denote by $\psi(t,x)$ the function coinciding with $g(x)$ for $t=0$, $x\in D$, and coinciding with $\psi(t,x)$ for $(t,x)\in\partial\tilde{\pi}$; $\Lambda = [\inf\psi(t,x), \sup\psi(t,x)]$. We suppose that $L_2^v \prec L_1$ in the domain $G = R^r\times\Lambda$.

By a *generalized solution* of problem (3.18), we mean a function $u(t,x)$ satisfying the relation

$$u(t,x) = E\,\psi(t-\tau_t^x, X^x_{\tau_t^x})\exp\left\{\int_0^t (\phi(X_s^x, u(t-s,X_s^x)), dW_s) - \frac{1}{2}\int_0^t |\phi(X_s^x, u(t-s,X_s^x))|^2\,ds\right\}, \tag{3.19}$$

where X_t^x is a solution of the stochastic equation

$$X_t^x - x = \int_0^t \sigma(X_s^x)\,dW_s + \int_0^t b(X_s^x)\,ds\,;$$
$$\tau^x = \inf\{t : X_t^x \notin D\},\quad \tau_t^x = \tau^x\wedge t\,. \tag{3.20}$$

We observe that the integration in the exponential factor in (3.19) may be carried out from 0 to τ_t^x. Then the right-hand side of (3.19) will not change.

THEOREM 3.3. *Suppose that* $L_2^v \prec L_1$ *in the domain* $G = R^r \times \Lambda$, *the functions* $\phi(x,v)$ *are Lipschitz continuous, and let the lateral boundary* $\partial\pi$ *of the cylinder* π *consist of two parts: of the set* $\partial\tilde{\pi}$ *of points which are regular for the operator* $L_1 - \frac{\partial}{\partial t}$, *and of the set* $\partial\pi \setminus \partial\tilde{\pi}$, *inaccessible for* $L_1 - \frac{\partial}{\partial t}$. *Then problem (3.18) has a unique generalized solution in the class of bounded measurable functions. This solution takes boundary values at every continuity point of the functions* $\psi(t,x)$ *which belongs to* $D \cup \partial\tilde{\pi}$. *If the projection of the set* $\partial\tilde{\pi}$ *on* ∂D *is uniformly t-regular for the operator* L_1 *in* D, *and the function* $\psi(t,x)$ *is continuous on the closure of the set* $D \cup \partial\tilde{\pi}$, *then the generalized solution is continuous for* $t \geq 0$, $x \in D$.

The proof of this theorem is arranged according to the same scheme as the proof of Theorem 3.1, so we shall only outline it here.

Let us consider the Banach space $B_{T,D}$ of bounded measurable functions $u(t,x)$, $0 \leq t \leq T$, $x \in D$, $\|u\| = \sup_{0 \leq t \leq T,\, x \in D} |u(t,x)|$. In the space $B_{T,D}$, we shall consider the operator

$$\tilde{F}[u] = E\,\psi(t-\tau_t^x, X^x_{\tau_t^x}) \exp\{\tilde{\mathfrak{J}}_0^t(\phi(u))\},$$

where X_t^x and τ_t^x are specified by equality (3.20), and

$$\tilde{\mathfrak{J}}_a^t(\phi(u)) = \int_a^t (\phi(X_s^x, u(t-s, X_s^x)), dW_s) - \frac{1}{2}\int_a^t |\phi(X_s^x, u(t-s, X_s^x))|^2\, ds\,.$$

Analogously to the way it was done when proving Theorem 3.1, one can verify that the operator $\tilde{F}$ is a contraction in the space $B_{t_0,D}$ for some $t_0 > 0$. The magnitude of t_0 depends on $\sup_{\partial\tilde{\pi} \cup D} |\psi(t,x)|$, $\sup_{(x,v) \in G} |\phi(x,v)|$, and on the Lipschitz constant of the function $\phi(x,v)$ in v. This immediately implies that the operator $\tilde{F}$ in the space $B_{T,D}$ has only one fixed point for any $T < \infty$, and therefore, the solution of problem (3.19) exists and is unique.

Let $(t_0,x_0) \in \partial\tilde{\pi}$ and $\psi(t,x)$ be continuous at a point (t_0,x_0). Since $(t_0,x_0) \in \partial\tilde{\pi}$, $\lim_{x\to x_0, x\in D} P_x\{\tau^x > t\} = 0$. From this, noting that the coefficients of stochastic equation (3.19) are bounded, we conclude that, for any $\delta > 0$.

$$\lim_{(t,x)\to(t_0,x_0)} P\{|\psi(t-\tau_t^x, X^x_{\tau_t^x}) - \psi(t_0,x_0)| > \delta\} = 0 .$$

Using this relation and the t-regularity of the point x_0 for the operator L_1 in D, we deduce from (3.19) that the generalized solution satisfies the boundary values at the point (t_0,x_0).

To prove the last claim of the theorem, it is sufficient to check that if the projection of the set $\partial\tilde{\pi}$ on ∂D is uniformly t-regular for L_1, and $\psi(t,x)$ is continuous, then the operator $\tilde{F}$ transforms the space of bounded continuous functions on $[0,T] \times (D \cup \partial D)$ into itself. This follows from the bounds given by Lemma 3.1 and from the fact that the trajectories of system (3.20) depend continuously on the initial conditions. □

One can prove that if the operator L_1 does not degenerate near the boundary and the functions $g(x)$, $\phi(x,u)$, and the coefficients of stochastic equation (3.20) are sufficiently smooth, then the solution of the mixed problem has any preassigned number of derivatives.

Completing this section we make a few remarks.

1. The reasonings of this section may readily be carried over to the case when terms of the form $c(x,u)u$ are involved in the operator L_2^u:

$$L_2^u u = \sum_{i=1}^{r} B^i(x,u) \frac{\partial u}{\partial x^i} + c(x,u)\, u .$$

In this case the equation defining the generalized solution, for example, in the case of Cauchy's problem, has the form

$$u(t,x) = Eg(X_t^x)\exp\left\{\int_0^t (\phi(X_s^x, u(t-s,X_s^x)), dW_s) - \right.$$

$$\left. -\frac{1}{2}\int_0^t |\phi(X_s^x, u(t-s,X_s^x))|^2 ds + \int_0^t c(X_s^x, u(t-s,X_s^x))\, ds\right\},$$

where X_t^x just as before is determined by equation (3.3). In particular, if $B(x,u) \equiv 0$, then the subordination condition holds, and the Cauchy problem $\frac{\partial u}{\partial t} = L_1 u + c(x,u)\,u$, $u(0,x) = g(x)$, has a generalized solution for an arbitrary operator L_1. If the coefficients of the operator L_1, the initial function, and the function $c(x,u)$ are smooth enough, then the generalized solution will be smooth as well. The next two chapters will deal with a number of problems for the equations of the form $\frac{\partial u}{\partial t} = L_1 u + c(x,u)\,u$.

2. One can consider the Dirichlet problem for equations with subordinate terms. Suppose that the projection on some axis x^1 of the domain in which the problem is considered, is small enough, and let the operator L_1 have, for example, a non-degenerate diffusion along the x^1-axis. Then it is possible to define the generalized solution with an equality similar to (3.4) or (3.19). If the boundary of the domain consists of a part inaccessible for the operator L_1 and of a part regular for L_1 in a sufficiently strong sense, then boundary conditions for the non-linear equation should be assigned only on the part of the boundary being regular for L_1. Under these conditions one can prove the existence of a unique generalized solution and examine its properties. The limit behavior as $t \to \infty$ of a solution of the mixed problem is closely related to the Dirichlet problem. A number of results in this direction can be found in the paper by Nissio [1].

§5.4 *On a class of systems of differential equations*

This section is concerned with some linear and quasi-linear systems of differential equations connected with diffusion processes. The important class of so-called diffusion-reaction equations are covered here. These equations describe the change with time of the concentration of some substances due to diffusion and to chemical reactions. We shall not strive for more generality and, as a rule, shall consider systems of two equations.

So, first we shall consider the linear system

$$\frac{\partial u_k(t,x)}{\partial t} = L_k u_k + c_{k1}(x)\, u_1 + c_{k2}(x)\, u_2 \,,$$
$$t > 0,\ x \in R^r;\ u_k(0,x) = g_k(x),\ k = 1,2 \,. \tag{4.1}$$

Here

$$L_k = \frac{1}{2} \sum a_k^{ij}(x) \frac{\partial^2}{\partial x^i \partial x^j} + \sum b_k^i(x) \frac{\partial}{\partial x^i}\,;\quad k = 1,2\,,$$

$$\sum a_k^{ij}(x)\lambda_i \lambda_j \geq 0\,.$$

The coefficients a_k^{ij}, b_k^i are assumed to be bounded and smooth enough. Usually, it is sufficient to suppose that $a_k^{ij}(x)$ have bounded second-order derivatives, and b_k^i have bounded first-order derivatives. The functions $c_{kj}(x)$ in (4.1) are assumed bounded from above and continuous. Moreover, let $c_{12}(x) \geq 0$ and $c_{21}(x) \geq 0$. We observe that unless both these functions are non-positive, then by going over to the new unknown functions $\tilde{u}_1 = u_1$, $\tilde{u}_2 = -u_2$ one can ensure that the corresponding coefficients are non-negative.

Let $\sigma_1(x)$, $\sigma_2(x)$ be square matrices with Lipschitz continuous elements such that $\sigma_1\sigma_1^* = (a_1^{ij}(x))$, $\sigma_2\sigma_2^* = (a_2^{ij}(x))$; $b_k = (b_k^1(x), \cdots, b_k^r(x))$, $k = 1,2$; W_t being a Wiener process in R^r. We shall consider the Markov process $(X_t, \nu_t, P_{x,i})$ in the state space $R^r \times \{1,2\}$, connected with system (4.1); and the corresponding Markov family $(X_t^{x,i}, \nu_t^{x,i}, P)$. If the indices x,i are written as superscripts on the trajectories, then the Markov family is considered; if the indices are subscripts on the probabilities or expectations, then the corresponding Markov process is. The transformation from a process to a family and conversely is carried out in a standard way. The family $(X_t^{x,i}, \nu_t^{x,i}, P)$ is defined via the stochastic equation

$$dX_t^{x,i} = \sigma_{\nu_t^{x,i}}(X_t^{x,i})dW_t + b_{\nu_t^{x,i}}(X_t^{x,i})dt\,, \quad X_0^{x,i} = x\,, \quad \nu_0^{x,i} = i\,, \tag{4.2}$$

where $\nu_t^{x,i}$ is a random process with two states 1 and 2 for which

$$P\{\nu_{t+\Delta}^{x,i} = k \mid \nu_s^{x,i}, X_s^{x,i}, s \in [0,t]\} = P\{\nu_{t+\Delta}^{x,i} = k \mid \nu_t^{x,i}, X_t^{x,i}\} =$$
$$= c_{\nu_t^{x,i},k}(X_t^{x,i})\,\Delta + o(\Delta)\,, \quad \Delta \downarrow 0\,, \quad k = 1,2\,; \quad k \neq \nu_t^{x,i}\,. \tag{4.3}$$

It is easily checked (we leave this to the reader) that, under the assumptions made on the coefficients, such a Markov process exists. We will compute its infinitesimal operator A. Let the function $f(x,i)$, $x \in R^r$, $i \in \{1,2\}$, be bounded and have uniformly continuous bounded first- and second-order derivatives in x. We shall denote by ζ the time of the first jump of the process $\nu_t : \zeta = \inf\{t : \nu_t \neq \nu_0\}$; and let χ_h be the indicator of the set $\{\zeta > h\}$, $h > 0$. By (4.2) and (4.3) we obtain:

$$Af(x,i) = \lim_{h \downarrow 0} h^{-1}[E_{x,i} f(X_h, \nu_h) - f(x,i)] =$$
$$= \lim_{h \downarrow 0} h^{-1} E_{x,i}(f(X_h, \nu_h) - f(x,i))\,\chi_h +$$
$$+ \lim_{h \downarrow 0} h^{-1} E_{x,i}(f(X_h, \nu_h) - f(x,i))\,(1 - \chi_h) = \tag{4.4}$$
$$= L_i f(x,i) - c_{ij}(x) f(x,i) + c_{ij}(x) f(x,j)\,, \quad i \neq j\,.$$

The convergence in this equality is uniform in $x \in R^r$, $i \in \{1,2\}$. Thus, the infinitesimal operator is defined on smooth functions $f(x,i)$ and has the above-indicated form.

The representation of the solution of problem (4.1) in the form of the mean value of some functional of the trajectories of the process $(X_t, \nu_t; P_{x,i})$ to be given in the following theorem, is a version of the Feynman-Kac formula.

THEOREM 4.1. *Suppose that the above cited assumptions on the coefficients of equations (4.1) are fulfilled. In addition, we shall assume that* $\sum_{i,j=1}^{r} a_k^{ij}(x)\lambda_i\lambda_j > c\sum \lambda_i^2$, $k = 1,2$, $c > 0$, *and let the initial functions* $g_1(x)$, $g_2(x)$ *be bounded and continuous. Then the solution of Cauchy problem (4.1) may be represented as follows*

$$u_k(t,x) = E_{x,k} g_{\nu(t)}(X_t) \exp\left\{\int_0^t c(\nu_s, X_s)\,ds\right\}, \tag{4.5}$$

where $c(i,x) = c_{ii}(x) + c_{ij}(x)$, $(i,j,k \in \{1,2\},\ i \neq j)$.

The proof of this theorem, provided one notes (4.4), does not differ in essence from the proof of the Feynman-Kac formula cited in §2.1, and we shall drop it. □

If the operator L_1 or L_2 is degenerate, then formula (4.5) gives the generalized solution of problem (4.1). One can prove that in the case when the coefficients and initial function are smooth enough, this generalized solution is a classical one, even if the operators L_k have degenerations. Formulae (4.5) define the solution of problem (4.1) also in the case when the initial functions have discontinuities, the initial conditions being satisfied, generally speaking, only at the continuity points of the initial functions.

Representations of the form (4.5) may be utilized for examining the behavior of the solution of system (4.1) as $t \to \infty$, when studying problems with small parameter and in problems with degenerations. An elegant remark due to Kac permits giving the representation of the solution of the telegrapher's equation through a random process of the $(X_t, \nu_t; P_{x,i})$ type.

Note that the solution of the first boundary value problem for the stationary system of equations in some domain $D \subset R^r$ can also be written down in the form of the mean value of proper functionals of the trajectories of the process $(X_t, \nu_t; P_{x,i})$:

$$\begin{gathered} L_1 u_1(x) + c_{11}(x) u_1(x) + c_{12}(x) u_2(x) = 0 , \\ L_2 u_2(x) + c_{21}(x) u_1(x) + c_{22}(x) u_2(x) = 0 , \\ x \in D;\ u_k(x)\big|_{x \in \partial D} = \psi_k(x),\ k = 1,2 . \end{gathered} \tag{4.6}$$

If $\tau_D = \inf\{t : X_t \notin D\}$, then the solution of problem (4.6) may be written down as follows:

$$u_k(x) = E_{x,k} \psi_{\nu_{\tau_D}}(X_{\tau_D}) \exp\left\{ \int_0^{\tau_D} c(\nu_s, X_s)\, ds \right\} , \quad k = 1,2 . \tag{4.7}$$

Here one should make some additional assumptions ensuring the finiteness with probability 1 of the variable τ_D, the regularity of the boundary points, and the finiteness of the expectation on the right-hand side of (4.7). For example, it is sufficient to assume that the operators L_k are uniformly elliptic, the domain D is bounded and has a smooth boundary, $c(i,x) \leq 0$, and the boundary functions $\psi_k(x)$ are continuous.

Formulae (4.7) may be used, for example, for examining various problems with small parameter in higher derivatives, for the study of system (4.6) when the operators L_1, L_2, have degenerations, for the analysis of problems in unbounded domains, and so on.

Now we proceed to study non-linear equations. We can consider a system like (4.1) where all coefficients depend on $u_k(t,x)$ and introduce the notion of the generalized solution just as in §5.1. Or we can restrict ourselves to a more special situation when only the coefficients in first-order derivatives and the functions c_{ij} depend on $u_k(t,x)$. If, in addition, we assume that the lower order terms are subordinate to the higher order terms (such is the case, for instance, when the operators L_k do not degenerate), then it is possible to proceed as in §5.3 (see M. Nissio [1]). Here we will consider an even more special situation when only the functions c_{ii}, $i \in \{1,2\}$ depend on $u_k(t,x)$.

So, let us consider the system

$$\frac{\partial u_1(t,x)}{\partial t} = L_1 u_1 + c_{11}(x,u_1,u_2)\,u_1 + c_{12}(x)\,u_2 \,,$$

$$\frac{\partial u_2(t,x)}{\partial t} = L_2 u_2 + c_{21}(x)\,u_1 + c_{22}(x,u_1,u_2)\,u_2 \,, \tag{4.8}$$

$$x \in R^r,\ t > 0;\ u_1(0,x) = g_1(x),\ u_2(0,x) = g_2(x)\,.$$

If the classical solution of problem (4.8) exists, then on account of Theorem 4.1 we arrive at the following equation for the functions u_1, u_2

$$u_k(t,x) = E_{x,k} g_{\nu_t}(X_t) \exp\left\{\int_0^t c(\nu_s, X_s, u_1(t-s,X_s), u_2(t-s,X_s))\,ds\right\}. \tag{4.9}$$

If the functions c_{ij} are continuous, bounded from above, and Lipschitz continuous in u_1, u_2, then system (4.9) has a unique solution in the class of bounded measurable functions. This solution may be constructed via the contraction mappings just as in Theorem 3.1. The solution of system (4.9) will be called a generalized solution of system (4.8).

In the next chapter we shall use formulae (4.9) for examining wave front propagation in systems of (4.8) type. Here we shall go into the question of the stabilization as $t \to \infty$ of the solutions of the following system

$$\begin{aligned} \frac{\partial u_1}{\partial t} &= L_1 u_1 + c_{11}(u_1,u_2)\,u_1 + c_{12} u_2,\ u_1(0,x) = g_1(x)\,, \\ \frac{\partial u_2}{\partial t} &= L_2 u_2 + c_{21} u_1 + c_{22}(u_1,u_2)\,u_2,\ u_2(0,x) = g_2(x)\,. \end{aligned} \tag{4.10}$$

Except for the assumption previously made we shall suppose that the functions c_{ij} do not depend directly on the space variables and $L_1 = L_2$.

We shall put $f_1(u_1,u_2) = c_{11}(u_1,u_2)u_1 + c_{12}u_2$, $f_2(u_1,u_2) = c_{21}u_1 + c_{22}(u_1,u_2)u_2$ and consider the system

$$\dot{u}_1 = f_1(u_1,u_2), \quad \dot{u}_2 = f_2(u_1,u_2) \,. \tag{4.11}$$

We shall assume that (u_1^0,u_2^0) is a stable equilibrium position of system (4.11).

THEOREM 4.2. *Suppose that* $L_1 = L_2$ *and in some convex domain* $D \ni (u_1^0,u_2^0)$ *there exists a continuous convex function* $\mathcal{H}(u_1,u_2)$ *(the Lyapunov function) such that* $\mathcal{H}(u_1^0,u_2^0) = 0$, $\mathcal{H}(u_1,u_2) > 0$ *for* $(u_1,u_2) \in D' = D\setminus(u_1^0,u_2^0)$, $\mathcal{H}(u_1,u_2)$ *being differentiable in* D. *Let, for* $(u_1,u_2) \in D'$, *the derivative of the function* $\mathcal{H}$ *by virtue of system (4.11) satisfy the inequality*

$$\frac{d\mathcal{H}}{dt}(u_1,u_2) = \frac{\partial\mathcal{H}}{\partial u_1} f_1 + \frac{\partial\mathcal{H}}{\partial u_2} f_2 < -\rho[(u_1-u_1^0)^2 + (u_2-u_2^0)^2] < 0 \,, \tag{4.12}$$

where $\rho(z)$, $z \geq 0$, $\rho(0) = 0$, *is a continuous increasing function.*

Suppose that $[G] \subset D$, *where* $[G]$ *is the closure of the set* $G = \{(u_1,u_2) : u_1 = g_1(x), u_2 = g_2(x), x \in R^r$. *Then, uniformly in* $x \in R^r$,

$$\lim_{t\to\infty} u_1(t,x) = u_1^0, \quad \lim_{t\to\infty} u_2(t,x) = u_2^0 \,.$$

Proof. First of all, we observe that in this special case (c_{ij} = const. for $i \neq j$ and $L_1 = L_2$) the random process $(X_t^{x,i}, \nu_t^{x,i})$ consists of the two independent processes $X_t^{x,i} = X_t^x$ and $\nu_t^{x,i} = \nu_t^i$, $x \in R^r$, $i \in \{1,2\}$. We shall denote by E^W and $\widetilde{E}$ the expectations corresponding to X_t^x and ν_t^i respectively.

Let us introduce the notation:

$$\mathcal{T}_0^\Delta(c,i) = \int_0^\Delta c(\nu_s^i, u_1(t+\Delta-s,X_s^x), u_2(t+\Delta-s,X_s^x))\,ds \,.$$

From (4.9), relying on the Markov property we see that

$$u_k(t+\Delta,x) = Eu_{\nu_\Delta^k}(t,X_\Delta^x)\exp\{\mathcal{T}_0^\Delta(c,k)\}\,.$$

Next, by representing the expectation with respect to the measure P in the form of the iterated integral and taking into account the convexity of the function $\mathcal{H}$ we get

$$\mathcal{H}(u_1(t+\Delta,x),\, u_2(t+\Delta,x)) =$$

$$= \mathcal{H}(E^W\widetilde{E}u_{\nu_\Delta^1}(t,X_\Delta^x)\exp\{\mathcal{T}_0^\Delta(c,1)\},\, E^W\widetilde{E}u_{\nu_\Delta^2}(t,X_\Delta^x)\exp\{\mathcal{T}_0^\Delta(c,2)\}) \leq \tag{4.13}$$

$$\leq E^W\mathcal{H}(\widetilde{E}u_{\nu_\Delta^1}(t,X_\Delta^x)\exp\{\mathcal{T}_0^\Delta(c,1)\},\, \widetilde{E}u_{\nu_\Delta^2}(t,X_\Delta^x)\exp\{\mathcal{T}_0^\Delta(c,2)\})\,.$$

Since the functions $u_k(t,x)$ and $c(i,u_1,u_2)$ are continuous and $P\{\nu_\Delta^i \neq i\} = c_{ij}\Delta + o(\Delta)$, $\Delta \downarrow 0$, the right-hand side of (4.12), as $\Delta \downarrow 0$, may be rewritten in the form

$$E^W\mathcal{H}(\widetilde{E}u_{\nu_\Delta^1}(t,X_\Delta^x)\exp\{\overline{\mathcal{T}}_0^\Delta(c,1)\},\, \widetilde{E}u_{\nu_\Delta^2}(t,X_\Delta^x)\exp\{\overline{\mathcal{T}}_0^\Delta(c,2)\} + o(\Delta)\,, \tag{4.14}$$

where

$$\overline{\mathcal{T}}_0^\Delta(c,i) = \int_0^\Delta c(\nu_s^i, u_1(t,X_\Delta^x), u_2(t,X_\Delta^x))\,ds\,.$$

Let us consider the transformation Q_Δ of the plane (u_1,u_2) into itself which is defined as follows

$$Q_\Delta : (u_1,u_2) \to (u_1^\Delta, u_2^\Delta)\,;$$

$$u_i^\Delta = \widetilde{E}u_{\nu_\Delta^i}\exp\left\{\int_0^\Delta c(\nu_s^i,u_1,u_2)\,ds\right\},\quad i\in\{1;2\}\,.$$

Note that up to an infinitesimal of a higher order as $\Delta \downarrow 0$, the transformation Q_Δ is a translation along the trajectories of dynamical system (4.11). This assertion results from the equality

$$u_i^\Delta = u_i \exp\{\Delta \cdot c(i,u_1,u_2)\}(1 - c_{ij}\Delta) + \\ + u_j c_{ij}\Delta + o(\Delta) = \\ = u_i + (u_i c_{ii}(u_1,u_2) + u_j c_{ij})\Delta + o(\Delta) = \\ = u_i + f_i(u_1,u_2)\Delta + o(\Delta), \; i \neq j, \; \Delta \downarrow 0 .$$

The argument of the function $\mathcal{H}$ in (4.14) can be written down in the form

$$Q_\Delta(u_1(t,X_\Delta^x), u_2(t,X_\Delta^x)) .$$

Let us put

$$H(t) = \sup_{x \in R^r} \mathcal{H}(u_1(t,x), u_2(t,x)) .$$

We suppose that $H(t) > h > 0$. Further, note that, up to an infinitesimal of a higher order as $\Delta \downarrow 0$, the argument of the function $\mathcal{H}$ in (4.14) is a translation of the point $(u(t,X_\Delta^x), u(t,X_\Delta^x))$ along the trajectories of system (4.11). In view of this, relations (4.12)-(4.14) imply that one can find $q = q(h) > 0$ such that for sufficiently small Δ

$$H(t+\Delta) < H(t) - q\Delta .$$

Noting that the point (u_1^0, u_2^0) is the only zero of the function $\mathcal{H}(u_1,u_2)$, the foregoing inequality implies the claim of the theorem. □

We will provide a simple example illustrating the importance of the conditions of convexity type in Theorem 4.2. Let us consider the system

$$\frac{\partial u_1}{\partial t} = \frac{\partial^2 u_1}{\partial x^2} + cu_1 , \quad \frac{\partial u_2}{\partial t} = \frac{\partial^2 u_2}{\partial x^2} + cu_2 , \\ u_1(0,x) = \sin x, \; u_2(0,x) = \cos x, \; t > 0, \; x \in R^1 . \tag{4.15}$$

The point $(0,0)$ of the plane (u_1,u_2) is an unstable equilibrium point of the corresponding system (4.11) for $c > 0$. All trajectories of this system other than the point $(0,0)$ run to infinity as $t \to \infty$. The pair of functions $u_1(t,x) = e^{(c-1)t} \sin x$, $u_2(t,x) = e^{(c-1)t} \cos x$ is the solution of system (4.15). If $c \in (0,1)$, then for any $x \in R^1$, the point $(u_1(t,x), u_2(t,x))$ is attracted to the unstable equilibrium point $(0,0)$ as $t \to \infty$.

§5.5 *Parabolic equations and branching diffusion processes*

Up to now we have been considering quasi-linear parabolic equations by using the representations of solutions of linear problems in the form of functional integrals. These representations gave us the corresponding integral equations for solutions of non-linear problems. This section will present an explicit representation for solutions of a class of quasi-linear equations. Such a representation can be obtained by considering a branching diffusion process connected with the equation. We shall not strive for more generality and only show the construction in the simplest case.

So, we shall consider the Cauchy problem

$$\frac{\partial u(t,x)}{\partial t} = \frac{1}{2}\frac{\partial^2 u}{\partial x^2} + \sum_{k=1}^{\infty} p_k u^k - u \,, \qquad u(0,x) = g(x), \; x \in R^1, \; t > 0 \,. \tag{5.1}$$

Here $p_1, p_2, \cdots, p_r, \cdots \geq 0$, $\sum_1^\infty p_k = 1$. A *branching diffusion process* is connected with equation (5.1) which is arranged as follows. At the initial time, there is one particle which travels along the line R^1 and, at a random time τ, it splits into a random number ν of particles. Each of the newborn particles goes on travelling independently of the past travel of the particles and of the behavior of other descendants and again splits in a random time τ_1. The new particles also travel and split and so on. The travel of the particles is described by a Wiener process (this is due to the fact that equation (5.1) involves the generator of the Wiener process $\frac{1}{2}\frac{\partial^2}{\partial x^2}$). The distribution of the time τ between the birth of a particle and the first splitting is the same for all particles. Namely, it is the exponential distribution: $P\{\tau > t\} = e^{-t}$, the random variable τ being independent of the travel of this particle and of the behavior of other particles. The number ν of descendants of a particle in one splitting has the distribution $\{p_k\}$ and also does not depend on the past travel of the process and on the behavior of other particles. By time t there will be a random number $n(t) = 1, 2, \cdots$ of the particles which will occupy positions $X_t^1, \cdots, X_t^{n(t)}$.

Let us consider the function

$$u(t,x) = E_x\, g(X_t^1)\, g(X_t^2) \cdots g(X_t^{n(t)}) . \tag{5.2}$$

Here the index x in the expectation sign indicates the initial position of the initial particle. Generally speaking, the expectation in (5.2) may become infinite for some t. If $\sup_{x \in R^1} |g(x)| < 1$, then the expectation in (5.2) is always finite.

THEOREM 5.1. *Suppose that* $\sup_{x \in R^1} |g(x)| < 1$. *Then the function* $u(t,x)$ *defined by equality (5.2) is a solution of problem (5.1).*

Proof follows the work by McKean [3]. Let τ be the time at which the initial particle splits. Then we can decompose the expectation in (5.2) into two summands:

$$u(t,x) = E_x\, g(X_t^1) \cdots g(X_t^{n(t)}) = E_x \chi_{\tau > t}\, g(X_t^1) + \tag{5.3}$$
$$+ \int_0^t p_\tau(s)\, ds \int_{R^1} p(s,x,y) \sum_{k=1}^{\infty} p_k\, u^k(t-s,y)\, dy ,$$

where $\chi_{\tau<t}$ is the indicator of the set $\{\tau < t\}$, $p_\tau(s)$ is the density function of the random variable τ, $p(s,x,y)$ is the transition density function of the Wiener process. Next, from the definition of the branching diffusion process it results that the right-hand side of (5.3) may be transformed into the form

$$u(t,x) = e^{-t} E_x g(X_t^1) + \sum_{k=1}^{\infty} p_k \int_0^t e^{-s} ds \int_{-\infty}^{\infty} p(s,x,y) u^k(t-s,y) dy =$$

$$= e^{-t} E_x g(X_t^1) + \sum_1^{\infty} p_k \int_0^t e^{-(t-s)} ds \int_{-\infty}^{\infty} p(t-s,x,y) dy\, u^k(s,y) . \tag{5.4}$$

The right-hand side of this equality is continuous in (t,x) and differentiable in t; therefore, the function $u(t,x)$ is continuous and differentiable in t. Differentiating (5.4) in t yields

$$\frac{\partial u}{\partial t}(t,x) = -e^{-t} E_x g(X_t^1) + e^{-t} \frac{\partial E_x g(X_t^1)}{\partial t} +$$

$$+ \sum_{k=1}^{\infty} p_k u^k(t,x) - \sum_{k=1}^{\infty} p_k \int_0^t e^{-(t-s)} ds \int_{-\infty}^{\infty} p(t-s,x,y) dy\, u^k(s,y) + \tag{5.5}$$

$$+ \sum_{k=1}^{\infty} p_k \int_0^t e^{-(t-s)} ds \int_{-\infty}^{\infty} \frac{\partial p(t-s,x,y)}{\partial t} u^k(s,y) dy .$$

Now we shall use the following relations

$$\frac{\partial E_x g(X_t^1)}{\partial t} = \frac{1}{2} \frac{\partial^2 E_x g(X_t^1)}{\partial x^2}, \quad \frac{\partial p(t-s,x,y)}{\partial t} = \frac{1}{2} \frac{\partial^2 p(t-s,x,y)}{\partial x^2} .$$

Noting these equalities, (5.5) gives

$$\frac{\partial u}{\partial t}(t,x) = -e^{-t} E_x g(X_t^1) + \frac{1}{2}\frac{\partial^2(e^{-t}E_x g(X_t^1))}{\partial x^2} + \sum_{k=1}^{\infty} p_k u^k(t,x) -$$

$$- \sum_1^{\infty} p_k \int_0^t e^{-(t-s)} ds \int_{-\infty}^{\infty} p(t-s,x,y)\, dy\, u^k(s,y) + \tag{5.6}$$

$$+ \frac{1}{2}\frac{\partial^2}{\partial x^2}\left(\sum_{k=1}^{\infty} p_k \int_0^t e^{-(t-s)} ds \int_{-\infty}^{\infty} p(t-s,x,y)\, u^k(s,y)\, dy\right).$$

Now (5.4) implies that the sum of the first and the fourth summands on the right-hand side of (5.6) is equal to $-u(t,x)$, and the sum of the second and the fifth summands equals $\frac{1}{2}\frac{\partial^2 u(t,x)}{\partial x^2}$. Finally, we get for $u(t,x)$ the equation

$$\frac{\partial u}{\partial t}(t,x) = \frac{1}{2}\frac{\partial^2 u}{\partial x^2} + \sum_{k=1}^{\infty} p_k u^k(t,x) - u(t,x) .$$

Relying on the property of the Wiener process and taking into account that $P\{n(t)=1\} = 1-t+o(t)$ as $t \downarrow 0$, one can easily check that $\lim_{t \downarrow 0} u(t,x) = g(x)$ at every point of continuity of the initial function.

Formula (5.2) takes on an especially elegant form when the initial function is as follows

$$g(x) = \begin{cases} 1, & x \le 0, \\ 0, & x > 0. \end{cases} \tag{5.7}$$

Since $X_t^k = x + W_t^k$, where W_t^k is a standard Wiener process starting from zero, we conclude that, in the case when $g(x)$ is a step-function, formula (5.2) takes the form

$$u(t,x) = P\{x + \max_{1 \le k \le n(t)} W_t^k < 0\} = P\{\min_{1 \le k \le n(t)} W_t^k > x\} , \tag{5.8}$$

where W_t^k is the Wiener process starting from the point 0. One should bear in mind that the variables W_t^k in (5.8) are not mutually independent for different k (with large probability, recently split particles occupy close positions). McKean [3] and then Bramson [1] successfully used formula (5.8) for examining the Kolmogorov-Petrovskii-Piskunov equation (see Chapter VI). Another set of problems where it turned out to be useful to represent solutions of equations of (5.1) type with the help of some branching diffusion process is as follows. These are problems on finding necessary and sufficient conditions for the solution of Cauchy's problem to exist for all $t > 0$ (so-called explosion criterion). The works of Ikeda and Watanabe [1] are devoted to this question. The general construction of the branching processes with diffusion is available in Ikeda, Nagasawa and Watanabe [1].

Chapter VI

QUASI-LINEAR PARABOLIC EQUATIONS WITH SMALL PARAMETER. WAVE FRONTS PROPAGATION

§6.1 *Statement of problem*

Since the 1930's, a number of articles have been emerging in which the diffusion equation with the non-linear term

$$\frac{\partial u(t,x)}{\partial t} = \frac{D}{2}\frac{\partial^2 u(t,x)}{\partial x^2} + f(u(t,x)) \tag{1.1}$$

or systems of equations of a similar type, are used for describing some physical, chemical, or biological processes.

Such models has been used with success, for example, in population genetics (Fisher [1], Aronson and Weinberger [1,2]), in the theory of excitable media (see e.g. Romanovskii, Vasiliev, Yahno [1]), in chemical kinetics (Frank-Kameneckii [1]) and other fields. In particular, the function $u(t,x)$ may be interpreted as the concentration of some particles at the point $x \in R^1$ at time t. These particles diffuse with the diffusion coefficient $\frac{D}{2}$ in time and multiply themselves (or perish). The function $f(u)$ characterizes the rate of variation of concentration in the absence of diffusion. Typical forms of the function $f(u)$ may be seen among those represented in Fig. 1. In the case 1-a, the function $f(u)$ is characterized by the fact that $f'(0) = \sup_{0<u<1} u^{-1}f(u)$. In the case 1-b $0 \leq f'(0) < \sup_{0<u<1} u^{-1}f(u)$, but $f(u) > 0$ for $0 < u < 1$. For instance, the function $f(u)$ represented in Fig. 1-a means that, for small u, the rate of concentration growth is approximately proportional to the concentration, and, for

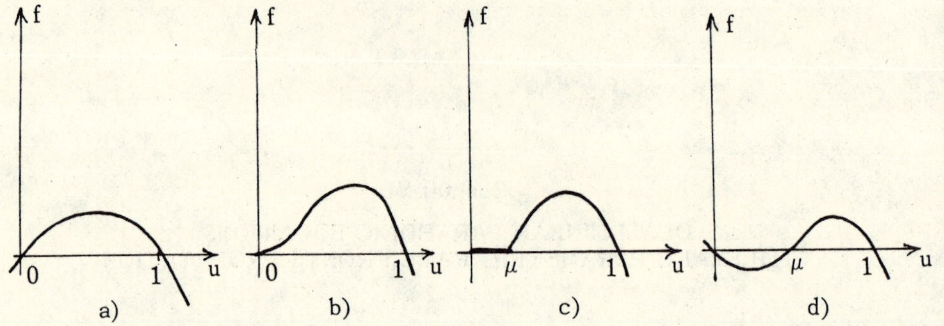

Fig. 1

the concentrations close to 1, the multiplication slows down. For $u > 1$, the particles are "killed" with the rate $|f(u)|$.

To select a unique solution of problem (1.1), one should assign initial conditions. Let

$$u(0,x) = \chi_{x<0}(x) = \begin{cases} 1, & \text{for } x < 0 \\ 0, & \text{for } x \geq 0 . \end{cases} \tag{1.2}$$

This means that, at the initial moment, the particles occupied the region $x < 0$. The evolution of the function $u(t,x)$ in time is defined by two processes (for the sake of simplicity, we will speak of the case 1-a; qualitative behavior of $u(t,x)$ in the other cases is similar). In the first place, a particle will diffuse along the x-axis. Secondly, at the expense of the non-linear term, the multiplication of particles will occur at those points where the concentration does not exceed 1. The interaction of these factors causes the initial step-function to smooth out; and this smooth, monotonically decreasing profile will, with some velocity, travel as a wave from left to right. After a large time, a certain profile and a certain velocity of this wave will be attained, namely, the function $u(t,x)$ will be, in a sense, close to $V(x-\alpha t)$, where $V(\xi)$ is the profile of the limit wave, and α is its velocity. To derive the equation for the function $V(\xi), -\infty < \xi < \infty$, one should substitute the function $V(x-\alpha t)$ in equation (1.1):

$$\frac{D}{2} V''_{\xi\xi} + \alpha V'_{\xi} + f(V) = 0, \quad -\infty < \xi < \infty . \tag{1.3}$$

If (1.2) is taken as the initial function, then the condition at infinity

$$\lim_{\xi \to -\infty} V(\xi) = 1, \quad \lim_{\xi \to \infty} V(\xi) = 0 \tag{1.4}$$

should be joined to equation (1.3). In equation (1.3), the velocity α is still unknown. To determine the wave velocity α, we shall suppose, for the present, that $f(u) = cu$, where $c = f'_u(0)$ (we recall that now we are studying the function $f(u)$ presented in Fig. 1-a, so that $c = f'(0) = \sup_{0<u<1} u^{-1} f(u) > 0$). We will consider, along with problem (1.1)-(1.2), the following linear problem

$$\frac{\partial \tilde{u}(t,x)}{\partial t} = \frac{D}{2} \frac{\partial^2 \tilde{u}}{\partial x^2} + c\tilde{u}, \quad \tilde{u}(0,x) = \chi_{x<0}(x) .$$

As it follows from §2.1,

$$\tilde{u}(t,x) = e^{ct} P\{x + \sqrt{D}\, W_t \le 0\} , \tag{1.5}$$

where W_t is a Wiener process on the line, $W_0 = 0$. This formula implies that

$$\lim_{t\to\infty} \tilde{u}(t, \beta t) = \lim_{t\to\infty} e^{ct} \int_{\beta t}^{\infty} \frac{1}{\sqrt{2\pi D t}} e^{-\frac{z^2}{2tD}} dz = \begin{cases} 0, & \text{for } \beta > \sqrt{2cD} \\ \infty, & \text{for } \beta < \sqrt{2cD} . \end{cases} \tag{1.6}$$

Equality (1.6) may be interpreted as follows: for large t, the velocity of propagation of the large values domain of the function $\tilde{u}(t,x)$ is asymptotically equal to $\sqrt{2cD}$. Now we shall return to problem (1.1)-(1.2). The Feynman-Kac formula implies the following equation for $u(t,x)$:

$$u(t,x) = E\,\chi_{x<0}(x+\sqrt{D}\,W_t)\exp\left\{\int_0^t c(u(t-s,x+\sqrt{D}\,W_s))\,ds\right\}, \tag{1.7}$$

where $c(u) = u^{-1}f(u)$. According to assumption, $c = f_u'(0) \geq c(u)$. Noting this we conclude from (1.5) and (1.7) that $u(t,x) \leq \tilde{u}(t,x)$, and thus by (1.6) $\lim_{t\to\infty} u(t,\beta t) = 0$ for $\beta > \sqrt{2cD}$. Therefore, for large t, the domain of small values of the function $u(t,x)$ contracts no more rapidly than with velocity $\sqrt{2Dc}$. Of course, in the domain of the values of $u(t,x)$ remote from zero, the behavior of the solution of problem (1.1)-(1.2) differs from that of $\tilde{u}(t,x)$, because cu approximates $f(u)$ only near zero. The fact that the function $f(u)$ is negative for $u > 1$ does not allow the solution of problem (1.1)-(1.2) to exceed 1. Using equation (1.7) and the strong Markov property of the Wiener process, one can prove, that $\lim_{t\to\infty} u(t,\beta t) = 1$ for $\beta < \sqrt{2cD}$ (see the proof of Theorem 1.2). This means that for large t, the domain of the values close to 1, spreads with velocity not smaller than $\sqrt{2Dc}$. If, for large t, the function $u(t,x)$ is close to a solution of the running wave $V(x-\alpha t)$ type, then the above reasoning implies that $\alpha = \sqrt{2cD}$.

Equation (1.1)-(1.2) with the function $f(u)$ represented in Fig. 1-a have been examined in the article by Kolmogorov, Petrovskii and Piskunov [1]. Henceforth equation (1.1) with such a function $f(u)$ will be referred to as KPP equation. In the foregoing article, it is established that, for every $t > 0$, the solution $u(t,x)$ of problem (1.1)-(1.2) monotonically decreases in x from 1 to 0. Let us denote by $x^* = x^*(t)$ the only value x for which $u(t,x^*) = \frac{1}{2}$. Then

$$\lim_{t\to\infty} t^{-1}x^*(t) = \sqrt{2cD}, \quad \lim_{t\to\infty} u(t,x^*(t)+z) = V(z),$$

where $V(z)$ is a solution of problem (1.3)-(1.4) for $\alpha = \sqrt{2cD}$ which obeys the condition $V(0) = \frac{1}{2}$. The solution of problem (1.3)-(1.4) with the extra condition $V(0) = \frac{1}{2}$ exists and is unique for any $\alpha \geq \sqrt{2cD}$.

In the case when the non-linear term in (1.1) has the form represented in Fig. 1-d, it turns out that only one value α^* of the parameter α exists for which problem (1.3)-(1.4) is solvable. In this case, for large t, the solution of problem (1.3)-(1.4) is close to $V(x-\alpha^* t)$, where $V(\xi)$ is a solution of problem (1.3)-(1.4) for $\alpha = \alpha^*$ (see Kanel' [1], Aronson and Weinberger [1,2], Fife and McLeod [1]). The velocity α^* may be both positive and negative. The sign of α^* coincides with that of $\int_0^1 f(u)\,du$. When the behavior of the function $f(u)$ is "intermediate" (see Fig. 1-b and 1-c), then, for large t, the solution of problem (1.1)-(1.2) also approaches a solution of the running wave type having a proper velocity α and a profile defined by problem (1.3)-(1.4).

Hence, two questions are involved in the study of problem (1.1)-(1.2). The first one is about the profile of the wave which will be reached in a large time, the second one is about the velocity at which this wave travels. The first question is more delicate and, presumably, less interesting from the experimenter's point of view. In this connection, it is desirable to separate these two questions. It turns out that this may be done, at least, if the non-linear term has the form represented in Fig. 1-a.

To carry out such a separation, let us consider the function $u^\varepsilon(t,x) = u(t/\varepsilon, x/\varepsilon)$. For large t, this function is to be close to $V\left(\frac{x-\alpha t}{\varepsilon}\right)$. For small positive ε, $V\left(\frac{x-\alpha t}{\varepsilon}\right)$ is close to 1 for $x < \alpha t$, and is close to 0 otherwise. Since $u(t,x)$ is a solution of problem (1.1)-(1.2), we have for $u^\varepsilon(t,x)$:

$$\frac{\partial u^\varepsilon}{\partial t} = \frac{\varepsilon D}{2}\frac{\partial^2 u^\varepsilon}{\partial x^2} + \frac{1}{\varepsilon} f(u^\varepsilon), \quad u^\varepsilon(0,x) = \chi_{x<0}(x). \tag{1.8}$$

For any $t > 0$, $x \in R^1$, the solution of problem (1.8) tends to a step function:

$$u^\varepsilon(t,x) \approx V(\varepsilon^{-1}(x-\alpha t)) \to \chi_{x<0}(x-\alpha t)$$

as $\varepsilon \downarrow 0$. This step function travels from left to right with velocity α. Thus, when studying, (in the preliminary examination) the behavior of the solution of problem (1.8) as $\varepsilon \downarrow 0$, only the question concerning the velocity of the wave propagation remains. The wave profile has the standard form of a step function.

Going over to equation (1.8) is also convenient in the respect that this enables us to extend the problem to the case of variable coefficients and allows us to examine a number of new effects related to space non-homogeneity. In this chapter, we shall be concerned with the following problem with small parameter

$$\frac{\partial u^{\varepsilon}(t,x)}{\partial t} = \frac{\varepsilon}{2} \sum_{i,j=1}^{r} \frac{\partial}{\partial x^i}\left(a^{ij}(x) \frac{\partial u^{\varepsilon}}{\partial x^j}\right) + \frac{1}{\varepsilon} f(x,u^{\varepsilon}) ,$$
$$t > 0 , \ x \in R^r , \ u^{\varepsilon}(0,x) = g(x) . \tag{1.9}$$

Here the $a^{ij}(x)$ are bounded functions having bounded second-order derivatives and such that the form $\sum_{i,j=1}^{r} a^{ij}(x)\lambda_i\lambda_j$ does not degenerate uniformly in R^r. For every x, the function $f(x,u)$ has the form of one of the functions represented in Fig. 1. Hypotheses on $f(x,u)$ will be refined later on. The initial function $g(x)$ is always assumed to be non-negative, bounded and continuous everywhere in R^r except for, possibly, a finite number of smooth manifolds where $g(x)$ has gaps. Moreover, we will assume that the support G_0 of the function $g(x)$ does not coincide with $R^r : G_0 = \{x \in R^r : g(x) > 0\} \neq R^r$.

How will the solution of problem (1.9) behave for small ε? First, we will go into this question far from the zeros of the function $f(x,u)$, i.e. far from the points $u = 0$, $u = 1$ (and also far from $u = \mu \in (0,1)$ for the case represented in Fig. 1-d, and far from $[0,\mu]$ for the case in Fig. 1-c). Far from the zeros of the function $f(x,u)$, the function $u^{\varepsilon}(t,x)$ will basically change in accordance with the ordinary differential equation

$$\frac{d\tilde{u}^{\varepsilon}(t,x)}{dt} = \varepsilon^{-1} f(x,\tilde{u}^{\varepsilon}(t,x)), \tilde{u}^{\varepsilon}(0,x) = g(x) . \tag{1.10}$$

This implies that, for small ε, the function $u^\varepsilon(t,x)$ is close to some step-function $\kappa_t(x)$ taking two values: 0 and 1. Let G_t be the support of the function $\kappa_t(x)$: $G_t = \{x \in R^r : \kappa_t(x) = 1\}$. For small ε, the evolution of the function $u^\varepsilon(t,x)$, in the preliminary study, is described by the change of the set G_t with time. The boundary ∂G_t of the set G_t may be interpreted as the wave front. This wave front travels with velocity of order 1 as $\varepsilon \downarrow 0$. The advance of this front is an implication of the interaction between the diffusion term in equation (1.9) and the non-linear term accounting for the multiplication and killing of particles.

As will be seen in §6.4, in the case represented in Fig. 1-d, the advances of the front are, in a sense, of local nature. Namely, for small Δ, the set $\partial G_{t+\Delta}$ is close to G_t; the variation of the front near a point $x \in \partial G_t$ in a small time is determined by the set G_t and by the behavior of the coefficients of equation (1.9) near the point x. In the case when, for every x, the function $f(x,u)$ has the form represented in Fig. 1-c, the set G_t changes in a similar fashion.

If, for some x, the function $f(x,u)$ has the form represented in Fig. 1-a, then generally speaking the evolution of the front is no longer of local nature: namely, for small Δ, the set $G_{t+\Delta}$ may contain points which are at a finite distance from G_t. Here we have an "appearance of new sources" such that a wave starts propagating in all directions away from each of these sources. If $f'_u(x,0) = c = \text{const.}$ does not depend on x, then this effect does not occur in the case 1-a either. Problem (1.9) with the non-linear term of the form 1-a will be examined in the next section. The effect of "appearing new sources" takes place in the case of the non-linear term of the form 1-b too.

Equation (1.1) and systems of equations of similar kind are used, in particular, to describe the spreading of excitation in an excitable medium. In such a model, the excitation is associated with diffusion and chemical transmutations of some molecules and ions. Parallel to such a chemical description, a phenomenological model of spreading of excitation also exists. According to the latter, every point in a medium may be in one of two states: excited and non-excited. In the domain where the excitation spreads a velocities field is assigned. To be more exact, one indicates the velocity $v(x,e)$ of spreading at the point x in the direction of the vector e. Suppose that at time $t = 0$, the excitation occupied a domain $G_0 \subset R^r$. Then, by time t, it will occupy a domain $\tilde{G}_t^v$, consisting of the points at which the excitation has arrived when spreading in all directions in accordance with the assigned velocities field:

$$\tilde{G}_t^v = \left\{ x \in R^r : \inf_{\phi_0 \in G_0, \phi_1 = x} \int_0^1 \frac{|\dot{\phi}_s| ds}{v(\phi_s, \dot{\phi}_s)} \leq t \right\}.$$

The questions suggest themselves, first, how one can go over from the kinetic description of spreading the excitation to the phenomenological one via the velocities field, and the second question is: when is such a transition possible?

It is the limit passage as $\varepsilon \downarrow 0$ in equation (1.9) which leads to a separation of the space R^r into two sets: a set G_t where $u^\varepsilon(t,x)$ is close to 1 and the complement to G_t where $u^\varepsilon(t,x)$ is close to 0. If the non-linear term has the form represented in Fig. 1-d, then, as will be seen later on, the spreading of the excitation may be described through a proper velocity field $v(x,e)$: $G_t = \tilde{G}_t^v$. We shall show how this field may be determined by the equation.

If the non-linear term is of the form 1-a, it turns out that the expansion of the domain of excitation cannot, generally speaking, be described through a velocity field.

When studying problems with non-linear boundary conditions, it is also suitable to go over to problem (1.9) with small parameter. Such models are useful for studying the spread of excitation when multiplication occurs only near the boundary (see, e.g. Freidlin and Sivak [1]). Section 6 is devoted to these points.

An alternative possible generalization is as follows. One can examine wave propagation when the random particle motion is not a diffusion process, but some other random process. This generalization may be performed especially simply in the case where the particles motion is described by a Markov process. In §6.5, the case will be treated when the particles motion is described by one of the components of a diffusion process.

In this chapter, we shall also examine certain systems of diffusion-reaction equations similar to KPP equation, and compute the velocity of wave propagation (§6.7).

Almost all our constructions in this chapter are underlain by the Feynman-Kac formula and by the estimates for probabilities of large deviations for various classes of random processes.

§6.2 *Generalized KPP equation*

In this section, we shall consider equation (1.9) and some of its generalizations with the non-linear term $f(x,u)$ which, for every $x \in R^r$, has the form presented in Fig. 1-a. To be more exact, we suppose that $f(x,0) = f(x,1) = 0$ for $x \in R^r$; $f(x,u) > 0$ for $u \in (0,1)$, $x \in R^r$; $f(x,u) < 0$ for $u \notin [0,1]$, $x \in R^r$. Put $c(x,u) = u^{-1}f(x,u)$ for $u > 0$ and $c(x,0) = \lim_{u \downarrow 0} u^{-1}f(x,u)$ (Fig. 2). We assume that the function $c(x,u)$, $x \in R^r$, $u \in [0,\infty)$, is continuous and satisfies a Lipschitz condition in u. Let $\max_{0 \le u \le 1} c(x,u) = c(x,0) = c(x)$. The class of all such functions $f(x,u)$ will be denoted by $\mathfrak{A}$. When considering examples, it will sometimes be convenient for us to allow the function $c(x,u)$ to have gaps in x. These gaps bring about no essential changes in our reasonings, and we shall not dwell on the case of discontinuous functions $c(x,u)$ specially. All the results of this section are taken from Freidlin [15].

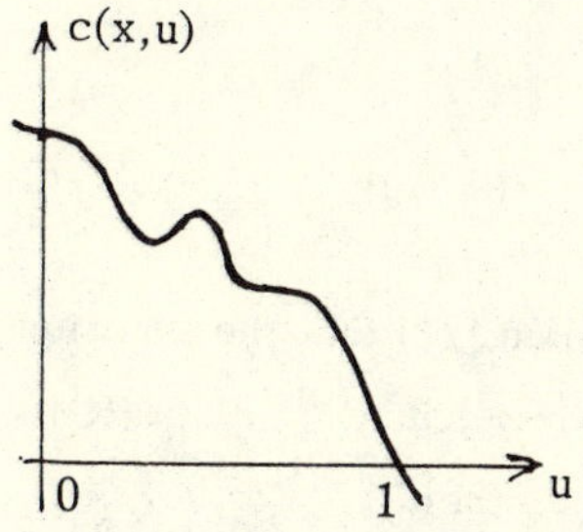

Fig. 2

So, consider the Cauchy problem

$$\frac{\partial u^\varepsilon(t,x)}{\partial t} = \frac{\varepsilon}{2} \sum_{i,j=1}^{r} \frac{\partial}{\partial x^i}\left(a^{ij}(x)\frac{\partial u^\varepsilon}{\partial x^j}\right) + \frac{1}{\varepsilon} f(x,u^\varepsilon) = L^\varepsilon u^\varepsilon + \varepsilon^{-1} f(x,u^\varepsilon)\,, \tag{2.1}$$

$$t > 0\,,\ x \in R^r\,,\ u^\varepsilon(0,x) = g(x) \geq 0\,.$$

By $(X_t^\varepsilon, P_x^\varepsilon)$ we will denote the diffusion process in R^r corresponding to the operator L^ε. As was explained in §1.7, the action functional $\lambda(\varepsilon) S_{0T}(\phi)$ in the space $C_{0,T}(R^r)$ is associated with the family of the processes $(X_t^\varepsilon, P_x^\varepsilon)$ as $\varepsilon \downarrow 0$. For absolutely continuous $\phi \in C_{0T}(R^r)$

$$S_{0T}(\phi) = \frac{1}{2}\int_0^T \sum_{i,j=1}^{r} a_{ij}(\phi_s)\dot\phi_s^i \dot\phi_s^j\, ds,\ (a_{ij}(x)) = (a^{ij}(x))^{-1}\,,$$

and $S_{0T}(\phi) = +\infty$ for the other $\phi \in C_{0T}(R^r)$. The normalizing coefficient $\lambda(\varepsilon)$ is equal to ε^{-1}. By the definition of the action functional, the following relations hold: for any function $\phi \in C_{0T}(R^r), \phi_0 = x$, and arbitrary $\gamma, \delta > 0$ there is an $\varepsilon_0 > 0$ such that, for $0 < \varepsilon_0 < \varepsilon$

$$P_x^\varepsilon\{\rho_{0T}(X^\varepsilon, \phi) < \delta\} \geq \exp\{-\varepsilon^{-1}(S_{0T}(\phi) + \gamma)\}\,. \tag{2.2}$$

Moreover, for any $s < \infty$ the set $\Phi_s = \{\phi \in C_{0T}(R^r) : \phi_0 = x, S_{0T}(\phi) \leq s\}$ is compact in $C_{0,T}(R^r)$, and also, for arbitrary $\gamma, \delta > 0$ one can find $\varepsilon_0 > 0$ such that for $\varepsilon < \varepsilon_0$

$$P_x^\varepsilon\{\rho_{0T}(X^\varepsilon, \Phi_s) \geq \delta\} \leq \exp\{-\varepsilon^{-1}(s - \gamma)\}\,. \tag{2.3}$$

It is possible to prove (see Wentzell and Freidlin [2]) that the functional $S_{0T}(\phi)$ is lower semi-continuous in $C_{0T}(R^r)$, i.e. if $\phi^{(n)} \in C_{0T}(R^r)$ converges to ϕ uniformly on $[0,T]$ as $n \to \infty$, then

$$S_{0T}(\phi) \leq \varliminf_{n\to\infty} S_{0T}(\phi^{(n)})\,.$$

Let us introduce the functional

$$R_{0T}(\phi) = \int_0^T c(\phi_s)\,ds - S_{0T}(\phi)\,.$$

LEMMA 2.1. *Suppose that* $g(x)$, $x \in R^r$, *is a non-negative bounded function. We shall put* $G_0 = \{x \in R^r : g(x) > 0\}$ *and assume that* G_0 *belongs to the closure of the set* (G_0) *of its interior points, the function* $g(x)$ *being continuous for* $x \in (G_0)$. *Let the function* $c(x)$, $x \in R^r$, *be bounded and uniformly continuous. Then*

$$\lim_{\varepsilon \downarrow 0} \varepsilon \ln E_x^\varepsilon g(X_t^\varepsilon) \exp\left\{\varepsilon^{-1} \int_0^t c(X_s^\varepsilon)\,ds\right\} = \qquad (2.4)$$

$$= \sup\{R_{0T}(\phi) : \phi_0 = x, \phi_t \in G_0\}\,.$$

Proof. We shall denote by m the upper bound on the right-hand side of (2.4). Since $c(x)$ is bounded, we obtain that $m < \infty$. Remembering the properties of the function $g(x)$ and the fact that the functional $R_{0T}(\phi)$ is upper semi-continuous, we conclude that, for any $\gamma > 0$, one can find $\hat{\phi} \in C_{0T}(R^r)$ such that $\hat{\phi}_0 = x$, $\rho(\hat{\phi}_t, R^r \setminus G_0) = \delta_1 > 0$ and $R_{0t}(\hat{\phi}) > m - \gamma$. Let us denote by κ a positive number so small that $\int_0^t |c(\phi_s) - c(\psi_s)|\,ds < \gamma/2$, provided $\rho_{0t}(\phi, \psi) < \kappa$; $\delta_2 = \kappa \wedge \frac{\delta_1}{2}$.

Using bound (2.2) and the continuity of $g(x)$ on the set (G_0), we get for sufficiently small ε:

$$E_x^\varepsilon g(X_t^\varepsilon) \exp\left\{\varepsilon^{-1} \int_0^t c(X_s^\varepsilon)\,ds\right\} \ge E_x^\varepsilon \chi_{\rho_{0,t}(X^\varepsilon, \hat{\phi}) < \delta_2} g(X_t^\varepsilon) \times \qquad (2.5)$$

$$\exp\left\{\varepsilon^{-1} \int_0^t c(X_s^\varepsilon)\,ds\right\} \ge \min_{x : |x - \hat{\phi}_t| < \delta_2} |g(x)| \exp\left\{\varepsilon^{-1} \int_0^t c(\hat{\phi}_s)\,ds - \frac{\gamma}{2\varepsilon}\right\} \times$$

$$\times P_x^\varepsilon\{\rho_{0t}(X^\varepsilon,\hat{\phi}) < \delta_2\} \geq \exp\left\{-\frac{2\gamma}{\varepsilon} + \varepsilon^{-1}\int_0^t c(\hat{\phi}_s)\,ds - \varepsilon^{-1} S_{0,t}(\hat{\phi})\right\} \geq$$

$$\geq \exp\{\varepsilon^{-1}(m-3\gamma)\}\,.$$

Now we proceed to derive the upper bound. We put $s = |m| + t\cdot\sup_{x\epsilon R^r}|c(x)| + 1$. We have

$$E_x^\varepsilon g(X_t^\varepsilon)\exp\left\{\varepsilon^{-1}\int_0^t c(X_s^\varepsilon)\,ds\right\} = E_x^\varepsilon g(X_t^\varepsilon)\chi_{\rho_{0,t}(X^\varepsilon,\Phi_s)\geq\kappa/2} \times$$
$$\times \exp\left\{\varepsilon^{-1}\int_0^t c(X_s^\varepsilon)\,ds\right\} + E_x^\varepsilon g(X_t^\varepsilon)\chi_{\rho_{0,t}(X^\varepsilon,\Phi_s)<\kappa/2}\exp\left\{\varepsilon^{-1}\int_0^t c(X_s^\varepsilon)\,ds\right\}, \tag{2.6}$$

where $\Phi_s = \{\phi\,\epsilon\,C_{0,t}(R^r) : \phi_0 = x, S_{0,t}(\phi)\leq s\}$, $\kappa > 0$ being defined above.

The first summand on the right-hand side of (2.6) is bounded by inequality (2.3): no matter what the $\gamma > 0$, for sufficiently small ε, this summand is smaller than

$$\exp\{\varepsilon^{-1}t\cdot\sup_{x\epsilon R^r}|c(x)| - \varepsilon^{-1}(s-\gamma)\} \leq$$
$$\leq \exp\{-\varepsilon^{-1}(|m|+1-\gamma)\}\,. \tag{2.7}$$

To bound the second summand, we observe that in view of the compactness of the set Φ_s it is possible to choose in it a finite $\kappa/2$ – net: $\phi^{(1)},\cdots,\phi^{(N)}$. Then the second summand is bounded from above by the quantity

$$\sup_{x\epsilon R^r}|g(x)|\sum_{i=1}^N E_x^\varepsilon\chi_{\rho_{0,t}(\phi^{(i)},X^\varepsilon)<\kappa}\exp\left\{\varepsilon^{-1}\int_0^t c(X_s^\varepsilon)\,ds\right\} \leq$$
$$\leq \sup_{x\epsilon R^r}|g(x)|\sum_{i=1}^N \exp\left\{\varepsilon^{-1}\left(\int_0^t c(\phi_s^{(i)})\,ds + \gamma/2\right)\right\} P_x^\varepsilon\{\rho_{0,t}(X^\varepsilon,\phi^{(i)}) < \kappa\}\,. \tag{2.8}$$

Put $a_i = \inf\{S_{0,t}(\phi) : \rho_{0,t}(\phi,\phi^{(i)}) < \kappa\} - \gamma/4$, $i = 1,\cdots,N$. Since $S_{0,t}(\phi)$ is semi-continuous, one can find $\mu > 0$ such that $\rho_{0,t}(\Phi_{a_i}, \phi^{(i)}) > \kappa + \mu$. Thus, by bound (2.3), we have for ε small enough:

$$P_x^\varepsilon\{\rho_{0,t}(X^\varepsilon, \phi^{(i)}) < \kappa\} \le P_x^\varepsilon\{\rho_{0,t}(X^\varepsilon, \Phi_{a_i}) \ge \mu\} \le \le \exp\{-\varepsilon^{-1}(a_i - \gamma/4)\} . \tag{2.9}$$

From (2.8) and (2.9), it follows that the second summand on the right-hand side of inequality (2.6) is bounded from above by the quantity

$$\sum_{i=1}^{N} \exp\{\varepsilon^{-1} [\sup\{R_{0t}(\phi) : \rho_{0,t}(\phi, \phi^{(i)}) < \kappa\} + 2\gamma]\} \tag{2.10}$$

for ε small enough. Finally, from (2.6), (2.7) and (2.10) we conclude that, for sufficiently small ε,

$$E_x^\varepsilon g(X_t^\varepsilon) \exp\left\{\varepsilon^{-1} \int_0^t c(X_s^\varepsilon)\, ds\right\} \le \exp\{\varepsilon^{-1}(m+3\gamma)\} . \tag{2.11}$$

Since γ is arbitrarily small, from (2.5) and (2.11) follows the claim of Lemma 2.1. □

Let us denote

$$V(t,x,y) = \sup\{R_{0,t}(\phi) : \phi \in C_{0,t}(R^r), \phi_0 = x, \phi_t = y\} ,$$

$$V(t,x) = \sup_{y \in G_0} V(t,x,y), Q = \{(t,x) : t > 0, x \in R^r, V(t,x) < 0\} ,$$

$$K(t,x) = \begin{cases} 1, & \text{for } (t,x) \notin Q , \\ 0, & \text{for } (t,x) \in Q . \end{cases}$$

It is easily checked that $V(t,x,y)$ and $V(t,x)$ are continuous functions increasing in t; $V(t,x,y) = V(t,y,x)$.

The following assertion holds for a solution $u^\varepsilon(t,x)$ of problem (2.1).

THEOREM 2.1. *Suppose that the function* $f(x,u)$ *belongs to* $\mathfrak{A}$, *and the function* $c(x)$ *is bounded and uniformly continuous in* R^r. *Let* $g(x)$ *be a non-negative, bounded function whose support* G_0 *belongs to the closure of the set* (G_0) *of its interior points. Besides, the function* $g(x)$ *is assumed to be continuous for* $x \in (G_0)$. *Suppose that, for* $(t,x) \in Q$

$$V(t,x) = \sup \{R_{0,t}(\phi) : \phi \in C_{0,t}(R^r), \phi_0 = x, \phi_t \in G_0, V(t-s, \phi_s) < \\ < 0 \text{ for } s \in (0,t)\} . \tag{2.12}$$

Then, for any $T, \delta > 0$, *for* (t,x) *such that* $V(t,x) \neq 0$, $\lim_{\varepsilon \downarrow 0} u^\varepsilon(t,x) = K(t,x)$ *uniformly in* $(t,x) \in \{(t,x) : t \in [0,T],\ |x| < T,\ |V(t,x)| \geq \delta\} = Q_\delta$.

Proof. The Feynman-Kac formula implies that the function $u^\varepsilon(t,x)$ obeys the relation

$$u^\varepsilon(t,x) = E_x^\varepsilon g(X_t^\varepsilon) \exp\left\{\varepsilon^{-1} \int_0^t c(X_s^\varepsilon, u^\varepsilon(t-s, X_s^\varepsilon))\, ds\right\}, \tag{2.13}$$

where $c(x,u) = u^{-1} f(x,u)$. Whence, noting that by the assumption $c(x) = c(x,0) \geq c(x,u)$, we conclude:

$$0 \leq u^\varepsilon(t,x) \leq E_x^\varepsilon g(X_t^\varepsilon) \exp\left\{\varepsilon^{-1} \int_0^t c(X_s^\varepsilon)\, ds\right\} . \tag{2.14}$$

From (2.1) and (2.14) it results that

$$\overline{\lim_{\varepsilon \downarrow 0}}\ \varepsilon \ln u^\varepsilon(t,x) \leq V(t,x) .$$

Thus, $\lim_{\varepsilon \downarrow 0} u^\varepsilon(t,x) = 0$ provided $(t,x) \in Q$, where $V(t,x) < 0$. This convergence is uniform on the set $Q_\delta \cap Q$.

Now we shall show that $\lim_{\varepsilon \downarrow 0} u^\varepsilon(t,x) = 1$ whenever $V(t,x) > 0$. Along with the Markov process $(X_t^\varepsilon, P_x^\varepsilon)$ in the state space R^r, we shall consider the Markov process $(Y_s^\varepsilon, P_{t,x}^\varepsilon)$ corresponding to the operator $L^\varepsilon - \frac{\partial}{\partial t}$ in the state space $(-\infty,\infty)\times R^r$. The trajectories $Y_s^\varepsilon = (t_s, X_s^\varepsilon)$ of this new process are deterministic motions along the line $(-\infty,\infty)$ with constant velocity -1 as their first component t_s. The second component is the process X_t^ε in R^r. The process $(Y_s^\varepsilon, P_{t,x}^\varepsilon)$ is easily seen to be a strong Markov process.

Let λ be a small positive number. We shall introduce Markov times:

$$\tau_1^{\varepsilon,\lambda} = \tau_1 = \inf\{s : u^\varepsilon(t_s, X_s^\varepsilon) \geq 1-\lambda\},$$

$$\tau_2^{\varepsilon,\lambda} = \tau_2 = \inf\{s : V(t_s, X_s^\varepsilon) = 0\}, \quad \tau^{\varepsilon,\lambda} = \tau = \tau_1^{\varepsilon,\lambda} \wedge \tau_2^{\varepsilon,\lambda}.$$

The strong Markov property of the process $(Y_s^\varepsilon, P_{t,x}^\varepsilon)$ together with (2.13) leads to the following relation

$$u^\varepsilon(t,x) = E_{t,x}^\varepsilon u^\varepsilon(t_\tau, X_\tau^\varepsilon) \exp\left\{\varepsilon^{-1}\int_0^\tau c(X_s^\varepsilon, u^\varepsilon(t_s, X_s^\varepsilon))\,ds\right\} =$$

$$= E_{t,x}^\varepsilon \chi_{\tau=\tau_1} u^\varepsilon(t_{\tau_1}, X_{\tau_1}^\varepsilon) \exp\left\{\varepsilon^{-1}\int_0^{\tau_1} c(X_s^\varepsilon, u^\varepsilon(t_s, X_s^\varepsilon))\,ds\right\} + \tag{2.15}$$

$$+ E_{t,x}^\varepsilon \chi_{\tau\neq\tau_1} u^\varepsilon(t_{\tau_2}, X_{\tau_2}^\varepsilon) \exp\left\{\varepsilon^{-1}\int_0^{\tau_2} c(X_s^\varepsilon, u^\varepsilon(t_s, X_s^\varepsilon))\,ds\right\}.$$

Since the function $c(x,u)$ is non-negative in the domain $0 \leq u \leq 1-\lambda$, the first summand on the right-hand side of (2.15) is bounded from below by the quantity

$$(1-\lambda)\, E_{t,x}^\varepsilon \chi_{\tau=\tau_1} = (1-\lambda)\, P_{t,x}^\varepsilon\{\tau=\tau_1\}. \tag{2.16}$$

To bound the second summand, we observe that, at the points (t_0, x_0) for which $V(t_0, x_0) = 0$ for any $\delta > 0$, we have

$$u^\varepsilon(t_0, x_0) > \exp\left\{-\frac{\delta}{\varepsilon}\right\}, \tag{2.17}$$

whenever ε is small enough. Indeed, let $\bar\phi \in C_{0,t_0}(R^r)$, $\bar\phi_0 = x_0$, $\bar\phi_{t_0} \in (G_0)$. Moreover, suppose that for $s \in [\theta, t_0-\theta]$, the point $(t-s, (\bar\phi_s)$ is at a positive distance κ from the complement to the set Q, and $R_{0t}(\bar\phi) > -\delta/4$ (see Fig. 3). Such a function $\bar\phi$ may be chosen by virtue of hypothesis (2.12). The number θ can be selected arbitrarily small; of course in doing so, the distance κ will also be a small, but positive number. Note that the curve $(t_0-s, \bar\phi_s)$ goes inside the domain Q. And moreover, except for small parts near $s = 0$ and $s = t_0$, it goes at a positive distance κ from the complement of the domain Q. This observation implies that $u^\varepsilon(t_0-s, \bar\phi_s)$ is close to zero for $s \in [\theta, t_0-\theta]$ and small ε. Therefore,

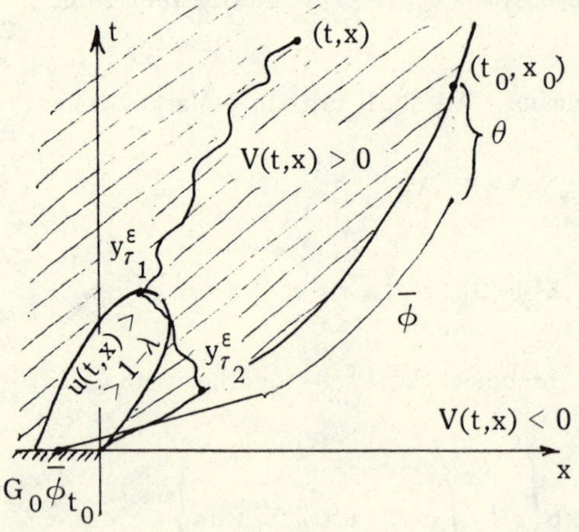

Fig. 3

$$\sup_{\theta<s<t_0-\theta} [c(\bar\phi_s) - c(\bar\phi_s, u^\varepsilon(t-s, \bar\phi_s))] < \frac{\delta}{4},$$

provided ε is small enough. From this it follows that one can find θ and κ_0 so small that

$$u^\varepsilon(t_0, x_0) = E^\varepsilon_{x_0} g(X^\varepsilon_{t_0}) \exp\left\{\varepsilon^{-1} \int_0^t c(X^\varepsilon_s, u^\varepsilon(t_0-s, X^\varepsilon_s))\, ds\right\} \geq$$

$$\geq E^\varepsilon_{x_0} g(X^\varepsilon_s)\, \chi_{\rho_{0,t_0}(\bar\phi, X^\varepsilon) < \kappa_0} \cdot \exp\left\{\varepsilon^{-1} \int_0^t c(X^\varepsilon_s, u^\varepsilon(t_0-s, X^\varepsilon_s))\, ds\right\} \geq$$

$$\geq P^{\varepsilon}_{x_0}\{\rho_{0,t_0}(\bar{\phi},X^{\varepsilon}) < \kappa_0\} \exp\left\{\varepsilon^{-1}\left(\int_0^t c(\bar{\phi}_s)ds - \delta/2\right)\right\} \geq$$

$$\geq \exp\{\varepsilon^{-1}(R_{0,t_0}(\bar{\phi}) - 3\delta/4)\} > \exp\left\{-\frac{\delta}{\varepsilon}\right\}$$

for ε small enough. Here we have used inequality (2.2) for bounding $P^{\varepsilon}_{x_0}\{\rho_{0,t_0}(X^{\varepsilon},\bar{\phi}) < \kappa\}$ from below. So, (2.17) is established.

We shall denote $V_0 = V(t,x) > 0$ and choose $h > 0$ such that $\inf\{V(s,y): |s-t| < h, |x-y| < h\} > \frac{1}{2} V_0$. Let us select $\delta \in (0, \alpha/2)$, where $\alpha = h \cdot \min_{|y-x|\leq h,\, 0\leq u\leq 1-\lambda} c(y,u)$.

By (2.17), fcr ε small enough

$$u^{\varepsilon}(t_{\tau_2}, X^{\varepsilon}_{\tau_2}) > \exp\left\{-\frac{\delta}{\varepsilon}\right\}.$$

We shall denote $\tau^{\varepsilon}_3 = \tau_3 = \inf\{s : |X^{\varepsilon}_s - x| = h\}$. It is clear that $P^{\varepsilon}_x\{\tau_3 < b\} \to 0$ as $\varepsilon \downarrow 0$ for any $b > 0$. The second summand on the right-hand side of (2.15) may be bounded from below by the quantity

$$E^{\varepsilon}_{t,x}\chi_{\tau=\tau_2<\tau_3} \exp\left\{\frac{-\delta+\alpha}{\varepsilon}\right\} - E^{\varepsilon}_{t,x}\chi_{\tau_3\leq\tau_2=\tau} \geq$$
$$\geq \exp\left\{\frac{\alpha}{2\varepsilon}\right\} P^{\varepsilon}_{t,x}\{\tau=\tau_2<\tau_3\} - P^{\varepsilon}_{t,x}\{\tau_3\leq\tau_2\}. \tag{2.18}$$

Noting that $\lim_{\varepsilon\downarrow 0} P^{\varepsilon}_{t,x}\{\tau_3\leq\tau_2\} = 0$ we conclude from (2.15), (2.16) and (2.18) that $u^{\varepsilon}(t,x) > 1-\lambda$ for ε small enough.

Finally, let us show that $\overline{\lim_{\varepsilon\downarrow 0}}\, u^{\varepsilon}(t,x) \leq 1$. For this, we shall pick a small $\lambda > 0$. Let us denote $D^{\varepsilon} = \{(t,x): t\geq 0, u^{\varepsilon}(t,x) \geq 1+\lambda\}$ and denote by $\tau^{\varepsilon}_4 = \tau_4 = \inf\{s : Y^{\varepsilon}_s \notin D^{\varepsilon}\}$ the first exit time of the process Y^{ε}_s from D^{ε}. Equation (2.13) yields that

$$u^\varepsilon(t,x) = E^\varepsilon_{t,x} u^\varepsilon(t-\tau_4, X^\varepsilon_{\tau_4}) \exp\left\{\varepsilon^{-1}\int_0^{\tau_4} c(X^\varepsilon_s, u^\varepsilon(t-s, X^\varepsilon_s))\,ds\right\} =$$

$$= E^\varepsilon_{t,x} u^\varepsilon \chi_{\tau_4<t} \exp\left\{\varepsilon^{-1}\int_0^{\tau_4} c\,ds\right\} + E^\varepsilon_{t,x}\chi_{\tau_4=t} g(X^\varepsilon_t)\exp\left\{\varepsilon^{-1}\int_0^t c\,ds\right\} \le$$

$$\le (1+\lambda) P^\varepsilon_{t,x}\{\tau_4 < t\} + \|g\| \exp\left\{-\frac{t}{\varepsilon} \min_{2+\|g\|\ge u\ge 1+\lambda,\, |y-x|\le h} |c(y,u)|\right\} \times \tag{2.19}$$

$$\times P^\varepsilon_{t,x}\{\tau_4 = t\} + P^\varepsilon_x\{\inf\{s : |X^\varepsilon_s - x| = h\} < t\}\,.$$

The last summand in the right-hand side of (2.19) tends to zero together with ε. All the remaining summands taken together are bounded from above by the quantity $(1+2\lambda)$, for ε small enough. This implies that $\overline{\lim}_{\varepsilon\downarrow 0}\, u^\varepsilon(t,x) \le 1$. The above cited bounds yield that $\lim_{\varepsilon\downarrow 0} u^\varepsilon(t,x) = 1$ uniformly in the domain Q_δ. □

REMARK. The assertion of Theorem 2.1 clearly remains true if, in the definition of the function $V(t,x)$, the upper bound is taken solely over $y \in \partial G_0$, that is, over the boundary of the set G_0. In the definition of the function $V(t,x,y)$, for $x \notin G_0$ and $y \in \partial G_0$, the upper bound may be taken only over the set of the functions $\{\phi \in C_{0t}(R^r), \phi_0 = x, \phi_t = y, \phi_s \notin G_0 \text{ for } s \in [0,t)\}$.

We shall define in the space R^r the Riemannian metric $d(x,y)$ via the metric form $ds^2 = \sum_{i,j=1}^r a_{ij}(x)\,dx^i dx^j$, $(a_{ij}(x)) = (a^{ij}(x))^{-1}$:

$$d(x,y) = \inf\left\{\int_0^1 \sqrt{\sum_{i,j=1}^r a_{ij}(\phi_s)\dot\phi^i_s\dot\phi^j_s}\,ds : \phi \in C_{0,1}(R^r), \phi_0 = x, \phi_1 = y\right\}.$$

THEOREM 2.2. *Suppose that the function* $f(x,u)$ *belongs to* $\mathfrak{A}$, *and let* $c(x) = c(x,0) = c = \text{const.} > 0$. *Suppose that* $g(x)$ *satisfies the conditions listed in Theorem 2.1 and put* $G_0 = \{x \in R^r, g(x) > 0\}$. *Then* $\lim_{\varepsilon \downarrow 0} u^\varepsilon(t,x) = 1$, *if* $d(x,G_0) < t\sqrt{2c}$, *and* $\lim_{\varepsilon \downarrow 0} u^\varepsilon(t,x) = 0$, *if* $d(x,G_0) > t\sqrt{2c}$.

To prove Theorem 2.2, we shall need the following

LEMMA 2.2. *Put* $\mathcal{H}(t,x,y) = \{\phi \in C_{0,t}(R^r) : \phi_0 = x, \phi_t = y\}$. *Then*

$$\inf\left\{\int_0^t \sum_{i,j=1}^r a_{ij}(\phi_s)\dot\phi_s^i \dot\phi_s^j \, ds : \phi \in \mathcal{H}(t,x,y)\right\} = \frac{1}{t}\, d^2(x,y)\,.$$

This lower bound is attained on the set of minimal geodesics which connect the points x *and* y, *provided the parameter along these geodesics is proportional to the arc length.*

Proof. Let us denote by $\ell_{0t}(\phi)$ the length of the curve ϕ between the points ϕ_0 and ϕ_t: $\ell_{0,t}(\phi) = \int_0^t \sqrt{\Sigma\, a_{ij}(\phi_s)\dot\phi_s^i \dot\phi_s^j}\, ds$. Suppose that γ_s, $0 \le s \le t$, is the minimal geodesic connecting the points $x,y \in R^r$. Let the parameter s along the curve be proportional to the arc length. Then,

$$\sum_{i,j=1}^r a_{ij}(\gamma_s)\dot\gamma_s^i \dot\gamma_s^j = \frac{d^2(x,y)}{t^2} = \frac{1}{t^2}\, \ell_{0,t}^2(\gamma)\,, \quad s \in [0,t]\,;$$

$$\int_0^t \sum_{i,j=1}^r a_{ij}(\gamma_s)\dot\gamma_s^i \dot\gamma_s^j\, ds = \frac{1}{t}\, d^2(x,y)\,.$$

Now we shall consider an arbitrary, absolutely continuous function $\phi \in C_{0t}(R^r)$ for which $\phi_0 = x$, $\phi_t = y$. Then by the Schwarz inequality

$$\int_0^t \sum_{i,j=1}^r a_{ij}(\phi_s)\dot\phi_s^i\dot\phi_s^j ds \geq \frac{1}{t}\,\ell^2(\phi) \geq \frac{1}{t}\, d^2(x,y)\,. \tag{2.20}$$

The first inequality in (2.20) turns into equality only provided the parameter along the curve ϕ is proportional to the arc length (then $\Sigma\, a_{ij}(\phi_s)\dot\phi_s^i\dot\phi_s^j = \text{const.}$). The last inequality becomes equality, if ϕ is the minimal geodesic connecting the points x and y. □

Now we proceed to prove Theorem 2.2. Let us evaluate the function $V(t,x,y)$. On the basis of Lemma 2.2 we arrive at

$$V(t,x,y) = \sup\{R_{0t}(\phi) : \phi \in C_{0t}(R^r), \phi_0 = x, \phi_t = y\} =$$

$$= ct - \inf\left\{\frac{1}{2}\int_0^t \sum a_{ij}(\phi_s)\dot\phi_s^i\dot\phi_s^j ds : \phi \in C_{0,t}(R^r), \phi_0 = x, \phi_t = y\right\} =$$

$$= ct - \frac{1}{2t}\, d^2(x,y)\,.$$

Whence,

$$V(t,x) = ct - \frac{1}{2t}\, d^2(x,G_0);\ Q = \{(t,x) : V(t,x) < 0\} = \{(t,x) : d(x,G_0) > t\sqrt{2c}\}\,.$$

The upper bound involved in the definition of the function $V(t,x,y)$, is attained on the minimal geodesics which connect the points x and y, and are equipped with the parametrization proportional to arc length. This implies condition (2.12) of Theorem 2.1 to be valid. On the basis of this theorem we infer that $\lim_{\varepsilon\downarrow 0} u^\varepsilon(t,x) = 1$ if $d(x,G_0) < t\sqrt{2c}$, and $\lim_{\varepsilon\downarrow 0} u^\varepsilon(t,x) = 0$ if $d(x,G_0) > t\sqrt{2c}$. □

The manifold $\{x \in R^r : d(x,G_0) = t\sqrt{2c}\}$ separates the sets where $u^\varepsilon(t,x)$ is close to 1 and to 0 for small ε. This manifold is a *wave front* at time t. The geodesics of the metric $d(x,y)$ play the role of *rays*, and the front travels in accordance with the Huygens principle. The domain where $u^\varepsilon(t,x)$ is close to 1, may be interpreted as the domain occupied by the excitation. Then the statement of Theorem 2.2 means that, for a constant $c(x) = c$, the velocity $v(x,e)$ of spreading of the excitation at a point x in the direction of the unit vector e, is equal to

$$v(x,e) = \left(\sum_{i,j=1}^{r} a_{ij}(x) e^i e^j \right)^{-1/2} \sqrt{2c} .$$

In particular, if equation (2.1) has the form $\frac{\partial u^\varepsilon}{\partial t} = \frac{\varepsilon D}{2} \Delta u^\varepsilon + \varepsilon^{-1} f(x,u^\varepsilon)$ then $v(x,e) = \sqrt{2Dc}$. Therefore, for $c(x) = \text{const.}$ and for small ε, we arrive at the phenomenological account of spreading excitations. Below, we shall show that, for variable $c(x)$, such a transition is, generally speaking, impossible.

The function $V(t,x,y)$ playing the basic role when evaluating the wave front, satisfies the corresponding Hamilton-Jacobi equation. To write down the Hamilton-Jacobi equation for the variational problem

$$\sup \left\{ \int_0^t \left[c(\phi_s) - \frac{1}{2} \sum_{i,j=1}^{r} a_{ij}(\phi_s) \dot\phi_s^i \dot\phi_s^j \right] ds : \phi \in C_{0t}(R^r), \phi_0 = x, \phi_t = y \right\} ,$$

one must take the Legendre transform $H(y,\alpha)$ of the function $F(y,p) = c(y) + \frac{1}{2} \sum_1^r a_{ij}(y) p^i p^j$ with respect to the variables $p = (p^1, \cdots, p^r)$: (see, e.g. Gel'fand and Fomin [1]). In our case $H(y,\alpha) = c(y) - \frac{1}{2} \sum_{i,j=1}^{r} a^{ij}(y) \alpha_i \alpha_j$, where $(a^{ij}(y)) = (a_{ij}(y))^{-1}$, and the Hamilton-Jacobi equation for the function $V(t,x,y)$ with respect to the variables t, y takes the form

$$\frac{\partial V}{\partial t} = c(y) - \frac{1}{2} \sum_{i,j=1}^{r} a^{ij}(y) \frac{\partial V}{\partial y^i} \frac{\partial V}{\partial y^j} .$$

The set of points $x \in R^r$ defined by the equation $V(t,x) = \sup_{y \in G_0} V(t,x,y) = 0$ for a fixed t, may be looked upon as the wave front at time t. Even in the simplest case treated in Theorem 2.2, this set can have a complicated topological structure. However, if $c(x) = c =$ const., then the set $G_t = \{x \in R^r : V(t,x) > 0\}$, where $u(t,x)$ is close to 1, changes, in a sense, continuously: for small Δ, the set $G_{t+\Delta}$ contains only points close to G_t. On the other hand, if the function $c(x)$ is not constant, then, generally speaking, new effects appear: at certain moments, new components of the connectedness of the set G_t arise. We shall illustrate this effect with the following example.

EXAMPLE 2.1. Let $x \in R^1$, $a^{11}(x) = 1$, $f(x,u) \in \mathfrak{A}$, $g(x) = \chi_{x \le 0}(x)$. In this case

$$R_{0,t}(\phi) = \int_0^t \left[c(\phi_s) - \frac{1}{2}\dot{\phi}_s^2\right] ds\,, \quad V(t,x) = \sup\{R_{0,t}(\phi) : \phi \in C_{0t}(R^1), \phi_0 = x, \phi_t$$

The Euler equation for the extremals of the functional $R_{0,t}(\phi)$ is as follows

$$\ddot{\phi}_s = c'(\phi_s)$$

This equation has the first integral

$$\frac{1}{2}\dot{\phi}_s^2 + c(\phi_s) = H = \text{const.} \tag{2.21}$$

We shall return to considering functions $c(x)$ of general form later on, and now, to carry the computation through, we will assume the function $c(x)$ to be piecewise constant. Let $c(x) = c_1 > 0$ for $x < h$, and $c(x) = c_2 > 2c_1$ for $x \ge h > 0$. Inside each of the domains $\{x < h\}$ and $\{x > h\}$, the Euler equation takes the form $\ddot{\phi} = 0$. So, the extremals of the functional $R_{0,t}(\phi)$ will be either segments of lines or broken lines with vertices on the line $x = h$ (Fig. 4). Let us compute $V(t,x)$ for the

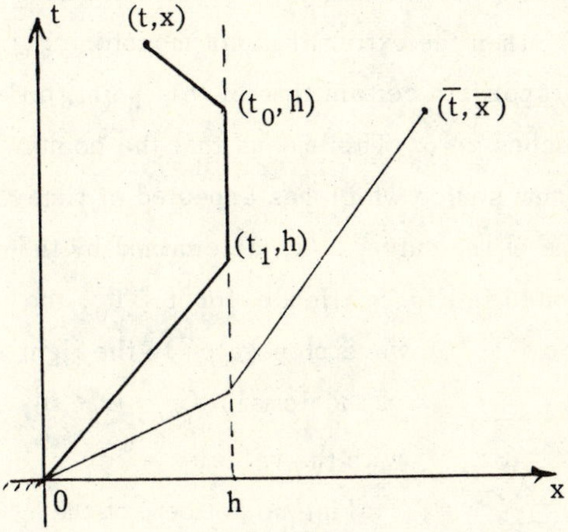

Fig. 4

points of the line $x = h$. On the broken line connecting the points (t_0, h), (t_1, h) and $(0,0)$, the functional $R_{0,t_0}(\phi)$ takes the value

$$R^{t_1} = c_2(t_0 - t_1) + c_1 t_1 - \frac{h^2}{2t_1} = c_2 t_0 - (c_2 - c_1) t_1 + \frac{h^2}{2t_1}$$

Let us find t_1, for which this quantity is maximal. It is easily seen that

$$\max_{t_1} R^{t_1} = c_2 t_0 - h\sqrt{2(c_2 - c_1)} .$$

This maximum is attained for $t_1 = \hat{t}_1 = \dfrac{h}{\sqrt{2(c_2 - c_1)}}$. Thus, if $t_0 > \hat{t}_1$, then the absolute maximum is attained on the broken line (rather than on the segment connecting the points (t_0, h) and $(0,0)$), and

$$V(t_0, h) = R^{\hat{t}_1} = c_2 t_0 - h\sqrt{2(c_2 - c_1)} .$$

The condition $V(t,x) = 0$ yields that the wave front reaches the point $x = h$ at the time $T_0 = \dfrac{h\sqrt{2(c_2 - c_1)}}{c_2}$. We observe that $T_0 > \hat{t}_1$, since $c_2 > 2c_1$.

So, $\lim_{\varepsilon \downarrow 0} u^\varepsilon(t,h) = 1$ for $t > T_0$. It is not difficult to check that, for $x < \frac{1}{2}[h + T_0\sqrt{2c_1}] = \bar{x}$, the upper bound involved in the definition of the function $V(t,x)$, is attained on the linear segments connecting the points (t,x) and $(0,0)$. Therefore, $V(t,x) = c_1 t - \dfrac{x^2}{2t}$ for $x < \bar{x}$, and the wave front in this domain travels according to the law $x = t\sqrt{2c_1}$. For $x > \bar{x}$, the upper bound is attained on broken lines having vertices on the line $x = h$. In particular, if $x \in (\bar{x}, h)$, then the extremal is not monotonic: first, it reaches the point $x = h$, spends a certain time at this point, and then, in the remaining time, it reaches zero. This means that the points $x \in (\bar{x}, h)$ will be excited by the new source which has appeared at time T_0 at the point $x = h$. The shape of the curve $t^*(x)$ determined by the equation $V(t^*, x) = 0$, is represented in Fig. 5. Hence, for $t < T_0$, the wave propagates to the right of the domain $G_0 = \{x < 0\}$ with the velocity $\sqrt{2c_1}$ "taking no notice" of the fact that after $x = h$, the coefficient $c(x)$ takes a larger value c_2. But at time $T_0 = \dfrac{h\sqrt{2(c_2 - c_1)}}{c_2}$, a "new source" arises at the point $x = h$, away from which a wave starts propagating in both directions: to the left

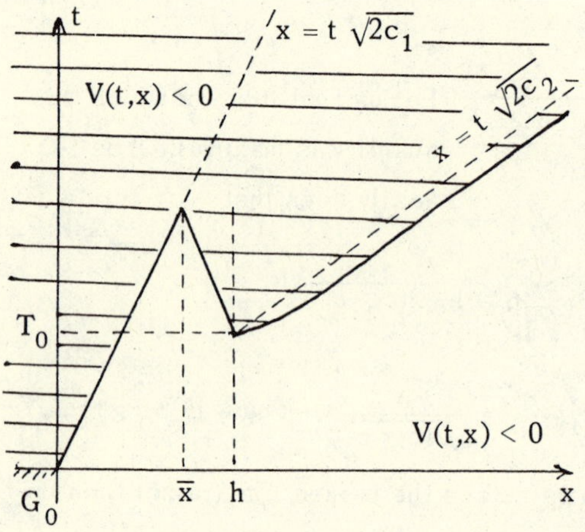

Fig. 5

with the velocity $\sqrt{2c_1}$, and to the right with the velocity close to $\sqrt{2c_2}$. On the right of the point $x = h$, the equation $V(t^*, x) = 0$ takes the form

$$\sup_{t_0}\left[c_2(t^* - t_0) + c_1 t_0 - \frac{(x-h)^2}{2(t^* - t_0)} - \frac{h^2}{2t_0}\right] = 0 .$$

It is not hard to verify that condition (2.12) is fulfilled here. In the shaded domain in Fig. 5, $\lim_{\varepsilon \downarrow 0} u^\varepsilon(t,x) = 1$. Below the curve $t^*(x)$, this limit vanishes.

This example shows that the domain $G_t = \{x \in R^1 : V(t,x) > 0\}$ can expand in a non-continuous way. For arbitrarily small Δ, the set $G_{T_0+\Delta}$ contains points (a neighborhood of the point $x = h$) which are at a positive distance from the set G_{T_0}. Certainly, this kind of expansion cannot be obtained in the phenomenological theory, where everything is determined by the velocity field.

It is readily seen that this effect also arises when the function $c(x)$ is smooth, provided it grows rapidly enough. Let us go into more detail in this question.

So, suppose just as before that $G_0 = \{x \in R^1, x < 0\}$; and let $c(x)$ be a continuously differentiable function increasing from $A = c(0)$ to $B = \lim_{x\to\infty} c(x)$, $c'(x) > 0$ for $x > 0$, (see Fig. 6). We shall consider the Cauchy problem

$$\frac{\partial u^\varepsilon}{\partial t} = \frac{\varepsilon}{2}\frac{\partial^2 u^\varepsilon}{\partial x^2} + \frac{1}{\varepsilon} c(x, u^\varepsilon)\, u^\varepsilon ,\, u^\varepsilon(0,x) . \tag{2.22}$$

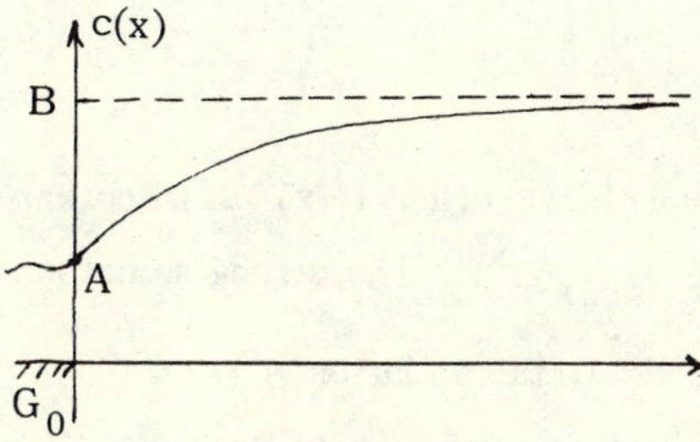

Fig. 6

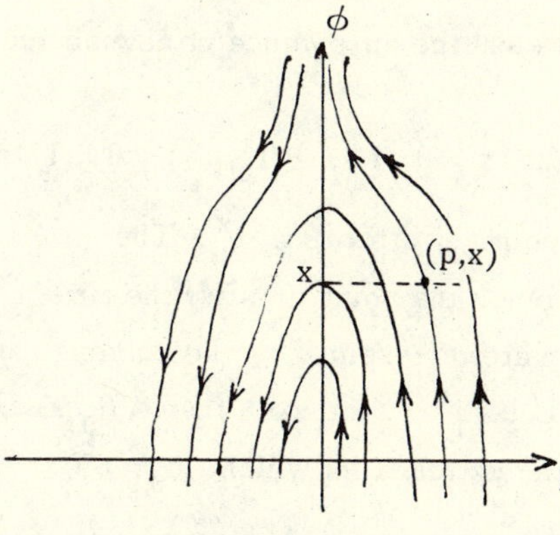

Fig. 7

Let us put $\dot{\phi} = p$. As follows from (2.21), the extremals of the functional $R_{0,t}$, corresponding to problem (2.22), are described by the level curves of the function $F(p,\phi) = \frac{1}{2}p^2 + c(\phi)$, (Fig. 7). Let us examine in more detail the extremal ϕ^x starting from the point x with zero initial velocity: $\dot{\phi}^x_0 = p_0 = 0$. Such an extremal satisfies equation (2.21) for $H = c(x)$. Integrating this equation with the initial condition $\phi^x_0 = x$ we derive that ϕ^x_s is determined by the following equality:

$$\int_{\phi^x_s}^{x} \frac{dz}{\sqrt{2(c(x) - c(z))}} = s . \tag{2.23}$$

This extremal reaches the point $x = 0$ at the time

$$t_0^x = \int_0^x \frac{dz}{\sqrt{2(c(x)-c(z))}} . \tag{2.24}$$

Let us compute the value of the functional R_{0,t_0^x} on the function ϕ^x:

$$R_{0,t_0^x}(\phi^x) = \int_0^{t_0^x} \left[c(\phi_s^x) - \frac{1}{2}(\dot{\phi}_s^x)^2 \right] ds =$$

$$= 2 \int_0^{t_0^x} c(\phi_s^x)\, ds - c(x)\, t_0^x .$$

Here we have made use of equality (2.21).

Now we shall show that whether or not the function $t^*(x)$ is monotonically increasing, depends on the sign of $R_{0,t_0^x}(\phi^x)$. The strong monotonicity of this function means that, for any $t > 0$ the equation $t^*(x) = t$ defines in a unique way the position of the front $x = x(t)$ at the time t. Hence it appears clear that in this case $x(t)$ changes continuously, and no new sources arise. If the function $t^*(x)$ has regions where it decreases or is not strictly monotonic, then this means the appearance of new sources or jumps of the wave front.

Suppose that $R_{0,t_0^x}(\phi^x) < 0$ for some $x > 0$, i.e. $V(t_0^x, \phi^x) < 0$. This means that the wave front reaches the point x after time t_0^x. The extremals which, starting from $x > 0$, reach the point 0 after the time t_0^x, are to be non-monotone. In Fig. 7, they are represented as the curves starting from the point (p,x), $p > 0$. Let $t_1 > t_0^x$ be such that $V(t_1,x) = 0$; and suppose that $\hat{\phi}_x$, $0 \le s \le t_1$, is the extremal for which $R_{0,t_1}(\hat{\phi}) = V(t_1,x) = 0$.

We will denote $a = \max_{0 \le s \le t_1} \hat{\phi}_s = \hat{\phi}_{\hat{t}}$ (see Fig. 8). Suppose $x_1 \in (x,a)$; and let the number δ be such that $\hat{\phi}_\delta = x_1$ and $\hat{\phi}_s < x_1$ for $s \in [0,\delta)$. We will consider the function $\tilde{\phi}_s$, $0 \le s \le t_1$,

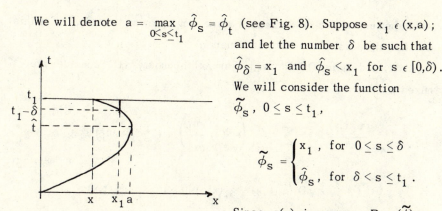

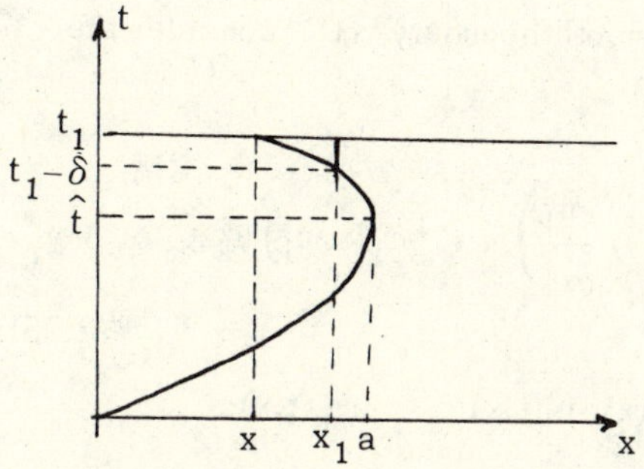

Fig. 8

$$\tilde{\phi}_s = \begin{cases} x_1, & \text{for } 0 \le s \le \delta \\ \hat{\phi}_s, & \text{for } \delta < s \le t_1. \end{cases}$$

Since $c(x)$ increases, $R_{0,\delta}(\tilde{\phi}) > R_{0,\delta}(\hat{\phi})$. The functions $\tilde{\phi}_s$ and $\hat{\phi}_s$ coincide on $[\delta, t_1]$. Thus, $R_{0,t_1}(\tilde{\phi}) > R_{0,t_1}(\hat{\phi})$. This implies that one can find $t < t_1$ such that $V(t,x_1) = 0$, i.e. the wave front reaches the point $x_1 > x$ before the point x. This means that the function $t^*(x)$ is not monotone: $t^*(x_1) < t^*(x)$. Thereby, if for some $x > 0$

$$\int_0^{t_0^x} c(\phi_s^x)\, ds < t_0^x \frac{c(x)}{2},$$

then new sources are "igniting" ahead of front. Here ϕ_s^x and t_0^x are specified by formulae (2.23), (2.24).

It is not difficult to show that the converse is also true, namely: if for all $x \in R^1 \setminus G_0$

$$\int_0^{t_0^x} c(\phi_s^x)\, ds \ge t_0^x \frac{c(x)}{2},$$

then the function $t^*(x)$ is monotone non-decreasing. The proof of this assertion is dropped.

Now, we proceed to examine the wave front propagation when condition (2.12) is not fulfilled. In particular, this enables us to consider the case when the function $c(x)$ decreases while moving away from G_0. For this, we shall need some auxiliary statements.

Given a domain $D \subset R^r$ with a smooth boundary ∂D, consider the mixed problem

$$\frac{\partial u_k^\varepsilon(t,x)}{\partial t} = \frac{\varepsilon}{2} \sum_{i,j=1}^{r} \frac{\partial}{\partial x^i}\left(a^{ij}(x) \frac{\partial u_k^\varepsilon}{\partial x^j}\right) + \varepsilon^{-1} c_k(x, u_k^\varepsilon) u_k^\varepsilon ,$$

$$t > 0,\ x \in D;\ u_k^\varepsilon(0,x) = g_k(x),\ u_k^\varepsilon(t,x)\Big|_{x \in \partial D} \quad \psi_k(t,x) . \tag{2.25}$$

Here $k = 1,2$; $c_k(x,u)$ are bounded, Lipschitz continuous functions.

LEMMA 2.3. *Let* $c_1(x,u) \le c_2(x,u)$ *for* $x \in D$, $-\infty < u < \infty$, $0 \le g_1(x) \le g_2(x) \le \bar{g} < \infty$, $0 \le \psi_1(t,x) \le \psi_2(t,x) < \bar{\psi} < \infty$. *Then* $u_1^\varepsilon(t,x) \le u_2^\varepsilon(t,x)$ *for* $t \ge 0$, $x \in D$.

Proof. For $t > 0$, $x \in D$, the following relation holds for the difference $w^\varepsilon(t,x) = u_2^\varepsilon(t,x) - u_1^\varepsilon(t,x)$:

$$\frac{\partial w^\varepsilon}{\partial t} = \frac{\varepsilon}{2} \sum_{1}^{r} \frac{\partial}{\partial x^i}\left(a^{ij}(x) \frac{\partial w^\varepsilon}{\partial x^j}\right) + \varepsilon^{-1} \hat{c}^\varepsilon(t,x) w^\varepsilon + \varepsilon^{-1} F(t^\varepsilon,x) \tag{2.26}$$

$$w^\varepsilon(0,x) = \hat{g}(x) \ge 0,\ w^\varepsilon(t,x)\Big|_{x \in \partial D} = \hat{\psi}(t,x) , \tag{2.27}$$

where $\hat{c}^\varepsilon(t,x) = (u_2^\varepsilon - u_1^\varepsilon)^{-1}(c_2(x,u_2^\varepsilon) u_2^\varepsilon - c_2(x,u_1^\varepsilon) u_1^\varepsilon)$, $F^\varepsilon(t,x) = c_2(x,u_1^\varepsilon) u_1^\varepsilon - c_1(x,u_1^\varepsilon) u_1^\varepsilon$, $\hat{g}(x) = g_2(x) - g_1(x)$ and $\hat{\psi}(t,x) = \psi_2(t,x) - \psi_1(t,x)$. Let $(X_t^\varepsilon, P_x^\varepsilon)$ be a diffusion process in R^r corresponding to the differential operator $L^\varepsilon = \frac{\varepsilon}{2} \sum_1^r \frac{\partial}{\partial x^i}\left(a^{ij}(x) \frac{\partial}{\partial x^j}\right)$. We shall denote by $\tau^\varepsilon = \inf\{s : X_s^\varepsilon \notin D\}$ the first exit time from the domain D; and by $G(t,x)$ we denote a function on $D \cup (\partial D \times [0,\infty))$ which is equal to $\hat{g}(x)$ for $t = 0$, and equal to $\hat{\psi}(t,x)$ for $x \in D$, $t > 0$. From (2.26) - (2.27) it follows that $w^\varepsilon(t,x)$ can be written in the form

$$w^\varepsilon(t,x) = E_x^\varepsilon\, G(t-(\tau^\varepsilon \wedge t), X^\varepsilon_{t\wedge\tau^\varepsilon}) \exp\left\{\frac{1}{\varepsilon}\int_0^{t\wedge\tau^\varepsilon} \hat{c}(t-s, X_s^\varepsilon)\,ds\right\} +$$

$$+\frac{1}{\varepsilon} E_x^\varepsilon \int_0^{\tau^\varepsilon\wedge t} F^\varepsilon(t-s, X_s^\varepsilon)\exp\left\{\frac{1}{\varepsilon}\int_0^s \hat{c}(t-s_1, X^\varepsilon_{s_1})\,ds_1\right\}ds\ .$$

This implies the claim of the lemma: $w^\varepsilon(t,x) = u_2^\varepsilon(t,x) - u_1^\varepsilon(t,x) \geq 0$. □

Note that, this lemma is surely an implication of the maximum principle for linear parabolic equations.

We will introduce the velocity field

$$v(x,e) = \left(\sum_1^r a_{ij}(x)e^i e^j\right)^{-1/2} \sqrt{2c(x)}\ ,\quad x \in R^r\ ,$$

$$e = (e^1,\cdots,e^r)\ ,\quad \sum_{k=1}^r (e^k)^2 = 1\ .$$

Suppose that $g(x)$ is the initial function in problem (2.1), and let the conditions listed in Theorem 2.1 be fulfilled for it, $G_0 = \{x \in R^r : g(x) > 0\}$. We set

$$\tau_{G_0}(x) = \inf_{\phi \in C_{0,t}(R^r), \phi_0 = x, \phi_t \in G_0} \int_0^t \frac{|\dot\phi_s|\,ds}{v\left(\phi_s, \frac{\dot\phi_s}{|\dot\phi_s|}\right)} =$$

$$= \inf\left\{\int_0^t \sqrt{\sum_{i,j=1}^r a_{ij}(\phi_s)\dot\phi_s^i\dot\phi_s^j (2c(\phi_s))^{-1}}\,ds : \phi \in C_{0,t}(R^r), \phi_0 = x, \phi_t \in G_0\right\}.$$

LEMMA 2.4. *Let the conditions of Theorem 2.1, except for condition (2.12), be fulfilled. Then* $\lim_{\varepsilon\downarrow 0} u^\varepsilon(t,x) = 1$ *for* $z \in G_t = \{x \in R^r, \tau_{G_0}(x) < t\}$.

Proof. Suppose that $z \in G_t$, and let $\hat{\phi} \in C_{0,t}(R^r)$, $\hat{\phi}_0 = z, \hat{\phi}_t \in [G_0]$, be such that $\tau_{G_0}(z) = \int_0^t \sqrt{\Sigma a_{ij}(\hat{\phi}_s)\dot{\hat{\phi}}_s^i \dot{\hat{\phi}}_s^j (2c(\hat{\phi}_s))^{-1}} ds$. We shall choose a small number $\delta_1 > 0$ and consider a finite covering of the curve $\hat{\phi}$ by Riemannian open balls $U_1, U_2, \cdots, U_N$ of radius δ_1: the ball U_1 has its center at the point $\hat{\phi}_t = O_1$; the center of the ball U_k is at the point O_k at which the sphere ∂U_{k-1} intersects the curve $\hat{\phi}$. One should take the first intersection point after $\hat{\phi}$ passes through the center of U_{k-1} (we move along the curve $\hat{\phi}$ from $\hat{\phi}_t$ towards $\hat{\phi}_0$!); $U_k = \{y \in R^r : d(O_k, y) < \delta_1\}$ (Fig. 9). Along with the solution $u^\varepsilon(t,x)$ of problem (2.1), we shall consider the following functions: $u_1^\varepsilon(t,x)$ —the solution of problem (2.25) for $D = U_1$, $c_1(u) = \inf_{x \in U_1} c(x,u)$, $g_1(x) = g(x)$, $\psi_1(t,x) = 0$ for $x \in \partial D$ and $t > 0$; $u_2^\varepsilon(t,x)$ —the solution of problem (2.25) in $D = U_1$ with $g_2(x) = g(x), \psi_2(t,x) = 0$ for $x \in \partial D$, $t > 0$, and with $c_2(x,u) = c(x,u)$; and, finally, $\overline{u}_1^\varepsilon(t,x)$ —the solution of (2.1) for $c(x,u) = c_1(u)$, with the initial function $\tilde{g}(x) = g(x)$ for $x \in G_0 \cap U_1$ and $g(x) = 0$ for $x \notin G_0 \cap U_1$.

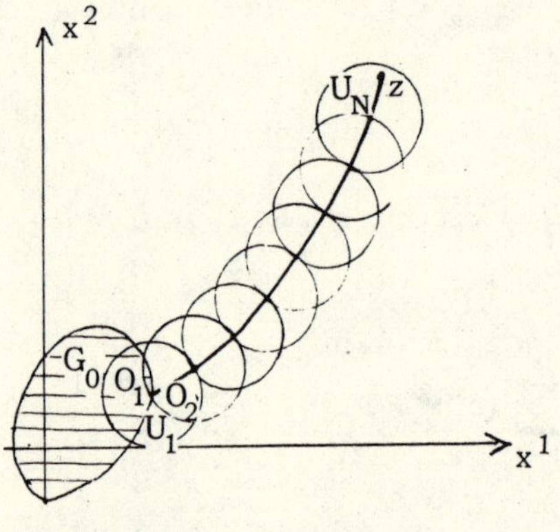

Fig. 9

Noting that $u^\varepsilon(t,x)$ is non-negative, Lemma 2.3 implies that $u^\varepsilon(t,x) \geq u_2^\varepsilon(t,x)$ for $x \in U_1$. Next, Lemma 2.3 yields that $u_2^\varepsilon(t,x) \geq u_1^\varepsilon(t,x)$ for $x \in U_1$. With the aid of the reasoning given when proving Lemma 2.1, one may deduce that $u_1^\varepsilon(t,x) \geq \overline{u}_1^\varepsilon(t,x) - o_\varepsilon(1)$ as $\varepsilon \downarrow 0$, $x \in U_1$. At last, relying on Theorem 2.2 we conclude that

$$\lim \overline{u}_1^\varepsilon(t,x) = 1 \quad \text{for} \quad d(x, G_0 \cap U_1) < t\sqrt{2c_1(0)},$$

where $c_1(0) = \inf_{x \in U_1} c(x,0)$. All these bounds together yield that

$\underline{\lim}_{\varepsilon \downarrow 0} u^{\varepsilon}(t,x) \geq 1$ on the set $\{x \in U_1, d(x,G_0) < t\sqrt{2c_1(0)}\}$. Next we apply this procedure in the domain U_2, then in U_3 and so on. If $\tau_{G_0}(z) < t$, then for δ_1 small enough, we obtain that before the time t, the point z will be reached by the excitation spreading in the domain U_k with the velocity corresponding to $\inf_{x \in U_k} c(x,0)$, This results in $\underline{\lim}_{\varepsilon \downarrow 0} u^{\varepsilon}(t,z) \geq 1$. Just as was done when proving Theorem 2.1, it is possible to check that $\overline{\lim}_{\varepsilon \downarrow 0} u^{\varepsilon}(t,x) \leq 1$. Hence, $\lim_{\varepsilon \downarrow 0} u^{\varepsilon}(t,x) = 1$ for $z \in G_t$. □

REMARK. It is worthwhile noting that one cannot derive a similar upper bound for $u^{\varepsilon}(t,x)$ via "local" arguments, that is, by means of replacing $c(x,u)$ in a neighborhood of a point x_0 by a larger function $\overline{c}_{x_0}(u)$. This follows, in particular, from Example 2.1. To get the upper bound, it would be necessary, at each subsequent step, to take as the initial function not a step function, but a positive function which depends on ε and takes into account what value (of order of $\exp\{-\varepsilon^{-1}\alpha(x)\}$) has already been accumulated at the point x at the expense of the preceding steps (compare with item 1 of the next section).

The following theorem enables us to examine the solution of problem (2.1) in those cases when condition (2.12) does not hold.

THEOREM 2.3. *Suppose that a function* $f(x,u)$ *belongs to* $\mathfrak{A}$ *and the function* $c(x,u) = u^{-1}f(x,u)$ *is Lipschitz continuous,* $c(x,0)$ *being bounded. Let a function* $g(x)$ *satisfy the conditions formulated in Theorem 2.1. We shall denote* $G_t = \{x \in R^r : \tau_{G_0}(x) < t\}, H = \{(t,x) : t > 0, x \in G_t\}$, $[H]$ *designates the closure of the set* H *in* $[0,\infty) \times R^r$, *and*

$$\tilde{V}(t,x) = \sup\{R_{0,s}(\phi) : \phi \in C_{0,s}(R^r), s \leq t, \phi_0 = x, (s,\phi_s) \in [H],$$

$$(\sigma,\phi_\sigma) \notin [H] \ \textit{for} \ \sigma \in [0,s)\}.$$

1. *Suppose that* $\widetilde{V}(t,x) < 0$ *for* $(t,x) \notin [H]$. *Then* $\lim_{\varepsilon \downarrow 0} u^{\varepsilon}(t,x) = 1$ *for* $(t,x) \in H$ *and* $\lim_{\varepsilon \downarrow 0} u^{\varepsilon}(t,x) = 0$ *for* $(t,x) \notin [H]$.

2. *Let us denote* $\widetilde{H} = H \cup \{(t,x) : \widetilde{V}(t,x) > 0\}$, *and let* $[\widetilde{H}]$ *be the closure of* $\widetilde{H}$ *in* $[0,\infty) \times R^r$. *We shall assume that, for* $(t,x) \notin [\widetilde{H}]$

$$\widetilde{V}(t,x) = \sup \{R_{0,s}(\phi) : \phi \in C_{0,s}(R^r), \phi_0 = x, (s,\phi_s) \in [\widetilde{H}], \\ (\sigma,\phi_\sigma) \notin [\widetilde{H}] \text{ for } \sigma \in [0,s), s \le t\}. \tag{2.28}$$

Then $\lim_{\varepsilon \downarrow 0} u^{\varepsilon}(t,x) = 1$ *for* $(t,x) \in \widetilde{H}$ *and* $\lim_{\varepsilon \downarrow 0} u^{\varepsilon}(t,x) = 0$ *for* $(t,x) \notin [\widetilde{H}]$.

Proof. First we shall show that $\lim_{\varepsilon \downarrow 0} u^{\varepsilon}(t,x) = 0$ provided $\widetilde{V}(t,x) < 0$. We shall put $\tau_t = \inf\{s : (s, X_s^{\varepsilon}) \in [H]\} \wedge t$. This random variable is a Markov time, so the relation holds

$$u^{\varepsilon}(t,x) = E_x^{\varepsilon} u^{\varepsilon}(t-\tau_t, X_{\tau_t}^{\varepsilon}) \exp\left\{\frac{1}{\varepsilon}\int_0^{\tau_t} c(X_s^{\varepsilon}, u^{\varepsilon}(t-s, X_s^{\varepsilon}))\, ds\right\} \le \\ \le E_x^{\varepsilon}\, G(t-\tau_t, X_{\tau_t}^{\varepsilon}) \exp\left\{\frac{1}{\varepsilon}\int_0^{t} c(X_s^{\varepsilon})\, ds\right\} = \zeta^{\varepsilon}, \tag{2.29}$$

where $c(x) = c(x,0)$, and $G(t,x)$ is a function vanishing for $t = 0$, $x \in R^r \setminus G_0$, and equal to $A = 1 \vee \sup\{g(x) : x \in R^r\}$ on the lateral boundary Γ of the domain H (see Fig. 10). Analogously to the way Lemma 2.1

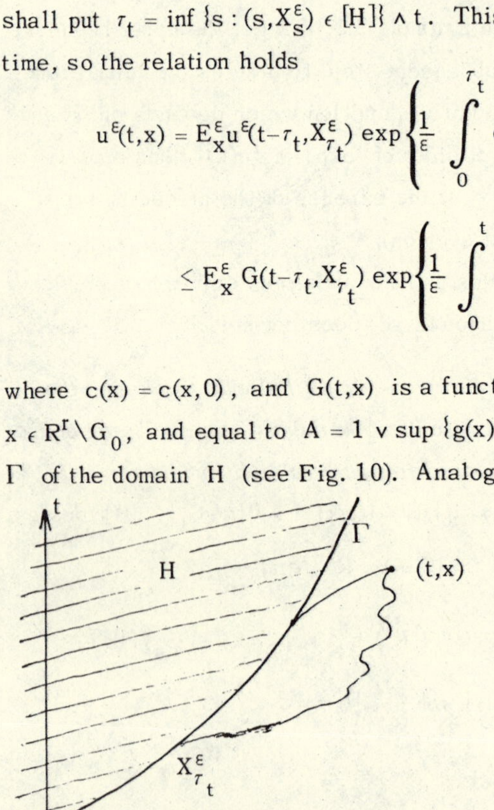

Fig. 10

was proved, it is easily checked that the expression ζ^{ε} on the right-hand side of (2.29) satisfies the relation $\lim_{\varepsilon \downarrow 0} \varepsilon \ln \zeta^{\varepsilon} = \widetilde{V}(t,x)$. This relation, along with (2.29), implies that $\lim_{\varepsilon \downarrow 0} u^{\varepsilon}(t,x) = 0$ provided $\widetilde{V}(t,x) < 0$.

Now, to complete the proof of the first claim of Theorem 2.3, it suffices to refer to Lemma 2.4, according to which $\lim_{\varepsilon \downarrow 0} u^{\varepsilon}(t,x) = 1$ for $(t,x) \in H$. Finishing the proof of the second claim is carried out just as the final part of the proof of Theorem 2.1 was arranged, equality (2.28) playing the role of relation (2.12). □

EXAMPLE 2.2. Let $x \in R^1$, $a^{11}(x) = 1$, $f(x,u) \in \mathfrak{A}$, $g(x) = \chi_{x \leq 0}(x)$. The function $c(x) = c(x,0)$ is assumed to be monotonically decreasing for $x > 0$. We shall consider the function ψ_s, $s \geq 0$, defined by the differential equation

$$\dot{\psi}_s = \sqrt{2c(\psi_s)}, \quad \psi_0 = 0. \tag{2.30}$$

This function increases monotonically; $\dot{\psi}_s$ decreases monotonically, remaining positive (Fig. 11). Let us show that the shaded domain in the plane (t,x) in Fig. 11 is none other than $H = \{(t,x) : t > 0, \tau_{G_0}(x) < t\}$ in our example. In fact, let us consider a point (t_0, x_0) in the shaded domain, and let (s_0, x_0) be a point on the curve $\psi : \psi_{s_0} = x_0$. Clearly, $s_0 < t_0$. We put $\tilde{\phi}_s = \psi_{s_0 - s}$, $s \in [0, s_0]$. The definition of ψ_s implies that $|\dot{\tilde{\phi}}_s| \, (2c(\tilde{\phi}_s))^{-1/2} = 1$. Therefore,

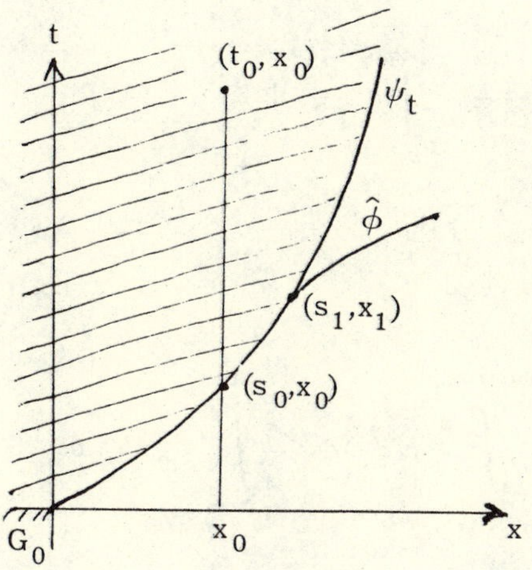

Fig. 11

$$\tau_{G_0}(x_0) = \inf \left\{ \int_0^t |\dot{\phi}_s| (2c(\phi_s))^{-1/2} ds : \phi \in C_{0t}(R^1), \phi_0 = x, \phi_t \leq 0, t \geq 0 \right\} \leq$$

$$\leq \int_0^{s_0} |\dot{\tilde{\phi}}_s| (2c(\tilde{\phi}_s))^{-1/2} ds = s_0 < t_0$$

and thereby, $(t_0, x_0) \in H$. It is readily verified that the points lying on the right of the curve ψ in Fig. 11, do not belong to the domain H.

Now we shall show that $\tilde{V}(t,x) < 0$ for $x > \psi_t$. Let the functional involved in the definition of $\tilde{V}(t,x)$ attain its upper bound on an extremal $\hat{\phi}_s$, $s \in [0, t-s_1]$:

$$\tilde{V}(t,x) = \int_0^{t-s_1} \left[c(\hat{\phi}_s) - \frac{1}{2} \dot{\hat{\phi}}_s^2 \right] ds .$$

Along with $\hat{\phi}_s$, we shall consider the function $\hat{\hat{\phi}}_s$ which is just a segment of $\psi : \hat{\hat{\phi}} = \psi(t-s), s \in [0, t-s_1]$. Since $c(x)$ monotonically decreases with the growth of x, and $\hat{\hat{\phi}}_s < \hat{\phi}_s$, $s \in [0, t-s_1]$, we conclude that $c(\hat{\hat{\phi}}_s) > c(\hat{\phi}_s)$. Next, we observe, that

$$\int_0^{t-s_1} \frac{1}{2} \dot{\hat{\phi}}_s^2 \, ds \geq \frac{(x-x_1)^2}{2(t-s_1)} > \int_0^{t-s_1} \frac{\dot{\hat{\hat{\phi}}}_s^2}{2} \, ds .$$

In the last inequality, we have made use of the fact that $\dot{\hat{\hat{\phi}}}_s < \frac{x-x_1}{t-s_1}$, for $s \in [0, t-s_1]$. The above cited bounds yield:

$$\tilde{V}(t,x) = \int_0^{t-s_1} \left[c(\hat{\phi}_s) - \frac{1}{2} \dot{\hat{\phi}}_s^2 \right] ds < \int_0^{t-s_1} c(\hat{\hat{\phi}}_s) \, ds - \int_0^{t-s_1} \frac{1}{2} \dot{\hat{\hat{\phi}}}_s^2 \, ds = 0$$

for $x > \psi_t$. Hence, on account of the first claim of Theorem 2.3 we arrive at

$$\lim_{\varepsilon \downarrow 0} u^{\varepsilon}(t,x) = \begin{cases} 1, & \text{for } x < \psi_t, t > 0 , \\ 0, & \text{for } x > \psi_t, t > 0 , \end{cases}$$

where ψ_t is the solution of problem (2.30). Therefore, in this example, wave propagation may be described via the phenomenological model of exitation spreading.

§6.3 *Some remarks and refinements*

This section concerns some generalizations and refinements of the results of the preceding section.

1. We shall consider problem (2.1) with an initial function depending on small parameter. For brevity, we shall confine ourselves to the one-dimensional case. So, let $g^{\varepsilon}(x) = 1$ for $x < 0$, $g^{\varepsilon}(x) = \exp\{-\varepsilon^{-1}\alpha(x)\}$ for $x \geq 0$, where $\alpha(x)$ is a positive function increasing with x. We put

$$\overline{R}_{0,t}(\phi) = -\alpha(\phi_t) + R_{0,t}(\phi) ,$$

$$\overline{V}(t,x) = \sup\{\overline{R}_{0,t}(\phi) : \phi \in C_{0,t}(R^1), \phi_0 = x, \phi_t \in R^1\} .$$

By the properties of action functionals, noting that $c(x) = c(x,0) = \max_{0 \leq u \leq 1} c(x,u)$, it is easily deduced that $\lim_{\varepsilon \downarrow 0} u^{\varepsilon}(t,x) = 0$ provided $\overline{V}(t,x) < 0$. For the equality $\lim_{\varepsilon \downarrow 0} u^{\varepsilon}(t,x) = 1$ to be established at the points (t,x) for which $\overline{V}(t,x) > 0$, it is necessary, in addition, to make an assumption, similar to (2.12):

$$\overline{V}(t,x) = \sup\{\overline{R}_{0,t}(\phi) : \phi \in C_{0,t}(R^1), \phi_0 = x , \quad \overline{V}(s,\phi_s) < 0 \text{ for } s \in [0,t)\} \text{ if } \overline{V}(t,x) < 0 . \tag{3.1}$$

If (3.1) holds, then the wave front is assigned by the equality $\overline{V}(t,x) = 0$.

Suppose, for example, that $a^{11}(x) = D = \text{const.}$, $\alpha(x) = \alpha \cdot x$, $\alpha > 0$, and $c(x) = c = \text{const.}$ Then for $x > \alpha Dt$

$$\overline{V}(t,x) = \sup\left\{-\alpha\phi_t + ct - \int_0^t \frac{\dot{\phi}_s}{2D} ds : \phi \in C_{0,t}(R^1), \phi_0 = x\right\} =$$

$$= ct - \inf_{y \geq 0}\left\{\alpha y + \frac{(x-y)^2}{2Dt}\right\} = ct - \alpha x + \frac{\alpha^2 Dt}{2}$$

Whence we find that, at time t, for $\alpha\sqrt{D} < \sqrt{2c}$ the wave front is at the point $x = x(t) = \frac{t}{\alpha}\left(c + \frac{1}{2}\alpha^2 D\right)$. If $\alpha > 0$ is small enough, then the front velocity may be arbitrarily large. Its smallest value $\min\left(\frac{c}{\alpha} + \frac{\alpha D}{2}\right) = \sqrt{2cD}$ is the velocity which is obtained in the case when the initial function is equal to $\chi_{x<0}(x)$.

Thus, the possible velocities of wave front propagation lie in the interval $[\sqrt{2cD}, \infty)$.

2. Let us see how our results will change, if equation (2.1) involves terms with first-order derivatives. The arguments used in §6.1 show that the coefficients in first-order derivatives must be of order 1 as $\varepsilon \downarrow 0$. So, instead of (2.1) we shall consider the following problem:

$$\frac{\partial u^\varepsilon}{\partial t} = \frac{\varepsilon}{2} \sum_{i,j=1}^{r} \frac{\partial}{\partial x^i}\left(a^{ij}(x) \frac{\partial u^\varepsilon}{\partial x^j}\right) + \sum_{i=1}^{r} b^i(x) \frac{\partial u^\varepsilon}{\partial x^i} + \frac{1}{\varepsilon} f(x,u^\varepsilon) =$$

$$= L^\varepsilon u^\varepsilon + \frac{1}{\varepsilon} f(x,u^\varepsilon), \quad u^\varepsilon(0,x) = g(x) . \tag{3.2}$$

The assumptions concerning $f(x,u)$ and $g(x)$ will be the same as those in Theorem 2.1. We shall introduce the functional $\hat{R}_{0,t}(\phi)$ which, for absolutely continuous $\phi \in C_{0t}(R^r)$, is assigned by the formula

$$\hat{R}_{0,t}(\phi) = \int_0^t \left[c(\phi_s) - \frac{1}{2} \sum_{i,j=1}^{r} a_{ij}(\phi_s)(\dot{\phi}_s^i - b^i(\phi_s))(\dot{\phi}_s^j - b^j(\phi_s))\right] ds ,$$

$$(a_{ij}(x)) = (a^{ij}(x))^{-1} ;$$

for the remaining $\phi \in C_{0t}(R^r)$, we put $\hat{R}_{0,t}(\phi) = -\infty$. Using the limit theorems for probabilities of large deviations of the random process connected with the operator L^ε, one can verify that the claim of Theorem 2.1 is true for the solution of problem (3.2) whenever the functional $\hat{R}_{0,t}(\phi)$ is taken instead of $R_{0,t}(\phi)$.

The presence of the drift causes a number of new effects in advance of the wave front. As an example, we shall consider the one-dimensional case: $x \in R^1$, $a^{11}(x) = 1$, $g(x) = \chi_{x<0}(x)$, $b(x) = b = \text{const}$.

It is easy to check that if $c(x) = c = \text{const.}$, then the velocity of the wave front is equal to $\sqrt{2c} - b$. If $b < \sqrt{2c}$, then the front will travel to the right. For $b > \sqrt{2c}$, the domain of large values of the function $u^\varepsilon(t,x)$ (for $\varepsilon << 1$) will contract: $\lim_{\varepsilon \downarrow 0} u^\varepsilon(t,x) = 1$, provided $x < t(\sqrt{2c} - b)$, and $\lim_{\varepsilon \downarrow 0} u^\varepsilon(t,x) = 0$ for $x > t(\sqrt{2c} - b)$.

Let now $c(x)$ decrease monotonically for $x > 0$, $c(0) > b$, $c(h) = b$. Then the front will travel from left to right until it reaches the point h. At the time t, the front will take the position $\psi_t < h$, where ψ_t is a solution of the equation

$$\dot{\psi}_t = \sqrt{2c(\psi_t)} - b\,, \quad \psi_0 = 0\,.$$

The wave does not propagate to points on the right of the point $x = h$.

If after some $x_0 > h$ the function $c(x)$ starts increasing, then the same effect as in Example 2.1 is possible: for certain $x_1 > x_0$, a new source may arise away from which a wave will propagate in both directions.

We note that, for positive $b(x)$, the velocity of the wave propagation from left to right decreases rather than increases. This is because we are considering the backward Kolmogorov equation. The physical drift has its sign opposite to that of $b(x)$.

Finally, we observe that, in a similar way one can consider the problem of the propagation of the wave front related to the equation

$$\frac{\partial u^\varepsilon}{\partial t} = \mathcal{L}^\varepsilon u^\varepsilon + \frac{1}{\varepsilon} f(x,u^\varepsilon)\,. \tag{3.3}$$

Here $\mathcal{L}^\varepsilon$ is the generator of some Markov process $(X_t^\varepsilon, P_x^\varepsilon)$ to which corresponds the action functional $\varepsilon^{-1}\tilde{S}_{0,t}(\phi)$. For the process $(X_t^\varepsilon, P_x^\varepsilon)$, one can, for example, take a family of purely discontinuous Markov processes. Then $\mathcal{L}^\varepsilon$ will be an integral operator of convolution type. The corresponding action functionals are evaluated in Wentzell and Freidlin [2].

3. After going over to the equation with small parameter one can consider the wave propagation in a bounded domain with some given boundary conditions.

Given a domain $D \subset R^r$ with the sufficiently smooth boundary, consider the equation

$$\frac{\partial v^\varepsilon}{\partial t} = \frac{\varepsilon}{2} \sum_{i,j=1}^{r} \frac{\partial}{\partial x^i}\left(a^{ij}(x)\frac{\partial v^\varepsilon}{\partial x^j}\right) + \frac{1}{\varepsilon} f(x,v^\varepsilon) = L^\varepsilon v^\varepsilon + \varepsilon^{-1} f(x,v^\varepsilon) ,$$

$$t > 0,\ x \in D,\ v^\varepsilon(0,x) = g(x) \tag{3.4}$$

in the cylinder $(0,\infty) \times D$. The assumptions concerning the coefficients $a^{ij}(x)$ and the functions $f(x,v)$ and $g(x)$ will be the same as at the beginning of section 6.2. Boundary conditions should be adjoined to (3.4). We shall consider conditions of one of the following types

$$v^\varepsilon(t,x)\Big|_{x \in \partial D} = 0 , \tag{3.5}$$

$$\frac{\partial v^\varepsilon}{\partial \ell(x)}\Big|_{x \in \partial D} = 0 . \tag{3.6}$$

We shall assume $\ell(x)$ to be a smooth vector field on ∂D nowhere tangent to the manifold ∂D.

In the case of problem (3.4)-(3.5), the limit behavior of $v^\varepsilon(t,x)$ as $\varepsilon \downarrow 0$ is defined by the same functional $R_{0,t}(\phi)$. However, the upper bound involved in the definition of $V(t,x,y)$ should be taken over the set of the functions $\phi \in C_{0t}(R^r)$, $\phi_0 = x$, $\phi_t = y$, which do not leave the domain D during the time interval $(0,t)$. Allowing for these modifications, Theorem 2.1 remains valid for $v^\varepsilon(t,x)$. The proof follows the same scheme. Theorem 2.2 also holds true, whenever the distance between points $x,y \in D$ is defined as the lower bound of the Riemannian length of the curves which connect the points x and y without leaving D.

In the case of problem (3.4)-(3.6), one should change the functional R. When considering the Cauchy problem or problem (3.4)-(3.5), the functional R was constructed as the difference between $\int_0^t c(\phi_s)\,ds$ and the normed action functional corresponding to the family of the processes

$(X_t^\varepsilon, P_x^\varepsilon)$ in R^r, governed by the operator L^ε. In the case of problem (3.4)-(3.6), one should subtract from $\int_0^t c(\phi_s)ds$ the normed action functional $\tilde{S}_{0,t}(\phi)$ corresponding to the family of the processes $(X_t^\varepsilon, P_x^\varepsilon)$ which, inside the domain D, are governed by the operator L^ε, and, on the boundary, are subject to reflection in the direction of the field $\ell(x)$. As was explained in §2.5, the solution of problem (3.4)-(3.6) is written in terms of the process $(X_t^\varepsilon, P_x^\varepsilon)$. In the general case, the functional $\tilde{S}_{0,t}(\phi)$ has been calculated by Anderson and Orey [1] (see also Wentzell and Freidlin [2]). If the functional $R_{0,t}(\phi)$ is replaced by $\tilde{R}_{0,t}(\phi) = \int_0^t c(\phi_s)ds - \tilde{S}_{0,t}(\phi)$, and the upper bound in the definition of $V(t,x,y)$, $(x,y \in D \cup \partial D)$ is taken over the functions not leaving $D \cup \partial D$, then Theorem 2.1 may be applied to the solution of problem (3.4)-(3.6).

4. Now we shall discuss the non-linear term in equation (2.1). We have supposed that, for every $x \in R^r$, the function $c(x,u) = u^{-1}f(x,u)$ (see Fig. 2) vanishes at the point $u = 1$. It is not difficult to generalize the results of §6.2 to the case when the only non-negative root of the function $c(x,u)$ (for a fixed x) depends on $x : c(x,z(x)) = 0$, $c(x,u) > 0$ for $u < z(x)$, $c(x,u) < 0$ for $u > z(x)$, and $c(x,0) = \max\limits_{0 \le u \le z(x)} c(x,u)$. A slight modification in the proof of Theorem 2.1 shows that, in this case, $\lim\limits_{\varepsilon \downarrow 0} u^\varepsilon(t,x) = z(x)$, provided $V(t,x) > 0$. If $V(t,x) < 0$, then just as before $\lim\limits_{\varepsilon \downarrow 0} u^\varepsilon(t,x) = 0$.

Equality (2.13) and estimation with the action functional are convenient tools for studying the limit behavior of the solutions of equations such as

$$\frac{\partial u^\varepsilon(t,x)}{\partial t} = \frac{\varepsilon}{2} \frac{\partial^2 u^\varepsilon(t,x)}{\partial x^2} + \frac{1}{\varepsilon} c[x; u^\varepsilon(s,x), s \le t] u^\varepsilon(t,x),$$

where $c[x; u(s,x), s \le t]$ is a functional depending not only on the value of $u(t,x)$ at time t, but on the behavior of $u(s,x)$ for $s \le t$ as well. Examination of such non-linear terms enables one to model, via one equation, some effects which are usually described by a system of equations.

5. Let $D = \{(x,y) \in R^2 : -\infty < x < \infty, |y| < a\}$, $g(x,y) = \chi_{x<0}(x)$, $\varepsilon > 0$. We will consider the following problem

$$\frac{\partial u^\varepsilon(t,x,y)}{\partial t} = \frac{1}{2\varepsilon}\frac{\partial^2 u^\varepsilon}{\partial y^2} + \frac{\varepsilon}{2}\frac{\partial^2 u^\varepsilon}{\partial x^2} + \frac{1}{\varepsilon} f(u^\varepsilon),\ t > 0,\ x \in D,$$
$$u^\varepsilon(0,x,y) = g(x,y),\ u^\varepsilon(t,x,\pm a) = 0\,. \tag{3.7}$$

In regard to the function $f(u)$, we shall assume that $f(0) = 0$, $c(u) = u^{-1}f(u)$ decreases monotonically for $u \geq 0$ from $c = c(0) = \lim_{u\downarrow 0} u^{-1}f(u)$, $0 < c < \infty$, to $c(\infty) = \lim_{u\to\infty} c(u) \geq -\infty$. Let $(X_t^\varepsilon, P_x^\varepsilon)$ be a process governed by the operator $\frac{\varepsilon}{2}\frac{d^2}{dx^2}$, and $(Y_t^\varepsilon, \tilde{P}_y^\varepsilon)$ be a process governed by the operator $\frac{1}{2\varepsilon}\frac{d^2}{dy^2}$, independent of $(X_t^\varepsilon, P_x^\varepsilon)$. We put $\tau^\varepsilon = \inf\{t : |Y_t^\varepsilon| = a\}$.

Using the probability representation of solutions of linear parabolic equations (see Chapter II), one can write down the following equation for $u^\varepsilon(t,x,y)$:

$$u^\varepsilon(t,x,y) = E_{x,y}^\varepsilon g(X_t^\varepsilon)\chi_{\tau^\varepsilon > t} \exp\left\{\frac{1}{\varepsilon}\int_0^t c(u^\varepsilon(t-s, X_s^\varepsilon, Y_s^\varepsilon))\,ds\right\}, \tag{3.8}$$

where $E_{x,y}^\varepsilon$ denotes the integration with respect to the measure $P_x^\varepsilon \times \tilde{P}_y^\varepsilon$. From (3.8), on account of $c(u) \leq c(0) = c$, we get

$$u^\varepsilon(t,x,y) \leq e^{\frac{ct}{\varepsilon}} P_x\{X_t^\varepsilon \leq 0\} \cdot \tilde{P}_y^\varepsilon\{\tau^\varepsilon > t\}\,. \tag{3.9}$$

We shall denote by λ^ε the largest eigenvalue of the problem

$$\frac{1}{2\varepsilon}\phi''_{yy}(y) = \lambda^\varepsilon \phi(y),\ -a < y < a,\ \phi(\pm a) = 0\,.$$

A simple calculation shows that $\lambda^\varepsilon = \frac{\pi^2}{8a^2\varepsilon}$, the corresponding eigenfunction $\phi(y)$ being equal to $\cos\frac{\pi y}{2a}$. It is readily checked that $\tilde{P}_y^\varepsilon\{\tau^\varepsilon > t\} \sim$

$\phi(y) \cdot \exp\{-\lambda^{\varepsilon} t\}$ as $\varepsilon \downarrow 0$. Whence, remembering (3.9) we conclude:

$$u^{\varepsilon}(t,x,y) \leq \exp\left\{\frac{1}{\varepsilon}\left(ct - \frac{x^2}{2t} - \frac{\pi^2 t}{8a^2}\right)\right\}.$$

From this it results that if $x > t\sqrt{2c - \frac{\pi^2}{4a^2}}$ and $c > \frac{\pi^2}{8a^2}$, then $\lim_{\varepsilon \downarrow 0} u^{\varepsilon}(t,x,y) = 0$.

We shall denote by u^* a positive solution of the equation $c(u^*) = \frac{\pi^2}{8a^2}$. Such a u^* exists, provided $c(\infty) < \frac{\pi^2}{8a^2} < c$, and it is always unique. We note that $c(u) < c$ for $u > u^*$. Whence, analogous to the proof of Theorem 2.1, it is possible to verify that $\lim_{\varepsilon \downarrow 0} u^{\varepsilon}(t,x,y) = u^*$ for $x < t\sqrt{2c - \frac{\pi^2}{4a^2}}$.

Therefore, for $c < \frac{\pi^2}{8a^2}$, the solution $u^{\varepsilon}(t,x,y)$ of problem (3.7) for $\varepsilon \ll 1$ has the shape of a wave in the form of a step function with the height u^* which travels along the x-axis from left to right with the velocity $v = \sqrt{2c - \frac{\pi^2}{4a^2}}$. If $c < \frac{\pi^2}{8a^2}$, then $\lim_{\varepsilon \downarrow 0} u^{\varepsilon}(t,x,y) = 0$ for all x,y and $t > 0$. In particular, if the band is narrow enough, $\left(a < \frac{\pi}{2\sqrt{2c}}\right)$, then the wave does not propagate.

Certainly, a similar problem may be considered for equation with variable coefficients, with a non-linear term depending on x and y, and with variable band width. In this case a phenomenon of stopping the wave near some point may occur. For example, if equation (3.7) is considered in a band of the variable width $2a(x)$, then the wave will stop in front of the domain where $a(x) < \frac{\pi}{2\sqrt{2c}}$. Then the wave will continue its propagation along the x-axis where the band width is large enough.

6. Let us consider the simplest KPP equation

$$\frac{\partial u}{\partial t} = \frac{1}{2}\frac{\partial^2 u}{\partial x^2} + f(u),\ t > 0,\ x \in R^1, \tag{3.10}$$

$$u(0,x) = \chi_{x<0}(x) . \tag{3.11}$$

For the equations, homogeneous in x, our results may be easily formulated without introducing a small parameter. Namely, Theorem 2.2 implies the following simple assertion:

$$\begin{aligned} &\lim_{t\to\infty} u(t,vt) = 0 \quad \text{for} \quad v > \sqrt{2f'(0)} , \\ &\lim_{t\to\infty} u(t,vt) = 1 \quad \text{for} \quad v < \sqrt{2f'(0)} . \end{aligned} \tag{3.12}$$

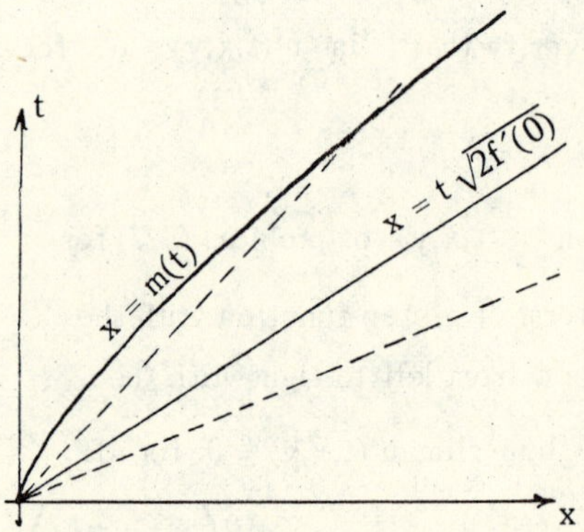

Fig. 12

In other words, the function $u(t,x)$ tends to 0 along any straight line passing through the origin which lies below the line $x = t\sqrt{2f'(0)}$ in the first quadrant. On the other hand, $u(t,x)$ tends to 1 along any straight line passing through the origin and lying above $x = t\sqrt{2f'(0)}$ (see Fig. 12). In Section 6.1, this result has been formulated in a slightly different way. First of all, we observe that, for every t, the function $u(t,x)$ monotonically decreases in x. For example, this follows from the monotonicity of the initial function and from the fact that $v(t,x) = \frac{\partial u(t,x)}{\partial x}$ satisfies the equation

$$\frac{\partial v}{\partial t} = \frac{1}{2}\frac{\partial^2 v}{\partial x^2} + f'(u(t,x))\,v, t > 0, x \in R^r .$$

for which the maximum principle holds. We shall denote by $m = m(t)$ a number for which $u(t,m(t)) = \frac{1}{2}$. On account of the above remark and of the fact that, for every t, $u(t,x)$ varies from 1 to 0 for $x \in (-\infty, \infty)$, we conclude that such an $m(t)$ exists and is unique. Hence, $m(t)$ is the coordinate of a point of the "fixed phase of a travelling wave." True, for every finite value of t, generally speaking, there is no wave, because the profile of the function $u(t,m(t)+z) = U_t(z)$ depends on t. However, as it has been shown in the work of KPP [1], this profile tends to some fixed profile as $t \to \infty$.

Statement (3.12) means that

$$m(t) = t\sqrt{2f'(0)} + o(t),\ t \to \infty .$$

The article by McKean [3] was followed by a series of works refining the behavior of $m(t)$ as $t \to \infty$ (Bramson [1], Uchiyama [1], Gartner [4]). The Feynman-Kac formula implies that

$$u(t,x) = E_x\, \chi_{x<0}(X_t) \exp\left\{\int_0^t c(u(t-s,X_s))\,ds\right\} \leq$$

$$\leq e^{ct} P_x\{X_t<0\} = \frac{e^{ct}}{\sqrt{2\pi t}} \int_x^\infty e^{-\frac{z^2}{2t}} dz = \frac{e^{ct}}{\sqrt{\pi}} \int_{\frac{x}{\sqrt{2t}}}^\infty e^{-u^2}\, du . \tag{3.13}$$

Here $c = f'(0) \geq \max\limits_{0\leq u\leq 1} c(u)$, $c(u) = u^{-1}f(u)$; (X_t, P_x) is a Wiener process on the line. Let $x = t\sqrt{2c} - \dfrac{\ln t}{2\sqrt{2c}}$. With the aid of L'Hopital's Rule it is readily checked that

$$\lim_{t\to\infty} \frac{e^{ct}}{\sqrt{\pi}} \int_{\sqrt{tc}-\frac{\ln t}{4\sqrt{ct}}}^{\infty} e^{-u^2}\, du = \frac{1}{\sqrt{2\pi}}. \tag{3.14}$$

From (3.13) and (3.14) it follows that

$$u\left(t, t\sqrt{2c} - \frac{\ln t}{2\sqrt{2c}}\right) < \frac{1}{2\pi} + o(1) < \frac{1}{2}$$

for t large enough. This implies that

$$m(t) \leq t\sqrt{2c} - \frac{1}{2\sqrt{2c}} \ln t \tag{3.15}$$

for large t. Bound (3.15) has incidentally been obtained in the above cited article by McKean.

The most precise result, among known ones to date, belongs to Bramson [1]. He showed that

$$m(t) = t\sqrt{2c} - \frac{3}{2\sqrt{2c}} \ln t + O(1),\ t \to \infty\,. \tag{3.16}$$

Recently Uchiyama [1] has shown that (3.16) remains true, if (3.11) is replaced by the weaker condition

$$0 < u(0,x) = o(e^{-\beta x}),\ x \to \infty\,,$$

for some $\beta > \sqrt{2c}$.

We will also mention the following result by McKean: If the limit $\lim\limits_{x\to\infty} u(0,x)\exp\{\beta x\} = a$ exists for some $\beta \in (0, \sqrt{2c})$, and $0 < a < \infty$, then

$$m(t) = \left(\frac{\beta}{2} + \frac{c}{\beta}\right) t + o(1),\ t \to \infty\,.$$

In their researches, McKean and Bramson used an elegant representation of the solution to the KPP equation in the form of the mean value of some functional of a branching process with diffusion (see §5.5). Garther [4] derived bound (3.16) and some of its generalizations using the asymptotic behavior of the first exit time of a Wiener process.

§6.4 *Other forms of non-linear terms*

We shall denote by $\mathfrak{D}$ the set of functions $f(x,y)$, $x \in R^r$, $-\infty < u < \infty$, such that, for every $x \in R^r$, the function $f(x,u)$ has the form represented in Fig. 1-d: $f(x,0) = f(x,\mu(x)) = f(x,1) = 0$, $0 < \mu(x) < 1$, $f(x,u) < 0$ for $u \in (0,\mu(x)) \cup (1,\infty)$, $f(x,u) > 0$ for $u \in (-\infty, 0) \cup (\mu(x), 1)$, $f'_u(x,0) < 0$,

$f'_u(x,1) < 0$. As is customary, we shall designate $c(x,u) = u^{-1}f(x,u)$. In the case considered here, the graph of $c(x,u)$ as a function of u for a fixed $x \in R^r$ has the form represented in Fig. 13. The function $c(x,u)$ will always be thought of as continuously differentiable.

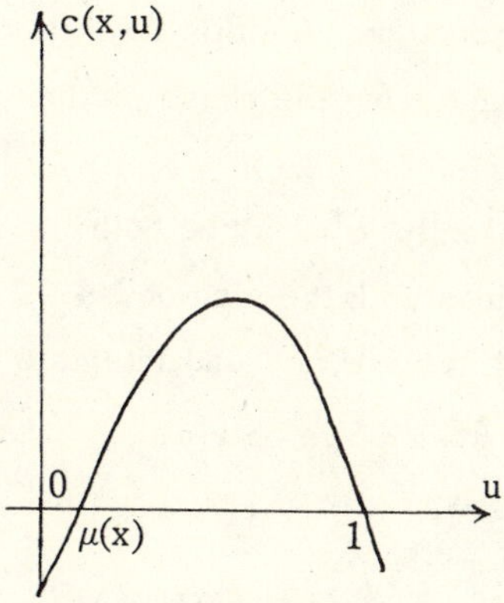

Fig. 13

As will be seen later on, the wave front propagation in the case when $f \in \mathfrak{D}$ is, in essence, a local process. The behavior of the front near a point $x_0 \in R^r$ is defined by the equation with "frozen" coefficients $a^{ij}(x_0)$ and non-linear term $f(x_0, u)$. Moreover, the front propagation in a direction e at a point x_0 is determined, in a certain sense, only by that part of the operator which acts in the direction e, i.e. it is essentially a one-dimensional process. In this connection, we will first recall results on the one-dimensional case, homogeneous in space.

So let us consider the Cauchy problem

$$\frac{\partial u(t,x)}{\partial t} = \frac{a}{2}\frac{\partial^2 u}{\partial x^2} + f(u),\ t > 0,\ x \in R^1,$$
$$u(0,x) = \chi_{x<0}(x)\,. \tag{4.1}$$

As it was shown by Aronson and Weinberger [1,2] and Fife and McLeod [1], the solution of problem (4.1) converges to a "traveling wave" solution: $U(x-\alpha t)$. In the case when $f \in \mathfrak{D}$, both the wave shape and its velocity α are entirely determined by the equation for $U(\xi)$. If $u(t,x) = U(x-\alpha t)$ is substituted in equation (4.1), then we arrive at the following equation for $U(\xi)$:

$$\frac{a}{2}U''_{\xi\xi} + \alpha U'_{\xi} + f(U) = 0,\ -\infty < \xi < \infty,$$
$$\lim_{\xi\to-\infty} U(\xi) = 1,\quad \lim_{\xi\to\infty} U(\xi) = 1\,. \tag{4.2}$$

It turns out that, for $f \in \mathfrak{D}$, problem (4.2) has a solution for one and only one $\alpha = \alpha^* = \alpha^*[a,f]$ which is determined by the coefficient a and the function f (we will remind that, for $f \in \mathfrak{A}$, problem 4.2 was solvable for any $\alpha \geq \sqrt{2af'(0)}$. It is easily checked that $\alpha^*[a,f] = \sqrt{a}\,\alpha^*[1,f] = \sqrt{a}\,\alpha^*[f]$. For $\alpha = \alpha^*$, the solution of problem (4.2) is unique up to the drift of the argument. The function $U(\xi)$ is monotone. All this information on problem (4.2) is obtained by means of analyzing the phase picture of equation (4.2) in the (U,U')-plane.

Unlike the KPP equation, for $f \in \mathfrak{D}$ the velocity α^* may be both positive and negative. The sign of α^* coincides with the sign of $\int_0^1 f(u)\,du$. Indeed, let us multiply equation (4.2) by $U'(\xi)$ and integrate from $-\infty$ to $+\infty$. Noting that $U'(\xi) \to 0$ as $\xi \to \pm\infty$ we obtain

$$0 = \frac{a}{2}\int_{-\infty}^{\infty} U''(\xi)U'(\xi)\,d\xi + \alpha\int_{-\infty}^{\infty}[U'(\xi)]^2\,d\xi + \int_{-\infty}^{\infty} f(U(\xi))\,U'(\xi)\,d\xi =$$

$$= \frac{a}{4}\int_{-\infty}^{\infty} d(U'(\xi))^2 + \alpha\int_{-\infty}^{\infty}[U'(\xi)]^2\,d\xi + \int_{0}^{1} f(U)\,dU =$$

$$= \alpha\int_{-\infty}^{\infty}[U'(\xi)]^2\,d\xi - \int_0^1 f(u)\,du\,.$$

Whence we find

$$\alpha = \int_0^1 f(u)\,du\left(\int_{-\infty}^{\infty}[U'(\xi)]^2\,d\xi\right)^{-1} \tag{4.3}$$

and, thereby, $\operatorname{sign}\alpha = \operatorname{sign}\int_0^1 f(u)\,du$. This means that, depending on the sign of $\int_0^1 f(u)\,du$, either the domain where $u(t,x)$ is close to 1 will expand (the sign $+$), or the domain where $u(t,x)$ is close to 0 will expand (the sign $-$) as t increases. In the case when $f \in \mathfrak{A}$, equality (4.3) surely holds too, but there $\int_0^1 f(u)\,du$ is always positive.

We proceed now to consider the corresponding problem with small parameter and variable coefficients. As before, we shall assume for the present that $x \in R^1$:

$$\frac{\partial u^\varepsilon(t,x)}{\partial t} = \frac{\varepsilon}{2}\, a(x)\, \frac{\partial^2 u^\varepsilon}{\partial x^2} + \frac{1}{\varepsilon}\, f(x, u^\varepsilon)\,,$$
$$u^\varepsilon(0,x) = \chi_{x<0}(x)\,. \tag{4.4}$$

The Feynman-Kac formula implies that

$$u^\varepsilon(t,x) = E_x\, \chi_{x<0}(X_t^\varepsilon)\, \exp\left\{\frac{1}{\varepsilon} \int_0^t c(X_s^\varepsilon, u^\varepsilon(t-s, X_s^\varepsilon))\, ds\right\}, \tag{4.5}$$

where $(X_t^\varepsilon, P_x^\varepsilon)$ is a process in R^r, corresponding to the operator $\frac{\varepsilon a(x)}{2} \frac{d^2}{dx^2}$. Since $f(u) < 0$ for $u > 1$, one can easily deduce from (4.5) that $0 < u^\varepsilon(t,x) \le 1$, one can easily deduce from (4.5) that $0 \le u^\varepsilon(t,x) \le 1$ for all $t \ge 0$, $x \in R^1$. It is also seen from (4.5) that, for small ε, the half-plane $\{(t,x) : t > 0,\ x \in R^1\}$ may be divided into a domain D_ε^1 where $u^\varepsilon(t,x)$ is close to 1, a domain D_ε^2 where $u^\varepsilon(t,x)$ is close to 0, and a narrow transition area $\mathcal{E}_\varepsilon$ near the front diminishing as $\varepsilon \downarrow 0$. Since $c(x,u) < 0$ for small u and $c(x,u)$ is close to 0 for small $1-u$, one can conclude that, for small ε, the expectation of (4.5) is mainly contributed to by those trajectories which go in the domain of the function $u^\varepsilon(t,x)$ values far from 0 and 1. Such values are taken by $u^\varepsilon(t,x)$ only near the wave front. This leads to the following two conclusions. First, for $f \in \mathfrak{D}$, the front travels in a continuous way. No jumps and no sources appear. This is due to the fact that $c(x,u) < 0$ for small u. Second, for $f \in \mathfrak{D}$, one cannot separate the question about the velocity and that about the shape of the wave, because the expectation of (4.5) is mainly contributed to by trajectories passing in the domain where $u^\varepsilon(t,x)$ changes fast. We will remind that in the case when $f \in \mathfrak{A}$, the basic contribution was from trajectories passing in the domain where $u^\varepsilon(t,x)$ is close to zero.

Therefore, for $f \in \mathfrak{D}$, in order to evaluate the low of the advance of the limit step function as $\varepsilon \downarrow 0$, it is necessary to take into consideration the shape of the wave in the transition area. However, the circumstance that the propagation is of a local nature, simplifies the problem. This enables one to examine equation (4.4) as $\varepsilon \downarrow 0$ relying on the comparison theorems (i.e. in essence, on the maximum principle) and on the properties of problem (4.1), homogeneous in x.

Let us clarify this for the case when $a(x) = a = \text{const}$. We will let $\alpha(x) = \sqrt{a}\,\alpha^*[f(x,\cdot)]$, and let for the sake of definiteness, $\alpha(x) > 0$ for $x \in R^1$. Repeating the argument involved in the proof of Lemma 2.4, one can make sure that $\lim_{\varepsilon \downarrow 0} u^\varepsilon(t,x) = 1$ for $x < \psi_t$, where the function ψ_t is defined by the equation

$$\dot{\psi}_t = \alpha(\psi_t),\ \psi_0 = 0\,. \tag{4.6}$$

On the other hand, the function $\tilde{u}^\varepsilon = 1 - u^\varepsilon$ obeys the equation

$$\frac{\partial \tilde{u}^\varepsilon}{\partial t} = \frac{\varepsilon a}{2}\frac{\partial^2 \tilde{u}^\varepsilon}{\partial x^2} + \frac{1}{\varepsilon}\tilde{f}(x,\tilde{u}^\varepsilon),\ \tilde{u}^\varepsilon(0,x) = 1 - \chi_{x<0}(x)\,,$$

where $\tilde{f}(x,u) = -f(x,1-u)$. It is readily seen that $\tilde{f}(x,u)$ also belongs to $\mathfrak{D}$, and thus, with the same reasoning as above, we can obtain a lower bound for $\tilde{u}^\varepsilon = 1 - u^\varepsilon$. This bound implies that $\lim_{\varepsilon \downarrow 0} \tilde{u}^\varepsilon(t,x) = 1 - \lim_{\varepsilon \downarrow 0} u^\varepsilon(t,x) = 1$, provided $x > \psi(t)$. Therefore the function $\psi(t)$ defined by equation (4.6), represents the coordinate of the front at time t. The velocity of the front propagation at a point x is defined by the formula $\alpha^*[f(x,\cdot)]\sqrt{a}$. For the foregoing reasoning to become a stringent proof, it is necessary, of course, to check that the functional $\alpha^*[f(x,\cdot)]$ depends smoothly on f.

Now we shall consider the multi-dimensional case and variable diffusion coefficients:

$$\frac{\partial u^{\varepsilon}(t,x)}{\partial t} = \frac{\varepsilon}{2} \sum_{i,j=1}^{r} \frac{\partial}{\partial x^i}\left(a^{ij}(x)\frac{\partial u^{\varepsilon}}{\partial x^j}\right) + \frac{1}{\varepsilon} f(x,u^{\varepsilon}) , \tag{4.7}$$

$$t > 0,\ x \in R^r;\ u^{\varepsilon}(0,x) = g(x) \geq 0 .$$

Let a point x_0 belong to the front at some moment. We shall make a linear coordinate transformation which turns the form $\Sigma\, a^{ij}(x_0)\lambda_i\lambda_j$ into a sum of squares. If in the new coordinates $(\tilde{x}^1, \cdots, \tilde{x}^r)$ the plane $\tilde{x}^1 = 0$ is the tangent plane of the front at the point x_0, then the subsequent advance of the front near the point x_0 will be determined by the operator $\frac{\varepsilon}{2}\frac{\partial^2 u}{(\partial\tilde{x}^1)^2} + \frac{1}{\varepsilon} f(x_0,u)$. Therefore, in the new coordinates the velocity at the point x_0 is $\alpha^*[f(x_0,\cdot)]$. If we now return to the old coordinates, then for the velocity $v(x,e)$ at a point x in a direction $e = (e^1, \cdots, e^r)$, $\sum_1^r (e^i)^2 = 1$ we obtain the formula

$$v(x,e) = \left(\sum_{i,j=1}^{r} a_{ij}(x)e^ie^j\right)^{-1/2} \alpha^*[f(x,\cdot)] . \tag{4.8}$$

This formula means that, if the velocity is measured in the Riemannian metric induced by the form $ds^2 = \sum_1^r a_{ij}(x)dx^i dx^j$, then the velocities field will be isotropic.

Formula (4.8) for the velocities field of the wave front relying on the above cited reasoning, had been suggested by Freidlin and Sivak [1]. The exact proof was then given by Gärtner [2].

In order to formulate the result derived by Gärtner, we shall introduce some notations. For any $x,y \in G_-^0 = \{x \in R^r : \alpha(x) < 0\}$, $\alpha(x) = \alpha^*[f(x,\cdot)]$, we set

$$\rho_-^0(x,y) = \inf \int_{T_1}^{T_2} \frac{|\dot{\phi}_s|\,ds}{v\left(\phi_s, \frac{\dot{\phi}_s}{|\dot{\phi}_s|}\right)},$$

where $v(x,e)$ is defined by formula (4.8). The lower bound in the definition of $\rho^0(x,y)$ is taken over all absolutely continuous curves ϕ_s, $T_1 \le s \le T_2$ connecting the points x and y and lying entirely in G_-^0. The distance $\rho_+^0(x,y)$ is defined by the same formula on the set $G_+^0 = \{x \in R^r : \alpha(x) \ge 0\}$, however the lower bound should be taken over curves lying entirely in G_+^0. In a similar fashion we shall define the distances $\rho_-^1(x,y)$ and $\rho_+^1(x,y)$ on the sets $G_-^1 = \{x \in R^r : \alpha(x) \le 0\}$ and $G_+^1 = \{x \in R^r : \alpha(x) > 0\}$ respectively. Since $\alpha(x) = \alpha^*[f(x,\cdot)]$ may vanish, these distances may, generally speaking, take the value $+\infty$.

Next, we introduce the following sets:

$$A^0 = \{x \in R^r : g(x) < \mu(x)\},\ A^1 = \{x \in R^r : g(x) > \mu(x)\},$$

$$Q_+^0 = \{(t,x) : t > 0,\ x \in G_+^0,\ \rho_+^0(x, G_+^0 \setminus A^0) > t\},$$

$$Q_-^0 = \{(t,x) : t > 0,\ x \in G_-^0,\ \rho_-^0(x, G_-^0 \cap A^0) < t\},$$

$$Q_+^1 = \{(t,x) : t > 0,\ x \in G_+^1,\ \rho_+^1(x, G_+^1 \cap A^1) < t\},$$

$$Q_-^1 = \{(t,x) : t > 0,\ x \in G_-^1,\ \rho_-^1(x, G_-^1 \setminus A^1) > t\},$$

$$Q^0 = Q_+^0 \cap Q_-^0,\ Q^1 = Q_+^1 \cup Q_-^1.$$

THEOREM 4.1. *Suppose that* $f \in \mathfrak{D}$, *and let* $|c(x,u)| = |u^{-1}f(x,u)|$ *and* $(1-u)^{-1}f(x,u)$ *be uniformly bounded for* $x \in R^r$, $u \in (0,1)$. *Then the solution* $u^\varepsilon(t,x)$ *of problem (4.7) converges to* 0 *uniformly in* (t,x) *belonging to any compact subset of the set* Q^0 *as* $\varepsilon \downarrow 0$, *and it converges to* 1 *uniformly in* (t,x) *belonging to any compact subset of the set* Q^1.

We restrict ourselves to the outline of the proof given above. The proof in detail may be found in the above cited article by Gärtner.

Hence, if one considers as excited all points x at which the function $u^{\varepsilon}(t,x)$ is close to 1, then in the case when $f \in \mathfrak{D}$, for small ε, we arrive at a phenomenological account of excitation spreading (for $\alpha(x) > 0$). The corresponding velocities field is assigned by formula (4.8). For $\alpha(x) < 0$, the domain, where there is no excitation, will expand with velocity $|v(x,e)|$.

EXAMPLE 4.1. Let us consider equation (4.7) for $f(x,u) = u(1-u)(u-\mu(x))$. For $\mu(x)$ taking values from $(0,1)$, such a function $f(x,u)$ belongs to the class $\mathfrak{D}$. It is easy to see that

$$\int_0^1 f(x,u)\,du = \frac{1}{12} - \frac{1}{6}\mu(x) .$$

Therefore, $\mu = \frac{1}{2}$ is a boundary value at which the velocity sign changes. For brevity, we will assume that $0 < \mu(x) < \frac{1}{2}$ for $x \in R^r$. To compute $\alpha(x) = \alpha^*[f(x,\cdot)]$, let us consider the equation for the wave profile:

$$\begin{gathered} U''_{\xi\xi} + \tilde{\alpha} U'_{\xi}(\xi) + f(x,U(\xi)) = 0,\ -\infty < \xi < \infty , \\ U(-\infty) = 1,\ U(+\infty) = 0 . \end{gathered} \tag{4.9}$$

It turns out that, for the above indicated function $f(x,u)$, problem (4.9) may be solved explicitly. By direct substitution in the equation one can verify that problem (4.9) is solvable for

$$\tilde{\alpha} = \tilde{\alpha}(x) = \frac{1}{\sqrt{2}} - \mu(x)\sqrt{2} .$$

For such an $\tilde{\alpha}$, the function

$$U(\xi) = \left[1 + \exp\left\{\frac{1}{\sqrt{2}}\xi\right\}\right]^{-1}$$

is a solution of problem (4.9). For the velocities field $v(x,e)$, we obtain the expression

$$v(x,e) = \frac{1}{\sqrt{2}}\left(\sum_{i,j=1}^{r} a_{ij}(x)\, e^i e^j\right)^{-\frac{1}{2}}\left(\frac{1}{\sqrt{2}} - \mu(x)\sqrt{2}\right) =$$

$$= \left(\sum_{i,j=1}^{r} a_{ij}(x)\, e^i e^j\right)^{-\frac{1}{2}}\left(\frac{1}{2} - \mu(x)\right).$$

In particular, in the case of homogeneous, isotropic diffusion $(a^{ij}(x) = D\,\delta^{ij})$, we have $v(x,e) = \sqrt{D}\left(\frac{1}{2} - \mu(x)\right)$.

Finally, we pause for a few words on the non-linearities represented in Fig. 1-b and Fig. 1-c.

If for every x the function $f(x,u)$ has the form represented in Fig. 1-b, then we face, generally speaking, two difficulties simultaneously: first, the evaluation of the velocity of the front wave propagation cannot be separated from taking into account the wave shape (just as in the case when $f \in \mathfrak{D}$); secondly, in this case (similar to the case when $f \in \mathfrak{A}$) new sources may appear. The latter appear, if in some domain of variation of x, $f'(x,0)$ takes values essentially larger than outside this domain. The situation gets simpler substantially, if, for example, $f'_u(x,0) = \text{const.}$ is assumed to be independent of $x \in R^r$. In this case, new sources do not appear, and the wave front propagation is of a local nature. For $f'_u(x,0) = \text{const.}$, it seems, the wave front propagation is described by velocities field (4.8) ($g(x)$ is looked upon as continuous and having a compact support). Here as $\alpha^*[f(x,\cdot)]$, one should take the expression

$$\alpha^*[f(x,\cdot)] = \frac{1}{\sqrt{2}} \inf_{\rho} \sup_{0<u<1} \left\{\rho'(u) + \frac{f(x,u)}{\rho(u)}\right\},$$

where $\rho(u)$ is a continuously differentiable function on $[0,1]$, positive for $u \in (0,1)$, for which $\rho(0) = 0$, $\rho'(0) > 0$. Such an assertion suggests itself as an implication of the results of the work by Hadeler and Rothe [1], where the one-dimensional problem, homogeneous in x, is considered.

We observe that if for every $x \in R^r$ the function $f(x,u)$ has the form represented in Fig. 1-b, then just as in the case when $f \in \mathfrak{A}$, the propagation is of a non-local nature. Non-local effects may be described with the functional $R^b_{0,t}(\phi) = \int_0^t [f'_u(\phi_s, 0) - \frac{1}{2} \sum_{i,j=1}^r a_{ij}(\phi_s)\dot\phi^i_s \dot\phi^j_s]ds$.

In the case when the functions $f(x,u)$, for every $x \in R^r$, have the form represented in Fig. 1-c, new sources do not appear. Here the process is of a local nature, and may be examined in a way similar to the case when $f \in \mathfrak{D}$.

It is also possible to consider non-linearities which are of a different form in different domains of the variation of x. Some results in this direction were obtained by Gärtner [2].

§6.5 *Other kinds of random movements*

As it has already been said, the wave front propagation is due to two factors; to the random motion of particles and to their multiplication. In the preceding sections, a diffusion process appears in the role of the random motion. Now, following the works by Sarafian and Safarian [1,2], we shall consider an alternative kind of random motion. Unlike diffusion, this motion has a finite velocity and is a "diffusion with inertia." We go straight to the problem with small parameter. In the case when the velocities field is homogeneous in space, the results of this section may be formulated without small parameter; namely, as some statement on the behavior of a solution of some boundary value problem as $t \to \infty$. Here we shall not strive for generality, and consider the simplest situation in which the effects we are interested in, still do appear.

So, we will consider the problem

$$\frac{\partial u^\varepsilon(t,x,y)}{\partial t} = \frac{1}{2\varepsilon}\frac{\partial^2 u^\varepsilon}{\partial x^2} + b(x,y)\frac{\partial u^\varepsilon}{\partial y} + \frac{1}{\varepsilon} f(x,y,u^\varepsilon) =$$

$$= L^\varepsilon u^\varepsilon + \frac{1}{\varepsilon} f(x,y,u^\varepsilon),\ t > 0,\ x \in (-1,1),\ -\infty < y < \infty, \tag{5.1}$$

$$u^\varepsilon(0,x,y) = g(x,y) \geq 0,\ \left.\frac{\partial u^\varepsilon(t,x,y)}{\partial x}\right|_{x=\pm 1} = 0 .$$

Here $b(x,y)$ and $f(x,y,u)$ are sufficiently smooth functions, bounded for $|x| \le 1$, $y \in (-\infty, \infty)$, $u \in [0,1]$ (for example, let them have bounded first-order derivatives). For brevity, we shall assume that $g(x,y)$ is continuous and has the set $G_0 \subseteq \{(x,y): |x| \le 1, y \le 0\}$ as its support.

Moreover, for every $x \in [-1,1]$ and $y \in (-\infty, \infty)$, the function $f(x,y,u)$ is assumed to have the form represented in Fig. 1-a; $f(x,y,0) = f(x,y,1) = 0$, $f(x,y,u) > 0$ for $u \in (0,1)$; $f(x,y,u) < 0$ for $u \notin [0,1]$. We shall let $c(x,y,u) = u^{-1} f(x,y,u)$, $c(x,y) = c(x,y,0) = \lim_{u \to 0} u^{-1} f(x,y,u)$, and suppose that the function $c(x,y,u)$ is Lipschitz continuous and

$$\max_{0 \le u \le 1} c(x,y,u) = c(x,y) .$$

Let $(X_t^\varepsilon, Y_t^\varepsilon; P_{x,y}^\varepsilon)$ be a diffusion process in the band $\pi = \{(x,y) \in R^2 : |x| \le 1, y \in R^1\}$, which, at the interior points of the band π, is governed by the operator L^ε and, on the boundary $\partial\pi = \{(x,y) \in R^2 : |x| = 1, y \in R^1\}$ it is subject to normal reflection. We have already considered such a process in Chapter IV. We will remind that the process $(X_t^\varepsilon, Y_t^\varepsilon; P_{x,y}^\varepsilon)$ may be constructed starting from the Wiener process $(\overline{W}_t, \overline{P}_x)$ on the segment $[-1,1]$ with reflection at the end-points of this segment. Namely, one should put $X_t^\varepsilon = W_{t/\varepsilon}$, Y_t^ε being a solution of the ordinary differential equation $\dot{Y}_t^\varepsilon = b(X_t^\varepsilon, Y_t^\varepsilon)$.

Equation (5.1) is a degenerate one, so the question about the existence of a classical solution of problem (5.1) is not trivial. If such a solution exists, then the Feynman-Kac formula yields for it the following equation

$$u^\varepsilon(t,x,y) = E_{x,y}^\varepsilon \, g(X_t^\varepsilon, Y_t^\varepsilon) \exp\left\{\frac{1}{\varepsilon} \int_0^t c(X_s^\varepsilon, Y_s^\varepsilon, u^\varepsilon(t-s, X_s^\varepsilon, Y_s^\varepsilon))\, ds\right\} . \tag{5.2}$$

The function $u^\varepsilon(t,x,u)$ which satisfies equation (5.2) for all $t \ge 0$, $x \in [-1,1]$, $y \in R^1$, will be called a generalized solution of problem (5.1). This definition is correct. Since the function $c(x,y,u)$ is assumed to be

Lipshitz continuous, it is not hard to prove that a generalized solution exists (compare with the results of §5.3). In general, if the functions $g(x,y)$, $c(x,y,u)$, and $b(x,y)$ are assumed smooth enough, then it is possible to demonstrate that the generalized solution is smooth and satisfies equation (5.1). But we will not do this and provide all our results for the generalized solution of problem (5.1).

We shall consider for brevity the case when $b(x,y) = b(x)$ does not depend on y, $f(x,y,u) = f(u) = u \cdot c(u)$. Besides, we shall assume that $1 \geq g(x,y) = g(y) \geq 0$, $G_0 = \text{supp}\, g(x,y) = \{(x,y) \in R^2 : |x| \leq 1,\ y \leq 0\}$.

Let us rewrite problem (5.1) for this case

$$\frac{\partial u^\varepsilon(t,x,y)}{\partial t} = \frac{\varepsilon}{2}\frac{\partial^2 u^\varepsilon}{\partial x^2} + b(x)\frac{\partial u^\varepsilon}{\partial y} + \frac{1}{\varepsilon}c(u^\varepsilon)u^\varepsilon ,$$

$$u^\varepsilon(0,x,y) = g(y), \quad \left.\frac{\partial u^\varepsilon(t,x,y)}{\partial x}\right|_{x=\pm 1} = 0 .$$

It is supposed that

$$c = c(0) = \sup_{0\leq u\leq 1} c(u) . \tag{5.4}$$

To describe the limit behavior of $u^\varepsilon(t,x,y)$ as $\varepsilon \downarrow 0$, we shall need some constructions.

Let us consider the eigenvalue problem

$$\frac{1}{2}\phi''(x) + \beta b(x)\phi(x) = \lambda\phi(x),\ x \in (-1,1) ,$$

$$\phi'(-1) = \phi'(1) = 0 ,$$

where β is a parameter. As is known, this problem has a discrete spectrum. The eigenvalue $\lambda = \lambda(\beta)$ having the maximal real part, is real and has multiplicity one. The corresponding eigenfunction is positive. From the fact that $\lambda(\beta)$ has multiplicity one, it follows that $\lambda(\beta)$ is differentiable in β (see Kato [1]). It is readily checked that the function

$\lambda(\beta)$ is convex (see e.g. Freidlin and Wentzell [2]). We will denote by $L(\alpha)$, $\alpha \in R^1$ the Legendre transform of the function $\lambda(\beta)$:

$$L(\alpha) = \sup_{\beta} [\alpha\beta - \lambda(\beta)] ;$$

$L(\alpha)$ is a non-negative, convex function, equal to $+\infty$ outside some finite interval. It is straightforward that $L(\alpha) = 0$ for $\alpha = \frac{1}{2}\int_{-1}^{1} b(x)\,dx$.

We put $\alpha_0 = \sup\{\alpha : L(-\alpha) < \infty\}$, $L_0 = \lim_{\alpha \uparrow \alpha_0} L(-\alpha)$. It is clear that, for $\alpha \in (0, \alpha_0)$, the function $L(-\alpha)$ is continuous, α_0 being a positive number.

THEOREM 5.1. *Suppose that the function* $f(u) = c(u) \cdot u$ *satisfies the conditions previously given in this section. Let us put* $\overline{b} = \frac{1}{2}\int_{-1}^{1} b(x)\,dx$. *If* $c(0) = c < L_0$, *then a unique positive solution* α^* *of the equation* $L(-\alpha^*) = c$ *exists, and the relations*

$$\lim_{\varepsilon \downarrow 0} u^{\varepsilon}(t,x,y) = \begin{cases} 1 \ \textit{for} \ |x| \le 1,\ y < (\alpha^* - \overline{b})t , \\ 0 \ \textit{for} \ |x| \le 1,\ y > (\alpha^* - \overline{b})t \end{cases} \tag{5.5}$$

hold for the solution of problem (5.3). If $c \ge L_0$, *then relations (5.5) hold for* $\alpha^* = \alpha_0$. *The convergence in (5.5) is uniform on every compact subset of* $(0,\infty) \times \pi$ *which does not contain points* (t,x,y) *such that* $y = \alpha^* t - \overline{b}t$.

The proof of this theorem is arranged in accordance with the same scheme as the proof of Theorem 2.1. At some parts, it is easier because this equation is invariant with respect to translations along the y-axis.

Without any loss of generality, one can think that $\overline{b} = 0$. If $\overline{b} \neq 0$, then one should go over to the new unknown function $\tilde{u}^{\varepsilon}(t,x,y) = u^{\varepsilon}(t,x,y+\overline{b}t)$.

In the case under consideration, equality (5.2) takes the form:

$$u^{\varepsilon}(t,x,y) = E^{\varepsilon}_{x,y}\, g(Y^{\varepsilon}_t) \exp\left\{\frac{1}{\varepsilon}\int_0^t c(u^{\varepsilon}(t-s, X^{\varepsilon}_s, Y^{\varepsilon}_s))\,ds\right\} . \tag{5.6}$$

From (5.6) and (5.4) it follows that

$$0 \le u^{\varepsilon}(t,x,y) \le E^{\varepsilon}_{x,y}\, g(Y^{\varepsilon}_t) \exp\left\{\frac{ct}{\varepsilon}\right\} \le \sup_{z \in R^1} g(z) e^{\frac{ct}{\varepsilon}} \cdot P^{\varepsilon}_{x,y}\{Y^{\varepsilon}_t \le 0\}\,. \tag{5.7}$$

To estimate the probability on the right-hand side, we shall make use of the action functional for the family of processes Y^{ε}_t. We shall denote by $S_{0,t}(\phi)$ the functional on the space $C_{0,t}(R^1)$ which, for absolutely continuous functions, is defined by the equality

$$S_{0,t}(\phi) = \int_0^t L(\dot{\phi}_s)ds \,;$$

for the other elements of $C_{0,t}(R^1)$ we put $S_{0,t}(\phi) = +\infty$. The functional $S_{0,t}(\phi)$ is lower semi-continuous. It is the normed action functional for the family of the processes Y^{ε}_t as $\varepsilon \downarrow 0$ (see §1.7; there are further references there). The corresponding normalizing coefficient is equal to ε^{-1}. We will recall that this means that, for any $\delta_1 h,\ s > 0$ and $\phi \in C_{0,t}(R^1)$, $\phi_0 = y$, for ε small enough, the following relation holds:

$$\begin{aligned} &P^{\varepsilon}_{x,y}\{\sup_{0 \le s \le t} |Y^{\varepsilon}_s - \phi_s| < \delta\} \ge \exp\{-\varepsilon^{-1}(S_{0,t}(\phi) + h)\}\,,\\ &P^{\varepsilon}_{x,y}\{\rho_{0,t}(Y^{\varepsilon}, \Phi^y_s) \ge \delta\} \le \exp\{-\varepsilon^{-1}(s-h)\}\,, \end{aligned} \tag{5.8}$$

where $\Phi^y_s = \{\phi \in C_{0,t}(R^1) : S_{0,t}(\phi) \le s, \phi_0 = y\}$, $\rho_{0,t}(\cdot,\cdot)$ is a metric in $C_{0,t}(R^1)$. Relations (5.8) are fulfilled uniformly in $x \in [-1,1]$, and y belonging to any compact set $F \in R^1$ and $\phi \in \Phi^y_s$.

The following lemma plays the same role as Lemma 2.1 did when proving Theorem 2.1.

LEMMA 5.1. *Let* $0 \le yt^{-1} = a < a_0$. *Then uniformly in* $x \in [-1,1]$

$$\lim_{\varepsilon \downarrow 0} \varepsilon \ln P^{\varepsilon}_{x,y}\{Y^{\varepsilon}_t \le 0\} = -tL(-a)\,.$$

If $ty^{-1} = a > a_0$, *then this limit is equal to* $-\infty$.

Proof. From (5.8), noting that the functional $S_{0,t}(\phi)$ is semi-continuous, we get

$$- \inf \{S_{0,t}(\phi) : \phi \in C_{0,t}(R^1), \phi_0 = y, \phi_t < 0\} \leq$$

$$\leq \varliminf_{\varepsilon \downarrow 0} \varepsilon \ln P^{\varepsilon}_{x,y}\{Y^{\varepsilon}_t \leq 0\} \leq \varlimsup_{\varepsilon \downarrow 0} \varepsilon \ln P^{\varepsilon}_{x,y}\{Y^{\varepsilon}_t \leq 0\} \leq \tag{5.9}$$

$$\leq - \inf \{S_{0,t}(\phi) : \phi \in C_{0,t}(R^1), \phi_0 = y, \phi_t \leq 0\} .$$

If $yt^{-1} = \alpha > \alpha_0$, then on some set of a positive measure from $[0, t]$, the absolutely continuous functions, for which $\phi_0 = y$, $\phi_t \leq 0$, must have a derivative not exceeding $-yt^{-1} = -\alpha < -\alpha_0$. On this set, $L(\dot{\phi}_s) = +\infty$, and thus the right-hand side of (5.9) becomes equal to $-\infty$. This implies that $\lim_{\varepsilon \downarrow 0} \varepsilon \ln P^{\varepsilon}_{x,y}\{Y^{\varepsilon}_t \leq 0\}$ exists and is equal to $-\infty$, provided $yt^{-1} = \alpha > \alpha_0$.

Now let $yt^{-1} = \alpha < \alpha_0$. Since the integrand in $S_{0,t}(\phi)$ depends only on $\dot{\phi}$, we deduce that all extremals are straight lines. Remembering the boundary conditions we find that the infimum on the right-hand side of (5.9) is attained on the function $\hat{\phi}_s = y - \frac{sy}{t}$ and is equal to $t \cdot L(-yt^{-1}) = tL(-\alpha)$. Taking into account that the function $L(-\alpha)$ is continuous for $\alpha \in [0, \alpha_0)$, we obtain that the left-hand side in (5.9) also is equal to $tL(-\alpha)$. This yields that $\lim_{\varepsilon \downarrow 0} \varepsilon \ln P^{\varepsilon}_{x,y}\{Y^{\varepsilon}_t \leq 0\} = -t \cdot L(-\alpha)$. □

Applying Lemma 5.1, we conclude from (5.7) that

$$\varlimsup_{\varepsilon \downarrow 0} \varepsilon \ln u^{\varepsilon}(t,x,y) \leq t(c - L(-yt^{-1})) . \tag{5.10}$$

For $yt^{-1} > \alpha_0$, the right-hand side of (5.10) is equal to $-\infty$.

First consider $c < L_0$. Note that, for $\alpha \in [0, \alpha_0)$, the function $L(-\alpha)$ is continuous and monotonically increasing, $L(0) = 0$ (we assume that $\bar{b} = 0$), and $L_0 = \lim_{\alpha \uparrow \alpha_0} L(-\alpha)$. Hence, it appears clear that the equation $L(-\alpha) = c < L_0$ has a unique solution α^* belonging to the interval $(0, \alpha_0)$. From (5.10) it follows that

$$\lim_{\varepsilon \downarrow 0} u^{\varepsilon}(t,x,y) = 0 \quad \text{for} \quad y > \alpha^* t, \ |x| \leq 1 , \tag{5.11}$$

because $L(-yt^{-1}) > L(-\alpha^*) = c$. This convergence is uniform in the domain $\{(x,y) \in \pi : |x| \leq 1, \ y > (\alpha^*+\delta)t, t \leq T\}$ for any $\delta > 0$, $T < \infty$.

If $c \geq L_0$, then we assume $\alpha^* = \alpha_0$. The validity of equality (5.11) for $\alpha^* = \alpha_0$ follows from Lemma 5.1.

Now let us show that

$$\lim_{\varepsilon \downarrow 0} u^{\varepsilon}(t,x,y) = 1 , \quad \text{for} \quad y < \alpha^* t , \ |x| \leq 1 . \tag{5.12}$$

Here one also can apply arguments quite similar to those used when proving Theorem 2.1. Let $y_0 < \alpha^* t_0$, $h = t_0 - y_0(\alpha^*)^{-1} > 0$. We will introduce the domain D in the space of variables (t,x,y): $D = D^{\varepsilon, y_0}_{\delta_1,\delta_2} = \{(t,x,y) : t > 0, |x| \leq 1, \ y < \alpha^* t, \ u^{\varepsilon}(t,x,y) < 1 - \delta_2, \ |y - y_0| < \delta_1\}$, where δ_1, δ_2 are small positive numbers. Fig. 14 represents the projection of this domain onto the (y,t)-plane.

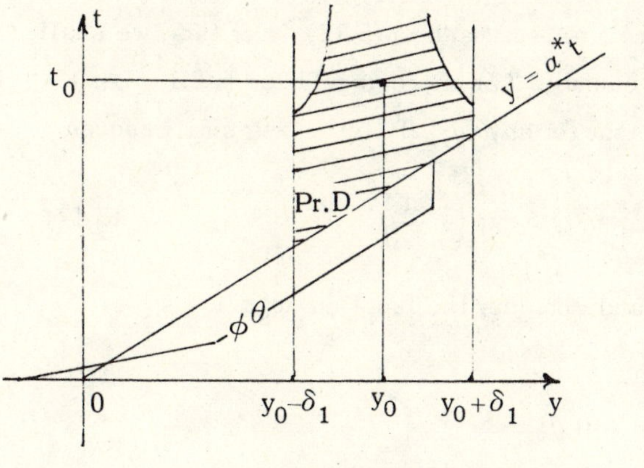

Fig. 14

Let us consider the Markov process $(t_s, X^{\varepsilon}_s, Y^{\varepsilon}_s; P^{\varepsilon}_{t,x,y})$ on the set $\{(t,x,y) : -\infty < t < \infty, |x| \leq 1, y \in R^1\}$, which is governed by the operator

$$-\frac{\partial}{\partial t} + \frac{1}{2\varepsilon}\frac{\partial^2}{\partial x^2} + b(x,y)\frac{\partial}{\partial y}$$

at the interior points of this set and is subject to the reflection along the normal on its boundary. For brevity, we denote $Z^{\varepsilon}_s = (t_s, X^{\varepsilon}_s, Y^{\varepsilon}_s)$.

Let us denote by $\tau = \tau^2$ the first exit time of the process $(Z^{\varepsilon}_s, P^{\varepsilon}_{t,x,y})$ from the domain $D : \tau = \inf\{s : Z^{\varepsilon}_s \notin D\}$. Clearly, $P^{\varepsilon}_{t,x,y}\{\tau \leq t\} = 1$ for $(t,x,y) \in D$. From relation (5.6) for $(t_0, x_0, y_0) \in D$, applying the strong Markov property, we derive:

$$u^{\varepsilon}(t_0,x_0,y_0) = E^{\varepsilon}_{t_0,x_0,y_0} u^{\varepsilon}(Z^{\varepsilon}_{\tau}) \exp\left\{\frac{1}{\varepsilon}\int_0^{\tau} c(u^{\varepsilon}(Z^{\varepsilon}_s))\,ds\right\} =$$

$$= \sum_{i=1}^{3} E_{t_0,x_0,y_0} \chi_i\, u^{\varepsilon}(Z^{\varepsilon}_{\tau}) \exp\left\{\frac{1}{\varepsilon}\int_0^{\tau} c(u^{\varepsilon}(Z^{\varepsilon}_s))\,ds\right\}. \tag{5.13}$$

Here we denoted by χ_1 the indicator of the set of those trajectories of the process Z^{ε}_s, which leave D across that part of the boundary where $y = a^*t$. By χ_2 we denoted the indicator of the set of trajectories leaving D across that part of the boundary where $|y-y_0| = \delta_1$. Finally, χ_3 is the indicator of the set of trajectories leaving D across that part of the boundary, where $u^{\varepsilon}(t,x,y) = 1-\delta_2$.

We observe that $c(u) \geq 0$ for $u \in [0,1]$, and $u^{\varepsilon}(\tau, X^{\varepsilon}_{\tau}, Y^{\varepsilon}_{\tau}) = 1-\delta_2$, whenever $\chi_3 = 1$. From this we conclude that the third summand on the right-hand side of (5.13) is not smaller than $(1-\delta_2) P^{\varepsilon}_{t_0,x_0,y_0}\{\chi_3 = 1\}$.

Now we shall estimate the first summand in (5.13). For this, we shall need an assertion similar to bound (2.17) in proving Theorem 2.1. Let $0 < a^* < a_0$. We shall show that for any $\mu > 0$, for $\varepsilon > 0$ small enough

$$u^{\varepsilon}(t,x,a^*t) \geq \exp\left\{-\frac{\mu t}{\varepsilon}\right\}. \tag{5.14}$$

Let us choose $\beta \in (a^*, a_0)$ and consider the function ϕ^{θ}_s, $0 \leq s \leq t$, θ being a small parameter:

$$\phi^{\theta}_s = \begin{cases} a^*t \text{ for } s \in [0,\theta], \\ -a^*s + a^*(t+\theta) \text{ for } s \in [\theta, t-\sqrt{\theta}], \\ -\beta(s-t) + a^*(\theta+\sqrt{\theta}) - \beta\sqrt{\theta} \text{ for } s \in [t-\sqrt{\theta}, t]. \end{cases}$$

This piece-wise linear function connects the point $a^*t = \phi^{\theta}_0$ and the point $\phi^{\theta}_t = (a^*-\beta)\sqrt{\theta} + a^*\theta$, which lies on the left of zero for sufficiently small θ. It is easily verified that $S_{0,t}(\phi^{\theta}) = \int_0^t L(\dot{\phi}^{\theta}_s)\,ds$ may be made arbitrarily small at the expense of choosing small θ. We note that for

$s \in (0, t-\sqrt{\theta})$, the function ϕ^θ lies in the domain where $u^\varepsilon(t,x,y)$ is close to zero for small ε. This, along with the first of bounds (5.8), leads to (5.14): for sufficiently small $\varepsilon > 0$:

$$u^\varepsilon(t,x,y) = E^\varepsilon_{x,\alpha^* t}\, g(Y^\varepsilon_t) \exp\left\{\int_0^t c(u^\varepsilon(t-s, X^\varepsilon_s, Y^\varepsilon_s))\, ds\right\} \geq$$

$$\geq \frac{1}{2}\, g((\alpha^*-\beta)\sqrt{\theta}+\alpha^*\theta) \exp\left\{\left[\varepsilon^{-1}\left(c-\frac{1}{2}\mu\right) - \varepsilon^{-1}\left(L(-\alpha^*)+\frac{\mu}{3}\right)\right] t\right\} \geq$$

$$\geq \exp\left\{-\frac{\mu t}{\varepsilon}\right\}.$$

To leave D, the trajectories, for which $\chi_1 = 1$, take longer than $h - \frac{2\delta_1}{\alpha^*} = h_1$. We shall assume that δ_1 is so small that $h_1 > 0$. Along these trajectories, during the time spent in D,

$$c(u^\varepsilon(t-s, X^\varepsilon_s, Y^\varepsilon_s) > \lambda > 0\,, \tag{5.15}$$

where $\lambda = \lambda(\delta_2) = \min\limits_{0 \leq u \leq 1-\delta_2} c(u) > 0$. Selecting $\mu < \frac{\lambda h_1}{t_0}$, from (5.14) and (5.15) we obtain

$$E^\varepsilon_{t_0,x_0,y_0} \chi_1\, u^\varepsilon(Z^\varepsilon_\tau) \exp\left\{-\varepsilon^{-1}\int_0^\tau c(u^\varepsilon(Z^\varepsilon_s))\, ds\right\} \geq$$

$$\geq \exp\left\{-\frac{1}{\varepsilon}(2\mu t_0 - \lambda h_1)\right\} P^\varepsilon_{t_0,x_0,y_0}\{\chi_1 = 1\} \geq P^\varepsilon_{t_0,x_0,y_0}\{\chi_1 = 1\}.$$

Finally, we observe that $P^\varepsilon_{t_0,x_0,y_0}\{\chi_1+\chi_3 = 1\} \to 1$ as $\varepsilon \downarrow 0$, because the probability that in time $[0,t_0]$ the component Y^ε_t moves off from the initial point y_0 by the value δ_1, tends to zero together with ε. This results from our assumption $\overline{b} = \frac{1}{2}\int_{-1}^1 b(x)\, dx = 0$.

Gathering the bounds of all summands in (5.13) we conclude that

$$\underline{\lim}_{\varepsilon \downarrow 0} u^{\varepsilon}(t_0,x_0,y_0) \geq (1-\delta_2)\, P^{\varepsilon}_{t_0,x_0,y_0}\{\chi_1+\chi_3=1\} - o_{\varepsilon}(1) \to 1-\delta_2$$

provided $a^* t_0 > y_0$. On the other hand, $\overline{\lim}_{\varepsilon \downarrow 0} u^{\varepsilon}(t,x,y) \leq 1$, since $c(u) < 0$ for $u \notin [0,1]$. Therefore, noting that δ_2 is arbitrarily small, for $y < a^* t$, we obtain

$$\lim_{\varepsilon \downarrow 0} u^{\varepsilon}(t,x,y) = 1 . \tag{5.16}$$

This convergence is uniform in the domain $\{(t,x,y) : \delta \leq t \leq T,\ |x| \leq 1,\ y < a^* t - \delta\}$ for any $\delta > 0$, $T < \infty$.

To complete the proof of the theorem, (5.16) should also be established for the case $c \geq L_0$. To this end, let us consider an increasing sequence of functions $c_n(u)$ of the same form as $c(u)$:

$$c_n(1) = 0,\ c_n(u) > 0 \text{ for } u < 1;\ c_n(u) < 0 \text{ for } u > 1 ,$$

$$c_n(0) = \max_{0 \leq u \leq 1} c_n(u) < L_0,\quad \lim_{n \to \infty} c_n(0) = L_0 .$$

Let $u_n^{\varepsilon}(t,x,y)$ be a solution of problem (5.3), where the function $c_n(u)$ is taken instead of $c(u)$. Moreover let a_n^* be a solution of the equation $c_n(0) = L(-a)$. The condition $c_n(0) \to L_0$ implies that $a_n^* \to a_0$. Since $c_n(0) < L_0$, we have that $\lim_{\varepsilon \downarrow 0} u_n^{\varepsilon}(t,x,y) = 1$ for $y < a_n^* t$. Whence, taking into account that due to the maximum principle $u^{\varepsilon}(t,x,y) \geq u_n^{\varepsilon}(t,x,y)$, we conclude that equality (5.16) is also valid in the case $c \geq L_0$. Here one should assume $a^* = a_0$. □

Now we shall go into more detail in examining equations with a non-linear term depending on x. Let us consider the problem

$$\frac{\partial u^{\varepsilon}(t,x,y)}{\partial t} = \frac{1}{2\varepsilon}\frac{\partial^2 u^{\varepsilon}}{\partial x^2} + b(x)\frac{\partial u^{\varepsilon}}{\partial y} + \frac{1}{\varepsilon}\, c(x,u^{\varepsilon})\, u^{\varepsilon} ,$$

$$t > 0,\ |x| < 1,\ -\infty < y < \infty,\ u^{\varepsilon}(0,x,y) = g(y),\ \left.\frac{\partial u^{\varepsilon}}{\partial x}(t,x,y)\right|_{|x|=1} = 0. \tag{5.17}$$

We suppose that the function $c(x,u)$ is continuously differentiable and let, for every x, it decrease monotonically in u, $c(x,0) > 0$, $c(x,1) = 0$. In regard to the initial function, we shall make the same assumptions as in Theorem 5.1. For brevity, we shall assume that $\overline{b} = \frac{1}{2} \int_{-1}^{1} b(x)\,dx = 0$, $b(x) \not\equiv 0$.

The fact that the non-linear term depends on the fast variable x, causes the following complication. In order to calculate the law of the wave propagation, it is already insufficient to know the asymptotics of the probabilities of large deviations for the particles motion, i.e. for Y_t^ε. Now it is necessary to know the asymptotics of the probabilities of large deviations for the two-dimensional process $(Y_t^\varepsilon, Z_t^\varepsilon)$, where $Z_t^\varepsilon = \int_0^t c(X_s^\varepsilon)\,ds$, $c(x) = c(x,0)$.

To describe this asymptotic behavior let us consider the eigenvalue problem (β_1, β_2 are number parameters):

$$\frac{1}{2}\phi''(x) + (\beta_1 b(x) + \beta_2 c(x))\phi(x) = \lambda(\beta_1, \beta_2)\phi(x),$$

$$|x| < 1,\ \phi'(1) = \phi'(-1) = 0.$$

Let $\lambda(\beta_1, \beta_2)$ be the eigenvalue of this problem with the largest real part, $L(a^1, a^2)$ being the Legendre transform of the convex function $\lambda(\beta_1, \beta_2)$:

$$L(a^1, a^2) = \sup_{\beta_1, \beta_2} [a^1\beta_1 + a^2\beta_2 - \lambda(\beta_1, \beta_2)].$$

The function $L(a^1, a^2)$ is equal to $+\infty$ outside some bounded domain. Inside the domain of finiteness, the function $L(a^1, a^2)$ is non-negative, continuous and convex.

Let the functional $S_{0,t}(\phi^1, \phi^2)$ be equal to $\int_0^t L(\dot\phi_s^1, \dot\phi_s^2)\,ds$ for absolutely continuous $\phi^1, \phi^2 \in C_{0,t}(R^1)$ and let it be equal to $+\infty$ for remaining pairs $\phi^1, \phi^2 \in C_{0,t}(R^1)$. It follows from Wentzell and Freidlin [2], that this functional is the action functional for the family of two-dimensional processes $(Y_t^\varepsilon, Z_t^\varepsilon)$ as $\varepsilon \downarrow 0$.

Let us put $\mathcal{L}(a) = \sup_{\gamma \in R^1} [\gamma - L(-a, \gamma)]$, $a \in R^1$.

THEOREM 5.2. *Suppose that there is a unique positive root* a^* *of the equation* $\mathcal{L}(a) = 0$, *the function* $\mathcal{L}(a)$ *being continuous at the point* a^*. *Then the relation*

$$\lim_{\varepsilon \downarrow 0} u^\varepsilon(t,x,y) = \begin{cases} 0, & \text{if } y > a^* t \\ 1, & \text{if } y < a^* t \end{cases}$$

is fulfilled for the solution $u^\varepsilon(t,x,y)$ *of problem (5.17).*

Let us outline *the proof* of this theorem. From (5.2) it follows that

$$0 \le u^\varepsilon(t,x,y) \le E_{x,y}\, g(Y_t^\varepsilon) \exp\{\varepsilon^{-1} Z_t^\varepsilon\} = H^\varepsilon(t,x,y) . \tag{5.18}$$

By the properties of the action functional we get the inequality

$$\sup\{\phi_t^2 - S_{0,t}(\phi^1,\phi^2) : \phi_0^1 = y, \phi_t^1 < 0, \phi_0^2 = 0\} \le \varliminf_{\varepsilon \downarrow 0} \epsilon \ln H^\varepsilon(t,x,y) \le$$
$$\le \varlimsup_{\varepsilon \downarrow 0} \epsilon \ln H^\varepsilon(t,x,y) \le \sup\{\phi_t^2 - S_{0,t}(\phi^1,\phi^2) : \phi_0^1 = y, \phi_t^1 \le 0, \phi_0^2 = 0\} . \tag{5.19}$$

Taking into account that the integrand in $S_{0,t}(\phi_1,\phi_2)$ depends only on derivatives, it is easily checked that the upper bound on the right-hand side of (5.19) is

$$t \sup_{\gamma \in R^1} [\gamma - L(-yt^{-1}, \gamma)] = t\, \mathcal{L}(yt^{-1}) .$$

This and (5.18) imply that $\lim_{\varepsilon \downarrow 0} u^\varepsilon(t,x,y) = 0$ for $y > a^* t$, $|x| \le 1$.

We leave it to the reader to prove that $\lim_{\varepsilon \downarrow 0} u^\varepsilon(t,x,y) = 1$ in the domain $\{|x| \le 1,\ y < a^* t\}$. The proof is arranged along the same lines as in Theorem 5.1. We only observe that the continuity of the function $\mathcal{L}(a)$ at the point a^* turns out to be helpful when constructing the bound similar to bound (5.14) in the proof of Theorem 5.1. □

§6.6 *Wave front propagation due to non-linear boundary effects*

As was previously said, wave front propagation in non-linear diffusion equations is the result of the interaction of two processes: random motion of particles and multiplication. In many problems, it is natural to think that the multiplication occurs not everywhere in domain, but only on the boundary or on some surfaces of co-dimension 1 lying interior to the domain. This must result in the wave propagating, basically, along these surfaces, or, at least, in the wave propagation all over the volume being defined by the surface phenomena.

Taking into account surface effects leads to a non-linear term in the equation taking the form $f(u)\delta_\Gamma(x)$, where $\delta_\Gamma(x)$ is the δ-function spread over the surface Γ. The availability of such a term in the equation is equivalent to some boundary condition, if Γ is the boundary of the domain under consideration, or it is equivalent to a condition of piecing together solutions along the surface Γ. Hence a problem arises with non-linear boundary conditions or non-linear glueing conditions. Here we shall consider the problem with non-linear boundary conditions, following the work by Korostelev and Freidlin [1].

Given a domain $D \subset R^r$ with smooth boundary ∂D, consider the mixed boundary value problem

$$\frac{\partial u^\varepsilon(t,x)}{\partial t} = \frac{\varepsilon}{2} \sum_{i,j=1}^{r} \frac{\partial}{\partial x^i}\left(a^{ij}(x)\,\frac{\partial u^\varepsilon}{\partial x^j}\right) = L^\varepsilon u^\varepsilon,\ t > 0,\ x \in D\,,$$

$$u^\varepsilon(0,x) = g(x),\ \frac{\partial u^\varepsilon}{\partial \ell(x)}(t,x) + \varepsilon^{-1} f(x,u^\varepsilon(t,x))\Big|_{x\in\partial D,\,t>0} = 0\,. \tag{6.1}$$

Here $a^{ij}(x)$ are bounded, twice continuously differentiable functions forming a positive definite matrix, and $\ell(x)$ is a smooth vector field on ∂D. In regard to the function $f(x,u)$, we assume that it is continuously differentiable for $x \in R^r$, $-\infty < u < \infty$, and belongs to the class $\mathfrak{A}$ (see §6.2). The initial function $g(x)$ is assumed to be continuous, non-negative and to have a compact support G_0.

We remind that, for $f(x,u) = c(x) \cdot u$, the solution of problem (6.1) may be represented in the form of the mean value of some functional of the process $(X_t^\varepsilon, P_x^\varepsilon)$ in $D \cup \partial D$, which, inside the domain, is governed by the operator L^ε, and, on the boundary, is subject to reflection in the direction $\ell(x)$ (see §2.5).

For brevity, we shall dwell on a special case of problem (6.1). Namely, here we shall assume that $D = R_+^r = \{x \in R^r : x^1 > 0\}$, $\ell(x) = (1, 0, \cdots, 0)$. Besides, we shall suppose that $\lim\limits_{u \to 0} u^{-1} f(u) = c = \text{const}$.

For the process $(X_t^\varepsilon, P_x^\varepsilon)$ in R_+^r with reflection on the boundary to be constructed, we have introduced in §1.6, following Anderson and Orey [1], the mapping $\Gamma : C_{0,T}(R^r) \to C_{0,T}(R_+^r)$: if $\zeta = \zeta_t = (\zeta_t^1, \cdots, \zeta_t^r)$, then $\Gamma_t(\zeta) = (\zeta_t^1 - [\min\limits_{0 \le s \le t} \zeta_s^1 \wedge 0], \zeta_t^2, \zeta_t^3, \cdots, \zeta_t^r)$. The mapping $\xi : C_{0,T}(R^r) \to C_{0,T}^+$, where $C_{0,T}^+$ is the space of non-decreasing non-negative continuous functions, has been defined by the formula: $\Gamma(\zeta) - \zeta = (\xi(\zeta), 0, 0, \cdots, 0)$. The space $C_{0,T}^+$ has been equipped with the topology of uniform convergence. Let $\sigma(x)$ be a matrix with Lipschitz continuous elements for which $\sigma(x)\sigma^*(x) = (a^{ij}(x))$; $b(x) = (b^1(x), \cdots, b^r(x))$, $b^j(x) = \frac{1}{2} \sum_{i=1}^r \frac{\partial a^{ij}(x)}{\partial x^i}$.

We will consider the stochastic equation

$$Y_t^{\varepsilon,x} = x + \sqrt{\varepsilon} \int_0^t \sigma(\Gamma_s(Y^{\varepsilon,x})) dW_s + \varepsilon \int_0^t b(\Gamma_s(Y^{\varepsilon,x})) ds . \tag{6.2}$$

As it was explained in §1.6, this equation has a unique solution, and the family of processes $X_t^{\varepsilon,x} = \Gamma_t(Y^{\varepsilon,x})$ induces the Markov process $(X_t^\varepsilon, P_x^\varepsilon)$ in R_+^r with reflection along the field $\ell(x) = (1, 0, \cdots, 0)$ which is governed by the operator L^ε inside R_+^r. The local time $\xi_t^{\varepsilon,x} = \xi_t(Y^{\varepsilon,x})$ on the boundary is associated with this process.

Theorem 2.5.1 and the Remark following this theorem imply that the solution $u^\varepsilon(t,x)$ of problem (6.1) for $D = R_+^r$, $\ell(x) = (1, 0, \cdots, 0)$ satisfies the following relation

$$u^{\varepsilon}(t,x) = E_x^{\varepsilon} g(X_t^{\varepsilon}) \exp\left\{\frac{1}{\varepsilon} \int_0^t c(X_s^{\varepsilon}, u^{\varepsilon}(t-s, X_s^{\varepsilon}))\, d\xi_s^{\varepsilon,x}\right\}, \qquad (6.3)$$

where $c(x,u) = u^{-1} f(x,u)$.

To examine the behavior of $u^{\varepsilon}(t,x)$ as $\varepsilon \downarrow 0$, we shall need the action functional for the pair of processes $(X_t^{\varepsilon}, \xi_t^{\varepsilon,x})$. Let us give the construction of this functional for our case.

First of all, we observe that the action functional for the family of the processes $Y_t^{\varepsilon,x}$ defined by equality (6.2) has the form:

$$\varepsilon^{-1} S_{0,T}^{Y}(\phi) = \frac{1}{2\varepsilon} \int_0^T \sum_{i,j=1}^{r} a_{ij}(\Gamma_s \phi)\, \dot{\phi}_s^i \dot{\phi}_s^j ds, \ \phi \in C_{0T}(R^r), (a_{ij}(x)) = (a^{ij}(x))^{-1}.$$

This assertion may be established just as Theorem 1.7.4 (see Anderson and Orey [1], Wentzel and Freidlin [2]). Now note that the pair of processes $(X_t^{\varepsilon,x}, \xi_t^{\varepsilon,x})$ is obtained from $Y_t^{\varepsilon,x}$ using the continuous transformation $B = (\Gamma, \xi): (X_t^{\varepsilon,x}, \xi_t^{\varepsilon,x}) = B_t(Y^{\varepsilon,x}) = (\Gamma_t(Y^{\varepsilon,x}), \xi_t(Y^{\varepsilon,x}))$. The transformation B is invertible, and

$$\psi_t = B^{-1}(\phi, \mu) = \phi_t - \int_0^t \chi_{\partial D}(\phi_s)\, \ell(\phi_s)\, d\mu_s .$$

In accordance with Property 1 of action functionals (see §1.7), the action functional for the pair $(X_t^{\varepsilon,x}, \xi_t^{\varepsilon,x})$ has the form

$$\frac{1}{\varepsilon} S_{0,T}^{x,\xi}(\phi, \mu) = \frac{1}{\varepsilon} S_{0,T}^{Y}(B^{-1}(\phi, \mu)) =$$

$$= \frac{1}{2\varepsilon} \int_0^T \sum_{i,j=1}^{r} a_{ij}(\phi_s)(\dot{\phi}_s^i - \chi_{\partial D}(\phi_s)\ell^i(\phi_s)\dot{\mu}_s)(\dot{\phi}_s^j - \chi_{\partial D}(\phi_s)\ell^j(\phi_s)\dot{\mu}_s) ds .$$

If the mapping B^{-1} is not defined for some pair (ϕ, μ), we assume $S_{0T}^{x,\xi}(\phi, \mu) = +\infty$.

Let us denote

$$R_{0,t}^{+}(\phi, \mu) = c\,\mu_t - S_{0,t}^{x,\xi}(\phi, \mu),$$

where $c = \lim_{u\to 0} u^{-1} f(u) = \text{const}$. Since $f(x,u) \in \mathcal{A}$, $\max_{0\leq u\leq 1} c(x,u) \leq c$. Hence, noting that $\dot{\xi}_s^{\varepsilon,x} \geq 0$, (6.3) yields

$$0 \leq u^{\varepsilon}(t,x) \leq E_x\, g(X_t^{\varepsilon}) \exp\left\{\frac{c\xi_t^{\varepsilon}}{\varepsilon}\right\}. \tag{6.4}$$

By Property 3 of action functional (§1.7),

$$\begin{aligned}\lim_{\varepsilon\downarrow 0} \varepsilon \ln E_x^{\varepsilon}\, g(X_t^{\varepsilon}) \exp\left\{\frac{c}{\varepsilon}\,\xi_t^{\varepsilon}\right\} &= \\ = \sup\{R_{0,t}^{+}(\phi, \mu) : \phi_0 = x,\ \phi_t \in G_0,\ \dot{\mu}_s \geq 0 \text{ for } s \leq t\}&.\end{aligned} \tag{6.5}$$

It is not difficult to check that the maximal value of the right-hand side of (6.5) is attained for

$$\dot{\mu}_s = \frac{1}{a_{11}(\phi_s)}\left[\left(c + \sum_{i=1}^{r} a_{i1}(\phi_s)\dot{\phi}_s^i\right) \vee 0\right]$$

and is

$$V(t,x) = \sup_{\phi_0 = x,\, \phi_t \in G_0} \left\{\frac{1}{2}\int_0^t \left(\frac{\chi_{\partial D}(\phi_s)}{a_{11}(\phi_s)}\left[\left(c + \sum_{i=1}^{r} a_{i1}(\phi_s)\dot{\phi}_s^i\right) \vee 0\right]^2 - \sum_{i,j=1}^{r} a_{ij}(\phi_s)\dot{\phi}_s^i \dot{\phi}_s^j\right) ds\right\}.$$

From (6.4) and (6.5) it follows that

$$\lim_{\varepsilon\downarrow 0} u^{\varepsilon}(t,x) = 0,$$

provided $V(t,x) < 0$. It is easy to prove (we leave this to the reader) that $\lim_{\varepsilon \downarrow 0} u^{\varepsilon}(t,x) = 0$ for the interior points of the domain R^r_+ non-belonging to the closure $[G_0]$ of the set G_0. Note that for such points, $V(t,x)$ may be positive. However, if $x \in \{x \in R^r : x^1 = 0\}$, then the condition $V(t,x) > 0$ implies that $\lim_{\varepsilon \downarrow 0} u^{\varepsilon}(t,x) = 1$. The proof of the last claim is similar to the final part of the proof of Theorem 2.1, and we drop it.

It is worthwhile noting that the assumption about $c(x,0) = c = \text{const.}$ saves us from assumptions of (2.12) type. However, in the case considered here, in spite of the assumption $c(x,0) = \text{const.}$, the appearance of new sources ahead of the front as well as jumps of the wave front are possible. Later on this will be illustrated by examples.

So, we arrive at the following result.

THEOREM 6.1. *Suppose that* $D = R^r_+$, $\underline{\ell}(x) = (1,0,\cdots,0)$, $c(x,u) = c =$ const. *Then* $\lim_{\varepsilon \downarrow 0} u^{\varepsilon}(t,x) = 0$ *for any point* x *lying inside* R^r_+ *and non-belonging to* $[G_0]$. *If* $x \in \partial R^r_+$ *and* $V(t,x) > 0$, *then* $\lim_{\varepsilon \downarrow 0} u^{\varepsilon}(t,x) = 1$. *If* $x \in \partial R^+$ *and* $V(t,x) < 0$ *then* $\lim_{\varepsilon \downarrow 0} u^{\varepsilon}(t,x) = 0$.

Let us consider a number of examples illustrating the peculiarities of the wave front propagation due to the boundary conditions. In all these examples, the diffusion coefficients are assumed to be constant, $c(x,0) = c = \text{const.}$, with the domain being either a half-plane or a strip. Thus, as is readily verified, the extremals are always either segments of straight lines or broken lines, and thereby all necessary extremums may be easily computed.

EXAMPLE 6.1. Suppose that $D = R^2_+$, $L^{\varepsilon} = \frac{\varepsilon\sigma^2}{2}\Delta$, $\sigma = \text{const.}$, $\ell(x) = (1,0)$, $c(x,u) = c = \text{const.}$

Let $G_0 = \{(x^1,x^2) : |x^2| < h, 0 \le x^1 < \delta\}$. In this case, the upper bound involved in the definition of $V(t,0,x^2)$ for $x^2 > h$ is attained by the function $\phi_s = (\phi^1_s, \phi^2_s)$, $0 \le s \le t$, $\phi^1_s \equiv 0$, $\phi^2_s = x^2 - \frac{x^2-h}{t}s$:

$$V(t,0,x^2) = -\frac{t}{2\sigma^2}\frac{(x^2-h)^2}{t^2} + \frac{t\sigma^2 c^2}{2}.$$

Whence, equating $V(t,0,x^2)$ to zero, we find that at time t the wave front travelling along the boundary, is at the points $x^2(t) = \pm(h + t\sigma^2 c)$. Hence, in this example

$$\lim_{\varepsilon \downarrow 0} u^\varepsilon(t,0,x^2) = \begin{cases} 1, & \text{if } |x^2| < h + t\sigma^2 c \\ 0, & \text{if } |x^2| > h + t\sigma^2 c. \end{cases}$$

In the interior points, $\lim_{\varepsilon \downarrow 0} u^\varepsilon(t,x) = 0$, provided $x \notin [G_0]$.

Now let the domain $G_0 = \{(x^1,x^2) : 0 < a < x^1 < 2a,\ |x^2| < h\}$ have no points in common with the x-axis. Then for small t the function $V(t,x)$ is negative for any $x \notin [G_0]$. Initially, there will be no wave. At the time $T_0 = \frac{2a}{c\sigma^2}$, simultaneously on the entire interval $[-h, h]$ of the x^2-axis, an excited area arises which starts propagating in both directions along the x^2-axis. At the first moment, the velocity of this wave is infinite. Then it will decrease and becomes equal to $c\sigma^2$.

Now let the support of the initial function have the form represented in Fig. 15. From the point $(0,0)$ the wave starts propagating to the right with velocity $c\sigma^2$. In the time $T_1 = \frac{\beta}{c\sigma^2}$, the excitation from the point $(0,0)$ reaches the point $(0,\beta)$ on the x^2-axis. However, if β is large enough, $(\beta > 2h)$, then the excitation at the point $(0,\beta)$ will be induced at the time $T_2 = \frac{2h}{c\sigma^2}$ by the segment MN of the shaded domain G_0 in Fig. 15. Therefore, a new source may also appear for constant coefficients and $c(x,0) = \text{const}$. The appearance of new sources

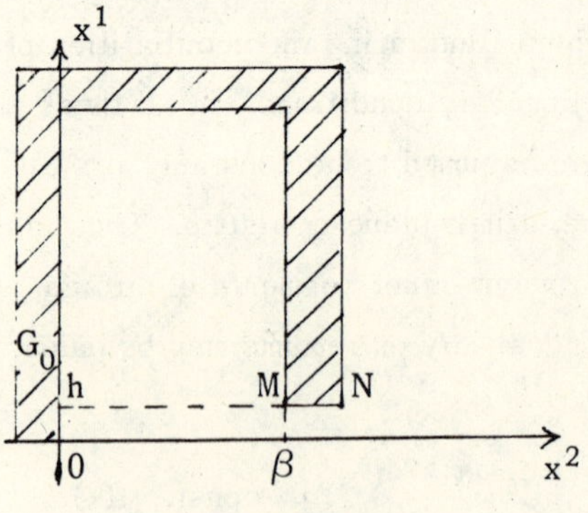

Fig. 15

and jumps of the wave front on the boundary may be caused by the mutual disposition of the support of the initial function and the boundary of the domain.

EXAMPLE 6.2. Let the operator L be the same as in the preceding example, the domain D being the band: $\{(x^1,x^2)\epsilon R^2 : 0<x^1<h, -\infty<x^2<\infty\}$; $\ell(0,x^2) = (1,0), \ell(h,x^2) = (-1,0)$; $c(0,x^2) = c_1$, $c(h,x^2) = c_2$. Our results may be readily reformulated for the case of the band. The proof undergoes no substantial changes. Let us suppose that at the initial moment, there is a point source at the point (0,0). If the domain occupied by the excitation is small enough, then the law of the propagation will be close to that given by a point source. In any case, the domain occupied by the excitation at time t, is obtained as the union of the excitation domains of all possible point sources from G_0.

So, let at the initial moment, a point source be at the point (0,0). The picture of the wave front propagation depends on the relation between c_1 and c_2.

If $c_1 \geq c_2$, then a wave starts propagating along the lower boundary of the band with velocity $v_1 = c_1\sigma^2$. This wave propagates "taking no notice" of the upper boundary. At the time $T_0 = \frac{2h}{c_2\sigma^2}$, at the point (h,0) of the upper boundary, a new source takes fire, and a wave starts propagating along the upper boundary in both directions off this source. In the initial region, the velocity of this wave is prescribed by the formula $v(z) = \frac{c\sigma^2\sqrt{h^2-z^2}}{2|z|}$ After reaching the points $(\pm z_0, h)$, where z_0 is the root of the equation $v(z_0) = v_1 = c_1\sigma^2$, the wave continues propagating in both directions along the upper boundary with velocity $v_1 = c_1\sigma^2$ (rather than with $v_2 = c_2\sigma^2$). In the stationary movement (for large t), the front on the upper boundary is $h\sqrt{3}$ behind the front on the lower boundary.

In the case when $c_2 > c$, a somewhat more complicated picture takes place in which the wave front on the upper boundary first lags behind the front on the lower boundary, and then leaves it behind. The velocity of the stationary movement on the both boundaries is $v_2 = c_2\sigma^2$.

EXAMPLE 6.3. Let $D = R^2_+$, $\ell(x^1,0) = (\operatorname{ctg} \alpha, 1)$, where $\alpha (0 < \alpha < \pi)$ is the angle between the direction of the reflection and the positive direction of the x^2-axis. Just as before, $L^\varepsilon = \frac{\varepsilon\sigma^2}{2}\Delta$, $c(x,0) = c = \text{const}$. This case may be easily reduced to that considered in Theorem 6.1, with the aid of a linear change of variables. In doing so, the reflection becomes a normal one, and rather than the operator L^ε an operator appears containing terms with mixed derivatives. In the example under consideration, the wave from the point source on the boundary propagates to the left and to the right with different velocities. The velocity of the wave propagation to the right is $v_+ = c\sigma^2 \operatorname{ctg} \frac{\alpha}{2}$. While α varies from 0 to π, the velocity diminishes from $+\infty$ to 0. The velocity of the wave in the negative direction of the x^2 axis, on the contrary, varies from 0 to $+\infty$.

§6.7 *On wave front propagation in a diffusion-reaction system*

The behavior as $t \to \infty$ of solutions of some parabolic systems of differential equations similar to the KPP equation is treated in this section. Such systems describe the diffusion of several types of particles combined with the multiplication and mutual transmutation of these particles. They belong to the so-called class of diffusion-reaction equations (see, e.g. Fife [1,2]). Our results may also be formulated by introducing, in a proper way, a small parameter. But we are concerned only with the case homogeneous in the space variables, so there is no necessity in introducing a small parameter. The wave front propagation characterizes the behavior of solutions as $t \to \infty$.

So let us consider the Cauchy problem

$$\frac{\partial u_1(t,x)}{\partial t} = \frac{D_1}{2}\frac{\partial^2 u_1}{\partial x^2} + c_{11}(u_1,u_2)u_1 + c_{12}u_2 ,$$

$$\frac{\partial u_2(t,x)}{\partial t} = \frac{D_2}{2}\frac{\partial^2 u_2}{\partial x^2} + c_{21}u_1 + c_{22}(u_1,u_2)u_2 , \tag{7.1}$$

$$t > 0,\ x \in R^1,\ u_1(0,x) = g_1(x),\ u_2(0,x) = g_2(x) .$$

This system is a special case of system (5.4.8). Just as in the case of one equation, the behavior as $t \to \infty$ of the solutions of such a system depends essentially on the phase picture of the system of the ordinary differential equations (a point system) which is obtained when there is no diffusion, i.e. for $D_1 = D_2 = 0$:

$$\frac{d\bar{u}_1}{dt} = c_{11}(\bar{u}_1,\bar{u}_2)\bar{u}_1 + c_{12}\bar{u}_2 = f_1(\bar{u}_1,\bar{u}_2)$$

$$\frac{d\bar{u}_2}{dt} = c_{21}\bar{u}_1 + c_{22}(\bar{u}_1,\bar{u}_2)\bar{u}_2 = f_2(\bar{u}_1,\bar{u}_2) . \tag{7.2}$$

In the case of one equation, the corresponding point system reduced to the equation $\frac{d\bar{u}}{dt} = f(\bar{u})$. We made various assumptions about this equation considering various forms of the function $f(u)$ (see Fig. 1). In the case of system (7.2), similar assumptions are conveniently formulated in terms of the phase picture of system (7.2).

Let the following conditions be fulfilled.

1. The vector field $(f_1(u_1,u_2), f_2(u_1,u_2))$ has on the (u_1,u_2)-plane an unstable equilibrium position at the point $(0,0)$ and a stable equilibrium position at a point (a_1,a_2), $a_1 > 0$, $a_2 > 0$. All integral curves of system (7.2) starting at the points of the first quadrant $K_+ = \{(u_1,u_2) : u_1 \geq 0, u_2 \geq 0, u_1^2 + u_2^2 \neq 0\}$ are attracted to the point (a_1,a_2) as $t \to \infty$ (see Fig. 16).

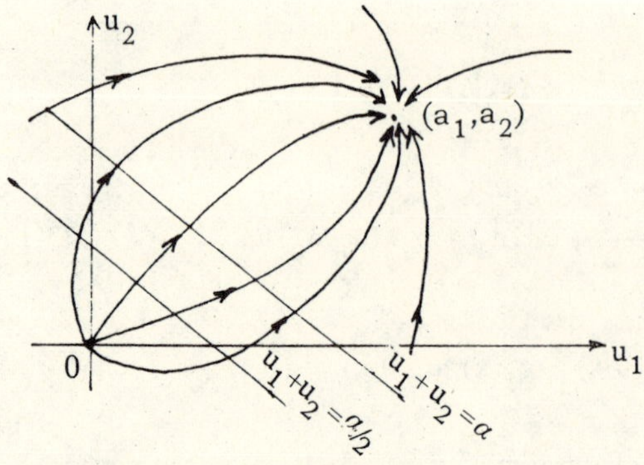

Fig. 16

2. $c_{11}(0,0) = \sup_{u_1,u_2\geq 0} c_{11}(u_1,u_2)$, $c_{22}(0,0) = \sup_{u_1u_2\geq 0} c_{22}(u_1,u_2)$.

3. One can find $a > 0$ and $N_1,N_2 > 0$ such that the following conditions hold:

(a) $c_{11}(u_1,u_2) + c_{12} > 0$, $c_{22}(u_1,u_2) + c_{21} > 0$ for $(u_1,u_2) \in B_a = \{(u_1,u_2) : u_1 \geq 0, u_2 \geq 0, u_1 + u_2 \leq a\}$;

(b) $c_{11}(N_1,u_2)N_1 + c_{12}u_2 \leq 0$ for $u_2 \in [0,N_2]$ and $c_{22}(u_1,N_2)N_2 + c_{21}u_1 \leq 0$ for $u_1 \in [0,N_1]$;

(c) the solution of the Cauchy problem for system (7.2) with the initial data $u_1(0,x) = g_1(x), u_2(0,x) = g_2(x)$ from $\mathfrak{A}_{a/2} = K_+ \cap \left\{(u_1,u_2) : u_1 + u_2 > \frac{a}{2}\right\}$ is attracted to the point (a_1,a_2). This means that for any $\delta > 0$ one can find $t_0 = t_0(\delta)$ such that $\sup_{x\in R^r} [|u_1(t,x) - a_1| + |u_2(t,x) - a_2|] < \delta$ for $t > t_0$, provided $\{(u_1,u_2) : u_1 = g_1(x), u_2 = g_2(x), x \in R^r\} \subset \mathfrak{A}_{a/2} \cap \{(u_1,u_2) : |u_1| + |u_2| < 1/\delta\}$;

4. D_1, D_2, c_{12}, and c_{21} are positive constants.

5. The functions $c_{11}(u_1,u_2)$ and $c_{22}(u_1,u_2)$ are bounded from above and Lipschitz continuous.

Of all possible systems of (7.1) type, conditions 1 and 2 single out systems similar to KPP equation.

For condition 3-(c) to be valid, one can give simple necessary conditions in terms of the coefficients, although there are a number of unstudied points here. If $D_1 = D_2$, then the conditions ensuring the validity of 3-(c) are given in Theorem 5.4.2. Here, Condition 3 enables us to use reasoning similar to the concluding part of the proof of Theorem 2.1.

The positiveness of c_{12} and c_{21} required in Condition 4 ensures, as will be seen in the sequel, the stabilization of the same wave front velocity for both components.

Let $\nu(t)$, $t \geq 0$, be a Markov process with two states 1 and 2 for which

$$P\{\nu(t+\Delta)=j \mid \nu(t)=i\} = c_{ij}\Delta + o(\Delta),\ \Delta \downarrow 0,\ i \neq j\ .$$

We will consider the random process X_t defined by the equality

$$dX_t = \sqrt{D_{\nu(t)}}\, dW_t,\ t > 0,\ X_0 = x \in R^1\ ,$$

where W_t is a Wiener process on the line, D_1 and D_2 being the coefficients of system (7.1). It is easily seen that the pair $(X_t, \nu(t))$ together with the family of probability measures $P_{x,i}$, $x \in R^1$, $i \in \{1,2\}$, forms a Markov process in the state space $R^1 \times \{1,2\}$. As was demonstrated in §5.4, the solution $(u_1(t,x), u_2(t,x))$ of system (7.1) satisfies the following relations

$$u_i(t,x) = E_{x,i}\, g_{\nu(t)}(X_t) \exp\left\{ \int_0^t c(\nu(s), u_1(t-s,X_s), u_2(t-s,X_s))\, ds \right\}, \tag{7.3}$$

where $c(i,u_1,u_2) = c_{ii}(u_1,u_2) + c_{ij}$, $i,j \in \{1,2\}$, $i \neq j$.

For the sake of simplicity, we assume that $g_i(x) = a_i \chi_{x<0}(x)$, $i = 1,2$. Just as in the case of the KPP equation, one can expect that for large t the solution $(u_1(t,x), u_2(t,x))$ of system (7.1) will have the form of a wave propagating from left to right with some velocity c^* :

$$(u_1(t,x), u_2(t,x)) \approx (U_1(x-c^*t), U_2(x-c^*t))\ .$$

This wave is defined by its two profiles $U_1(\xi), U_2(\xi)$ and by the velocity c^*. If c^* is known, the profiles may be found as the solutions of the system obtained from (7.1), by substituting $U_i(x-c^*t)$ for $u_i(t,x)$:

$$\frac{D_1}{2} U_1'' + c^*U_1' + f_1(U_1,U_2) = 0,\ \lim_{\xi\to-\infty} U_1(\xi) = a_1,\ \lim_{\xi\to-\infty} U_2(\xi) = a_2\ ,$$

$$\frac{D_2}{2} U_2'' + c^*U_2' + f_2(U_1,U_2) = 0,\ \lim_{\xi\to\infty} U_1(\xi) = \lim_{\xi\to\infty} U_2(\xi) = 0\ .$$

A weaker version of this assertion to be dealt with has the following form:*) there is a c^* such that for any $h_0 > 0$ uniformly in $h > h_0$

$$\lim_{t\to\infty} u_i(t,(c^*+h)t) = 0 \quad \text{for} \quad i = 1,2\ , \tag{7.4a}$$

$$\lim_{t\to\infty} u_i(t,(c^*-h)t) = a_i \quad \text{for} \quad i = 1,2\ . \tag{7.4b}$$

It is natural to call the constant c^* for which relations (7.4a)-(7.4b) hold, the velocity of the wave front. We shall compute c^* and establish relations (7.4).

Let $\lambda = \lambda(\alpha)$ be the largest eigenvalue of the following matrix whose elements depend on the parameter $\alpha \in R^1$:

$$\begin{pmatrix} -c_{12} + \alpha D_1 & c_{12} \\ c_{21} & -c_{21} + \alpha D_2 \end{pmatrix}$$

(the eigenvalues of this matrix are real). It is readily verified that $\lambda(\alpha)$ is a non-decreasing, convex function having an asymptote with slope $\max(D_1, D_2)$ as $\alpha \to \infty$, and an asymptote with the slope $\min(D_1, D_2)$ as $\alpha \to -\infty$.

Let us denote by $\ell(\beta)$, $\beta \in R^1$, the Legendre transform of the function $\lambda(\alpha) : \ell(\beta) = \sup_\alpha [\alpha\beta - \lambda(\alpha)]$.

For the sake of definiteness, let $D_1 < D_2$. From the above properties of the function $\lambda(\alpha)$ it follows that $\ell(\beta) = \infty$ outside $[D_1, D_2]$. On the interval $[D_1, D_2]$, the function $\ell(\beta)$ is finite, positive and convex. It is differentiable on (D_1, D_2), and its derivative tends to $+\infty$ as $\beta \to D_2$, and it tends to $-\infty$ as $\beta \to D_1$.

THEOREM 7.1 (Freidlin [18]). *Suppose that Conditions 1-5 are fulfilled and let* $D_1 < D_2$. *Then relations (7.4) hold for*

$$c^* = \frac{|\lambda(\alpha^*) + B|}{\sqrt{2(\alpha^* - A)}}\ ,$$

*)It is a version of the weak law of disturbance propagation (Aronson and Weinberger [2]).

where $A = \frac{c_{22}+c_{21}-c_{11}-c_{12}}{D_2-D_1}$, $B = \frac{D_2(c_{11}+c_{12})-D_1(c_{22}+c_{21})}{D_2-D_1}$,

$c_{11} = c_{11}(0,0)$, $c_{22} = c_{22}(0,0)$, a^* *being the root of the equation*

$$\lambda(a^*) + B = 2(a^*-A)\lambda'(a^*), \lambda'(a) = \frac{d\lambda(a)}{da} .$$

Proof. First we shall prove that, for any $h > 0$,

$$\lim_{t\to\infty} u_i(t,(c^*+h)t) = 0, \quad i = 1,2 .$$

We shall provide bounds only for the function $u_1(t,x)$; the proof for $u_2(t,x)$ is similar. Our reasoning is quite analogous to that carried out before (see, e.g. Theorem 2.1), but now, we have the small parameter t^{-1} rather than the small parameter ε.

Condition 2 and (7.3) imply that

$$u_1(t,x) \le E_{x,1}\, g_{\nu(t)}(X_t) \exp\left\{\int_0^t c(\nu_s)ds\right\}, \tag{7.5}$$

where $c(i) = c(i,0,0) = c_{ii}(0,0) + c_{ij}, (i \neq j)$. Next, we observe that $X_t = X_0 + \widetilde{W}(\int_0^t D_{\nu(s)}ds)$, where $\widetilde{W}_t$ is a proper Wiener process independent of $\nu(t), \widetilde{W}_0 = 0$. Relying on this, we get from (7.5) for $x = ct, c > 0$,

$$u_1(t,x) \le (a_1 \vee a_2) E\, P\{\widetilde{W}(\xi_t) > ct | \xi_t\} \exp\{A\xi_t + B\} . \tag{7.6}$$

Here we designate $\xi_t = \int_0^t D_{\nu(s)}ds$, the constants A and B being defined by the conditions $c(i) = AD_i + B$ for $i = 1,2$. Since

$$P\{\widetilde{W}(\xi_t) > ct|\xi_t\} \le \text{const.} \times \exp\left\{-\frac{c^2t^2}{2\xi_t}\right\},$$

from (7.6) one can infer that

$$u(t,ct) \le c_1 E \exp\left\{-\frac{c^2t^2}{2\xi_t} + A\xi_t + Bt\right\}, \tag{7.7}$$

where c_1 is some positive constant. Let us show that, for $c > c^*$, the right-hand side of (7.7) tends to zero as t grows. To this end, we shall compute the logarithmic asymptotics of the expectation as $t \to \infty$. This asymptotic behavior is determined by the large deviations of the random process $t^{-1}\xi_t$ from its limit as $t \to \infty$.

As it follows from Theorem 1.7.6 (see also Wentzell and Freidlin [2]), the deviations of order 1 for $t^{-1}\xi_t$ as $t \to \infty$ are described by the action function $\ell(\beta)$ defined above. Whence, relying on Property 3 of action functionals from §1.7, we conclude that

$$\lim_{t\to\infty} t^{-1} \ln E \exp\left\{-t\left(\frac{c^2}{2\frac{\xi_t}{t}} - A\frac{1}{t}\xi_t - B\right)\right\} = \tag{7.8}$$

$$= -\min_x \left(\frac{c^2}{2x} - Ax - B + \ell(x)\right).$$

Since the function $\ell(x)$ is $+\infty$ outside $[D_1, D_2]$, this minimum may be taken only over $[D_1, D_2]$. The foregoing properties of $\ell'(x)$ imply that the minimum is attained inside the interval (D_1, D_2).

Let us verify that the right-hand side of equality (7.8) vanishes for $c = c^*$. We will denote by $\tilde{a} = \tilde{a}(x)$ the solution of the equation $x = \lambda'(\tilde{a})$. Such an $\tilde{a}(x)$ is defined in a unique way for $x \in (D_1, D_2)$. Then $\ell(x) = x\tilde{a}(x) - \lambda(\tilde{a}(x))$, $x \in (D_1, D_2)$. It is not hard to check that $\ell'(x) = \tilde{a}(x)$. The minimum of the function

$$f(x) = \frac{(c^*)^2}{2x} - Ax - B + \ell(x)$$

is attained at some point $\bar{x} \in (D_1, D_2)$. We will denote $\bar{a} = \tilde{a}(\bar{x})$. Then remembering the definition of $\tilde{a}$, from the condition $f(\bar{x}) = 0$, we derive the equations

$$\frac{(\lambda(a^*) + B)^2}{4(a^* - A)\bar{x}^2} + A - \bar{a} = 0, \quad \lambda'(\bar{a}) = \bar{x}. \tag{7.9}$$

The equation for α^*

$$\frac{\lambda(\alpha^*)+B}{2(\alpha^*+A)} = \lambda'(\alpha^*) \tag{7.10}$$

should be joined to (7.9). From (7.9) and (7.10) it results that

$$[\lambda'(\bar{\alpha})]^2(\bar{\alpha}-A) = [\lambda'(\alpha^*)]^2(\alpha^*-A) .$$

Since $[\lambda'(\alpha)]^2(\alpha-A)$ is an increasing function of α, the last relation yields

$$\bar{\alpha} = \alpha^*, \ \bar{x} = \lambda'(\alpha^*) = \frac{\lambda(\alpha^*)+B}{2(\alpha^*-A)} .$$

Noting these equalities we arrive at

$$\min f(x) = f(\bar{x}) = \frac{(\lambda(\alpha^*)+B)^2}{4\bar{x}(\alpha^*-A)} - A\bar{x} - B + \bar{x}\alpha^* - \lambda(\alpha^*) = 0 .$$

So, the right-hand side of (7.8) vanishes for $c = c^*$. It is easily seen that the function

$$\gamma(c) = \min_x \left(\frac{c^2}{2x} - Ax - B + \ell(x)\right) \tag{7.11}$$

increases with c. Therefore, $\gamma(c) > 0$ for $c > c^*$, and according to (7.7) and (7.8)

$$u(t, ct) < c_1 \exp\{-t\gamma(c)\}, \ c > c^* ,$$

for sufficiently large t, which establishes relation (7.4a).

We proceed to prove (7.4b). Our reasoning, generally, follows the same scheme as the concluding part of the proof of Theorem 2.1. Condition 3 will help us to obtain the needed bounds.

The following statement results from (7.4a): for any $\delta > 0$, if $h > 0$ is small enough and t is sufficiently large, then the inequality

$$c(\nu_s, u_1(t-s, X_s), u_2(t-s, X_s)) > c(\nu_s, 0, 0) - \delta/2, \ 0 \le s \le t ,$$

is valid, with probability close to 1, for the trajectories X starting from the point $X_0 = (c^*+h)t$. On account of (7.8) this implies that

$$u_i(t,(c^*+h)t) > E_{(c^*+h)t,i}\, g_{\nu(t)}(X_t) \exp\left\{\int_0^t \left[c(\nu(s)) - \frac{\delta}{2}\right] ds\right\} > \tag{7.12}$$

$$> \exp\{-\delta t\}, \quad i = 1,2;$$

for sufficiently small $h > 0$ and sufficiently large t. This bound is a counterpart of inequality (2.17).

First of all, we will show that for every $h > 0$ and T sufficiently large

$$(u_1(t,x), u_2(t,x)) \in A_{a/2} \tag{7.13}$$

for $t > T$, $x < (c^*-h)t$. Let us put

$$D_\beta = \left\{(t,x) : -\infty < t < \infty,\ 0 < x < (c^*+\beta)t,\ u_1(t,x) + \right.$$

$$\left. + u_2(t;x) < \frac{3a}{4}\right\}, \quad \tau = \inf\{s : (t-s, X_s) \notin D_\beta\}.$$

Since $(X_s, \nu(s); P_{x,i})$ is a strong Markov process and τ is a Markov time, relations (7.3) imply the equality

$$u_i(t,x) =$$

$$= E_{x,i} u_{\nu(\tau)}(t-\tau, X_\tau) \exp\left\{\int_0^\tau c(\nu(s), u_1(t-s,X_s), u_2(t-s,X_s))\, ds\right\}, \tag{7.14}$$

$$i \in \{1;2\},\ (t,x) \in D_\beta .$$

From condition 3-(a) it follows that there is a $\kappa > 0$ for which $c(i,u_1(t,x),u_2(t,x)) > \kappa$ everywhere in the domain D_β. If $x < (c^*-h)t$, $h > 0$, then the time of reaching the set $\Gamma = \{(x,s) : s = \left(0 \vee \frac{x}{c^*+\beta}\right)$ by the "heat" process $(t-s, X_s)$ starting from the point (t,x), goes to infinity together with t. By (7.12), everywhere on the set Γ,

$$u_i(t,x) > K \exp\{-\delta t\}, \quad i \in \{1;2\},$$

for some $K > 0$. Moreover, by choosing β sufficiently small, $\delta/_K$ can be made arbitrarily small. From these bounds, with the help of (7.14) we conclude that for certain $T > 0$, (7.13) holds.

Now we shall show that for any $h_1 > 0$

$$(u_1(t,(c^*-h_1)t), u_2(t,(c^*-h_1)t)) \to (a_1,a_2), \; t \to \infty . \tag{7.15}$$

Let us pick $h \in (0,h_1)$ and $T_1 = T_1(h)$ so that (7.13) will hold for $t > T_1(h)$, $x \leq (c^*-h)t$. By condition 3-(b), one can assume that

$$u_1(t,x) < N_1, \quad u_2(t,x) < N_2 \tag{7.16}$$

for $t > 0$, $x \in R^1$. To finish the proof of the theorem, we shall need the following lemma.

LEMMA 7.1. *Suppose that the coefficients* $c_{ii}(u_1,u_2)$ *are Lipschitz continuous with a constant* K *and let* $c_{ii}(u_1,u_2) \leq M$, $i \in \{1,2\}$. *Denote by* $(u_1^g(t,x), u_2^g(t,x))$, $g = (g_1(x), g_2(x))$, $x \in R^1$, *the solution of the Cauchy problem for system (7.2) with the initial conditions* $u_1(0,x) = g_1(x)$, $u_2(0,x) = g_2(x)$. *Suppose that* $g_1(x), g_2(x), \tilde{g}_1(x)$, *and* $\tilde{g}_2(x)$ *are continuous functions in* R^1 *bounded in modulus by a constant* A *and let* $g = (g_1(x), g_2(x))$, $\tilde{g} = (\tilde{g}_1(x), \tilde{g}_2(x))$.

Then, for any $\delta > 0$, $x_0 \in R^1$, *and* $T > 0$, *one can find* $B = B(\delta,M,K,A\,T,x_0)$ *such that, for* $x \leq x_0$, $t \leq T$,

$$|u_1^g(t,x) - u_1^{\tilde{g}}(t,x)| + |u_2^g(t,x) - u_2^{\tilde{g}}(t,x)| < \delta$$

provided $g(x) = \tilde{g}(x)$ *for* $x < B$.

The *proof* of this lemma is easy to develop, for example, using the equality

$$u_i^g(t,x) = E_{x,i}\, g_{\nu(t)}(X_t) \exp\left\{ \int_0^t c(\nu_s, u_1^g(t-s,X_s), u_2^g(t-s,X_s))\, ds \right\}$$

and the similar representation for $u_i^{\tilde{g}}(t,x)$. So we drop the proof.

Consider the Cauchy problem

$$\frac{\partial v_1(t,x)}{\partial t} = \frac{D_1}{2}\frac{\partial^2 v_1}{\partial x^2} + c_{11}(v_1,v_2)\,v_1 + c_{12}v_2$$

$$\frac{\partial v_2(t,x)}{\partial t} = \frac{D_2}{2}\frac{\partial^2 v_2}{\partial x^2} + c_{21}v_1 + c_{22}(v_1,v_2)\,v_2 \tag{7.17}$$

$$t > T_1,\ x \in R^1,\ v_i(T_1,x) = \begin{cases} u_i(T_1,x) & \text{for } x < (c^*-h)\,T_1 \\ u_i(T_1,(c^*-h)\,T_1) & \text{for } x \geq (c^*-h)\,T_1\,. \end{cases}$$

The positive number T_1 in problem (7.17) will be chosen later. Now we will only assume that T_1 is large enough so that inclusion (7.13) holds for $t > T_1$, $x \leq (c^*-h)\,t$. Then by condition 3-(c), taking into account (7.16) we conclude that, for any $\delta > 0$, one can find T_2 such that

$$|v_1(t,x)-a_1| + |v_2(t,x)-a_2| < \delta \quad \text{for} \quad t > T_1 + \frac{T_2}{2}\ . \tag{7.18}$$

Next, by Lemma 7.1, it is possible to choose T_1 so that

$$|v_1(t,x)-u_1(t,x)| + |v_2(t,x)-u_2(t,x)| < \delta \tag{7.19}$$

for $t \in [T_1, T_1+T_2]$, $x < (c^*-h_1)(T_1+T_2)$.

From (7.18) and (7.19) we obtain (7.4b), which completes the proof of Theorem 7.1. □

REMARK. When proving (7.4a), we used that $c(i) = AD_i + B$, $i \in \{1;2\}$ for some A and B. If $D_1 = D_2$, then it may be impossible to find such

A and B. The above reasonings also do not work in the case of a system with more than two unknown functions. In these cases one can act as follows (for brevity we will clarify everything for a system of two equations). Denote by $\lambda(\alpha^1, \alpha^2), (\alpha^1, \alpha^2) \in R^2$ the maximal eigenvalue of the matrix

$$\begin{pmatrix} -c_{12} + \alpha^1 D_1 + \alpha^2 c(1) & c_{12} \\ c_{21} & -c_{21} + \alpha^1 D_2 + \alpha^2 c(2) \end{pmatrix}.$$

It is not difficult to verify that the function $\lambda(\alpha^1, \alpha^2)$ is convex. Let $\ell(\beta_1, \beta_2)$ be its Legendre transform. One can prove that $\ell(\beta_1, \beta_2)$ is the action function describing large deviations for the two-dimensional process $(\int_0^t D_{\nu(s)} ds, \int_0^t c(\nu_s) ds)$ as $t \to \infty$. Noting this fact, with the help of reasonings similar to those used above, one can prove that the wave propagation velocity c^* is the root of the equation

$$\min_{x,y} \left(\frac{c^{*2}}{2x} - y + \ell(x,y) \right) = 0 . \tag{7.20}$$

One should bear in mind that, in the case of the general situation, the function $\ell(\beta_1, \beta_2)$ differs from $+\infty$ only in some interval in the plane (β_1, β_2). This interval may degenerate into a point. For example, let $D_1 = D_2 = D$, $c(1,0,0) = c(2,0,0) = c$. Then simple calculation shows that $\lambda(\alpha^1, \alpha^2) = \alpha^1 D + \alpha^2 c$, $\ell(D,c) = 0$, $\ell(\beta_1, \beta_2) = +\infty$ for $(\beta_1, \beta_2) \neq (D,c)$. Equation (7.20) yields: $c^* = \sqrt{2Dc}$.

Chapter VII
WAVE FRONT PROPAGATION IN PERIODIC AND RANDOM MEDIA

§7.1 *Introduction*

Consider the KPP equation:

$$\frac{\partial u(t,x)}{\partial t} = \frac{D}{2}\frac{\partial^2 u(t,x)}{\partial x^2} + f(u(t,x)),\ t > 0,\ x \in R^1,$$
$$u(0,x) = \begin{cases} 1, & x \le 0 \\ 0, & x > 0 . \end{cases} \tag{1.1}$$

We assume that $f(0) = f(1) = 0$, $f(u) > 0$ for $u \in (0,1)$, $f(u) < 0$ for $u \notin [0,1]$, and $f'(0) = \sup_{0<u<1} u^{-1}f(u)$. As was said above, for large t, the function $u(t,x)$ is close to the solution $V(x-\alpha t)$ of equation (1.1) which is a wave propagating from left to right with the velocity $\alpha = \sqrt{2Df'(0)}$. The shape $V(\xi)$ of the wave is defined by the equation

$$\frac{D}{2}V''(\xi) + \alpha V'(\xi) + f(V(\xi)) = 0,\ -\infty < \xi < \infty,$$
$$V(-\infty) = 1,\ V(+\infty) = 0 .$$

In particular, this result implies that for any $h > 0$

$$\lim_{t\to\infty} u(t,(\alpha+h)t) = 0,\quad \lim_{t\to\infty} u(t,(\alpha-h)t) = 1 .$$

Therefore, the domain of large (close to 1) values of the function $u(t,x)$ propagates with velocity α.

This remark enables us to extract the simpler problem on the propagation velocity from the more delicate one concerned with the wave shape.

Furthermore, our investigation of the propagation velocity of the domain of large values of the function $u(t,x)$ may be considered in a more general situation, for example, in the case when both the diffusion coefficients and the multiplication rate depend on the space variables. Certainly, in the case of arbitrary diffusion coefficients and a non-linear term $f(x,u)$, there is no hope that, for large t, a constant propagation velocity of the domain of large values of the function $u(t,x)$ will be reached. For this constant velocity to be reached, it is necessary to make hypotheses on the homogeneity (in a sense) of the diffusion coefficients and multiplication rate. The following two types of such hypotheses are conceived as being natural. First, one can consider the periodic media case, i.e. look upon the diffusion coefficients and multiplication rate as functions periodic in the space variables. Secondly, one can consider random space-homogeneous media, i.e. think of these functions as random homogeneous fields. Some results concerning both cases without detailed proofs are in Gärtner and Freidlin [2] (see also Gartner [3] and Freidlin [17]). In this chapter, we consider both cases in detail. We do not touch on a number of interesting questions, such as the nature of the transition area between the domain of large values and the domain where $u(t,x)$ is close to zero (in the case of problem (1.1) this question reduces to that of the wave shape). We also omit the case of the function $f(u)$ having zeros inside the interval $(0,1)$.

Just as in the preceding chapter, our research is based on applying the Feynman-Kac formula and limit theorems for large deviations probabilities.

In Chapter VI, large deviations theorems in the space of continuous functions were used substantially when studying wave propagation for quasi-linear equations with small diffusion. To calculate the asymptotic wave propagation velocity for the equations to be considered in the present chapter, we apply large deviations theorems for some families of random vectors (finite-dimensional), rather than for diffusion processes with small parameter. Let us formulate the corresponding results to be used in the sequel.

Let $(\Omega_\theta^t, \mathcal{F}_\theta^t, P_\theta^t)$ be a family of probability spaces, where t runs over the positive half-line $(0, \infty)$ and the parameter θ varies over an arbitrary non-empty set Θ. Consider a family of r-dimensional random vectors η_θ^t defined on the corresponding measurable spaces $(\Omega_\theta^t, \mathcal{F}_\theta^t)$, $t \in (0, \infty)$, $\theta \in \Theta$. Suppose that for some positive function $\varepsilon(t)$, such that $\varepsilon(t) \to 0$ as $t \to \infty$, and for all $z \in R^r$, the limit

$$G(z) = \lim_{t \to \infty} \varepsilon(t) \ln E_\theta^t \exp\left\{\frac{1}{\varepsilon(t)} (z, \eta_\theta^t)\right\} \tag{1.2}$$

independent of $\theta \in \Theta$, exists uniformly in the parameter θ. Here E_θ^t denotes the expectation with respect to the probability measure P_θ^t, and $(\cdot, \cdot)$ designates scalar product in the space R^r. We also admit that the expectation in (1.2) as well as the limit $G(z)$ can take the value $+\infty$.

Let us introduce the action function $S: R^r \to [0, \infty]$ by the equality

$$S(y) = \sup_{z \in R^r} [(y,z) - G(z)], \quad y \in R^r .$$

The function S is convex and lower continuous. Let D(G) be the set of all points of the form $\nabla G(x)$. Here z runs over all points of the space R^r, where the function G is finite and differentiable. Denote by $\overline{D}(G)$ the set of all points $y \in R^r$, for which a sequence $(y_n) \subset D(G)$ exists such that simultaneously

$$y_n \to y \quad \text{and} \quad S(y_n) \to S(y)$$

as $n \to \infty$. If the function G is differentiable at all points of the space R^r, then $\overline{D}(G)$ coincides with the domain of finiteness of the action function S (Rockafellar [1]). For any $s \geq 0$, we set

$$\Phi(s) = \{y \in R^r : S(y) \leq s\} .$$

The sets $\Phi(s)$ are closed and convex. If 0 is an interior point of the domain of finiteness of the function G, then the sets $\Phi(s)$, $s \geq 0$, are bounded.

Let $\rho(\cdot,\cdot)$ be the Euclidean metric in the space R^r.

THEOREM 1.1. *Suppose that, for some* $s \geq 0$, *the set* $\Phi(s)$ *is non-empty and bounded. Then for any* $\delta > 0$, $h > 0$, *we can choose* $t_0 > 0$ *such that the bound*

$$P_\theta^t\{\rho(\eta_\theta^t, \Phi(s)) > \delta\} \leq \exp\left\{-\frac{1}{\varepsilon(t)}(s-h)\right\}$$

holds for $t > t_0$ *and all* $\theta \in \Theta$.

THEOREM 1.2. *For any* $\delta > 0$, $h > 0$, *and for all* $y \in \overline{D}(G)$, *a* $t_0 > 0$ *exists such that*

$$P_\theta^t\{\rho(\eta_\theta^t, y) < \delta\} \geq \exp\left\{-\frac{1}{\varepsilon(t)}(S(y)+h)\right\}$$

for $t > t_0$ *and all* $\theta \in \Theta$.

The proofs of Theorems 1.1 and 1.2 in essence follow from Theorem 1.7.2 (see also Wentzell and Freidlin [2], Chapter V).

In the next section, we shall make more precise the statement of the problem and obtain a number of auxiliary results for the case of periodic coefficients. In Section 7.3, we calculate the asymptotic wave front propagation velocity in the multi-dimensional problem with periodic coefficients. Next, in §7.4 and §7.5, we formulate the problem on wave front propagation in a one-dimensional random medium and give auxiliary results. The calculation of the wave front velocity for this case is carried out in §7.6. The last section, §7.7, is devoted to deriving more explicit expressions for the velocity and to considering examples.

§7.2 *Calculation of the action functional*

Consider the Cauchy problem for the quasi-linear equation

$$\frac{\partial u(t,x)}{\partial t} = Lu(t,x) + f(x,u(t,x)), (t,x) \in (0,\infty) \times R^r , \tag{2.1}$$

with the initial condition

$$u(0,x) = g(x) . \tag{2.2}$$

Here L denotes a linear elliptic second-order differential operator

$$L = \frac{1}{2} \sum_{i,j=1}^{r} a^{ij}(x) \frac{\partial^2}{\partial x^i \partial x^j} + \sum_{i=1}^{r} b^i(x) \frac{\partial}{\partial x^i} .$$

We shall assume the matrix $a(x) = (a^{ij}(x))$ to be non-singular, the coefficients $a^{ij}(x)$, $b^i(x)$, and the function $f(x)$, $x = (x^1,\cdots,x^r) \in R^r$, $u \in R^1$, satisfy the Lipschitz condition with respect to x and u and depend on the variables $x^1, \cdots, x^r$ periodically with period 1. From now on, the function $g(x)$ is supposed to be non-negative, continuous, not identically equal to zero and to have a compact support. Under these assumptions, problem (2.1)-(2.2) is known to have a unique classical solution $u(t,x)$.

We are going to examine the asymptotic behavior of the function $u(t,x)$ as $t \to \infty$. As it was seen in Chapter VI, if L is the Laplace operator and the function $f(x,u)$ does not depend on x, vanishes for $u = 0$ and $u = 1$, and has some extra properties, then there is a positive constant c^* such that the following assertion holds:

For any $c > c^*$, the function $u(t,x)$ tends to zero as $t \to \infty$ in the region $|x| > ct$, and for any $c \in (0,c^*)$, it tends to 1 in the region $|x| < ct$. Thus, c^* can be interpreted as an asymptotic wave front propagation velocity.

If one wants the function u to behave similarly in the case when the coefficients of equation (2.1) depend periodically on the space variable x, it is necessary to impose some additional assumptions on the function f. Namely, we assume that for any $x \in R^r$

$$\begin{gathered} f(x,0) = f(x,1) = 0 , \\ f(x,u) \geq 0 \text{ for } 0 < u < 1 , \; f(x,u) < 0 \text{ for } u > 1 , \end{gathered} \tag{2.3}$$

What is more, we suppose that for every $h \in (0,1)$, there is a point $x_h \in R^r$ such that

$$f(x_h, u) > 0 \quad \text{for} \quad u \in (0,h) \tag{2.4}$$

Furthermore, let the function

$$c(x,u) = u^{-1} f(x,u), \quad (x,u) \in R^r \times (0,\infty)$$

be extendible to the hyperplane $u = 0$, and let in this hyperplane, $c(x,u)$ as function of u attain its absolute maximum $c(x)$:

$$c(x) = c(x,0) = \sup_{u>0} c(x,u) . \tag{2.5}$$

From (2.3) and the properties of the initial function it follows in particular that the solution of Cauchy's problem (2.1)-(2.2) is non-negative, bounded from above by the quantity $1 \vee \max_{x \in R^r} g(x)$ and satisfies the relation:

$$\overline{\lim_{t \to \infty}} \sup_{x \in R^r} u(t,x) \leq 1 . \tag{2.6}$$

Henceforth we shall utilize the following integral equation for the function $u(t,x)$. Consider the Markov family (X_t^x, P) defined by the Ito stochastic differential equation

$$dX_t^x = \sigma(X_t^x)\, dW_t + b(X_t^x)\, dt, \; X_0^x = x \in R^r ,$$

where $b(x) = \{b^i(x)\}$, $\sigma(x) = (a(x))^{1/2}$, W_t is the r-dimensional Wiener process defined on a probability space (Ω, N, P). Let (X_t, P_x) be the corresponding Markov process. Then, as it has already been seen previously, the function $u(t,x)$ is the unique bounded solution of the equation:

$$u(t,x) = E_x\, g(X_t) \exp\left\{ \int_0^t c(X_s, u(t-s, X_s))\, ds \right\} =$$

$$\left(E g(X_t^x) \exp\left\{ \int_0^t c(X_s^x, u(t-s, X_s^x))\, ds \right\} \right).$$

For further considerations it will be helpful to note that due to the periodicity of the drift $b(x)$ and the diffusion matrix $a(x)$, for any $x \in R^r$ and any r-dimensional vector z with integer components, the distribution in the path space of the process (X_t, P_x) with respect to the probability measure P_{x+z} coincides with the distribution of the process $X_t + z$ with respect to P_x.

The equality

$$P_x^t(A) = \frac{E \exp\{\int_0^t c(X_s^x)\, ds\}\, \chi_A}{E \exp\{\int_0^t c(X_s^x)\, ds\}} \tag{2.8}$$

defines a family of probability measures P_x^t, $t > 0$, $x \in R^r$, in the measurable space $(\Omega, \mathcal{N})$. We will make use of large deviations theorems for the family of random vectors

$$\eta_x^t = \frac{1}{t}(x - X_t^x) \tag{2.9}$$

defined on the probability spaces $(\Omega, \mathcal{N}, P_x^t)$, $t > 0$, $x \in R^r$, respectively. Let C_π be the Banach space of all periodic continuous functions of period 1 in R^r with the uniform norm. To apply Theorems 1.1 and 1.2 to our case, we shall need the following

LEMMA 2.1. *For any* $z = (z_1, \cdots, z_r)$,

$$\lim_{t \to \infty} \frac{1}{t} \ln E_x^t \exp\{t(z, \eta_x^t)\} = \lambda(z) - \lambda(0) \tag{2.10}$$

exists uniformly in $x \in R^r$, *with* $\lambda(z)$ *being the eigenvalue of the differential operator*

$$L^z = L - \sum_{i,j=1}^{r} a^{ij}(x) z_i \frac{\partial}{\partial x^i} + c(x) - \sum_{i=1}^{r} b^i(x) z_i + \frac{1}{2} \sum_{i,j=1}^{r} a^{ij}(x) z_i z_j \quad (2.11)$$

in the space C_π *corresponding to a positive eigenfunction. The function* $\lambda(z)$, $z \in R^r$, *is differentiable.*

Proof. For any $z \in R^r$, we define a Markov family $(X_t^x(z), P)$ in R^r by the following stochastic differential equation:

$$dX_t^x(z) = \sigma(X_t^x(z))\, dW_t + [b(X_t^x(z)) - a(X_t^x(z))\, z]\, dt\,.$$

Let $(X_t(z), P_x^z)$ be the corresponding Markov process. The equality

$$(Q_t^z \psi)(x) =$$

$$= E \exp\left\{ \int_0^t \left[c(X_s^x(z)) - (b(X_s^x(z)), z) + \frac{1}{2} (a(X_s^x(z))\, z, z) \right] ds \right\} \psi(X_t^x(z)) \quad (2.12)$$

defines a continuous semi-group Q_t^z of linear bounded operators depending on the parameter z in the space C_π. Note that the operator L^z is the restriction of the infinitesimal operator of the semi-group Q_t^z to the set of all twice continuously differentiable periodic functions of period 1. The space C_π can be identified in a natural way with the Banach space C_{T^r} of all continuous functions on the r-dimensional unit torus T^r considered as a factor-group of R^r by the integer lattice. Denote by $\bar{a}^{ij}(x)$, $\bar{b}^i(x)$, and $\bar{c}(x)$ the restriction of the periodic functions $a^{ij}(x)$, $b^i(x)$, and $c(x)$ respectively to the unit torus T^r. Let $(\bar{X}_t^x(z), P)$ and $(\bar{X}_t(z), \bar{P}_x^z)$ be a Markov family and Markov process on the torus T^r governed by the differential operator:

$$\overline{L^z} = \frac{1}{2} \sum_{i,j=1}^{r} \bar{a}^{ij}(x) \frac{\partial^2}{\partial x^i \partial x^j} + \sum_{i,j=1}^{r} [\bar{b}^j(x) - \bar{a}^{ij}(x) z_i] \frac{\partial}{\partial x^j} .$$

We denote by $\bar{Q}_t^z$ the semi-group of linear bounded operators acting by the formula

$$\bar{Q}_t^z \bar{\psi}(x) =$$

$$= E \exp \left\{ \int_0^t \left[\bar{c}(\bar{X}_s^x(z)) - (b(\bar{X}_s^x(z)), z) + \frac{1}{2} (\bar{a}(\bar{X}_s^x(z)) z, z) \right] ds \right\} \bar{\psi}(\bar{X}_t^x(z)) ,$$

$$x \in T^r, \quad \bar{\psi} \in C_{T^r} .$$

It is straightforward to check that the "quasi-transition probabilities" $\bar{Q}^z(t,x,\Gamma) = (\bar{Q}_t^z \chi_\Gamma)(x)$ with $t > 0$, $x \in T^r$ and Γ being a Borel subset of T^r, are strictly positive and satisfy the Doeblin condition. So, there is a number $\lambda(z)$ such that, for all $t > 0$, $e^{t\lambda(z)}$ is a simple eigenvalue of the operator $\bar{Q}_t^z$ with the strictly positive eigenfunction $\bar{u}^z$ independent of t (Partzsch [1]). It is easily seen that $e^{t\lambda(z)}$ is an eigenvalue of the operator Q_t^z and

$$Q_t^z u^z = e^{t\lambda(z)} u^z , \tag{2.13}$$

where $u^z \in C_\pi$ corresponds to $\bar{u}^z \in C_{T^r}$. From this, by limit passage to the infinitesimal operator of the semi-group Q_t^z, we conclude that $\lambda(z)$ is the eigenvalue of the operator L^z in C_π corresponding to a positive eigenfunction.

The process $(X_t(z), P_x^z)$ differs from the process (X_t, P_x) only by a drift. Thus, the measures P_x^z and P_x are absolutely continuous, and the density of one measure with respect to another has the form:

$$\frac{dP_x^z}{dP_x}(X.) = \exp \left\{ - \int_0^t (z, \sigma(X_s) dW_s) - \frac{1}{2} \int_0^t (a(X_s) z, z) ds \right\} .$$

Using this and the definition of the random vectors η_x^t, (2.12) can be rewritten in the form:

$$(Q_t^z \psi)(x) = E \exp\left\{ \int_0^t c(X_s^x) ds + t(z, \eta_x^t) \right\} \psi(X_t^x) . \qquad (2.14)$$

Since the function u^z is strictly positive and periodic, one can deduce from (2.14) via equation (2.13) that:

$$\lim_{t \to \infty} \frac{1}{t} \ln E \exp\left\{ \int_0^t (c(X_s^x) + t(z, \eta_x^t) \right\} = \qquad (2.15)$$

$$= \lim_{t \to \infty} \frac{1}{t} \ln (Q_t^z 1)(x) = \lambda(z) ;$$

the convergence being uniform in $x \in R^r$. From this, recalling the definition of the probability measures P_x^t, we deduce the relation (2.10). It remains to prove that the function $\lambda(z)$ is differentiable. As it follows from the linear operators disturbancy theory (Kato [1]), for this purpose it is sufficient to make sure that the operator function $z \to Q_1^z$ is differentiable in the uniform operator topology. This can be done with representation (2.14). Thus, Lemma 2.1 is proved. □

Lemma 2.1 implies that, for probability measures (2.8) and random vectors (2.9), the exponential bounds of Theorems 1.1 and 1.2 are valid for all $s \geq 0$ and all $y \in R^r$ with $\varepsilon(t) = t^{-1}$. The sets $\Phi(s)$ are compact and the action function has the form:

$$S(y) = H(y) + \lambda(0) ,$$

where

$$H(y) = \sup_{z \in R^r} [(y,z) - \lambda(z)], \ y \in R^r .$$

From (2.12) and (2.15), we derive the estimate:

$$\lambda(z) \geq \min_{x \in R^r} \left[c(x) - (b(x),z) + \frac{1}{2}(a(x)z,z) \right].$$

This yields that $\frac{\lambda(z)}{|z|} \to \infty$ as $|z| \to \infty$. Hence (Rockafellar [1]), the function $H(y)$ is finite for all $y \in R^r$ and moreover, being convex, it is continuous. Furthermore, note that the set $\{y \in R^r : H(y) < 0\}$ is non-empty, since $\lambda(0) > 0$, and that the action functional $S(y)$ vanishes for $y = \nabla\lambda(0)$. The inequality $\lambda(0) > 0$ follows from relation (2.15) with $z = 0$ if one takes into account that by virtue of conditions (2.4) and (2.5), $c(x) \geq 0$ and $c(x) \not\equiv 0$.

§7.3 *Asymptotic velocity of wave front propagation in periodic medium*

We now proceed to study the asymptotic behavior of the solution $u(t,x)$ of problem (2.1)-(2.2).

LEMMA 3.1. *For any closed set* $F \subset \{y \in R^r : H(y) > 0\}$, $\overline{\lim_{t \to \infty}} \frac{1}{t} \ln \sup_{y \in F} u(t,ty) \leq -\min_{y \in F} H(y)$.

Proof. Let a number s be chosen in such a way that $0 < s < \min_{y \in F} H(y)$. Since the closed set F does not intersect the compact set $\Psi(s) = \{y \in R^r : H(y) \leq s\}$, the distance 2δ between them is positive. For t sufficiently large, the support of the initial function g is contained in the δt-neighborhood $U_{\delta t}(0)$ of the point 0. Therefore, on account of condition (2.5), for such t equation (2.7) implies the bound

$$u(t,ty) \leq \|g\| E_{ty} \exp\left\{ \int_0^t c(X_s)\, ds \right\} \chi_{X_t \in U_{\delta t}(0)} \tag{3.1}$$

with $\|g\| = \sup_{x \in R^r} |g(x)|$. From this, using definitions (2.8) and (2.9) of the probability measures P_x^t and random vectors η_x^t, we obtain

$$\sup_{y\epsilon F} u(t,ty) \leq \|g\| \sup_{y\epsilon F} P_{ty}^{t}\{\rho(\eta_{ty}^{t}, y) < \delta\} E_{ty} \exp\left\{\int_0^t c(X_s)\,ds\right\} \leq$$

$$\leq \|g\| \sup_{x\epsilon R^r} P_x^t\{\rho(\eta_x^t, \Psi(s)) > \delta\} \sup_{x\epsilon R^r} E_x \exp\left\{\int_0^t c(X_s)ds\right\} .$$

The expression on the right-hand side of this inequality can be estimated from below via relation (2.15) with $z = 0$ and the exponential bound contained in Theorem 1.1. As a result we arrive at the following inequality

$$\overline{\lim_{t\to\infty}} \frac{1}{t} \ln \sup_{y\epsilon F} u(t,ty) \leq -s .$$

Since s can be chosen arbitrarily close to $\min_{y\epsilon F} H(y)$, this fact leads to the assertion of Lemma 3.1. □

Let $Q = [0,1]^r$ be a unit cube and $U_\delta(y)$ the δ-neighborhood of a point y in the space R^r. Now we are going to cite an asymptotic bound for the function u which is, in a sense, opposite to (3.1).

LEMMA 3.2. *For all* $y \epsilon R^r$ *for which* $H(y) > 0$,

$$\underline{\lim_{t\to\infty}} \frac{1}{t} \ln \inf_{\tilde{y}\epsilon U_\delta(y)} u(t,t\tilde{y}) \geq$$

$$\geq \underline{\lim_{t\to\infty}} \frac{1}{t} \ln \inf_{x\epsilon Q,\tilde{y}\epsilon U_{2\delta}(y)} E_x \exp\left\{\int_0^t c(X_s)ds\right\} \chi_{X_t\epsilon U_{\delta t}(x-t\tilde{y})}$$

provided $\delta > 0$ *is small enough.*

Proof. Put

$$\ell = \underline{\lim_{t\to\infty}} \frac{1}{t} \ln \inf_{\tilde{y}\epsilon U_\delta(y)} u(t,t\tilde{y}) .$$

We first verify that $l > -\infty$. By the Markov property of the process (X_t, P_x), we obtain, for any $\tilde{y} \in R^r$, the following bound

$$E_{t\tilde{y}}g(X_t) \geq \prod_{k=1}^{[t]-1} \inf_{x \in U_\delta((t-k+1)\tilde{y})} P_x\{X_1 \in U_\delta((t-k)\tilde{y})\} \times$$

$$\times \inf_{x \in U_\delta((t-[t]+1)\tilde{y})} E_x g(X_{t-[t]+1}),$$

where $[t]$ is the integer part of t. Taking into account that the process (X_t, P_x) is "periodic", we conclude that

$$\inf_{\tilde{y} \in U_\delta(y)} E_{t\tilde{y}}g(X_t) \geq [\inf_{x \in Q, \tilde{y} \in U_{2\delta}(y)} P_x\{X_1 \in U_\delta(x-\tilde{y})\}]^{[t]-1} \times$$

$$\times \inf_{x \in U_{2|y|+3\delta}(0), 1 \leq s \leq 2} E_x g(X_s).$$

Since all the factors on the right-hand side are positive, we have

$$\varliminf_{t \to \infty} \frac{1}{t} \ln \inf_{\tilde{y} \in U_\delta(y)} E_{t\tilde{y}} g(X_t) > -\infty.$$

From this, noting equation (2.7), we derive that $\ell > -\infty$.

For all $x \in R^r$ and any positive numbers η and t, let us introduce the Markov times:

$$\sigma_\eta(t) = \min\{s \geq 0 : |X_s - (t-s)y| \geq \eta t\},$$

$$\tau_{x,\eta}(t) = \min\{s \geq 0 : |X_s - x + sy| \geq \eta t\}.$$

Pick $\varepsilon > 0$ so that the 2ε-neighborhood of the point y is contained entirely in the region in which the function H is positive, and fix $\delta \in \left(0, \frac{\varepsilon}{3}\right)$ and $h \in (0,1)$ in an arbitrary manner. If $\sigma_\varepsilon(t) > t$, then by Lemma 3.1, $u(t-s, X_s) \leq h$ from some time on. Hence, $c(X_s, u(t-s, X_s)) \geq c_h(X_s)$ for $s \in \left[0, \frac{1}{2}t\right]$ where

$$c_h(x) = \inf_{u \in [0,h]} c(x,u) .$$

By the Markov property, this and equation (2.7) together result in the following bound:

$$\inf_{\tilde{y} \in U_\delta(y)} u(t, t\tilde{y}) \geq \inf_{\tilde{y} \in U_\delta(y)} E_{t\tilde{y}} \Bigg[\exp\left\{ \int_0^{\kappa t} c_h(X_s)\, ds \right\} \times$$

$$\times \chi_{\sigma_\varepsilon(t) > \kappa t, X_{\kappa t} \in U_{(1-\kappa)\delta t}((1-\kappa)ty)} \Bigg] \cdot \inf_{\tilde{y} \in U_\delta(y)} u((1-\kappa)t, (1-\kappa)t\tilde{y}) \geq$$

$$\geq \inf_{x \in Q, \tilde{\tilde{y}} \in U_{2\delta}(y)} E_x \exp\left\{ \int_0^{\kappa t} c_h(X_s)\, ds \right\} \cdot \chi_{\tau_{x, \frac{\delta}{\kappa}}(\kappa t) > \kappa t, X_{\kappa t} \in U_{\kappa\delta t}(x - \kappa t\tilde{\tilde{y}})} \times \tag{3.2}$$

$$\times \inf_{\tilde{y} \in U_\delta(y)} u((1-\kappa)t, (1-\kappa)t\tilde{y})$$

for any $\kappa \in \left(0, \frac{1}{2}\right)$.

The second part of this bound is obtained by translating the trajectories of the process X_s by the integer part of the vector $t\tilde{y}$, if one makes use of "periodicity" of the process and the following inclusions:

$$U_{\varepsilon t}(x - t\tilde{y} + (t-s)y) \supset U_{\delta t}(x - sy) ,$$

$$U_{(1-\kappa)\delta t}(x - t\tilde{y} + (1-\kappa)ty) \supset U_{\kappa\delta t}(x - \kappa t\tilde{\tilde{y}})$$

where x is the fractional part of $t\tilde{y}$, $\tilde{\tilde{y}} = y + 2(\tilde{y} - y)$.

Since $\ell > -\infty$, (3.2) implies the inequality

$$\ell \geq \varliminf_{t\to\infty} \frac{p_{h,\delta\kappa^{-1}}(t)}{t} , \tag{3.3}$$

where

$$p_{h,\eta}(t) =$$

$$= \ln \inf_{x\epsilon Q, \widetilde{y}\epsilon U_{2\delta}(y)} E_x \exp\left\{\int_0^t c_h(X_s)\,ds\right\} \chi_{\tau_{x,\eta}(t)>t, X_t \epsilon U_{\delta t}(x-t\widetilde{y})} \cdot \tag{3.4}$$

Again by the Markov property and the "periodicity" of the process X_t, we establish that the function $p_{h,\eta}(t)$ is semi-additive:

$$p_{h,\eta}(s+t) \geq p_{h,\eta}(s) + p_{h,\eta}(t), \quad s,t > 0 .$$

Moreover, it is obvious that $p_{h,\eta}(t) \leq t \max_x c(x)$, $t > 0$. These properties of the function $p_{h,\eta}(t)$ are known (Hille and Phillips [1]) to yield the equality

$$\lim_{t\to\infty} \frac{p_{h,\eta}(t)}{t} = \sup_{t>0} \frac{p_{h,\eta}(t)}{t} .$$

Thus (3.3) leads to the bound:

$$\ell \geq \sup_{t>0} t^{-1} p_{h,\kappa^{-1}\delta}(t) . \tag{3.5}$$

We now introduce another function

$$p(t) = \ln \inf_{x\epsilon Q, \widetilde{y}\epsilon U_{2\delta}(y)} E_x \exp\left\{\int_0^t c(X_s)\,ds\right\} \chi_{X_t \epsilon U_{\delta t}(x-t\widetilde{y})} \cdot \tag{3.6}$$

In accordance with conditions (2.4) and (2.5), we have: $c_h(x) \uparrow c(x)$ as $h \downarrow 0$. Consequently, by Fatou's lemma, the expectation on the right-hand side of (3.4) converges monotonically to the expectation on the right-hand side of (3.6) as $h \downarrow 0$, $\eta \uparrow \infty$. With this expectation depending continuously on x and y, we have:

$$p_{h,\eta}(t) \uparrow p(t) \quad \text{as} \quad h \downarrow 0, \; \eta \uparrow \infty .$$

Therefore, passing in (3.5) to the limit as $h \downarrow 0$, $\kappa \downarrow 0$, we get

$$\ell \geq \sup_{t>0} \frac{p(t)}{t}$$

From this, the proof is immediate. □

LEMMA 3.3. *For any compact set* $K \subset \{y \in R^r : H(y) > 0\}$,

$$\varliminf_{t\to\infty} \frac{1}{t} \ln \inf_{y \in K} u(t,ty) \geq -\max_{y \in K} H(y) .$$

Proof. Due to the compactness of the set K, it is sufficient to show that, for any $y \in R^r$, for which $H(y) > 0$, and for any $\varepsilon > 0$, a number $\delta > 0$ exists such that

$$\varliminf_{t\to\infty} \frac{1}{t} \ln \inf_{\tilde{y} \in U_\delta(y)} u(t,t\tilde{y}) \geq -H(y) - \varepsilon .$$

Using Lemma 3.2 and noting that the function H is continuous, we conclude that, for the above inequality to be valid, it is sufficient, in its turn, that for small $\delta > 0$

$$\varliminf_{t\to\infty} \frac{1}{t} \ln \mathfrak{J}(t) \geq - \sup_{\tilde{y} \in U_{2\delta}(y)} H(\tilde{y}) , \tag{3.7}$$

where

$$\mathfrak{J}(t) = \inf_{x \in Q, \tilde{y} \in U_{2\delta}(y)} E_x \exp\left\{\int_0^t c(X_s)\, ds\right\} \chi_{X_t \in U_{\delta t}(x - t\tilde{y})} .$$

Taking into account definitions (2.8) and (2.9) of the measures P_x^t and random vectors η_x^t, we arrive at

$$\mathfrak{J}(t) \geq \inf_{\tilde{y} \in U_{2\delta}(y)} \inf_{x \in Q} P_x^t\{\eta_x^t \in U_\delta(\tilde{y})\} \cdot \inf_{x \in Q} E_x \exp\left\{\int_0^t c(X_s)\, ds\right\} .$$

The first factor on the right-hand side of this inequality may be estimated from below by Theorem 1.2, and the second one by limit relation (2.15) with $z = 0$. As a result we get the required bound (3.7). Lemma 4.3 is proved. □

We proceed now to formulate the main result concerning the periodic case.

THEOREM 3.1 (i) *For any closed set* $F \subset \{y \in R^r : H(y) > 0\}$,

$$\lim_{t \to \infty} u(t,ty) = 0$$

uniformly in $y \in F$.

(ii) *For any compact set* $K \subset \{y \in R^r : H(y) < 0\}$,

$$\lim_{t \to \infty} u(t,ty) = 1$$

uniformly in $y \in K$.

Proof. Assertion (i) follows immediately from Lemma 3.1. Let us prove assertion (ii). We set $\Psi(s) = \{y \in R^r : H(y) = s\}$, $\underline{\Psi}(s) = \{y \in R^r : H(y) \leq s\}$, $\overline{b} = \int_{T^r} \overline{b}(x)\overline{\mu}(dx)$, where $\overline{\mu}$ is the normalized invariant measure of the process $(\overline{X}_t, \overline{P}_x)$ on the torus T^r. Next, for any $\delta > 0$ and all $T > 1$, we introduce

$$\Gamma_T = [\{1\} \times \underline{\Psi}(\delta)] \cup [\bigcup_{1 \leq t \leq T} \{t\} \times (t\,\Psi(\delta))] .$$

Thus the set Γ_T consists of the lateral surface of the truncated cone and one of its bases, that is, the set $\underline{\Psi}(\delta)$. It follows easily from equation (2.7) that $u(1,x) > 0$ for all $x \in R^r$. Therefore, relying on Lemma 3.3, we have that for sufficiently large t,

$$u(s,x) \geq e^{-2\delta t} \quad \text{for all} \quad (s,x) \in \Gamma_t . \tag{3.8}$$

Next, for any positive t, h, η we introduce the Markov times:

$$\sigma_\Gamma(t) = \min\{s \geq 0 : (t-s, X_s) \in \Gamma_t\},$$

$$\sigma_h(t) = \min\{s \geq 0 : u(t-s, X_s) \geq h\}$$

$$\tau_\eta(t) = \min\{s \geq 0 : |X_s - X_0 - s\bar{b}| > \eta t\}.$$

If $u(t-s, X_s) < h$ for all $s \in [0,t]$, then we set $\sigma_h(t) = \infty$. Using Ito's equation one can easily prove that for any $\eta > 0$,

$$\sup_{x \in R^r} P_x\{\tau_\eta(t) \leq t\} \to 0 \tag{3.9}$$

as $t \to \infty$. Choose η so that the distance between the η-neighborhood of the set K and the set $\{H \geq 0\}$ is positive. Then there is a number $\kappa \in (0,1)$ such that for all $y \in K$

$$t - 1 \geq \sigma_\Gamma(t) > \kappa t \quad \text{if} \quad \tau_\eta(t) > t. \tag{3.10}$$

By the strong Markov property of the process X_t, we obtain from equation (2.7) that

$$u(t,x) = E_x \exp\left\{\int_0^{\tau \wedge t} c(X_s, u(t-s, X_s))\, ds\right\} \cdot u(t - (\tau \wedge t), X_{\tau \wedge t}) \tag{3.11}$$

for any Markov time τ. Relying on condition (2.3) and the definition of the Markov time $\sigma_h(t)$, we deduce from (3.11), with $\tau = \sigma_h(t)$, the bound:

$$u(t,x) \geq h P_x\{\sigma_h(t) \leq t\}.$$

On account of relations (2.6) and (3.9), we remark that in order to prove assertion (ii), it is sufficient to verify that for any $h \in (0,1)$, uniformly in $y \in K$,

$$P_{ty}\{\sigma_h(t) > t, \tau_\eta(t) > t\} \to 0 \tag{3.12}$$

as $t \to \infty$. Using (3.8) and (3.10) we get the bound

$$P_{ty}\{\sigma_h(t)>t, \tau_\eta(t)>t\} \le P_{ty}\{\kappa t < \sigma_\Gamma(t) \le \sigma_h(t)\} \le$$

$$\le e^{\delta t} E_{ty} \exp\left\{\frac{1}{2}\int_0^{\sigma_\Gamma(t)} c(X_s, u(t-s, X_s))\, ds\right\} [u(t-\sigma_\Gamma(t), X_{\sigma_\Gamma(t)})]^{1/2} \times$$

$$\times \exp\left\{-\frac{1}{2}\int_0^{\sigma_\Gamma(t)} c(X_s, u(t-s, X_s))\, ds\right\} \chi_{\kappa t < \sigma_\Gamma(t) \le \sigma_h(t)} \le$$

$$\le e^{\delta t} E_{ty} \exp\left\{\frac{1}{2}\int_0^{\sigma_\Gamma(t)} c(X_s, u(t-s, X_s))\, ds\right\} [u(t-\sigma_\Gamma(t), X_{\sigma_\Gamma(t)})]^{1/2} \times$$

$$\times \exp\left\{-\frac{1}{2}\int_0^{\kappa t} c_h(X_s)\, ds\right\},$$

where as before $c_h(x) = \inf_{u \in [0,h]} c(x,u)$. By the Hölder inequality and equality (3.11) with $\tau = \sigma_\Gamma(t)$, we deduce that

$$P_{ty}\{\sigma_h(t)>t, \tau_\eta(t)>t\} \le$$
$$\le e^{\delta t}[u(t,ty)]^{1/2} \left[E_{ty} \exp\left\{-\int_0^{\kappa t} c_h(X_s)\, ds\right\}\right]^{1/2}. \tag{3.13}$$

The function u is bounded. Since, by condition (2.4), the function $c_h(x)$ is non-negative and not identically equal to zero, we have

$$0 > \lambda_h = \lim_{t\to\infty} \frac{1}{t} \ln \sup_{x \in R^r} E_x \exp\left\{-\int_0^t c_h(X_s)\, ds\right\}. \tag{3.14}$$

Namely, as is seen from relation (2.15), λ_h is the eigenvalue of the operator $L - c_h(x)$ in C_π such that the corresponding eigenfunction is positive. Clearly, this eigenvalue cannot be equal to zero. If we now choose $\delta < \frac{\kappa}{2} |\lambda_h|$ then (3.13) and (3.14) together imply (3.12), the convergence being uniform in $y \in K$. This completes the proof. □

COROLLARY.

$$\lim_{t\to\infty} u(t,ty) = \begin{cases} 0, & \textit{for} \quad H(y) > 0 \\ 1, & \textit{for} \quad H(y) < 0 . \end{cases}$$

Hence, the set $t \cdot \{y \in R^r : H(y) = 0\}$ may be interpreted as the front of the wave which becomes stabilized for large t.

If

$$\min_{z \in R^r} \lambda(z) > 0 \tag{3.15}$$

then $H(0) < 0$. In this case, for all unit vectors $e \in R^r$, the equation $H(ve) = 0$ has a unique positive solution $v = v^*(e)$. The definition of the function H yields:

$$v^*(e) = \inf \frac{\lambda(z)}{(e,z)} ,$$

where the infimum is taken over all $z \in R^r$ for which $(e,z) > 0$. Here $v^*(e)$ can be thought of as the asymptotic wave propagation velocity in the direction of the vector e. It is easily seen that condition (3.15), in particular, holds if the differential operator has the self-adjoint form:

$$L = \frac{1}{2} \sum_{i,j=1}^{r} \frac{\partial}{\partial x^i} \left(a^{ij}(x) \frac{\partial}{\partial x^j} \right) .$$

§7.4 *Kolmogorov-Petrovskii-Piskunov equation with random multiplication coefficient*

In this section we start by studying the one-dimensional Cauchy problem

$$\frac{\partial u(t,x)}{\partial t} = \frac{1}{2}\frac{\partial^2 u(t,x)}{\partial x^2} + f(x,u(t,x)) ,$$

$$u(0,x) = g(x) , \quad (t,x) \in (0,\infty) \times R^1 , \tag{4.1}$$

for random functions f and g. To be more exact, let $f(x,u) = f(x,u;\hat{\omega})$ and $g(x) = g(x;\hat{\omega})$, x, $u \in R^1$, $\hat{\omega} \in \hat{\Omega}$, be measurable functions defined on a complete probability space $(\hat{\Omega}, \hat{\mathfrak{F}}, \hat{P})$. The function $f(x,u)$ is assumed to be stationary (in the narrow sense) with respect to the variable x. This means that for any $h \in R^1$, the distribution of the random function $f_h(x,u) = f(x+h, u)$ in the space of measurable functions of two variables with the σ-field generated by cylinder sets, coincides with the corresponding distribution of the function $f(x,u)$. As in the periodic case, we shall require the following conditions to be fulfilled:

(A 1) The function g is non-negative and bounded P-a.s.

(A 2) A random variable η exists such that the function g vanishes on the interval $[\eta,\infty)$, $\int_{R^1} g(x)\,dx > 0$ and

$$\int_{R^1} [g(x)-1]^+ dx < \infty , \tag{4.2}$$

where $z^+ = \max(z,0)$.

(A 3) For all $x \in R^1$, $\hat{P}$-a.s.

$$f(x,0) = f(x,1) = 0 ,$$

$f(x,u) \geq 0$ for $u \in (0,1)$ and $f(x,u) < 0$ for $u \in (1,\infty)$.

(A 4) For all $x \in R$, the function

$$c(x,u) = u^{-1} f(x,u) , \quad u > 0$$

is continuous in u for $u > 0$ P-a.s. The limit $\lim_{u\to 0} c(x,u) = c(x,0)$ exists with probability one and

$$\xi(x) = c(x,0) = \sup_{u>0} c(x,u) .$$

Let (W_t, P_x) be a one-dimensional Wiener process. We remind that the P_x are measures on the Borel σ-field $\mathfrak{N}$ in $C_{0,\infty}(R^1)$ which are obtained out of the standard Wiener measure by means of translations by $x \in R^1$. In particular, $P_x\{W_0 = x\} = 1$. As in §7.3, we shall make use of the fact that the (stochastic) solution $u(t,x) = u(t,x;\hat{\omega})$ of problem (4.1) satisfies the equation

$$u(t,x) = E_x \exp\left\{\int_0^t c(W_s, u(t-s, W_s))\, ds\right\} g(W_t) , \tag{4.3}$$

$(t,x) \in [0,\infty) \times R^1$. Any classical solution of problem (4.1) is a solution of equation (4.3). Conversely, every function $u(t,x)$, twice continuously differentiable in t and in x and satisfying (4.3), is a solution of problem (4.1). Thus, any measurable function $u(t,x)$ satisfying equation (4.3) is called a generalized solution of Cauchy's problem (4.1). If the initial function g is continuous and the function f is bounded and Lipschitz continuous in its two variables, then problem (4.1) has a classical solution. However, in the case where the functions f and g are random, it is desirable to get rid of such rigid conditions, but in return the solution is to be sought in the generalized form. Let us introduce in addition to (A 1) - (A 4), the following condition:

(A 5) With probability one, the function f satisfies the "stochastic" Lipschitz condition

$$|f(x,u_1) - f(x,u_2)| \le \zeta(x)\, |u_1 - u_2|, (x, u_1, u_2 \in R^1) , \tag{4.4}$$

where $\zeta(x) = \zeta(x, \hat{\omega})$ is a measurable random function such that for all $t \ge 0$ and $x \in R^1$

$$E_x \exp\left\{\int_0^t \zeta(W_s)\,ds\right\} < \infty \quad \hat{P}\text{-a.s.} \tag{4.5}$$

Since the function $f(x,u)$ is stationary, without any loss of generality the function $\zeta(x)$ can be thought of as stationary. By Jensen's inequality it is easily deduced that (4.5) is, in particular, fulfilled if $\zeta(x)$ obeys Cramer's condition:

$$\hat{E} \exp\{t\,\zeta(0)\} < \infty$$

for all real t.

Let us formulate the result on the existence and uniqueness of a generalized solution of Cauchy's problem (4.1).

THEOREM 4.1. *Let Conditions (A 1), (A 3) and (A 5) hold. Then, with probability one, Cauchy's problem (4.1) has a unique bounded generalized solution* $u(t,x)$, *continuous for* $(t,x) \in (0,\infty) \times R^1$ *and satisfying the inequality*:

$$0 \le u(t,x) \le 1 \vee \sup_{x \in R^1} g(x) . \tag{4.6}$$

If, moreover, (4.2) is fulfilled, then

$$\overline{\lim_{t\to\infty}} \sup_{x\in R^1} u(t,x) \le 1 \quad (\hat{P}\text{-a.s.}) \tag{4.7}$$

If the function f *is locally Hölder continuous and the function* g *is continuous, then the generalized solution* $u(t,x)$ *is continuous for all* $(t,x) \in [0,\infty) \times R^1$ *continuously differentiable in* t *and twice in* x *in the region* $(0,\infty) \times R^1$ *and satisfies problem (4.1).*

The proof of this theorem follows the argument given in §5.3, and we omit it.

§7.5 *The definition and basic properties of the function* $\mu(z)$

Suppose that $\mathcal{F}^f$ and $\mathcal{F}^\xi$ and σ-fields generated by the (stationary in x) random functions $f(x,u)$ and $\xi(x)$ respectively ($\xi(x)$ has been defined in Condition (A 4)). The corresponding group of translations on $\mathcal{F}^f$ will be denoted by (θ_x), $x \in R^1$. The operators θ_x, $x \in R^1$, map $\mathcal{F}^\xi$ onto itself. The translation operators acting on $\mathcal{F}^f$-measurable random variables will be denoted by the same symbols. In particular, for any $x, y, u \in R^1$, we have

$$\theta_y f(x,u) = f(x+y,u) \quad \text{and} \quad \theta_y \xi(x) = \xi(x+y) \ (\hat{P}\text{-a.s.}) .$$

The sub-σ-fields of $\mathcal{F}^f$ and $\mathcal{F}^\xi$ consisting of all events that are $\hat{P}$-a.s. invariant with respect to the group of translations (θ_x) will be designated by $\mathcal{F}^f_{\text{inv}}$ and $\mathcal{F}^\xi_{\text{inv}}$ respectively. Clearly, $\mathcal{F}^\xi_{\text{inv}} = \mathcal{F}^f_{\text{inv}} \cap \mathcal{F}^\xi$.

Let $\tau_y = \min\{t \geq 0 : W_t = y\}$ be the first hitting time of the Wiener process to a point $y \in R^1$. In §7.6, when studying asymptotics of a solution for Cauchy's problem (4.1), we shall rely on large deviations theorems for the random variable $\eta_t = t^{-1}\tau_0$ with respect to the family of probability measures of the form:

$$P^t(A) = \frac{E_t \exp\{\int_0^{\tau_0} [\xi(W_s) + z_0] ds\} \chi_A}{E_t \exp\{\int_0^{\tau_0} [\xi(W_s) + z_0] ds\}} , \quad A \in \mathcal{N} , \tag{5.1}$$

where z_0 is an appropriate random variable on $(\hat{\Omega}, \hat{\mathcal{F}}, \hat{P})$ such that the denominator in definition (5.1) does not go to infinity for large t. To specify the corresponding action function, we must for any $z \in R^1$ compute the limit

$$G(z) = \lim_{t\to\infty} \frac{1}{t} \ln E^t \exp\{z\tau_0\} =$$

$$= \lim_{t\to\infty} \frac{1}{t} \ln \frac{E_t \exp\{\int_0^{\tau_0} [\xi(W_s) + z + z_0] ds\}}{E_t \exp\{\int_0^{\tau_0} [\xi(W_s) + z_0] ds\}} = \mu(z+z_0) - \mu(z_0) . \tag{5.2}$$

We shall demonstrate in Theorem 5.1 that the function $\mu(z)$ on the right-hand side of (5.2) is given by the formula

$$\mu(z) = \hat{E}\left[\ln E_1 \exp\left\{\int_0^{\tau_0} [\xi(W_s)+z]\,ds\right\} |\mathcal{F}^{\xi}_{inv}\right] \tag{5.3}$$

for any $z \in R^1$. The function $\mu(z)$ takes real values including $+\infty$. Since the expression in the expectation in (5.3) is convex, lower semi-continuous and monotonically non-decreasing in z, for any $z \in R$ one can choose such a version of this conditional expectation such that the function $\mu(z)$ is convex, lower semi-continuous and monotonically non-decreasing P-a.s. Later on the function $\mu(z)$ will be assumed to have these properties.

Since $\xi(x) \geq 0$, $\mu(z) = \infty$ for $z > 0$. Thus, a non-positive $\mathcal{F}^{\xi}_{inv}$-measurable random variable $\overline{g}$ exists such that $\mu(z) < \infty$ for $z < \overline{g}$ and $\mu(z) = \infty$ for $z > \overline{g}$. And what is more, by the definition, $\overline{g} = -\infty$ if $\mu(z) = \infty$ for all $z \in R^1$.

If the random function $\xi(x)$ is metrically transitive, then the conditional expectation in (5.3) turns into the unconditional one and the function $\mu(z)$ and the variable $\overline{g}$ are non-random.

THEOREM 5.1. (i) *For all* $z \in R^1$,

$$\mu(z) = \lim_{t\to\infty} \frac{1}{t} \ln E_t \exp\left\{\int_0^{\tau_0} [\xi(W_s)+z]\,ds\right\}, \quad \hat{P}\text{-a.s.} \,. \tag{5.4}$$

(ii) *With probability one, the function* $\mu(z)$ *is continuously differentiable and the derivative* $\mu'(z)$ *is positive and monotonically non-decreasing for* $z < \overline{g}$; $\mu'(z) \to 0$ *as* $z \to -\infty$ *on the set* $\{\overline{g} > -\infty\}$.

(iii) $\mu(z) \leq 0$ *for* $z \leq \overline{g}$ $\hat{P}$-a.s. .

Proof. Let c be an arbitrary real number. We set

$$\overline{\mu}(z) = \begin{cases} \mu(z+\overline{g}), & \text{if } \overline{g} > c\,, \\ 0, & \text{otherwise.} \end{cases}$$

As the random variable g is $\mathcal{F}^{\xi}_{\mathrm{inv}}$-measurable, we have

$$\mu(z) = \hat{E}\,[\Psi(z)|\mathcal{F}^{\xi}_{\mathrm{inv}}]\,, \tag{5.5}$$

where

$$\Psi(z) = \chi_{\overline{g}>c} \ln E_1 \exp\left\{\int_0^{\tau_0} [\xi(W_s)+\overline{g}+z]\,ds\right\}.$$

(i) Using the strong Markov property of the Wiener process one can verify that the random function

$$\overline{\mu}_{s,t}(z) = \chi_{\overline{g}>c} \ln E_t \exp\left\{\int_0^{\tau_s} [\xi(W_{s_1})+\overline{g}+z]\,ds_1\right\}, \quad s < t\,,$$

is additive:

$$\overline{\mu}_{s,t}(z) = \overline{\mu}_{s,u}(z) + \overline{\mu}_{u,t}(z) \quad \text{for} \quad s < u < t\,.$$

Moreover, it is easily checked that

$$\overline{\mu}_{s,t}(z) \geq -K_z \cdot (t-s)\,, \quad \hat{P}\text{-a.s.}$$

and for all $h \in R^1$

$$\theta_h \overline{\mu}_{s,t}(z) = \overline{\mu}_{s+h,t+h}(z)\,,$$

where K_z is a constant. Hence, the ergodic theorem can be applied to the random functions $\overline{\mu}_{0,t}(z)$: for any $z \in R^1$ the following limit relation

$$\lim_{t\to\infty} t^{-1}\,\overline{\mu}_{0,t}(z) = \overline{\mu}(z) \tag{5.6}$$

holds $\hat{P}$-a.s.. Evidently, (5.6) is true, with probability one, simultaneously for all rational z, in particular, for the "discontinuity" point $z = 0$ as well. From this, taking into account that the functions $\overline{\mu}_{s,t}(z)$ and

$\overline{\mu}(z)$ are convex and monotone, we infer that (5.6) is valid simultaneously for all real z $\hat{P}$-a.s. (Rockafellar [1], Theorem 10.8). Since c is arbitrary, (5.4) has thereby been demonstrated to hold on the set $\{\overline{g} > -\infty\}$ for all $z \in R^1$, $\hat{P}$-a.s.. On the set $\{\overline{g} = -\infty\}$, assertion (5.4) is proved in a similar way, with the random functions $\mu(z)$ and

$$\mu_{s,t}(z) = \ln E_t \exp\left\{ \int_0^{\tau_s} [\xi(W_u) + z]du \right\}$$

being used rather than $\overline{\mu}(z)$ and $\overline{\mu}_{s,t}(z)$ respectively.

(ii) By definition, $\overline{\mu}(z) < \infty$ for $z < 0$. Thus, from (5.5) and the monotonicity of the function $\Psi(z)$ it follows that $\Psi(z) < \infty$ for all $z < 0$ $\hat{P}$-a.s.. From this, in turn, follows that the function $\Psi(z)$ is continuously differentiable in the interval $(-\infty, 0)$ $\hat{P}$-a.s. and its derivative has the form

$$\Psi'(z) = \chi_{\overline{g}>c} \frac{E_1 \tau_0 \exp\{\int_0^{\tau_0} [\xi(W_s) + \overline{g} + z]ds\}}{E_1 \exp\{\int_0^{\tau_0} [\xi(W_s) + \overline{g} + z]ds\}} .$$

Since the function $\Psi(z)$ is convex, $h^{-1}[\Psi(z+h) - \Psi(z)]$ is monotone non-decreasing in h. In particular, for $h > 0$

$$\Psi'(z) \leq \frac{\Psi(z+h) - \Psi(z)}{h} .$$

The conditional expectation with respect to the σ-field $\mathcal{F}^{\xi}_{inv}$ on the right-hand side is finite for $z + h < 0$, $\hat{P}$-a.s.. Therefore, for $z < 0$,

$$\nu(z) = \hat{E}[\Psi'(z) | \mathcal{F}^{\xi}_{inv}]$$

is finite $\hat{P}$-a.s.. Further, noting that the function $\overline{\mu}(z)$ is convex, we conclude that, for $z < 0$, the one-sided derivatives $\overline{\mu}^{+}(z)$ and $\overline{\mu}^{-}(z)$ exist. As $h^{-1}[\Psi(z+h) - \Psi(z)]$ is monotone non-decreasing in h, we obtain by Fatou's lemma that for any $z < 0$

$$\overline{\mu}^{+}(z) = \overline{\mu}^{-}(z) = \nu(z) \quad (\hat{P}\text{-a.s.}) . \tag{5.7}$$

Let us show that, for any z, a version of the conditional expectation $\hat{E}[\Psi'(z)|\mathcal{F}^{\xi}_{inv}]$ can be chosen such that the function $\nu(z)$ is continuous for $z < 0$ $\hat{P}$-a.s.. Since the function $\Psi'(z)$ is continuous and (in view of the convexity of $\Psi(z)$) is monotone non-decreasing for $z < 0$, we conclude that there exists a version of the function $\nu(z)$, non-decreasing with probability one. It turns out that this version is continuous for $z < 0$ $\hat{P}$-a.s.. To prove this fact, it is sufficient to verify that the restriction of $\nu(z)$ to the set of negative rational numbers is continuous $\hat{P}$-a.s.. This is easily done via Fatou's lemma remembering that $\Psi'(z)$ is continuous and monotone. Consequently, the function $\nu(z)$ can be chosen continuous for $z < 0$ $\hat{P}$-a.s.. Utilizing this and the monotonicity of the functions $\overline{\mu}^{+}(z)$ and $\overline{\mu}^{-}(z)$ we derive from (5.7) that the function $\overline{\mu}(z)$ is continuously differentiable for $z < 0$ $\hat{P}$-a.s., which implies the continuous differentiability of $\mu(z)$ for $z < \overline{g}$. That the derivative $\mu'(z)$ is positive and monotone, follows from the fact that $\mu(z)$ is convex and strictly monotone. The fact that $\mu'(z)$ vanishes as $z \to -\infty$ is not difficult to deduce from the bound

$$\mu(z) \geq \ln E_1 e^{z\tau_0} = -\sqrt{2|z|}, \quad z \leq 0 .$$

(iii) Without loss of generality, the function $\xi(x)$ can be assumed bounded. For any $z < \overline{g}$, the function

$$u^z(x) = E_x \exp\left\{ \int_0^{\tau_0} [\xi(W_s) + z]ds \right\}, \quad x \geq 0 ,$$

is finite and satisfies the equation

$$\frac{1}{2}\frac{d^2u^z}{dx^2} + [\xi(x)+z]u^z = 0 . \tag{5.8}$$

More precisely, the function $u^z(x)$ is continuously differentiable in x, the derivative $\frac{du^z}{dx}$ is absolutely continuous, and equation (5.8) holds a.e. with respect to the Lebesgue measure. Hence the function $\phi^z = \ln u^z$ satisfies the equation

$$\frac{1}{2}\frac{d^2\phi^z}{dx^2} + \frac{1}{2}\left(\frac{d\phi^z}{dx}\right)^2 = -\xi(x) - z . \tag{5.9}$$

The function ϕ^z is differentiable in z. The initial data $\phi^z(0) = 0$ and $\frac{d\phi^z}{dx}(0) = \lim_{x \downarrow 0} \frac{d\phi^z}{dx}(x)$ of equation (5.9) are convex in z and, thi are one-sidedly differentiable in z. So, one may differentiate equati. n (5.9) with respect to z. Then, for the function $\psi^z = \frac{d\phi^z}{dx}$ we obtain the equation

$$\frac{1}{2}\frac{d^2\psi^z}{dx^2} + \frac{d\phi^z}{dx}\frac{d\psi^z}{dx} = -1$$

with the general solution

$$\psi^z(x) = \psi^z(0) + \int_0^x \left[\frac{d\psi^z}{dx}(0) - 2\int_0^y (u^z(\tilde{y}))^2 d\tilde{y}\right] \frac{dy}{(u^z(y))^2} \tag{5.10}$$

Now let us assume that $\mu(z) > 0$ for some $z < \overline{g}$. Then, by assertion (i), $u(x)$ converges exponentially to $+\infty$ as $x \to \infty$. In view of this, (5.10) guarantees the function $\psi^z(x)$ to be bounded from above as $x \to \infty$. But this cannot hold. Namely, by assertion (i),

$$\lim_{x\to\infty} \frac{1}{x}\, \phi^z(x) = \mu(z) . \tag{5.11}$$

Since the functions $\phi^z(x)$ and $\mu(z)$ are convex and differentiable in z, $z < \overline{g}$, the limit relation (5.11) can be differentiated with respect to z.

As a result we have

$$\lim_{x\to\infty} \frac{1}{x} \psi^z(x) = \mu'(z)$$

(Rockafellar [1]). Since by assertion (ii), $\mu'(z) > 0$, we have that $\psi^z(x) \to \infty$ as $x \to \infty$, which contradicts the above. Hence, $\mu(z) \leq 0$ for $z \leq \overline{g}$, as was to be proved. □

Let $U_\delta(x)$ be, as before, the δ-neighborhood of a point $x \in R^1$. The following theorem characterizes the discontinuity point $\overline{g}$ of the function $\mu(z)$ as the limit of a random functional of the Wiener process.

THEOREM 5.2. *For any* $\delta > 0$

$$\overline{g} = -\lim_{t\to\infty} \frac{1}{t} \ln E_0 \exp\left\{\int_0^t \xi(W_s)\,ds\right\} \chi_{W_t \in U_\delta(0)}, \quad (\hat{P}\text{-a.s.}).$$

Proof. 1° By using the Markov property of the Wiener process it is easily seen that the function

$$p_\delta(t) = \ln \inf_{x\in U_\delta(0)} E_x \exp\left\{\int_0^t \xi(W_s)\,ds\right\} \chi_{W_t \in U_\delta(0)}, \quad t > 0,$$

is semiadditive: for any $s, t > 0$,

$$p_\delta(s+t) \geq p_\delta(s) + p_\delta(t).$$

Moreover,

$$\inf_{0<s<t} p_\delta(s) > -\infty$$

for all $t > 0$. Consequently, as is known, $t^{-1}p_\delta(t)$ converges to a limit $\ell_\delta \in (-\infty, \infty]$ as $t \to \infty$, and in addition

$$\ell_\delta = \lim_{t\to\infty} t^{-1}p_\delta(t) = \sup_{t>0} t^{-1}p_\delta(t). \qquad (5.12)$$

2° Let us prove that

$$\ell_\delta = \lim_{t\to\infty} \frac{1}{t} \ln E_0 \exp\left\{ \int_0^t \xi(W_s)\, ds \right\} \chi_{W_t \epsilon U_\delta(0)} \cdot \tag{5.13}$$

Noting that the random function $\xi(x)$ is non-negative and applying the strong Markov property of the Wiener process, we get, for any $x \epsilon R^1$ and $t > 0$, the bound:

$$E_x \exp\left\{ \int_0^{t+1} \xi(W_s)\, ds \right\} \chi_{W_{t+1} \epsilon U_\delta(0)} \geq$$

$$\geq E_x \chi_{\tau_0 \leq 1} \exp\left\{ \int_{\tau_0}^{t+\tau_0} \xi(W_s)\, ds \right\} \chi_{W_{t+\tau_0} \epsilon U_\delta(0)} \cdot \chi_{W_{t+1} \epsilon U_\delta(0)} =$$

$$= E_x \chi_{\tau_0 \leq 1} E_0 \exp\left\{ \int_0^t \xi(W_s)\, ds \right\} \chi_{W_t \epsilon U_\delta(0)} \times$$

$$\times P_{W_t} \{W_{1-s} \epsilon U_\delta(0)\}\Big|_{s=\tau_0} .$$

Whence it follows that

$$\inf_{x \epsilon U_\delta(0)} E_x \exp\left\{ \int_0^{t+1} \xi(W_s)\, ds \right\} \chi_{W_{t+1} \epsilon U_\delta(0)} \geq \inf_{x \epsilon U_\delta(0)} P_x\{\tau_0 \leq 1\} \times$$

$$\times E_0 \exp\left\{ \int_0^t \xi(W_s) ds \right\} \chi_{W_t \epsilon U_\delta(0)} \inf_{x \epsilon U_\delta(0),\, s \epsilon (0,1]} P_x\{W_s \epsilon U_\delta(0)\} .$$

The first and the third factors on the right-hand side of this inequality are positive, and by (5.12) the expression on the left-hand side is bounded from above by $\exp\{\ell_\delta(t+1)\}$. Therefore, we finally get the bound:

$$E_0 \exp\left\{\int_0^t \xi(W_s)\,ds\right\} \chi_{W_t \epsilon U_\delta(0)} \le c_\delta \cdot \exp\{(t+1)\ell_\delta\}, \tag{5.14}$$

where c_δ is some (non-random) constant independent of t. In particular, this ensures limit relation (5.13).

3° Let us show that the limit ℓ_δ in fact does not depend on δ. By the Markov property of the Wiener process, for arbitrary $\delta, \varepsilon > 0$, we derive the bound:

$$E_0 \exp\left\{\int_0^{t+1} \xi(W_s)\,ds\right\} \chi_{W_{t+1} \epsilon U_\delta(0)} \ge$$

$$\ge E_0 \exp\left\{\int_0^t \xi(W_s)\,ds\right\} \chi_{W_t \epsilon U_\varepsilon(0)} \cdot \inf_{x \epsilon U_\varepsilon(0)} P_x\{W_1 \epsilon U_\delta(0)\}$$

which, together with (5.13), immediately implies the inequality $\ell_\delta \ge \ell_\varepsilon$. Hence, the limit ℓ_δ does not depend on δ. In the sequel we will denote it simply by ℓ. Observe that $\ell \ge 0$.

4° For any $\eta > \delta > 0$, with $\hat{P}$ probability one,

$$\ell = \lim_{t\to\infty} \frac{1}{t} \ln E_0 \exp\left\{\int_0^t \xi(W_s)\,ds\right\} \chi_{\tau_{-\eta} > t,\, W_t \epsilon U_\delta(0)}. \tag{5.15}$$

To prove this, we introduce the functions

$$p_{\delta,\eta}(t) = \ln \inf_{x \epsilon U_\delta(0)} E_x \exp\left\{\int_0^t \xi(W_s)\,ds\right\} \chi_{\tau_{-\eta} > t,\, W_t \epsilon U_\delta(0)},$$

$t > 0$. By analogy to $1°$, it is possible to establish that $\frac{1}{t} p_{\delta,\eta}(t)$ tends to some limit $\ell_{\delta,\eta} \in (-\infty, \infty)$ as $t \to \infty$ and also

$$\ell_{\delta,\eta} = \lim_{t\to\infty} t^{-1} p_{\delta,\eta}(t) = \sup_{t>0} t^{-1} p_{\delta,\eta}(t) . \tag{5.16}$$

One can prove that the expectation

$$E_x \exp\left\{\int_0^t \xi(W_s)\, ds\right\} \chi_{\tau_{-\eta}>t, W_t \in U_\delta(0)} \tag{5.17}$$

is continuous in x $\hat{P}$-a.s.. It is monotone non-decreasing in η and converges (as $\eta \to \infty$) to the expectation

$$E_x \exp\left\{\int_0^t \xi(W_s)\, ds\right\} \chi_{W_t \in U_\delta(0)} \tag{5.18}$$

which also depends continuously on x. Therefore,

$$p_{\delta,\eta}(t) \uparrow p_\delta(t) \quad \text{as} \quad \eta \to \infty .$$

(Note that this can also be proved without using the fact that (5.17) and (5.18) are continuous in x.) From this, via (5.12) and (5.16) we deduce that

$$\ell_{\delta,\eta} \uparrow \ell_\delta = \ell \quad \text{as} \quad \eta \to \infty . \tag{5.19}$$

Next, for arbitrary $\alpha > 0$, $\eta > \delta > 0$, and $t > 0$, we obtain the bound

$$\theta_{-\alpha} \inf_{x\in U_\delta(0)} E_x \exp\left\{\int_0^{t+2} \xi(W_s)\, ds\right\} \chi_{\tau_{-\eta}>t+2, W_{t+2}\in U_\delta(0)} =$$

$$= \inf_{x\in U_\delta(-\alpha)} E_x \exp\left\{\int_0^{t+2} \xi(W_s)\, ds\right\} \chi_{\tau_{-\eta-\alpha}>t+2, W_{t+2}\in U_\delta(-\alpha)} \geq$$

$$\geq \inf_{x \in U_\delta(-a)} P_x\{\tau_{-\eta-a} > 1, W_1 \in U_\delta(0)\} \times$$

$$\times \inf_{x \in U_\delta(0)} E_x \exp\left\{\int_0^t \xi(W_s)\, ds\right\} \chi_{\tau_{-\eta-a} > t, W_t \in U_\delta(0)} \times$$

$$\times \inf_{x \in U_\delta(0)} P_x\{\tau_{-\eta-a} > 1, W_1 \in U_\delta(-a)\} .$$

Here we again have used the Markov property of the Wiener process. The first and the third factors on the right-hand side of the above inequality are positive. Hence, putting together this inequality and (5.16) we obtain

$$\theta_{-a} \ell_{\delta,\eta} \geq \ell_{\delta,\eta+a} \qquad (\hat{P}\text{-a.s.})$$

for all $a \geq 0$ and, consequently,

$$\frac{1}{T} \int_0^T \theta_{-a} \ell_{\delta,\eta}\, da \geq \frac{1}{T} \int_0^T \ell_{\delta,\eta+a}\, da .$$

By the ergodic theorem, the expression on the left-hand side of this inequality converges to $\hat{E}[\ell_{\delta,\eta} | \mathcal{F}^{\xi}_{inv}]$ as $T \to \infty$ and by (5.19), the right-hand side converges to ℓ. Therefore,

$$\hat{E}[\ell_{\delta,\eta} | \mathcal{F}^{\xi}_{inv}] \geq \ell .$$

On the other hand, obviously $\ell_{\delta,\eta} \leq \ell$. Thus, $\ell_{\delta,\eta} = \ell$ $\hat{P}$-a.s., which immediately yields assertion (5.15) and the $\mathcal{F}^{\xi}_{inv}$-measurability of ℓ as well.

5° Let us prove that $-\ell \leq g$ $\hat{P}$-a.s.. Remembering inequality (5.14), for any δ, $\varepsilon > 0$ and all natural numbers $n \geq 1$, we derive that:

$$E_0 \exp\left\{\int_0^{\tau_{-1}} [\xi(W_s) - \ell - \varepsilon]\,ds\right\} \chi_{\tau_{-1}\epsilon(n-1,n],\, W_n \epsilon U_\delta(0)} \le$$

$$\le \exp\{-(n-1)\,(\ell+\varepsilon)\}\, E_0 \exp\left\{\int_0^{n} \xi(W_s)\,ds\right\} \chi_{W_n \epsilon U_\delta(0)} \le \tag{5.20}$$

$$\le c_\delta \exp\{2\ell - (n-1)\,\varepsilon\}\,.$$

Further, by using the strong Markov property of the Wiener process we arrive at the bound:

$$E_0 \exp\left\{\int_0^{\tau_{-1}} [\xi(W_s) - \ell - \varepsilon]\,ds\right\} \chi_{\tau_{-1}\epsilon(n-1,n],\, W_n \epsilon U_\delta(0)} \ge$$

$$\ge E_0 \exp\left\{\int_0^{\tau_{-1}} [\xi(W_s) - \ell - \varepsilon]\,ds\right\} \chi_{\tau_{-1}\epsilon(n-1,n]} \times \tag{5.21}$$

$$\times \inf_{t\epsilon(0,1]} P_{-1}\{W_t \,\epsilon\, U_\delta(0)\}\,.$$

Let $\delta > 1$. Then $\inf_{0<t\le 1} P_{-1}\{W_t \,\epsilon\, U_\delta(0)\} = P_{-1}\{W_1 \,\epsilon\, U_\delta(0)\} > 0$. Combining inequalities (5.20) and (5.21) and summing over n, we have

$$E_0 \exp\left\{\int_0^{\tau_{-1}} [\xi(W_s) - \ell - \varepsilon]\,ds\right\} \le \frac{c_\delta e^{2\ell}}{P_{-1}\{W_1 \,\epsilon\, U_\delta(0)\}} \sum_{n=0}^{\infty} e^{-n\varepsilon}$$

for $\delta > 1$. The expression on the right-hand side of this inequality is finite and non-random. Taking into account that the random function $\xi(x)$ is stationary and that ℓ is $\mathcal{F}^{\xi}_{\text{inv}}$ -measurable, it is straightforward to infer that $\mu(-\ell-\varepsilon) < \infty$ on the set $\{\ell < \infty\}$ $\hat{P}$-a.s.. Hence, $-\ell-\varepsilon < \overline{g}$ and since $\varepsilon > 0$ is arbitrary, $-\ell \leq \overline{g}$ $\hat{P}$-a.s..

6° It remains to show that $-\ell \geq \overline{g}$ $\hat{P}$-a.s.. Let us pick positive numbers c and ε in an arbitrary fashion and take $\overline{\ell} = (\ell \wedge c) - \varepsilon$. By the Markov property, for any $\delta > 0$, $t > 0$, we have the bound

$$E_0 \exp\left\{\int_0^{\tau_{-1}} [\xi(W_s) - \overline{\ell}]ds\right\} \geq E_0 \exp\left\{\int_0^{\tau_{-1}} [\xi(W_s) - \overline{\ell}]ds\right\} \times$$

$$\times \chi_{\tau_{-1}>t, W_t \epsilon U_\delta(0)} = E_0 \exp\left\{\int_0^t [\xi(W_s) - \overline{\ell}]ds\right\} \chi_{\tau_{-1}>t, W_t \epsilon U_\delta(0)} \times$$

$$\times E_{W_t} \exp\left\{\int_0^{\tau_{-1}} [\xi(W_s) - \overline{\ell}]ds\right\} \geq \inf_{x \epsilon U_\delta(0)} E_x \exp\{-\overline{\ell}\tau_{-1}\} \times$$

$$\times E_0 \exp \left.\int_0^t [\xi(W_s) - \overline{\ell}]ds\right\} \chi_{\tau_{-1}>t, W_t \epsilon U_\delta(0)} .$$

The first factor on the right-hand side of this inequality is positive $\hat{P}$-a.s.. According to (5.15), for $\delta < 1$, the second factor is (up to logarithmic equivalence) equal to $\exp t(\ell - \overline{\ell})$ and, therefore, runs to infinity as $t \to \infty$. Consequently,

$$E_0 \exp\left\{\int_0^{\tau_{-1}} [\xi(W_s) - \overline{\ell}]ds\right. = \infty \quad (\hat{P}\text{-a.s.})$$

so that $\mu(-\overline{\ell}) = \infty$, that is $-\overline{\ell} = -(\ell \wedge c) + \varepsilon \geq \overline{g}$ ($\hat{P}$ a.s.). In view of the fact that $c > 0$ and $\varepsilon > 0$ are arbitrary, this implies that $-\ell \geq \overline{g}$ $\hat{P}$-a.s.. This completes the proof of Theorem 5.2. □

§7.6 *Asymptotic wave front propagation velocity in random media*

The present section evaluates the asymptotic wave front propagation velocity for Cauchy's problem (4.1). In outline we shall follow §7.3 where the periodic case was treated.

Throughout this section, Conditions (A 1)-(A 5) are assumed to be fulfilled (see §7.4). In addition we will suppose that $-\infty < \overline{g} < 0$ $\hat{P}$-a.s.. This is equivalent to the assumption that the function $\mu(z)$ is not identically equal to $+\infty$ and the function $\xi(x)$ specified in Condition (A 4) is not equal to zero a.e. with respect to the Lebesgue measure. The foregoing assertion is not difficult to check via Theorem 5.2 and the inequality $-\overline{g} \geq \hat{E}\,[\xi(0)|\mathcal{F}^{\xi}_{inv}]$ which comes from bound (5.10) by Jensen's inequality.

Let $I(y)$ designate the Legendre transform of the function $\mu(z)$:

$$I(y) = \sup_{z \leq \overline{g}} [yz - \mu(z)], \quad y \in R^1 . \tag{6.1}$$

Since $\overline{g} < 0$ and $\mu(z) \geq \ln E_1 e^{z\tau_0} = -\sqrt{2|z|}$ for $z < 0$, the function $I(y)$ is finite and strictly decreasing for $y > 0$. Then, since $\mu(z) \to -\infty$ as $z \to -\infty$, we have $I(y) = \infty$ for $y \leq 0$. Therefore, noting that the function $I(y)$ is lower semi-continuous, we have that $I(y) \to +\infty$ as $y \downarrow 0$.

Let us demonstrate that $I(y) \to -\infty$ as $y \to +\infty$. Since the function $\mu(z)$ is convex, $\mu'(z)$ is monotone non-decreasing and tends to a limit $\mu'(\overline{g})$ as $z \uparrow \overline{g}$. If $\mu'(\overline{g}) = \infty$, the claim follows from the equality

$$I(\mu'(z)) = \mu'(z)z - \mu(z)$$

with passing to the limit as $z \uparrow \overline{g}$ and remembering that $\overline{g} < 0$. If, on the contrary, $\mu'(\overline{g}) < \infty$, then for any $y \geq \mu'(\overline{g})$, the function $yz - \mu(z)$ is monotone non-decreasing in z, $z < \overline{g}$. Consequently,

$$I(y) = y\overline{g} - \mu(\overline{g}) \quad \text{for} \quad y \geq \mu'(\overline{g}) ,$$

which implies: $I(y) \to -\infty$ as $y \to +\infty$.

These just proven properties of the function $I(y)$ yield that a unique positive number v^* exists such that $I\left(\frac{1}{v^*}\right) = 0$. Taking advantage of definition (6.1) of the function $I(y)$ leads to the more explicit expression for v^*:

$$v^* = \inf_{z \leq g} \frac{z}{\mu(z)} . \tag{6.2}$$

In what follows we shall establish that v^* is the asymptotic wave front propagation velocity for problem (4.1).

We begin by obtaining the lower bound for a solution $u(t,x)$ of problem (4.1).

LEMMA 6.1. *For arbitrary* $v > 0$, *the inequality*

$$\overline{\lim_{t\to\infty}} \frac{1}{t} \ln \sup_{x \geq vt} u(t,x) \leq -vI\left(\frac{1}{v}\right) \tag{6.3}$$

is valid $\hat{P}$-a.s..

Proof. Without loss of generality, we shall suppose that $I\left(\frac{1}{v}\right) > 0$. By the strong Markov property of the Wiener process, by inequality (4.6) and on account of Conditions (A 2), (A 4), we derive from equation (4.3) that

$$u(t,x) = E_x \exp\left\{\int_0^{\tau_\eta} c(W_s, u(t-s, W_s))\, ds\right\} \chi_{\tau_\eta \leq t} \cdot u(t-\tau_\eta, W_{\tau_\eta}) \leq$$

$$\leq (1 + \|g\|) E_x \exp\left\{\int_0^{\tau_\eta} \xi(W_s)\, ds\right\} \chi_{\tau_\eta \leq \frac{x}{v}}$$

for all $x \geq vt$. Here $\|g\| = \sup_{x \in R^1} |g(x)|$ and η is a variable defined in Condition (A 2). Further, the above inequality implies that

$$u(t,x) \le (1+\|g\|)\, e^{-\frac{xz}{v}} E_x \exp\left\{ \int_0^{\tau_\eta} [\xi(W_s)+z]\, ds \right\}.$$

for any $z < 0$ and all $x \ge vt$. This together with assertion (i) of Theorem 5.1 leads to the bound

$$\varlimsup_{t\to\infty} \frac{1}{t} \ln \sup_{x \ge vt} u(t,x) \le -v\left[\frac{z}{v} - \mu(z)\right],$$

provided $zv^{-1} - \mu(z) > 0$. In view of definition (6.1), this shows that bound (6.3) is valid, which completes the proof. □

LEMMA 6.2. *For any* $\delta > 0$ *and all* $v > v^*$, $\hat{P}$-a.s.

$$\varliminf_{t\to\infty} \frac{1}{t} \ln u(t,vt) \ge \tag{6.4}$$

$$\ge \varliminf_{t\to\infty} \frac{1}{t} \ln E_{vt} \exp\left\{ \int_0^t \xi(W_s)\, ds \right\} \chi_{W_t \in U_\delta(0)} .$$

Proof. For any $h > 0$, put

$$\xi_h(x) = \inf_{u \in (0,h)} c(x,u)$$

(see §7.4, Condition (A 4)). We introduce the Markov times

$$\sigma_{t,\kappa} = \min\{s \ge 0 : W_s \le v\cdot(t-s) - \kappa\};\quad t,\kappa > 0 .$$

Let us pick $\kappa > \delta + 1$ and $h \in (0,1)$ in an arbitrary way and take for all $s, t \in R^1$, $s < t$,

$$H_{h,x}(s,t) = \ln \inf_{x \in U_\delta(vt)} E_x \exp\left\{ \int_0^{t-s} \xi_h(W_{s_1})\, ds_1 \right\} \chi_{W_{t-s} \in U_\delta(vs),\, \sigma_{t,\kappa} > t-s}$$

and

$$H(s,t) = \ln \inf_{x \in U_\delta(vt)} E_x \exp\left\{\int_0^{t-s} \xi(W_{s_1})\,ds_1\right\} \chi_{W_{t-s} \in U_\delta(vs)} \cdot$$

Since $I\left(\frac{1}{v}\right) > 0$ for $v > v^*$, by Lemma 6.1, there is a random variable $t_0 > 0$ such that for all $t \geq t_0$ and all $x \geq vt - \kappa$, the function $u(t,x)$ does not exceed h and hence $c(x, u(t,x)) \geq \xi_h(x)$. Therefore, on the basis of the Markov property of the Wiener process we deduce from equation (4.3) the bound:

$$u(t,vt) \geq E_{vt} \exp\left\{\int_0^{t-t_0} \xi_h(W_s)\,ds\right\} \chi_{W_{t-t_0} \in U_\delta(vt_0), \sigma_{t,\kappa} > t-t_0} \times$$

$$\times \inf_{x \in U_\delta(vt)} E_x\, g(W_{t_0}) \,.$$

Since by Condition (A 2) $\inf_{x \in U_\delta(vt_0)} E_x\, g(W_{t_0}) > 0$, we conclude from this that

$$\varliminf_{t \to \infty} \frac{1}{t} \ln u(t,vt) \geq \varliminf_{t \to \infty} \frac{1}{t} H_{h,\kappa}(t_0,t) \,. \tag{6.5}$$

The random function $H_{h,\kappa}(s,t)$ is a semi-additive process. More precisely, the function $H_{h,\kappa}(s,t)$ is continuous in s and t and possesses, with probability one, the following properties (Kingman [1]):

(a) A constant C_1 exists such that

$$H_{h,\kappa}(s,t) \geq -C_1(t-s) \,,$$

for all $s, t \in R^1$, $s < t$.

(b) For any $s, t, u \in R^1$ such that $s < u < t$,

$$H_{h,\kappa}(s,t) \geq H_{h,\kappa}(s,u) + H_{h,\kappa}(u,t) \,.$$

(c) For any $s, t, T \in R^1$, $s < t$,

$$\theta_{vT} H_{h,\kappa}(s,t) = H_{h,\kappa}(s+T, t+T) .$$

Property (a) is not difficult to obtain via the bound

$$H_{h,\kappa}(s,t) \geq \ln \inf_{x \in U_\delta(vt)} P_x\{W_{t-s} \in U_\delta(vs), \sigma_{t,\kappa} > t-s\} .$$

The semi-additivity (b) follows from the definition of the function $H_{h,\kappa}(s,t)$ if one makes use of the Markov property of the Wiener process.

Property (c) is obvious.

Properties (a)-(c) ensure that

$$\varliminf_{t\to\infty} \frac{1}{t} H_{h,\kappa}(t_0,t) = \varliminf_{t\to\infty} \frac{1}{t} H_{h,\kappa}(0,t) \quad (\hat{P}\text{-a.s.})$$

and the "ergodic theorem":

$$\varliminf_{t\to\infty} \frac{1}{t} H_{h,\kappa}(0,t) = \sup_{t>0} \hat{E}\left[\frac{1}{t} H_{h,\kappa}(0,t) | \mathcal{F}^f_{inv}\right] \tag{6.6}$$

holds $\hat{P}$-a.s.. Recall that $\mathcal{F}^f_{inv}$ is the σ-field of the events generated by the random function $f(x,u)$, that are invariant with respect to translations in x. With the help of inequality (6.5) and the foregoing relations we arrive at the bound

$$\varliminf_{t\to\infty} \frac{1}{t} \ln u(t,vt) \geq \sup_{t>0} \hat{E}\left[\frac{1}{t} H_{h,\kappa}(0,t) | \mathcal{F}^f_{inv}\right]$$

In a manner analogous to the proof of Lemma 3.3 for the (non-random) subadditive function $p_{h,\eta}(t)$, taking into account (A 4), we conclude via the limit passage $h \downarrow 0$, $\kappa \uparrow \infty$, that

$$\varliminf_{t\to\infty} \frac{1}{t} \ln u(t,vt) \geq \sup_{t>0} E\left[\frac{1}{t} H(0,t) | \mathcal{F}^f_{inv}\right] . \tag{6.7}$$

Since the function $H(s,t)$ has Properties (a)-(c) too, we conclude that relation (6.6) is also valid for the function $H(0,t)$ in place of $H_{h,\kappa}(0,t)$. Thus, (6.7) at last leads to the inequality

$$\varliminf_{t\to\infty} \frac{1}{t} \ln u(t,vt) \geq \varliminf_{t\to\infty} \frac{1}{t} H(0,t)$$

which easily implies (6.4), completing the proof. □

Considering the random variables $\eta^t = t^{-1}\tau_0$ with respect to the probability measures P^t defined by equality (5.1) for $z_0 = \overline{g}$, it is possible (by large deviations theorems) to estimate from below the expression on the right-hand side of (6.4) and thereby to obtain a bound opposite to bound (6.3) in Lemma 6.1. In this case limit (5.2) and the action function for (η^t, P^t) have the form

$$G(z) = \mu(z+\overline{g}) - \mu(\overline{g}), \quad z \in R^1, \tag{6.8}$$

$$S(y) = I(y) - y\overline{g} + \mu(\overline{g}), \quad y \in R^1, \tag{6.9}$$

respectively (see §7.1). We now proceed to state the precise result.

LEMMA 6.3. *For any* $v > v^*$, *the bound*

$$\varliminf_{t\to\infty} \frac{1}{t} \ln u(t,vt) \geq -vI\left(\frac{1}{v}\right) \tag{6.10}$$

holds $\hat{P}$-a.s..

Proof. First we establish that for any $v > 0$

$$\lim_{t\to\infty} \frac{1}{t} P^t\left\{\frac{\tau_0}{t} \leq \frac{1}{v}\right\} \geq -S\left(\frac{1}{v}\right). \tag{6.11}$$

By Theorem 5.1, the function $G(z)$ is differentiable for $z < 0$, and the derivative $G'(z) = \mu'(z+\overline{g})$ runs over all the points of the interval $(0, \mu'(\overline{g}))$, where by the definition $\mu'(\overline{g}) = \lim_{z\uparrow\overline{g}} \mu'(z)$. Therefore, in the case when $v^{-1} \le \mu'(\bar{g})$, statement (6.11) follows from Theorem 1.2. However, if $v^{-1} > \mu^1(\bar{g})$ then as it is to be demonstrated just now, $S(v^{-1}) = S(\mu'(\bar{g})) = 0$ and bound (6.11) results from the corresponding bound for $v^{-1} = \mu'(\bar{g})$. Statement ii) of Theorem 5.1 implies that for $v^{-1} \ge \mu'(\bar{g})$ the function $Q(z) = v^{-1}z - G(z) = v^{-1}z - \mu(z+\bar{g}) + \mu(\bar{g})$ is monotone non-decreasing for $z \le 0$ and hence actually

$$S(v^{-1}) = \sup_{z<0} Q(z) = Q(0) = 0 .$$

Taking into account the definition of probability measures P^t and using statement (i) of Theorem 5.1, relation (6.11) can be rewritten as follows

$$\varliminf_{t\to\infty} \frac{1}{t} \ln E_{vt} \exp\left\{\int_0^{\tau_0} [\xi(W_s)+\bar{g}]\,ds\right\} \chi_{\tau_0 \le t} \ge \ge -v\left[S\left(\frac{1}{v}\right) - \mu(\bar{g})\right] . \tag{6.12}$$

Next, by using the strong Markov property of the Wiener process we get for any $\delta > 0$ and $t > 0$ the bound

$$E_{vt} \exp\left\{\int_0^t \xi(W_s)\,ds\right\} \chi_{W_t \in U_\delta(0)} \ge$$

$$\ge e^{-\bar{g}t} E_{vt} \exp\left\{\int_0^{\tau_0} [\xi(W_s)+\bar{g}]\,ds\right\} \chi_{\tau_0 \le t} \times \tag{6.13}$$

$$\times \inf_{0\le s\le t} E_0 \exp\left\{\int_0^{t-s} [\xi(W_{s_1})+\bar{g}]\,ds_1\right\} \chi_{W_{t-s} \in U_\delta(0)} .$$

The second factor on the right-hand side of inequality (6.13) is bounded from below with limit relation (6.12), and the last factor (by Theorem 5.2) equals 1 up to logarithmic equivalence. Further, utilizing Lemma 6.2, as the final result we obtain the bound

$$\varliminf_{t\to\infty} \frac{1}{t} \ln u(t,vt) \geq -\bar{g} - v\left[S\left(\frac{1}{v}\right) - \mu(\bar{g})\right]$$

for any $v > v^*$. This and (6.9) together imply the claim of the lemma. □

As in Lemma 6.2, we take for any $h > 0$

$$\xi_h(x) = \inf_{0<u<h} c(x,u) .$$

We shall also need the following condition:

(B) For any $h \in (0,1)$ and all $v \in R^1$

$$\eta_{v,h} = \overline{\lim_{t\to\infty}} \frac{1}{t} \ln E_{vt} \exp\left\{-\int_0^t \xi_h(W_s)ds\right\} < 0, \quad (\hat{P}\text{-a.s.}) .$$

Now we shall state the main result on wave front propagation for problem (4.1).

THEOREM 6.1. (i) *For any* $v > v^*$,

$$\lim_{t\to\infty} \sup_{x\geq vt} u(t,x) = 0 \quad (\hat{P}\text{-a.s.}) .$$

(ii) *If Condition (B) is fulfilled, then for any* $v \in (0,v^*)$

$$\lim_{t\to\infty} \inf_{0\leq x\leq vt} u(t,x) = 1 \quad (\hat{P}\text{-a.s.}) .$$

Proof. Since $I\left(\frac{1}{v}\right) > 0$ for $v > v^*$, statement i) follows immediately from Lemma 6.1. The proof of statement ii) is basically arranged similarly to the demonstration of the corresponding statement of Theorem 3.1. We first prove that for any $v \in (0,v^*)$

$$\varliminf_{t\to\infty} u(t,vt) \geq 1 \quad (\hat{P}\text{-a.s.}) \tag{6.14}$$

Let us pick $\delta > 0$ in an arbitrary way. Then by Lemma 6.3, one can find (perhaps, random) numbers $v^+ > v^*$ and $t_0 > 0$ such that

$$u(t,v^+t) \geq e^{-\delta t} \quad \text{for} \quad t \geq t_0 .$$

We shall introduce for arbitrary possible numbers t, h, η the Markov times:

$$\sigma^+(t) = \min\{s \geq 0 : W_s = v^+ \cdot (t-s)\} ,$$

$$\sigma_h(t) = \min\{s \geq 0 : u(t-s, W_s) \geq h\} ,$$

$$\tau_\eta(t) = \min\{s \geq 0 : |W_s - W_0| \geq \eta t\} .$$

In a manner analogous to how it was done when proving Theorem 3.1, one can deduce from equation (4.3) the following bound

$$u(t,x) \geq h\, P_x\{\sigma_h(t) \leq t\} , \tag{6.15}$$

for any $h \in (0,1)$, $x \in R^1$ and $t > 0$. Thus it is sufficient to show that for any $h \in (0,1)$ and some $\eta > 0$

$$P_{vt}\{\sigma_h(t) > t, \tau_\eta(t) > t\} \to 0 \tag{6.16}$$

as $t \to \infty$. Choose η and κ so that $0 < \eta < (v^* - v) \wedge v$ and $0 < \kappa < v - \eta$. Just as in the proof of Theorem 3.1, for any $h \in (0,1)$ and for sufficiently large t, we derive the bound

$$P_{vt}\{\sigma_h(t) > t, \tau_\eta(t) > t\} \leq P_{vt}\{\kappa t < \sigma^+(t) \leq \sigma_h(t) \wedge t\} \leq$$

$$\leq e^{\frac{\delta t}{2}} [u(t,vt)]^{1/2} \left[E_{vt} \exp\left\{ - \int_0^{\kappa t} \xi_h(W_s)\, ds \right\} \right]^{1/2} .$$

This implies that (6.16) is valid on the set $\{\eta_{v\kappa^{-1}h} < -\delta\}$ ($\eta_{v,h}$ has been specified in Condition (B)). Since $\delta > 0$ can be chosen arbitrarily small and κ does not depend on δ, we conclude that by Condition (B) convergence (6.16) holds for any $h \in (0,1)$ and for the above selected η $\hat{P}$-a.s.. We have thereby established that relation (6.14) is correct for any $v \in (0,v^*)$ P-a.s.. In a similar way one can find a $v_* < 0$ such that (6.14) is true for $v \in (v_*, 0)$ as well. Let $\underline{v} \in (v_*, 0)$, $\overline{v} \in (0, v^*)$ and $h \in (0,1)$ be chosen in an arbitrary fashion. From (6.14) it follows that for some $t_1 > 0$ and all $t \geq t_1$

$$u(t,\underline{v}t) \geq h \quad \text{and} \quad u(t,\overline{v}t) \geq h \quad (\hat{P}\text{-a.s.}) .$$

Therefore, for $t > t_1$

$$\sup_{x\in[\underline{v}t,\overline{v}t]} P_x\{\sigma_h(t) > t\} \leq \leq \sup_{x\in[\underline{v}t,\overline{v}t]} P_x\{W_{t-t_1} \in (\underline{v}t_1, \overline{v}t_1)\} . \tag{6.17}$$

The expression on the right-hand side of this inequality tends to zero as $t \to \infty$. This, together with bound (6.15), ensures

$$\varliminf_{t\to\infty} \inf_{x\in[\underline{v}t,\overline{v}t]} u(t,x) \geq h \quad (\hat{P}\text{-a.s.}) .$$

As $\underline{v} \in (v_*, 0)$, $\overline{v} \in (0, v^*)$ and $h \in (0,1)$ are picked arbitrarily, this and (4.7) together establish the claim of Theorem 6.1. □

In conclusion we give a few remarks.

1. Theorem 6.1 treated the behavior of the function $u(t,x)$ only in the half-plane $x \geq 0$. If the initial function $g(x)$ of problem 4.1 equals 1 for $x \leq \tilde{\eta}$, where $\tilde{\eta}$ is a random variable, then a stronger result holds: for any $v < v^*$

$$\lim_{t\to\infty} \inf_{-\infty<x<vt} u(t,x) = 1 .$$

To check this one only must replace bound (6.17) in the proof of Theorem 6.1 by the inequality

$$\sup_{x\leq \bar{v}t} P_x\{\sigma_h(t)>t\} \leq \sup_{x\leq \bar{v}t} P_x\{W_{\sigma_\Gamma(t)} \in [\tilde{\eta}\wedge 0, \bar{v}t]\},$$

where $\sigma_\Gamma(t)$ is the first hitting time of the process $(t-s, W_s)$, $s \geq 0$, to the set $\Gamma = \{(t,\bar{v}t) : t \geq 0\} \cup \{(0,x) : x \leq 0\}$.

2. The asymptotic wave front velocity v^* is the tangent of the angle between the vertical axis and the tangent to the graph of the function $\mu(z)$, passing through the origin. We can give examples showing that the point of tangency may coincide with the discontinuity point $(\bar{g}, \mu(\bar{g}))$ of the function $\mu(z)$.

3. If $\xi(x) \equiv c > 0$, then $\mu(z) = -\sqrt{-2(c+z)}$ for $z \leq -c$ and $\mu(z) = \infty$ for $z > -c$ so that $v^* = \sqrt{2c}$. This is a known result by Kolmogorov-Petrovskii-Piskunov (see Chapter 6). Let now $\xi(x) = c(x+\eta)$, where $c(x)$ is a periodic continuous positive function of period 1 and η is a random variable uniformly distributed in the unit interval $[0,1]$. In this case the function $\mu(z)$ coincides with the inverse function of $-\lambda(-z)$, $z \leq 0$, where $\lambda(z)$ is the eigenvalue of the differential operator

$$L^z = \frac{1}{2}\frac{d^2}{dx^2} - z\frac{d}{dx} + \left[c(x) + \frac{z^2}{2}\right],$$

corresponding to a positive and periodic eigenfunction of period 1. This agrees with results of §7.3.

4. Rather than problem (4.1) we can consider the Cauchy problem for a more general equation

$$\frac{\partial u}{\partial t} = Lu + f(x,u)$$

where

$$L = \frac{1}{2}a(x)\frac{\partial^2}{\partial x^2} + b(x)\frac{\partial}{\partial x}$$

is an elliptic differential operator with stationary random coefficients. To be more exact, $\Theta(x,u) = (a(x), b(x), f(x,u))$ is a random vector function on a probability space $(\hat{\Omega}, \hat{\mathcal{F}}, \hat{P})$ which is measurable and stationary in x. Suppose that $a(x) \geq \text{const} > 0$ and let the functions $a(x)$ and $b(x)$ be bounded and Hölder continuous $\hat{P}$-a.s.. The results of §§7.4-7.6 can easily be carried over to this more general statement of the problem. Here the function $\mu(z)$ defining the asymptotic wave front propagation velocity has the form

$$\mu(z) = \hat{E}\left[\ln E_1 \chi_{\tau_0<\infty} \exp\left\{\int_0^{\tau_0} [\xi(X_s)+z]ds\right\} | \mathcal{F}^{\Theta}_{\text{inv}}\right],$$

where (X_t, P_x) is a diffusion process with a (random) generator L, τ_0 is the first hitting time of the process X_t to the point 0 and $\mathcal{F}^{\Theta}_{\text{inv}}$ is the σ-field of invariant events of the stationary vector function $\Theta(x) = (a(x), b(x), \xi(x))$. The function $\mu(z)$ may, generally speaking, take positive values and its discontinuity point $\bar{g}$ may be positive as well. Since in the general case the "average" drift of the process (X_t, P_x) is not equal to zero, we conclude that for any $c \in (0,1)$ the region $\bar{G}_c(t) = \{x \in R^1 : u(t,x) < c\}$ can expand both in the direction of $+\infty$ and in the direction of $-\infty$ as $t \to \infty$. In connection with this, the equation $I(v^{-1}) = 0$ may have either two or no positive solution $v = v^*$. These cases of course require some reformulations of the statements of §7.5 and §7.6.

§7.7 *The function $\mu(z)$ and the one-dimensional Schrödinger equation with random potential*

This section shows how in a number of cases (with the exception of those already treated at the end of the previous section) the function $\mu(z)$ defined by equality (5.3) can be calculated in a more explicit form. We will assume the non-negative random function $\xi(x)$ to be metrically transitive and bounded $\hat{P}$-a.s..

Let $z < \bar{g}$ ($\bar{g}$ is a discontinuity point of the function $\mu(z)$). Then the function

$$u(x) = E_x \exp\left\{ \int_0^{\tau_0} [\xi(W_s)+z]\,ds \right\}, \quad x \geq 0, \tag{7.1}$$

satisfies the equation

$$\frac{1}{2}\frac{d^2u(x)}{dx^2} + [\xi(x)+z]\,u(x) = 0. \tag{7.2}$$

Recall that just as in the proof of assertion (iii) of Theorem 5.1, the function $u(x)$ is continuously differentiable, and in general the derivative $\frac{du}{dx}$ is only absolutely continuous, equation (7.2) being merely fulfilled a.e. with respect to the Lebesgue measure. Now one could make use of some ideas of Furstenberg for the Schrödinger equation (7.2) with random potential (Furstenberg [1], Molchanov [2]) and compute $\mu(z)$ as the limit of the function $\frac{1}{x}\ln u(x)$ as $x \to \infty$. Since we can use the representation (7.1) of the desired solution in addition to equation (7.2), we will act in a slightly different way. This will enable us to get rid of unnecessary technical restrictions. For any $h > 0$, the function

$$\phi_h(x) = -\frac{1}{h}[\ln u(x+h) - \ln u(x)] = -\frac{1}{h} E_{x+h} \exp\left\{ \int_0^{\tau_x} [\xi(W_s)+z]\,ds \right\},$$

$x \geq 0$, is $\mathcal{F}^\xi$-measurable (and metrically transitive), and the vector function $(\phi_h(x), \xi(x))$ is stationary. Consequently, the same is true of the function $\phi(x) = -\frac{d}{dx}\ln u(x)$ in place of $\phi_h(x)$. With the help of this function, Schrodinger's equation (7.2) can be reduced to the Riccati equation

$$\frac{d\phi(x)}{dx} = [\phi(x)]^2 + 2[\xi(x)+z]. \tag{7.3}$$

The function $\phi(x)$ can be extended to a solution of equation (7.3) on the entire real line preserving the $\mathcal{F}^\xi$-measurability and stationarity of

$(\phi(x), \xi(x))$. However this solution considered for $x \geq 0$ is "unstable" with respect to perturbations of the initial point $\phi(0)$. Therefore it may turn out to be more convenient to treat the function $\bar{\phi}(x) = \phi(-x)$, $x \in R^1$, rather than $\phi(x)$. Then equation (7.3) turns into the equation

$$\frac{d\bar{\phi}(x)}{dx} = -[\bar{\phi}(x)]^2 - 2[\bar{\xi}(x) + z], \tag{7.4}$$

where $\bar{\xi}(x) = \xi(-x)$. By the definition of the function $\bar{\phi}(x)$,

$$\ln u(x) = -\int_{-x}^{0} \bar{\phi}(y)\, dy .$$

Therefore,

$$\mu(z) = \hat{E} \ln u(1) = -\hat{E}\bar{\phi}(0) .$$

We observe that $\hat{E}\bar{\phi}(0) > 0$ for $z < \bar{g}$.

Let now $\tilde{\phi}(x)$, $x \in R^1$, be a metrically transitive stationary solution of equation (7.4) with $\hat{E}\tilde{\phi}(0) > 0$. Then the function

$$\tilde{u}(x) = \exp\left\{-\int_{-x}^{0} \tilde{\phi}(y)\, dy\right\}, \quad x \geq 0$$

obeys equation (7.2). Since $\hat{E}\tilde{\phi}(0) > 0$, by the ergodic theorem, $\tilde{u}(x)$ is bounded from above. Hence we derive that

$$\tilde{u}(x) = E_x \exp\left\{\int_0^{\tau_0 \wedge t} [\xi(W_s) + z]\, ds\right\} \tilde{u}(W_{\tau_0 \wedge t}) \geq$$

$$\geq E_x \exp\left\{\int_0^{\tau_0} [\xi(W_s) + z]\, ds\right\} \chi_{\tau_0 \leq t}$$

for any $t > 0$. Thus $E_x \exp\{\int_0^{\tau_0} [\xi(W_s)+z]ds\}$ is finite $\hat{P}$-a.s.. Consequently, $z \le \overline{g}$ (see step 6 of the proof of Theorem 5.2). For such z, we have already constructed the solution $\overline{\phi}(x)$ of equation (7.4). Let us prove that the functions $\widetilde{\phi}(x)$ and $\overline{\phi}(x)$ coincide. The function $\psi(x) = \widetilde{\phi}(x) - \overline{\phi}(x)$ satisfies the equation:

$$\frac{d\psi(x)}{dx} = -[\widetilde{\phi}(x)+\overline{\phi}(x)]\psi(x) ,$$

so that

$$\psi(x) = \psi(0) \exp\left\{-\int_0^x [\widetilde{\phi}(y)+\overline{\phi}(y)]dy\right\} .$$

Since $\hat{E}[\widetilde{\phi}(y)+\overline{\phi}(y)] > 0$, we conclude on the basis of the ergodic theorem that $\psi(x)$ tends to zero (exponentially fast) $\hat{P}$-a.s. as $x \to \infty$. This and the fact that $\psi(x)$ is stationary, yield that $\psi(x) \equiv 0$, i.e. $\widetilde{\phi}(x) \equiv \overline{\phi}(x)$. We now proceed to summarize.

THEOREM 7.1. *For any* $z < \overline{g}$, *there exists a unique metrically transitive stationary solution* $\overline{\phi}(x) = \overline{\phi}^z(x)$, $x \in R^1$, *of equation (7.4) with* $\hat{E}\,\overline{\phi}^z(0) > 0$. *Moreover,*

$$\mu(z) = -\hat{E}\,\overline{\phi}^z(0)$$

There is no such a solution for $z > \overline{g}$.

Consider the example. Let $\eta(x)$, $x \ge 0$, be a Markov process on some state space $(\mathcal{E}, \mathcal{B})$ with a unique normalized invariant measure. Assume that the stationary process $\overline{\eta}(x)$, $x \in R^1$, corresponding to this measure, is metrically transitive. We set $\overline{\xi}(x) = f(\overline{\eta}(x))$, where f is a nonnegative bounded measurable function. For some $z \in R^1$, we consider the Markov process $(\overline{\phi}^z(x), \eta(x))\,;P_{\overline{\phi},\eta})$, on the state space $K \times \mathcal{E}$ (K is a bounded Borel subset of R^1, $\overline{\phi}^z(x)$ satisfying equation (7.4)). Suppose that this process has a normalized invariant measure $\nu^z(d\phi, d\eta)$

such that $\iint_{K\times\mathfrak{S}} \phi\,\nu^z(d\phi, d\eta) > 0$ and the corresponding stationary process is metrically transitive. Then by Theorem 7.1, $z < \overline{g}$ and

$$\mu(z) = -\iint_{K\times\mathfrak{S}} \phi\,\nu^z(d\phi, d\eta) \tag{7.5}$$

We shall now provide a particular case. Let $\overline{\xi}(x)$ be a stationary Markov process with two states a and b $(0 \leq a < b)$ and with the corresponding intensities λ and μ, and $f(x) \equiv x$. For $z < -b$, the random variable $(\overline{\phi}(x), \eta(x))$ has $[\beta(z), \alpha(z)] \times \{a,b\}$ as its state space, where $\alpha(z) = -\sqrt{2(a+z)}$, $\beta(z) = -\sqrt{2(b+z)}$, and it meets all the above requirements. The densities $p(\phi)$ and $q(\phi)$ defined by the equalities $p(\phi)\,d\phi = \nu^z\,(d\phi, a)$ and $q(\phi)\,d\phi = \nu^z\,(d\phi, b)$ are solutions of the Kolmogorov equations

$$\frac{\partial}{\partial\phi}\,[(\alpha^2(z)-\phi^2)\,p^z(\phi)] + \lambda\,p^z(\phi) - \mu q^z(\phi) = 0\,,$$

$$\frac{\partial}{\partial\phi}\,[(\phi^2-\beta^2(z))\,q^z(\phi)] + \lambda\,p^z(\phi) - \mu q^z(\phi) = 0\,, \tag{7.6}$$

$$\beta(z) < \phi < \alpha(z)\,,$$

with the boundary conditions

$$\lim_{\phi\downarrow\beta(z)} p^z(\phi) = 0, \quad \lim_{\phi\uparrow\alpha(z)} (\alpha(z)-\phi)\,p^z(\phi) = 0\,,$$
$$\lim_{\phi\downarrow\beta(z)} q^z(\phi)(\phi-\beta(z)) = 0, \quad \lim_{\phi\uparrow\alpha(z)} q^z(\phi) = 0\,. \tag{7.7}$$

Problem (7.6) - (7.7) is solved in the explicit form:

$$p^z(\phi) = \frac{D(z)}{\alpha^2(z)-\phi^2}\,H^z(\phi)\,, \quad q^z(\phi) = \frac{D(z)}{\phi^2-\beta^2(z)}\,H^z(\phi)\,, \tag{7.8}$$

where

$$H^z(\phi) = \left(\frac{\alpha(z)-\phi}{\alpha(z)+\phi}\right)^{\frac{\lambda}{2\alpha(z)}} \cdot \left(\frac{\phi-\beta(z)}{\phi+\beta(z)}\right)^{\frac{\mu}{2\beta(z)}} \tag{7.9}$$

and $D(z)$ is a normalization factor such that

$$\int_{\beta(z)}^{\alpha(z)} (p^z(\phi)+q^z(\phi))\,d\phi = 1 \ . \tag{7.10}$$

It is easily understood that for $z > -b$, any solution of equation (7.4) explodes, i.e. it goes to infinity in finite time. Therefore, using (7.5) we obtain finally

$$\mu(z) = \begin{cases} -\int_{\beta(z)}^{\alpha(z)} \phi\,[p^z(\phi)+q^z(\phi)]\,d\phi\,, & z \le -b \\ +\infty & , \ z > -b \end{cases} \tag{7.11}$$

the functions $p^z(\phi)$ and $q^z(\phi)$ being defined by (7.8)-(7.10). Moreover, note that the integral on the right-hand side of (7.11) can be expressed by hypergeometric functions (Ryzyk and Gradstein [1]).

LIST OF NOTATIONS

δ^i_j, δ^{ij}, δ_{ij} : the Kronecker delta

$[a]$: the integer part of a number a

$\downarrow$: decreasing convergence

$a \wedge b = \min(a,b)$; $a \vee b = \max(a,b)$

R^r : r-dimensional Euclidean space

$|x| = \sqrt{\sum_1^r (x^i)^2}$, $x = (x^1, \cdots, x^r) \in R^r$

$\rho(a,b) = |a-b|$; $a,b \in R^r$ $(a,b) = \sum_1^r a^i b^i$: scalar product $(a,b \in R^r)$.

$U_\delta(a) = \{x \in R^r : |x-a| < \delta\}$

∂D : the boundary of a domain $D \subset R^r$

σ^* : the matrix adjoint to a matrix σ

$\mathrm{Tr}\, a$: the trace of a matrix a

$\mathcal{B}^r$: the Borel σ-field in R^r

$C_A(B)$: the space of continuous functions on a topological space A with values in a topological space B.

$C_{0,T}(R^r)$: the space of continuous functions in $[0,T]$ with values in R^r; $C^x_{0,T}(R^r) = \{\phi \in C_{0,T}(R^r) : \phi_0 = x\}$, $x \in R^r$.

$\|f\| = \sup_{0 \le s \le T} |f_s|$: norm in $C_{0,T}(R^r)$

$\rho_{0T}(\phi, \psi) = \|\phi - \psi\|$; $\phi, \psi \in C_{0,T}(R^r)$

$\mathcal{N}$: σ-field generated by cylinder sets in $C_{0,\infty}(R^r)$

$\mathcal{N}_t$: σ-field generated by cylinder sets in $C_{0,t}(R^r)$

$C^{k,\delta} = C^{k,\delta}_{R^r}(R^1)$: the space of real-valued functions in R^r having partial derivatives up to k-th order inclusively, k-th order derivatives being Hölder continuous with the exponent δ.

$(\mathcal{E}, \mathcal{B})$: measurable space (a set $\mathcal{E}$ with an indicated σ-field of subsets $\mathcal{B}$)

B : the Banach space of bounded measurable functions on $(\mathcal{E}, \mathcal{B})$; $\|f\| = \sup_{x \in \mathcal{E}} |f(x)|$, $f \in B$

T_t, $t \geq 0$: a semi-group of operators in B

A : the infinitesimal operator of a semi-group

D_A : the domain of definition of an operator A

[G] : the closure of a set G

(G) : the totality of the interior points of a set G

∇f : the gradient of a function $f(x)$, $x \in R^r$

Δ : the Laplace operator

L : differential operator

$S_{0T}(\phi)$: normed action functional

Ω : the space of elementary events

$\chi_A(\omega)$: the indicator of a set $A \subset \Omega$; $\omega \in \Omega$

$\mathcal{F}, \mathcal{N}, \mathcal{N}_t$: σ-fields in Ω

P : probability measure in a space Ω

E : mathematical expectation with respect to a measure P

D : variance

P_x : probability measure in $C_{0,\infty}(\mathcal{E})$ concentrated on the functions ϕ for which $\phi_0 = x \in \mathcal{E}$

E_x : expectation with respect to the measure P_x

W_t, $t \geq 0$: Wiener process

(X_t^x, P) : Markov family

(X_t, P_x) : Markov process

τ_D^x : the first exit time of a Markov random function X_t^x from a domain D

τ_D : the first exit time of a process (X_t, P_x) from D

ξ_t^x : local time

$H^2([a,b]\times\Omega)$: the space of functions $f(s,\omega), s \in [a,b], \omega \in \Omega$, for which
$\int_a^b Ef^2(s,\omega)ds < \infty$

$\bar{H}^2_{a,b}$: the space of progressively measurable functions $f(x,\omega) \in H^2([a,b]\times\Omega)$; for the functions from $\bar{H}^2_{a,b}$, the stochastic integral is defined in the book.

$L^2(\Omega) = \{\xi(\omega) : E|\xi(\omega)|^2 < \infty\}$

$\Longrightarrow$: implication sign

$\square$: the end of a proof.

$\emptyset$: the empty set

P-a.a. (P-a.e.) : almost all (almost everywhere) with respect to the measure P.

REFERENCES

R. F. Anderson and S. Orey

[1] 1976. Small random perturbations of dynamical systems with reflecting boundary. Nagoya Math. J., 60, 189-216.

D. G. Aronson and H. F. Weinberger

[1] 1975. Non-linear diffusion in population genetics, combustion and nerve propagation. Proceedings of the Tulane program in partial differential equations (Lecture Notes in Mathematics), Springer-Verlag: Berlin-Heidelberg-New York.

[2] 1978. Multi-dimensional non-linear diffusion in population genetics. Advances in Math., 30, 33-76.

A. Bensoussan, J. L. Lions and G. C. Papanicolaou

[1] 1978. Asymptotic analysis for periodic structures. Amsterdam-New York-Oxford: North-Holland Publishing Company.

Yu. N. Blagoveščenskii

[1] 1964. The Cauchy problem for quasi-linear parabolic equations in the degenerate case. Prob. Theory and Appl., 378-382 (Russian).

Yu. N. Blagoveščenskii and M. I. Freidlin

[1] 1961. Some properties of diffusion processes depending on parameter. Dokl. Akad. Nauk SSSR., 138, 508-511 (Russian).

R. M. Blumenthal

[1] 1957. An extended Markov property. Trans. Amer. Math. Soc., 85, 52-72.

R. M. Blumenthal and R. K. Getoor

[1] 1968. Markov processes and potential theory. New York-London: Academic Press.

A. A. Borovkov

[1] 1967. Boundary problems for random walks and large deviations in functional spaces. Prob. Theory and Appl., 12, 635-654.

A. N. Borodin

[1] 1977. A limit theorem for solutions of differential equations with random right-hand side. Prob. Theory and Appl., 22, 498-512.

M. D. Bramson

[1] 1978. Maximal displacement of branching Brownian motion. Comm. Pure Appl. Math., 31, 531-582.

R. H. Cameron and W. T. Martin

[1] 1944. Transformation of Wiener integrals under translations. Ann. Math., 45, 386-396.

M. Cranston, S. Orey and U. Rösler

[1] 1982. Exterior Dirichlet problems and the asymptotic behavior of diffusion. Preprint.

C. Delacherie et P. A. Meyer

[1] 1980. Probabilités et Potentiel. Théorie des Martingales. Herman.

M. D. Donsker

[1] 1951. An invariant principle for certain probability limit theorems. Mem. Amer. Math. Soc., 6.

J. L. Doob

[1] 1953. Stochastic Processes. New York: John Wiley.

[2] 1957. Conditional Brownian motion and the boundary limits of harmonic function. Bull. Soc. Math. France, 85, 431-458.

[3] 1959. Discrete potential theory and boundaries. J. Math. Mech., 8, 433-458.

E. B. Dynkin

[1] 1960. Theory of Markov processes. Oxford: Pergamon Press.

[2] 1964. Martin boundary and non-negative solutions of a boundary value problem with oblique derivative. Uspehi Mat. Nauk, 19, 3-50.

[3] 1965. Markov processes. I, II. Berlin: Springer-Verlag.

[4] 1982. Minimal excessive measures and functions, Trans. Amer. Math. Soc., 258, 217-244.

E. B. Dynkin and A. A. Yushkevich

[1] 1969. Markov Processes. Theorems and Problems. New York: Plenum Press.

A. Einstein

[1] 1926. Investigations on the theory of the Brownian movement. London: Methuen.

R. P. Feynman and A. Hibbs

[1] 1965. Quantum Mechanics and Path Integrals. New York: McGraw-Hill.

G. Fichera

[1] 1956. Sulla equazioni differentiali lineari ellipticо-paraboliche del secondo ordine. Atti Acc. Naz. Lincei Mem., Ser. 8, 5, 1-30.

[2] 1960. On a Unified Theory of Boundary Value Problems for Elliptic-Parabolic Equations of Second Order. Boundary Value Problems in Differential Equations. Univ. of Wisconsin Press, Madison, 97-120.

R. A. Fisher

[1] 1937. The wave of advance of advantageous genes. Ann. of Eugenics, 7, 355-369.

P. C. Fife

[1] 1978. Asymptotic States of Equations of Reaction and Diffusion. Bull. Amer. Math. Soc., 84, 693-724.

[2] 1979. Mathematical aspects of reacting and diffusing systems. Lecture Notes in Biomath., 28, Springer-Verlag.

P. C. Fife and J. B. McLeod

[1] 1977. The approach of solutions of non-linear diffusion equations to travelling wave solutions. MRC Technical Summary Report No. 1736.

D. A. Frank-Kamenezkii

[1] 1969. Diffusion and heat transfer in chemical kinetics. New York: Plenum Press.

M. I. Freidlin

[1] 1962. On stochastic equations of K. Ito and degenerate elliptic equations. Izv. Akad. Nauk SSSR, Ser. Mat., 26, 653-676, (Russian).

[2] 1963. Diffusion processes with reflection and the third boundary value problem. Prob. Theory and Appl., 8, 80-88 (Russian).

[3] 1964. Dirichlet's problem for an equation with periodic coefficients. Prob. Theory and Appl., 9, 133-139 (Russian).

[4] 1964. On a priori bounds for solutions of degenerate elliptic equations. Dokl. Acad. Nauk SSSR, 158, 281-283 (Russian).

[5] 1966. On the formulation of boundary value problems for degenerate elliptic equations. Dokl. Acad. Nauk SSSR, 170, 282-285 (Russian).

[6] 1967. Elliptic equations in unbounded regions. Dokl. Acad. Nauk SSSR, 172, 1286-1289 (Russian).

[7] 1967. Markov processes and differential equations. Ser. "Itogi Nauki", Moscow (Russian).

[8] 1967. Quasi-linear parabolic equations and measures on a functional space. Functional Anal. i Prilozen., 1.3, 74-82 (Russian).

[9] 1967. A certain class of degenerate quasi-linear equations. Dokl. Acad. Nauk SSSR, 177, 1015-1018 (Russian).

[10] 1968. On factorization of a non-negative definite matrix. Prob. Theory and Appl., 13, 375-378 (Russian).

[11] 1968. The smoothness of the solutions of degenerate elliptic equations. Izv. Acad. Nauk SSSR., Ser. Mat., 32, 1391-1413, (Russian).

[12] 1969. Existence "in the large" of smooth solutions of degenerate quasi-linear equations. Mat. Sbornik, 78, 332-348 (Russian).

[13] 1975. On diffusion processes with a small parameter. Trans. of the third Japan-USSR Symposium on Probability theory, Tashkent.

[14] 1978. Averaging principle and large deviations. Uspehi Mat. Nauk, 33:5, 107-160 (Russian).

[15] 1979. Propagation of concentration waves by a random motion connected with growth. Dokl. Acad. Nauk, 246, 544-548.

[16] 1981. On elliptic differential equations with small parameter. C. R. Math. Acad. Sci. Canada, III, 4, 209-214.

[17] 1982. On wave propagation in periodic media. In "Stochastic Analysis and Applications," M.A. Pinsky, Ed., Marcel Decker, New York 1984, 147-166.

[18] 1983. On wave front propagation in multi-component media. Submitted to Trans. Amer. Math. Soc., 286, 1, 181-193.

M. I. Freidlin and S. A. Sivak

[1] 1979. Small parameter method in multidimensional reaction-diffusion problem. Studia Biophysica, 76, 129-136 (GDR).

A. Friedman

[1] 1964. Partial differential equations of parabolic type. Englewood Cliffs, N. J.: Prentice-Hall.

[2] 1975. Stochastic differential equations and applications. I, II. New York-San Francisco-London: Academic Press.

H. Furstenberg

[1] 1963. Noncommuting random products. Trans. Amer. Math. Soc., 108, 377-428.

J. Gärtner

[1] 1978. On action functional for diffusion processes with small parameter in a part of higher derivatives. Preprint. Zentralinstitut für Mathematic und Mechanik, GDR, Berlin (Russian).

[2] 1980. Non-linear diffusion equations and excitable media. Dokl. Acad. Nauk SSSR, 254, 1310-1314 (Russian).

[3] 1981. On wave front propagation in random media. Math. Nachr., 100, 271-296 (GDR) (Russian).

[4] 1983. Location of Wave Fronts for the Multidimensional KPP Equation and Brownian First Exit Densities. Submitted to Math. Nachr. (GDR).

J. Gärtner and M. I. Freidlin

[1] 1978. A new contribution to the question of large deviations for random processes. Vestnic Mosc. Univ. Ser. Mat., 5, 52-59 (Russian).

[2] 1979. On the propagation of concentration waves in periodic and random media. Dokl. Acad. Nauk SSSR, 249, 521-525 (Russian).

I. M. Gelfand

[1] 1959. Some problems in the theory of quasi-linear equations. Uspehi Mat. Nauk, 14:2, 87-158 (Russian).

I. M. Gelfand and S. V. Fomin

[1] 1961. Variation calculus. Moscow: Fizmatgiz (Russian).

I. I. Gihman and A. V. Skorohod

[1] 1965. Introduction in the theory of stochastic processes. Moscow: Nauka.

[2] 1972. Stochastic differential equations. Berlin: Springer-Verlag.

[3] 1979. The theory of stochastic processes. I, II, III. Berlin: Springer-Verlag.

I. V. Girsanov

[1] 1960. On transforming a certain class of stochastic processes by absolutely continuous substitution of measures. Prob. Theory and Appl., 5, 285-301 (Russian).

[2] 1961. On a stochastic integral Ito's equation. Dokl. Acad. Nauk. SSSR, 138, 18-21 (Russian).

I. S. Gradstein and I. M. Ryzik

[1] 1963. Tables of integrals, sums, series, and products. Moscow, Fizmatgiz.

K. P. Hadeler and F. Rothe

[1] 1975. Travelling fronts in non-linear diffusion equations. Jorn. of Math. Biology, 2, 251-263.

R. Z. Has'minskii

[1] 1958. Diffusion processes and elliptic differential equations degenerating on the boundary of the region. Prob. Theory and Appl., 3, 430-461 (Russian).

[2] 1960. Ergodic properties of recurrent diffusion processes and stabilization of the solution of the Cauchy problem for parabolic equations. Prob. Theory and Appl., 5, 179-196 (Russian).

[3] 1963. On diffusion processes with small parameter. Izvestia Acad. Nauk SSSR., Ser. Mat., 27, 1281-1300 (Russian).

[4] 1966. Limit theorem for solutions of differential equation with small parameter. Prob. Theory and Appl., 11, 444-462 (Russian).

[5] 1968. On the averaging principal for stochastic differential equations. Kybernetika, Academia, Praha, 4, 260-279 (Russian).

[6] 1969. Stability of the systems of differential equations under random perturbation of their parameters. Moscow: Nauka (Russian).

E. Hille and R. S. Phillips

[1] 1957. Functional Analysis and Semi-Groups. Amer. Math. Soc.

L. Hörmander

[1] 1967. Hypoelliptic second-order differential equations. Acta Math., 119, 147-171.

G. Hunt

[1] 1956. Some theorems concerning Brownian motion. Trans. Amer. Math. Soc., 81, 294-319.

N. Ikeda and S. Watanabe

[1] 1970. On uniqueness and non-uniqueness of solutions for a class of non-linear equations and the explosion problem for branching processes. Journal of the Faculty of Science, Univ. of Tokyo, Sec. 1, 17, 187-214.

[2] 1981. Stochastic Differential equations and Diffusion Processes. Amsterdam-Oxford-New York-Tokyo: North-Holland/Kodansha.

N. Ikeda, M. Nagasawa and S. Watanabe

[1] 1968. Branching Markov processes, I, II, III, J. Math. Kyoto Univ., 8, 233-277, 365-410; 9, 97-162.

A. D. Ioffe and V. M. Tihomirov

[1] 1979. Theory of extremal problems. Amsterdam-New York: North-Holland.

K. Ito

[1] 1951. On a formula concerning differentials. Nagoya Math. J., 3, 55-65.

K. Ito and H. P. McKean, Jr.

[1] 1965. Diffusion processes and their sample paths. Berlin: Springer-Verlag.

M. Kac

[1] 1951. On some connections between probability theory and differential and integral equations. Proc. 2nd Berkeley Symp. on Math. Stat. and Prob., Univ. of California Press: Berkeley and Los Angeles.

[2] 1959. Probability and Related Topics in the Physical Sciences, New York: Wiley (Interscience).

[3] 1966. Can you hear the shape of a drum? Amer. Math. Monthly, 73, 1-23.

Ja. Kanel'

[1] 1962. On stabilization of solutions of the Cauchy problem for equations encountered in combustion theory. Mat. Sbornik, 59, 245-288 (Russian).

T. Kato

[1] 1976. Perturbation Theory for Linear Operators. Berlin and New York, 2nd ed: Springer-Verlag.

M. V. Keldysh

[1] 1951. On some cases of degenerate elliptic equations. Dokl. Acad. Nauk SSSR, 77, 181-183.

J. P. C. Kingman

[1] 1973. Subadditive ergodic theory. Annals of Probability, 1, 883-909.

F. Knight

[1] 1962. On the random walk and Brownian motion. Trans. Amer. Soc., 103, 218-228.

A. N. Kolmogoroff

[1] 1931. Über die analytishen Methoden in Wahrscheinlichkeitsrechnung, Math Ann. 415-458.

A. Kolmogorov, I. Petrovskii, and N. Piskunov

[1] 1937. Étude de l'équation de la diffusion avec croissence de la matière et son application a un problème biologique. Moscow Univ., Bull. Math., 1, 1-25.

A. P. Korostelev

[1] 1973. Probabilistic representation of the solution of the problem with oblique derivative. Prob. Theory and Appl., 18, 172-176 (Russian).

A. P. Korostelev and M. I. Freidlin

[1] 1980. On propagation of concentration waves with the help of non-linear boundary conditions. In the book "Factors of variety in mathematical ecology and population genetics." Puschino, 149-160 (Russian).

N. V. Krylov

[1] 1966. On regular boundary points for Markov processes. Prob. Theory and Appl., II, 690-695 (Russian).

O. A. Iadyženskaja, V. A. Solonnikov and N. N. Ural'ceva

[1] 1967. Linear and quasi-linear parabolic equations. Moscow: Nauka (Russian).

N. Levinson

[1] 1950. The first boundary value problem for the equation $\varepsilon\Delta u + Au_x + Bu_y + cu = 0$ for small ε. Ann. of Math., 51, 428-425.

P. Lévy

[1] 1948. Processus stochastiques et mouvement brownien. Paris: Gauthier-Villars.

M. B. Maljutov

[1] 1969. On the Poincaré boundary problem. Trans. Moscow Math. Soc., 20, 173-203.

P. Malliavin

[1] 1978. Stochastic calculus of variation and hypoelliptic operators. Proc. Intern. Symp., SDE, Kyoto (ed. by K. Ito), 195-263, Kinokuniya, Tokyo.

[2] 1978. C^*-hypoellipticity and degeneracy. Stochastic Analysis (ed. by A. Friedman and M. Pinsky) 199-214, 327-340, Academic Press, New York.

C. Marchioro and M. Pulvirenti

[1] 1982. Hydrodynamics in Two Dimensions and Vortex Theory. Comm. Math. Phys., 84, 483-503.

R. S. Martin

[1] 1941. Minimal positive harmonic functions. Trans. Amer. Math. Soc., 49, 137-172.

H. P. McKean

[1] 1969. Stochastic integrals. New York: Academic Press.

[2] 1969. Propagation of chaos for a class of non-linear parabolic equations. Vol. 2, (ed. by A. K. Aziz), Princeton, N. J.: Van Nostrand.

[3] 1975. Application of Brownian motion to the equation of Kolmogorov-Petrovskii-Piskunov. Comm. Pure Appl. Math. 28:3, 323-331 (correction: Comm. Pure Appl. Math., 29:5, 553-554).

P. A. Meyer

[1] 1966. Probability and potentials. Blaisdel, Waltham, Massachusetts.

N. Meyers and J. Serrin

[1] 1960. The exterior Dirichlet problem for second order elliptic partial differential equations, J. Math. Mech., 9, 4, 513-538.

C. Miranda

[1] 1970. Partial differential equations of elliptic type. 2nd ed., Berlin-Heidelberg-New York: Springer-Verlag.

S. A. Molchanov

[1] 1964. On a problem in the theory of diffusion processes. Prob. Theory and Appl., 9, 523-528.

[2] 1978. Structure of the eigenfunctions of the one-dimensional non-ordered structures. Izv. Acad. Nauk SSSR, Ser. Mat., 42, 70-103 (Russian).

M. Nikunen

[1] 1978. Self-adjoint diffusion processes with rapidly oscillating coefficients. Moscow Univ. Math. Bull., 33:5, 24-28 (Russian).

M. Nisio

[1] 1972. On stochastic differential equations associated with certain quasi-linear parabolic equations. Trans. Second Japan.-USSR symposium on probability theory, Kyoto, Vol. 2, 135-138.

O. A. Oleinik

[1] 1949. On Dirichlet problem for equations of elliptic type. Mat. Sbornik, 24:1, 3-14 (Russian).

O. A. Oleinik and E. V. Radkevich

[1] 1971. Second-order differential equations with non-negative characteristic form. Ser. "Itogi Nauki. Mat. Analiz", Moscow (Russian).

R. Paley, N. Wiener and A. Zygmund

[1] 1933. Note on random functions. Math. Z., 37, 647-668.

L. Partzsch

[1] 1971. On an ergodic theorem for Markov process with finite time of life. Prob. Theory and Appl., 16, 711-714.

R. S. Phillips and L. Sarason

[1] 1968. Elliptic-parabolic equations of the second order. J. Math. Mech., 17, 891-917.

M. Pinsky

[1] 1969. A note on degenerate diffusion processes. Prob. Theory and Appl., 14, 502-506.

[2] 1977. Isolated Singularities of Degenerate Elliptic Equations in R^2. Journ. of Differential Equations, 24, 274-281.

Yu V. Prohorov

[1] 1956. Convergence of random processes and limit theorems in probability theory. Prob. Theory and Appl., 1, 157-214 (Russian).

R. T. Rockafellar

[1] 1970. Convex Analysis. Princeton University Press.

Yu. M. Romanovskii, V. A. Vasiliev and V. G. Yahno

[1] 1979. Autowave processes in distributed kinetic systems. Uspehi Fiz. Nauk, 128, 626-666.

R. G. Safarian

[1] 1980. Some boundary value problems for degenerate diffusion process and for corresponding differential equations. Izv. Acad. Nauk Armianskoi SSSR, 10, 4, 259-267 (Russian).

V. V. Sarafian

[1] 1972. Diffusion processes and differential equations degenerating at isolated points. Prob. Theory and Appl., 17, 699-703 (Russian).

[2] 1982. On asymptotic of the first eigenvalue of elliptic operator with small parameter. Dokl. Acad. Nauk Armianskoi USSR, 75, 3 (Russian).

[3] 1982. The small disturbances of diffusion processes and differential operators degenerating on the boundary. Abstracts of Communications of the 4 USSR-Japan Symposium on probability theory and mathematical statistics, Tbilisi, USSR, Vol. 2, 187-189.

V. V. Sarafian and R. G. Safarian

[1] 1980. Diffusion processes with averaging and propagation of concentration waves. Dokl. Acad. Nauk Armianskoi USSR, 20, 5, 274-278 (Russian).

[2] 1982. Propagation of concentration waves and large deviations from averaged system. Ukrainskii Mat. Zurnal, 4, 2, 177-184 (Russian).

[3] On Asymptotic behavior of the solution of the Dirichlet problem for differential operators with small parameter. Submitted to Ukrainskii Mat. Zurnal. (Russian).

V. V. Sarafian, R. G. Safarian and M. I. Freidlin

[1] 1978. Degenerate diffusion processes and differential equations with small parameter. Uspehi Mat. Nauk, 33:6, 233-234 (Russian).

K. Sato and T. Ueno

[1] 1965. Multi-dimensional diffusion and Markov processes on the boundary. Jorn. of Math. of Kyoto Univ., 4, 529-605.

V. A. Shaloumov

[1] 1979. Averaging for diffusion processes with boundary conditions. VINITI, Moscow, Dep. N1550-79, 1-22 (Russian).

M. Shilder

[1] 1966. Some asymptotic formulas for Wiener integrals. Trans. Amer. Math. Soc., 125, 63-85.

B. Simon

[1] 1979. Functional Integration and Quantum Physics. New York-San Francisco-London: Academic Press.

A. V. Skorohod

[1] 1961. Stochastic equations for diffusions in a bounded region. Prob. Theory and Appl., 6, 264-274.

R. L. Stratonovich

[1] 1968. Conditional Markov processes and their application to the theory of optimal control. New York: Amer. Elsevier.

D. W. Stroock and S. R. S. Varadham

[1] 1979. Multidimensional diffusion processes. Berlin: Springer-Verlag.

H. Tanaka

[1] 1967. Local solutions of stochastic differential equations associated with certain quasi-linear parabolic equations. Jorn. of the Faculty of Science, Univ. of Tokyo, Sec. I, 14, 313-326.

K. Uchiyama

[1] 1980. Brownian first exit from and sojourn over one side moving boundary and application. Z. Wahrscheinlichkeits theorie Verw. Gebiete, 54, 75-116.

T. Ueno

[1] 1966. A survey on the Markov processes on the boundary of multidimensional diffusion. Proc. of the Fifth Berkeley Symp. on Math. Stat. and Probability, Univ. of California Press.

S. R. S. Varadhan

[1] 1966. Asymptotic probabilities and differential equations. Comm. Pure and Appl. Math., 19:3, 261-286.

[2] 1967. On the behavior of the fundamental solution of the heat equation with variable coefficients. Comm. Pure Appl. Math., 20:2, 431-455.

[3] 1967. Diffusion processes in a small time interval. Comm. Pure Appl. Math., 20:4, 659-685.

S. Watanabe

[1] 1971. On stochastic differential equations for multi-dimensional diffusion processes with boundary conditions. J. Math. Kyoto Univ., I, 11, 169-180; II, 11, 545-551.

A. D. Wentzell

[1] 1981. A Course in the Theory of Stochastic Processes. McGraw-Hill.

A.D. Wentzell and M.I. Freidlin

[1] 1970. On small random perturbations of dynamical systems. Uspehi Mat. Nauk, 28:1, 3-55 (Russian).

[2] 1979. Fluctuations in dynamical systems caused by small random perturbations. Moscow: Nauka (Russian).

N. Wiener

[1] 1923. Differential space. J. Math. Phys., 2, 131-174.

V.V. Zikov, S.M. Kozlov, O.A. Oleinik and Ha Ten Ngoan

[1] 1979. Averaging and G-convergence of differential operators. Uspehi Mat. Nauk, 34:5, 65-133 (Russian).

Library of Congress Cataloging in Publication Data

Freidlin, M. I. (Mark Iosifovich)
Functional integration and partial differential equations.

Bibliography: p.
1. Differential equations, Partial.
2. Probabilities. 3. Integration, Functional.
I. Title.
QA377.F72 1985 515.3′53 84–42874
ISBN 0–691–08354–1
ISBN 0–691–08362–2 (pbk.)

ANNALS OF MATHEMATICS STUDIES

109. Functional Integration and Partial Differential Equations, *by* MARK FREIDLIN
108. Arithmetic Moduli of Elliptic Curves, *by* NICHOLAS M. KATZ *and* BARRY MAZUR
107. Characters of Reductive Groups over a Finite Field, *by* GEORGE LUSZTIG
106. Topics in Transcendental Algebraic Geometry, edited *by* PHILLIP GRIFFITHS
105. Multiple Integrals in the Calculus of Variations and Nonlinear Elliptic Systems, *by* MARIANO GIAQUINTA
104. Etale Homotopy of Simplicial Schemes, *by* ERIC M. FRIEDLANDER
103. Seminar on Minimal Submanifolds, *edited by* ENRICO BOMBIERI
102. Seminar on Differential Geometry, *edited by* SHING-TUNG YAU
101. Random Fourier Series With Applications to Harmonic Analysis, *by* MICHAEL B. MARCUS *and* GILLES PISIER
100. Recent Developments in Several Complex Variables, *edited by* JOHN E. FORNAESS
99. The Spectral Theory of Toeplitz Operators, *by* L. BOUTET DE MONVEL *and* V. GUILLEMIN
98. K-Theory of Forms, *by* ANTHONY BAK
97. Riemann Surfaces and Related Topics: Proceedings of the 1978 Stony Brook Conference, *edited by* IRWIN KRA *and* BERNARD MASKIT
96. Base Change for GL(2), *by* ROBERT P. LANGLANDS
95. C*-Algebra Extensions and K-Homology, *by* RONALD G. DOUGLAS
94. Continuous Cohomology, Discrete Subgroups, and Representations of Reductive Groups, *by* A. BOREL *and* N. WALLACH
93. Seminar on Micro-Local Analysis, *by* VICTOR W. GUILLEMIN, MASAKI KASHIWARA, *and* YAKAHIRO KAWAI
92. The Classifying Spaces for Surgery and Cobordism of Manifolds, *by* IB MADSEN *and* R. JAMES MILGRAM
91. Seminar on Singularities of Solutions of Linear Partial Differential Equations, *edited by* LARS HÖRMANDER
90. Infinite Loop Spaces, *by* J. F. ADAMS
89. Invariant Forms on Grassmann Manifolds, *by* WILHELM STOLL
88. Foundational Essays on Topological Manifolds, Smoothings, and Triangulations, *by* ROBION C. KIRBY *and* LAURENCE C. SIEBENMANN
87. Scattering Theory for Automorphic Functions, *by* PETER D. LAX *and* RALPH S. PHILLIPS
86. 3-Manifolds, *by* JOHN HEMPEL
85. Entire Holomorphic Mappings in One and Several Complex Variables, *by* PHILLIP A. GRIFFITHS
84. Knots, Groups, and 3-Manifolds, *edited by* L. P. NEUWIRTH
83. Automorphic Forms on Adele Groups, *by* STEPHEN S. GELBART
82. Braids, Links, and Mapping Class Groups, *by* JOAN S. BIRMAN
81. The Discrete Series of GL_n over a Finite Field, *by* GEORGE LUSZTIG
80. Smoothings of Piecewise Linear Manifolds, *by* M. W. HIRSCH *and* B. MAZUR
79. Discontinuous Groups and Riemann Surfaces, *edited by* LEON GREENBERG
78. Strong Rigidity of Locally Symmetric Spaces, *by* G. D. MOSTOW
77. Stable and Random Motions in Dynamical Systems: With Special Emphasis on Celestial Mechanics, *by* JÜRGEN MOSER
76. Characteristic Classes, *by* JOHN W. MILNOR *and* JAMES D. STASHEFF
75. The Neumann Problem for the Cauchy-Riemann Complex, *by* G. B. FOLLAND *and* J. J. KOHN
74. Lectures on *p*-Adic *L*-Functions, *by* KENKICHI IWASAWA
73. Lie Equations, *by* ANTONIO KUMPERA *and* DONALD SPENCER

A complete catalogue of Princeton mathematics and
science books, with prices, is available upon request.

PRINCETON UNIVERSITY PRESS
41 WILLIAM STREET
PRINCETON, NEW JERSEY 08540

ISBN 0-691-08362-2